Perfectoid Spaces

Lectures from the 2017 Arizona Winter School

Mathematical
Surveys
and
Monographs

Volume 242

Perfectoid Spaces

Lectures from the 2017 Arizona Winter School

Bhargav Bhatt
Ana Caraiani
Kiran S. Kedlaya
Jared Weinstein
With an introduction by Peter Scholze
Bryden Cais, Editor

AMS
AMERICAN
MATHEMATICAL
SOCIETY
Providence, Rhode Island

EDITORIAL COMMITTEE

Robert Guralnick, Chair
Natasa Sesum
Benjamin Sudakov
Constantin Teleman

2010 *Mathematics Subject Classification.* Primary 11F80, 11G25, 14C30, 14F30, 14F40, 14G20, 14G22, 14G35, 14G40, 14L05.

For additional information and updates on this book, visit
www.ams.org/bookpages/surv-242

Library of Congress Cataloging-in-Publication Data

Names: Bhatt, Bhargav, 1983- | Cais, Bryden, editor. | Arizona Winter School on Perfectoid Spaces (20th : 2017 : Tucson, Ariz.)
Title: Perfectoid spaces : lectures from the 2017 Arizona Winter School / Bhargav Bhatt [and three others] ; with an Introduction by Peter Scholze ; Bryden Cais, editor.
Description: Providence, Rhode Island : American Mathematical Society, [2019] | Series: Mathematical surveys and monographs ; volume 242 | Prepared for the twentieth Arizona Winter School on Perfectoid Spaces, held March 11-17, 2017, at the University of Arizona in Tucson. | Includes bibliographical references.
Identifiers: LCCN 2019016582 | ISBN 9781470450151 (alk. paper)| ISBN 9781470465100 (Softcover)
Subjects: LCSH: Topological fields. | AMS: Number theory – Discontinuous groups and automorphic forms – Galois representations. msc | Number theory – Arithmetic algebraic geometry (Diophantine geometry) – Varieties over finite and local fields. msc | Algebraic geometry – Cycles and subschemes – Transcendental methods, Hodge theory. msc | Algebraic geometry – (Co)homology theory – p-adic cohomology, crystalline cohomology. msc | Algebraic geometry – (Co)homology theory – de Rham cohomology. msc | Algebraic geometry – Arithmetic problems. Diophantine geometry – Local ground fields. msc | Algebraic geometry – Arithmetic problems. Diophantine geometry – Rigid analytic geometry. msc | Algebraic geometry – Arithmetic problems. Diophantine geometry – Modular and Shimura varieties. msc | Algebraic geometry – Arithmetic problems. Diophantine geometry – Arithmetic varieties and schemes; Arakelov theory; heights. msc | Algebraic geometry – Algebraic groups – Formal groups, p-divisible groups. msc
Classification: LCC QA247 .P4485 2019 | DDC 516.3/5–dc23
LC record available at https://lccn.loc.gov/2019016582

Copying and reprinting. Individual readers of this publication, and nonprofit libraries acting for them, are permitted to make fair use of the material, such as to copy select pages for use in teaching or research. Permission is granted to quote brief passages from this publication in reviews, provided the customary acknowledgment of the source is given.

Republication, systematic copying, or multiple reproduction of any material in this publication is permitted only under license from the American Mathematical Society. Requests for permission to reuse portions of AMS publication content are handled by the Copyright Clearance Center. For more information, please visit www.ams.org/publications/pubpermissions.

Send requests for translation rights and licensed reprints to reprint-permission@ams.org.

© 2019 by the American Mathematical Society. All rights reserved.
The American Mathematical Society retains all rights
except those granted to the United States Government.
Printed in the United States of America.
Reprinted by the American Mathematical Society, 2021.

∞ The paper used in this book is acid-free and falls within the guidelines
established to ensure permanence and durability.
Visit the AMS home page at https://www.ams.org/

11 10 9 8 7 6 5 4 3 2   27 26 25 24 23 22 21

Contents

Preface BY BRYDEN CAIS	vii
Introduction BY PETER SCHOLZE	ix
Adic spaces BY JARED WEINSTEIN	1
1. An introduction to adic spaces	1
2. Perfectoid fields	11
3. Perfectoid spaces and diamonds	22
4. Banach-Colmez spaces	32
5. Projects	39
References	42
Sheaves, stacks, and shtukas BY KIRAN S. KEDLAYA	45
1. Sheaves on analytic adic spaces	46
2. Perfectoid rings and spaces	93
3. Sheaves on Fargues–Fontaine curves	121
4. Shtukas	148
Appendix A. Project descriptions	179
References	184
The Hodge-Tate decomposition via perfectoid spaces BY BHARGAV BHATT	193
1. Lecture 1: Introduction	193
2. Lecture 2: The Hodge-Tate decomposition for abelian schemes	200
3. Lecture 3: The Hodge-Tate decomposition in general	203
4. Lecture 4: Integral aspects	218
5. Exercises	230
6. Projects	237
References	241
Perfectoid Shimura varieties BY ANA CARAIANI	245
1. Introduction	245
2. Locally symmetric spaces and Shimura varieties	250

3.	Background from p-adic Hodge theory	265
4.	The canonical subgroup and the anticanonical tower	270
5.	Perfectoid Shimura varieties and the Hodge–Tate period morphism	279
6.	The cohomology of locally symmetric spaces: conjectures and results	287
	References	295

Preface

The Southwest Center for Arithmetic Geometry (SWC) was founded in 1997 by a group of seven mathematicians working in the southwest United States, and has been continuously supported by the National Science Foundation since that time. In the beginning, the SWC served as a true *center* for Arithmetic Geometry in the Southwest, but survives today in name only, having been subsumed by its principal activity, the *Arizona Winter School* (AWS).

The AWS was started with the ambitious goal of creating an intense and immersive workshop in which graduate students—especially those who may not be studying at traditional centers of number theory—would work under the guidance of leading experts to solve research-level problems at the forefront of the field. The very first school was held in the Spring of 1998 under the title "Workshop on Diophantine Geometry Related to the ABC Conjecture." In the twenty-one years that have followed, the AWS has been held annually each March on a different topic in arithmetic geometry and related areas, and has become a pillar of graduate education and training in these subjects throughout the country and abroad.

Over the years, the Arizona Winter School model has been adjusted and refined to meet the needs of an ever-growing and increasingly diverse audience: the five-day meeting, organized around a different central topic each year, now features a set of four lecture series by leading and emerging experts. A month before the school begins, each speaker proposes one or more research projects related to their lectures, and is assigned 10–15 graduate students who will work on these projects. At that time, speakers also provide detailed lecture notes for their courses. During the school, students attend lectures daily from 9am to 5pm, and work each evening (often into the early hours of the morning!) with speakers and designated assistants on these research projects. Students not assigned to these *research project groups* have the option to join one of two *problem sessions* devoted to solving advanced exercises related to the lecture series, or one of four *study groups* which focus on understanding the course lecture notes in detail; these additional activities are supervised by young researchers and allow students not assigned to one of the research projects to meaningfully engage with the workshop material on many levels. On the last day, the students from each research project group present their work to the whole school. The result is an extremely focused and immersive five days of mathematical activity for all.

This volume is comprised of the lecture notes which were prepared for the twentieth Arizona Winter School on "Perfectoid Spaces," held March 11–17, 2017 at the University of Arizona in Tucson, and attended by over 367 participants, making it the largest Winter School to date. The speakers were Bhargav Bhatt, Ana Caraiani, Kiran Kedlaya, and Jared Weinstein. We are greatly indebted to these authors for their hard work in making both the twentieth AWS and this proceedings volume a reality. Peter Scholze gave opening and closing lectures at the Winter School, which provided historical background as well as insights into his pioneering efforts and visions for the future of the subject, and we thank him most heartily for furnishing the introduction to this volume. We extend our sincere gratitude to the anonymous reviewers for their expert and careful readings of the articles, and for the valuable feedback they provided. We thank the National Science Foundation for their longstanding and continued support of the Arizona Winter School, the Clay Mathematics Institute for their partnership in organizing the 2017 AWS, and the University of Arizona Department of Mathematics for their support. Finally, we owe a great deal to the other members of the *Southwest Center*, both past and present, for their effort, perseverance, and vision in running the Arizona Winter School for more than twenty years, and for helping it to become the one-of-a-kind workshop that it is.

Bryden Cais

Introduction

Peter Scholze

The Arizona Winter School 2017 was around the subject of perfectoid spaces. These spaces belong to the world of p-adic geometry, and despite their rather complicated definition and appearance, they have found applications in many different areas of mathematics, including arithmetic geometry, number theory, representation theory, commutative algebra and algebraic topology.

The problem that motivated the author to introduce perfectoid spaces was the *weight-monodromy conjecture*. This is a conjecture on the ℓ-adic cohomology of projective smooth varieties X over a p-adic field K, whose analogue over an equal-characteristic field $\mathbb{F}_q((t))$ was proved by Deligne, [**Del80**], as a consequence of his work on the Weil conjectures. Let us recall the statement. For any integer i, consider the étale cohomology group $V = H^i_{\text{ét}}(X_{\overline{K}}, \mathbb{Q}_\ell)$ for some prime $\ell \neq p$, which is a representation of the absolute Galois group $G_K = \text{Gal}(\overline{K}/K)$. For simplicity, let us fix an element $\Phi \in G_K$ projecting to the geometric Frobenius on the residue field. Similarly to the case over a finite field, the element Φ acts via Weil numbers on $H^i_{\text{ét}}(X_{\overline{K}}, \mathbb{Q}_\ell)$, but now different weights can appear:

$$V = \bigoplus_{j=0}^{2i} V_j \ ,$$

where Φ acts through Weil numbers of weight j on V_j, i.e. all eigenvalues $\alpha \in \overline{\mathbb{Q}}_\ell$ of Φ on V_j are algebraic numbers whose complex absolute value is $q^{j/2}$ under any embedding $\overline{\mathbb{Q}} \hookrightarrow \mathbb{C}$. This follows from de Jong's alterations, [**dJ96**], and the Rapoport-Zink spectral sequence, [**RZ82**].

On the other hand, there is a monodromy operator $N : V \to V$ given as the logarithm of the action of inertia (using Grothendieck's quasi-unipotence theorem). This satisfies $N\Phi = q\Phi N$, and so N maps V_j into V_{j-2}.

CONJECTURE 1 (Deligne, [**Del71**]). *For all $j = 0, \ldots, i$, the map $N^j : V_{i+j} \to V_{i-j}$ is an isomorphism.*

This is similar to the hard Lefschetz theorem, and is sometimes said to be "Mirror dual" to it. Let us quickly explain what happens for an elliptic curve E. If it has (potentially) good reduction, then $N = 0$, and $V = V_1$, as follows from proper smooth base change and the Weil conjectures for the special fibre. If it has (potentially) semistable reduction, then N is nonzero, and so as V is 2-dimensional, one must have $V = V_2 \oplus V_0$ with both V_0 and V_2 being 1-dimensional; and this is indeed what happens, as can be deduced from an analysis of the Néron model, or by using Tate's uniformization of elliptic curves. This shows that as a rigid-analytic variety, E is the quotient of $\mathbb{G}_m$ by multiplication by q for some $q \in K$ with $0 < |q| < 1$. Then the Hochschild-Serre spectral sequence gives

$$E_2^{ij} = H^i(\mathbb{Z}, H_{\text{ét}}^j(\mathbb{G}_{m,\overline{K}}, \mathbb{Q}_\ell)) \Rightarrow H_{\text{ét}}^{i+j}(E_{\overline{K}}, \mathbb{Q}_\ell) \ .$$

Terms are nonzero only if $i, j \in \{0, 1\}$, and so the spectral sequence degenerates, with

$$H^0(\mathbb{Z}, H_{\text{ét}}^1(\mathbb{G}_{m,\overline{K}}, \mathbb{Q}_\ell)) = \mathbb{Q}_\ell(-1)$$

and

$$H^1(\mathbb{Z}, H_{\text{ét}}^0(\mathbb{G}_{m,\overline{K}}, \mathbb{Q}_\ell)) = \mathbb{Q}_\ell$$

contributing, which are of weight 2 and 0, respectively.

The idea (that was suggested to the author by his advisor M. Rapoport) is to try to reduce the case of mixed characteristic to the case of equal characteristic. This requires at minimum a comparison of the Galois groups that appear in mixed and equal characteristic. One such comparison is afforded by the theorem of Fontaine-Wintenberger.

THEOREM ([**FW79**]). *The absolute Galois groups of $\mathbb{Q}_p(p^{1/p^\infty})$ and $\mathbb{F}_p((t))$ are canonically isomorphic.*

Basically, the aim of the theory of perfectoid spaces is to find a geometric counterpart to this arithmetic statement. More precisely, we want to have categories of spaces over $\mathbb{Q}_p(p^{1/p^\infty})$ resp. $\mathbb{F}_p((t))$ that are equivalent via a "tilting" functor. This is in fact possible.

THEOREM ([**Sch12**], [**KL15**]). *The category of perfectoid spaces over $\mathbb{Q}_p(p^{1/p^\infty})$ is equivalent to the category of perfectoid spaces over $\mathbb{F}_p((t))$, via $X \mapsto X^\flat$.*

What is a perfectoid space? This is the subject of this book, so we will not say much here. Suffice to say that it is a special type of adic space, with the key condition being that locally it is of the form $\text{Spa}(R, R^+)$ where the Frobenius is surjective on R^+/p. This ensures that there are many (approximate) p-th roots of functions, and in fact excludes all familiar noetherian spaces from the picture.

In particular, our smooth projective variety X itself does not define a perfectoid space. In [**Sch12**] we found a trick that made it possible in many cases to "approximate" it by perfectoid spaces that in turn came from smooth projective varieties over $\mathbb{F}_p((t))$; this way, we concluded that Conjecture 1 holds for complete intersections in projective toric varieties. Unfortunately, to this day, noone was able to settle Conjecture 1 in full!

Since then, perfectoid spaces have played a key role in a number of questions, including:

(1) *p*-adic Hodge theory: This started with the work of Kedlaya-Liu [**KL15**] on relative *p*-adic Hodge theory and the work of the author [**Sch13**] on *p*-adic Hodge theory for rigid-analytic varieties. The general strategy here is to cover a rigid-analytic variety by perfectoid spaces using the *pro-étale site*: Any rigid space is locally in the pro-étale topology perfectoid. Recently, progress on integral *p*-adic Hodge theory has also been made using perfectoid techniques, [**BMS16**].

(2) Shimura varieties: The inverse limit over all levels at p of a Shimura variety is a perfectoid space by [**Sch15**]. This was in fact first observed by Weinstein in the case of the Lubin-Tate tower, [**Wei16**], and then generalized to all Rapoport-Zink spaces in [**SW13**]. This way, one gets applications towards the global Langlands conjecture, i.e. the relations between automorphic forms and Galois representations as in [**Sch15**], [**CS17**], [**ACC**$^+$**17**].
(3) Local Langlands: Combining the previous two directions, one can establish a general theory of shtukas in the p-adic setting; this was the main theme of the author's Berkeley lectures, [**SW14**]. This makes it possible to attack the local Langlands conjectures for a general p-adic reductive group.
(4) Commutative algebra: After Bhatt applied the almost purity theorem towards a special case of the direct summand conjecture, [**Bha14**], André has used perfectoid techniques to settle the full conjecture, [**And16**], with a simplified proof by Bhatt, [**Bha16**].
(5) Algebraic topology: The work [**BMS16**] on integral p-adic Hodge theory was motivated by computations in topological Hochschild homology, and the relation between the subjects is worked out in [**BMS**]. This led to a different perspective on topological Hochschild homology itself that has been worked out in [**NS17**].

Moreover, we must also mention the Fargues-Fontaine curve [**FF17**] that plays a critical role in several of these developments, in particular in p-adic Hodge theory, and in local Langlands: By Fargues' conjecture [**Far17**], the local Langlands conjecture can be understood as a geometric Langlands conjecture on the Fargues-Fontaine curve.

At the Arizona Winter School, a magnificent effort was made to present these developments to a very large audience of graduate students and other interested mathematicians. The lecture series of Bhatt (mostly on (1)), Caraiani (mostly on (2)), Kedlaya (mostly on (1) and (3), but including many foundational results on adic and perfectoid spaces), and Weinstein (on tilts and untilts of perfectoid spaces and the Fargues-Fontaine curve) each took on the demanding task of presenting some aspects, referring to each other in critical ways. The resulting excellent lecture notes that this volume combines will be an invaluable resource for learning the theory of perfectoid spaces.

In the evenings, the students worked on research projects and exercises, and I was thrilled by their infinite energy lasting until the morning hours. During the student presentations, I was more than once surprised by the results obtained by the students; sometimes, I couldn't even believe that this was correct! They have truly reshaped my own understanding of perfectoid spaces, which really are very mysterious objects. The school was a very memorable event, and I hope it will motivate many students to further develop the theory of perfectoid spaces, and to use it in new and surprising ways.

References

[ACC$^+$17] Patrick Allen, Frank Calegari, Ana Caraiani, Toby Gee, David Helm, Bao Le Hung, James Newton, Peter Scholze, Richard Taylor, and Jack Thorne. Potential automorphy over CM fields. 2017. in preparation.

[And16] Yves André, *La conjecture du facteur direct* (French, with French summary), Publ. Math. Inst. Hautes Études Sci. **127** (2018), 71–93, DOI 10.1007/s10240-017-0097-9. MR3814651

[Bha14] Bhargav Bhatt, *Almost direct summands*, Nagoya Math. J. **214** (2014), 195–204, DOI 10.1215/00277630-2648180. MR3211823

[Bha16] Bhargav Bhatt, *On the direct summand conjecture and its derived variant*, Invent. Math. **212** (2018), no. 2, 297–317, DOI 10.1007/s00222-017-0768-7. MR3787829

[BMS] Bhargav Bhatt, Matthew Morrow, and Peter Scholze, *Integral p-adic Hodge theory*, Publ. Math. Inst. Hautes Études Sci. **128** (2018), 219–397, DOI 10.1007/s10240-019-00102-z. MR3905467

[BMS16] B. Bhatt, M. Morrow, and P. Scholze, *Integral p-adic Hodge theory—announcement*, Math. Res. Lett. **22** (2015), no. 6, 1601–1612, DOI 10.4310/MRL.2015.v22.n6.a3. MR3507252

[CS17] Ana Caraiani and Peter Scholze, *On the generic part of the cohomology of compact unitary Shimura varieties*, Ann. of Math. (2) **186** (2017), no. 3, 649–766, DOI 10.4007/annals.2017.186.3.1. MR3702677

[Del71] Pierre Deligne, *Théorie de Hodge. I* (French), Actes du Congrès International des Mathématiciens (Nice, 1970), Gauthier-Villars, Paris, 1971, pp. 425–430. MR0441965

[Del80] Pierre Deligne, *La conjecture de Weil. II* (French), Inst. Hautes Études Sci. Publ. Math. **52** (1980), 137–252. MR601520

[dJ96] A. J. de Jong, *Smoothness, semi-stability and alterations*, Inst. Hautes Études Sci. Publ. Math. **83** (1996), 51–93. MR1423020

[Far17] Laurent Fargues. Geometrization of the local Langlands correspondence: an overview. 2017. arXiv:1602.00999.

[FF17] Laurent Fargues and Jean-Marc Fontaine, with a preface by Pierre Colmez. *Courbes et fibrés vectoriels en théorie de Hodge p-adique* (French), Astérisque **406** (2018), xiii+382. MR3917141.

[FW79] Jean-Marc Fontaine and Jean-Pierre Wintenberger, *Extensions algébrique et corps des normes des extensions APF des corps locaux* (French, with English summary), C. R. Acad. Sci. Paris Sér. A-B **288** (1979), no. 8, A441–A444. MR527692

[KL15] Kiran S. Kedlaya and Ruochuan Liu, *Relative p-adic Hodge theory: foundations* (English, with English and French summaries), Astérisque **371** (2015), 239. MR3379653

[NS17] Thomas Nikolaus and Peter Scholze, *On topological cyclic homology*, Acta Math. **221** (2018), no. 2, 203–409. MR3904731

[RZ82] M. Rapoport and Th. Zink, *Über die lokale Zetafunktion von Shimuravarietäten. Monodromiefiltration und verschwindende Zyklen in ungleicher Charakteristik* (German), Invent. Math. **68** (1982), no. 1, 21–101, DOI 10.1007/BF01394268. MR666636

[Sch12] Peter Scholze, *Perfectoid spaces*, Publ. Math. Inst. Hautes Études Sci. **116** (2012), 245–313, DOI 10.1007/s10240-012-0042-x. MR3090258

[Sch13] Peter Scholze, *p-adic Hodge theory for rigid-analytic varieties*, Forum Math. Pi **1** (2013), e1, 77, DOI 10.1017/fmp.2013.1. MR3090230

[Sch15] Peter Scholze, *On torsion in the cohomology of locally symmetric varieties*, Ann. of Math. (2) **182** (2015), no. 3, 945–1066, DOI 10.4007/annals.2015.182.3.3. MR3418533

[SW13] Peter Scholze and Jared Weinstein, *Moduli of p-divisible groups*, Camb. J. Math. **1** (2013), no. 2, 145–237, DOI 10.4310/CJM.2013.v1.n2.a1. MR3272049

[SW14] Peter Scholze and Jared Weinstein. *p*-adic geometry. 2014. notes from lecture course at UC Berkeley, Fall 2014.

[Wei16] Jared Weinstein, *Semistable models for modular curves of arbitrary level*, Invent. Math. **205** (2016), no. 2, 459–526, DOI 10.1007/s00222-015-0641-5. MR3529120

Adic spaces

Jared Weinstein

1. An introduction to adic spaces

This year's AWS topic is *perfectoid spaces*, a difficult topic to treat in one week if there ever was one. But given the interest in the topic, and the huge amount of important work awaiting young mathematicians who want to work on this field, it is certainly a worthy effort. The lecture notes here are meant to be a motivated introduction to adic spaces, perfectoid spaces and diamonds, for the reader who knows some algebraic geometry.[1]

1.1. What is a "space"? Consider the different kinds of geometric "spaces" you know about. First you learned about topological spaces. Then came various sorts of manifolds, which are topological spaces which locally look like a *model space* (an open subset of $\mathbf{R}^n$). Then you learned that manifolds could carry different structures (differentiable, smooth, complex, ...). You could express these structures in terms of the transition functions between charts on your manifold. But this is a little awkward, thinking of everything in terms of charts. Later you learned a more efficient definition: a manifold with one of these structures is a *ringed space* $(X, \mathcal{O}_X)$, where X is a topological space and $\mathcal{O}_X$ is a sheaf of rings on X, such that locally on X the pair $(X, \mathcal{O}_X)$ is isomorphic to one of the model spaces, together with its sheaf of (differentiable, smooth, complex) functions. An advantage of this point of view is that it becomes simple to define a morphism $f \colon X \to Y$ between such objects: it is a continuous map of topological spaces together with a homomorphism $\mathcal{O}_Y \to f_*\mathcal{O}_X$ (in other words, *functions pull back*).

This formulation of spaces in terms of pairs $(X, \mathcal{O}_X)$ was good preparation for learning about schemes, the modern language of algebraic geometry. This time the model spaces are affine schemes, which are spectra of rings. For a ring A, the topological space $\operatorname{Spec} A$ may have initially seemed strange—in particular it is not generally Hausdorff. But then you learn some advantages of working with schemes. For instance, an integral scheme X has a generic point η. It is enormously useful to take an object associated with X (a morphism to X, an $\mathcal{O}_X$-module, an étale sheaf on X, ...) and pass to its generic fiber, which is associated with the function field of X. Usually if some property is true on the generic fiber, then it is also true "generically" on X (that is, on a dense open subset). Number theorists use this

---

[1] Special thanks to Johannes Anschütz, Shamil Asgarli, Tony Feng, María-Inés de Frutos Fernández, Nadir Hajouji, Sean Howe, Siyan Li, Jackson Morrow, David Savitt, Peter Scholze, Koji Shimizu, and David Zureick-Brown for their helpful comments.

language all the time in the setting of Spec $\mathbf{Z}$: if a property holds over the generic point Spec $\mathbf{Q}$, then it holds at almost all special points Spec $\mathbf{F}_p$.

The language of *formal schemes* is useful for studying what happens in an infinitesimal neighborhood of a closed subset of a scheme. Thus formal schemes often arise in deformation theory. This time, the model spaces are formal spectra Spf A, where A is an admissible topological ring. (Examples of such A include $\mathbf{Z}_p$ and $\mathbf{Z}[\![T]\!]$.) The notation Spf stands for "formal spectrum", and refers to the collection of open prime ideals of A. This can be given the structure of a topological space X, which is equipped with a sheaf $\mathcal{O}_X$ of topological rings.

In the theory of *complex-analytic spaces*, the model space is the vanishing locus of a collection of holomorphic functions on an open subset of $\mathbf{C}^n$. Thus it is like the theory of complex manifolds, except that some singularities are allowed. The theory of complex-analytic spaces has many nice interactions with the theory of schemes. If X is a finite-type scheme over Spec $\mathbf{C}$, then there is a complex-analytic space X^{an}, the *analytification* of X, which is universal for the property of admitting a morphism of ringed spaces $(X^{\mathrm{an}}, \mathcal{O}_{X^{\mathrm{an}}}) \to (X, \mathcal{O}_X)$. Conversely, if $\mathcal{X}$ is a complex-analytic space admitting a closed immerson into projective space, then $\mathcal{X}$ is the analytification of a projective complex variety X, and then X and $\mathcal{X}$ have equivalent categories of coherent sheaves, and the equivalence respects cohomology groups (Serre's *GAGA* theorem). In this situation there are *comparison isomorphisms* between the étale cohomology groups of X and $\mathcal{X}$. There are further relations known as *uniformizations*; most well-known of these is the phenomenon that if E is an elliptic curve over Spec $\mathbf{C}$, then there exists a lattice $L \subset \mathbf{C}$ such that $E^{\mathrm{an}} \cong \mathbf{C}/L$ as complex-analytic spaces.

1.2. Rigid-analytic spaces. Let us turn our attention from archimedean fields ($\mathbf{R}$ and $\mathbf{C}$) to non-archimedean fields ($\mathbf{Q}_p$, $\mathbf{C}_p$, $k((t))$ for any field k). Both are kinds of complete metric fields, so it is natural to expect a good theory of manifolds or analytic spaces for a non-archimedean field K. Which ringed spaces $(X, \mathcal{O}_X)$ should serve as our model spaces? The naïve answer is that (to define a manifold) X should be an open subset of K^n, and $\mathcal{O}_X$ should be its sheaf of continuous K-valued functions. The problem with this approach is that X is totally-disconnected, which makes it too easy to glue functions together. This problem will ruin an attempt to emulate the complex theory: if $X = \mathbf{P}^1$ defined this way, then $H^0(X, \mathcal{O}_X) \neq K$ (violating GAGA) and $H^0_{\mathrm{\acute{e}t}}(X, A) \neq A$ (violating the comparison isomorphism).

Nonetheless, Tate observed that some elliptic curves over K (those with multiplicative reduction) admit an explicit uniformization by $K^\times$, which hints that there should be a good theory of analytic varieties. Tate's uniformization involved power series which converged on certain sorts of domains in $K^\times$. Tate's theory of *rigid-analytic spaces* is a language which satisfies most of the desiderata of an analytic space, including GAGA and the comparison isomorphisms. A brief summary of the theory: we define the *Tate algebra* $K\langle T_1, \ldots, T_n \rangle$ to be the K-algebra of power series in $K[\![T_1, \ldots, T_n]\!]$ whose coefficients tend to zero. (Alternately, this is the completion of the polynomial ring $K[T_1, \ldots, T_n]$ with respect to the "Gauss norm".) The Tate algebra has various nice properties: it is Noetherian, all ideals are closed, and there is a bijection between the maximal spectrum Spm $K\langle T_1, \ldots, T_n \rangle$ and the closed unit disc in $\overline{K}^n$, modulo the action of $\mathrm{Gal}(\overline{K}/K)$. An *affinoid K-algebra* is a quotient of a Tate algebra.

The model spaces in the theory of rigid-analytic spaces are $\operatorname{Spm} A$, where A is an affinoid K-algebra, and Spm means the set of maximal ideals. But the topology Tate puts on $\operatorname{Spm} A$ is not the one coming from $\overline{K}^n$, and in fact is not a topology at all, but rather a *Grothendieck topology*, with a collection of "admissible opens" and a notion of "admissible open covering". With this topology, $\operatorname{Spm} A$ carries a sheaf of rigid-analytic functions, whose global sections recover A. Then a rigid-analytic space over K is a pair $(X, \mathcal{O}_X)$, where X is a set carrying a Grothendieck topology and $\mathcal{O}_X$ is a sheaf of K-algebras, which is locally isomorphic to a model space $\operatorname{Spm} A$.

Despite this quirk about Grothendieck topologies, the theory of rigid-analytic spaces has had spectacular successes as a non-archimedean analogue to complex-analytic spaces: there is a rigid-analytic GAGA theorem, comparison theorems, fascinating theorems about uniformization of curves and of Shimura varieties, new moduli spaces which are local analogues of Shimura varieties (implicated in the proof of the local Langlands correspondence for GL_n over a p-adic field).

1.3. A motivation for adic spaces. Despite these successes, the theory of rigid-analytic spaces has a few shortcomings, which are addressed by the more general theory of *adic spaces*. One is the problem with topologies, illustrated in the following examples:

EXAMPLE 1.3.1. Let $X = \operatorname{Spm} K\langle T \rangle$ be the rigid-analytic closed unit disk, and let Y be the disjoint union of the open unit disc U with the circle $S = \operatorname{Spm} K\langle T, T^{-1} \rangle$. There is an open immersion $Y \to X$, which is a bijection on the level of points. But it is not an isomorphism, because the two spaces have different Grothendieck topologies. (The trouble is that $\{U, S\}$ is not an admissible cover of X, because U is not a finite union of affinoid subdomains.)

Another example: let $X = \operatorname{Spm} K\langle T \rangle$, let α be an element of the completion of $\overline{K}$ which is transcendental over K, and let $Y \subset X$ be the union of all affinoid subdomains U which do not "contain" α, in the sense that α does not satisfy the collection of inequalities among power series which define U. Then the open immersion $Y \to X$ is once again a bijection on points. Indeed, a point $x \in X$ is a Galois orbit of roots of an irreducible polynomial $f(T) \in K[T]$. Since $f(\alpha) \neq 0$, we have $|f(\alpha)| > |t|$ for some nonzero $t \in K$, and then x belongs to the rational subdomain defined by $|f| \leq |t|$, hence it belongs to Y. However, $Y \to X$ cannot be an isomorphism: the collection of affinoid subdomains U used to define Y does not admit a finite subcover of Y, whereas (since X is affinoid), any admissible cover of X by affinoid subdomains admits a finite subcover.

In both examples there was an open immersion $Y \to X$ which is a bijection on points but which is not an isomorphism. This suggests that there are certain hidden "points" in X which Y is missing. In fact in the world of adic spaces, Y is simply the complement in X of a single point.

Another shortcoming, if we may be so greedy as to point it out, is that rigid-analytic spaces are too narrowly tailored to the class of K-affinoid algebras studied by Tate. Whereas the category of adic spaces encompasses the categories of rigid-analytic spaces, formal schemes, and even ordinary schemes. This allows to pass between these categories very easily. For instance, if X is a formal scheme over $\operatorname{Spf} \mathbf{Z}_p$ (satisfying certain finiteness assumptions), then there should be a corresponding rigid space X^{rig}, its *rigid generic fiber*. This was worked out by Berthelot

[**Ber91**], but is rather subtle: if $X = \operatorname{Spf} \mathbf{Z}_p[\![T]\!]$, then X^{rig} is the rigid-analytic open unit disc, which is not even affinoid. Whereas in the adic world, there is a formal unit disc fibered over a two-point space $\operatorname{Spa} \mathbf{Z}_p$, and its generic fiber is simply the open subset lying over the generic point $\operatorname{Spa} \mathbf{Q}_p$.

1.4. Huber rings. The model spaces in the theory of adic spaces are associated to certain topological rings A. In light of our desiderata, A should be allowed to be $\mathbf{Z}_p[\![T]\!]$, or $\mathbf{Q}_p\langle T\rangle$, or even any ring whatsoever with its discrete topology. In the first and third case, the topology of A is generated by a finitely-generated ideal. In the second case, the topology of $\mathbf{Q}_p\langle T\rangle$ certainly isn't generated by p (since this is invertible in A), but rather there is an open subring $\mathbf{Z}_p\langle T\rangle$ whose topology *is* generated by p.

DEFINITION 1.4.1. A *Huber ring*[2] is a topological ring A containing an open subring A_0 carrying the linear topology induced by a finitely generated ideal $I \subset A_0$. The ring A_0 and the ideal I are called a *ring of definition* and an *ideal of definition*, respectively. (The data of A_0 and I are part of the data of a Huber ring; only their existence is.)

A Huber ring A is *Tate* if it contains a topologically nilpotent unit. Such an element is called a *pseudo-uniformizer*.

EXAMPLE 1.4.2.
(1) Any ring A can be given the discrete topology; then A is a Huber ring with $A_0 = A$ and $I = 0$.
(2) Let K be a nonarchimedean field: this means a topological field which is complete with respect to a nontrivial nonarchimedean real-valued metric $|\ |$. Since $|\ |$ is nontrivial, K contains an element ϖ with $0 < |\varpi| < 1$, which is then a pseudo-uniformizer of K. Then K is a Huber ring, $K^\circ = \{|x| \le 1\}$ is a ring of definition, and (ϖ) is an ideal of definition.
(3) Continuing with the previous example, we have the Tate K-algebra $A = K\langle T_1, \ldots, T_n\rangle$. This is a Tate Huber ring. The subring $K^\circ\langle T_1, \ldots, T_n\rangle$ is a ring of definition, and (ϖ) is an ideal of definition.
(4) Let R be any ring with its discrete topology. Then the power series ring $A = R[\![T_1, \ldots, T_n]\!]$ is a Huber ring which is not Tate. Then A itself is a ring of definition, and $(T_1, \ldots, T_n)$ is an ideal of definition.
(5) Similarly, if K is a nonarchimedean field with pseudouniformizer ϖ, then $A = K^\circ[\![T_1, \ldots, T_n]\!]$ is a Huber ring. Then A itself is a ring of definition, and $(\varpi, T_1, \ldots, T_n)$ is an ideal of definition.
(6) Let K be a nonarchimedean field which is perfect of characteristic p. The ring of Witt vectors $A = W(K^\circ)$ is a Huber ring, A itself is a ring of definition, and $(p, [\varpi])$ is an ideal of definition.
(7) Let $A = \mathbf{Q}_p[\![T]\!]$. It is tempting to say that A is a Huber ring, with a ring of definition $A_0 = \mathbf{Z}_p[\![T]\!]$ and an ideal of definition (p, T). But in fact one cannot put a topology on A which makes this work. Indeed, in such a topology $T^n \to 0$, and since multiplication by p^{-1} is continuous, $p^{-1}T^n \to 0$ as well. But this sequence never enters A_0, and therefore $A_0 \subset A$ is not open. (It is fine to say that $\mathbf{Q}_p[\![T]\!]$ is a Huber ring with ring of definition $\mathbf{Q}_p[\![T]\!]$ and ideal of definition (T), but then you are

---

[2]called an f-adic ring by Huber [**Hub94**].

artificially suppressing the topology of $\mathbf{Q}_p$, so that the sequence p^n does not approach 0.) There is a similar obstruction to $\mathbf{Z}_p[\![T]\!][1/p]$ being a Huber ring.

We need a few more basic definitions.

DEFINITION 1.4.3. A subset S of a topological ring A is *bounded* if for all open neighborhoods U of 0, there exists an open neighborhood V of 0 such that $VS \subset U$. An element $f \in A$ is *power-bounded* if $\{f^n\} \subset A$ is bounded. Let A° be the subset of power-bounded elements. If A is linearly topologized (for instance if A is Huber) then $A^\circ \subset A$ is a subring.

A Huber ring A is *uniform* if $A^\circ \subset A$ is bounded.

All of the Huber rings in Example 1.4.2 are uniform. A non-uniform Huber ring is $A = \mathbf{Q}_p[T]/T^2$, because $A^\circ = \mathbf{Z}_p + \mathbf{Q}_p T$ is unbounded.

REMARK 1.4.4. In a uniform Huber ring A, the power-bounded subring $A_0 \subset A$ serves as a ring of definition. Complete uniform Huber rings which are Tate are especially convenient because they are Banach rings. Indeed, suppose A is a uniform Tate Huber ring, and let $\varpi \in A$ be a pseudo-uniformizer. Then the topology on A is induced from the norm
$$|a| = 2^{\inf\{n:\, \varpi^n a \in A^\circ\}}.$$

1.5. Continuous valuations. The idea now is to associate to a Huber ring A a ringed space $\operatorname{Spa} A = (X, \mathcal{O}_X)$, which will serve as the model space for the theory of adic spaces. The points of X are quite interesting: they correspond to continuous valuations on the ring A.

Recall that an *ordered abelian group* is an abelian group Γ endowed with a translation-invariant total order $\leq$. These will be written multiplicatively. Examples include $\mathbf{R}_{>0}$ and any subgroup thereof. Another example is $\Gamma = \mathbf{R}_{>0} \times \mathbf{R}_{>0}$ under its *lexicographical ordering*: $(a,b) \leq (c,d)$ means that either $a < c$ or else $a = c$ and $b \leq d$. A feature of this Γ is that it contains $\mathbf{R}_{>0}$ (embedded along the first coordinate) together with, for each $a \in \mathbf{R}_{>0}$, elements (such as $(a, 1/2)$, respectively $(a, 2)$) which are between a and every real number less than (respectively, greater than) a. This concept easily generalizes to finite products $\mathbf{R}_{>0}^n$, or even infinite products of $\mathbf{R}_{>0}$ indexed by an ordinal.

DEFINITION 1.5.1. For an ordered abelian group Γ, a subgroup $\Gamma' \subset \Gamma$ is *convex* if any element of Γ lying between two elements of Γ' itself lies in Γ'.

It is a nice exercise to show that if $\Gamma', \Gamma'' \subset \Gamma$ are two convex subgroups then either $\Gamma' \subset \Gamma''$ or $\Gamma'' \subset \Gamma'$. Therefore the set of nontrivial convex subgroups forms a totally ordered set with respect to inclusion. The cardinality of this set is called the *rank* of Γ. Thus the rank of $\mathbf{R}_{>0}^n$ is n.

The condition for Γ to be rank 1 is equivalent to the following archimedean property: given $a, b \in \Gamma$ with $a > 1$, then there exists $n \in \mathbf{Z}$ with $b < a^n$. We remark that a rank 1 ordered abelian group can always be embedded into $\mathbf{R}_{>0}$.

DEFINITION 1.5.2. Let A be a topological ring. A *continuous valuation* on A is a map
$$|\cdot| : A \to \Gamma \cup \{0\},$$

where Γ is a totally ordered abelian group, and $\Gamma \cup \{0\}$ is the ordered monoid with least element 0. It is required that
- $|ab| = |a|\,|b|$,
- $|a+b| \leq \max(|a|,|b|)$,
- $|1| = 1$,
- $|0| = 0$,
- (Continuity) For all $\gamma \in \Gamma$, $\left\{ a \in A \,\middle|\, |a| < \gamma \right\}$ is open in A.

Two continuous valuations $|\ |: A \to \Gamma \cup \{0\}$ and $|\ |': A \to \Gamma' \cup \{0\}$ are *equivalent* if for all $a,b \in A$ we have $|a| \geq |b|$ if and only if $|a|' \geq |b|'$. In that case, after replacing Γ by the subgroup generated by the image of A, and similarly for Γ', there exists an isomorphism $\iota \colon \Gamma \cong \Gamma'$ such that $\iota(|a|) = |a|'$ for all $a \in A$.

Note that the kernel of $|\cdot|$ is a prime ideal of A which only depends on the equivalence class of $|\cdot|$.

DEFINITION 1.5.3. Let $\mathrm{Cont}(A)$ denote the set of equivalence classes of continuous valuations of A. For an element $x \in \mathrm{Cont}(A)$, we use the notation $f \mapsto |f(x)|$ to denote a continuous valuation representing x. We give $\mathrm{Cont}(A)$ the topology generated by subsets of the form $\left\{ x \,\middle|\, |f(x)| \leq |g(x)| \neq 0 \right\}$, with $f, g \in A$. For $x \in \mathrm{Cont}(A)$, the rank of x is the rank of the ordered abelian group generated by the image of a continuous valuation representing x.

Some remarks on the topology of $\mathrm{Cont}(A)$: Note that sets of the form $\{|g(x)| \neq 0\}$ are open, as are sets of the form $\{|f(x)| \leq 1\}$. This blends features of the Zariski topology on schemes and topology on rigid spaces. Furthermore, $\mathrm{Cont}(A)$ is quasi-compact, just as the spectrum of a ring is quasi-compact.

When A is a Huber ring, the set $\mathrm{Cont}(A)$ is a good candidate for the model space we want to build. For instance if A is a discrete ring, then $\mathrm{Cont}(A)$ contains one point x for each prime $\mathfrak{p} \in \mathrm{Spec}\,A$, namely the valuation pulled back from the trivial valuation on the residue field of $\mathfrak{p}$. The set $\mathrm{Cont}(\mathbf{Q}_p)$ is a single point, namely the equivalence class of the usual p-adic valuation on $\mathbf{Q}_p$.

Now consider $\mathrm{Cont}(\mathbf{Q}_p\langle T\rangle)$, which is our hypothetical "adic closed unit disc". For each maximal ideal $\mathfrak{m} \in \mathrm{Spm}\,\mathbf{Q}_p\langle T\rangle$, we do get a point in $\mathrm{Cont}(A)$ by pulling back the valuation on the nonarchimedean field $\mathbf{Q}_p\langle T\rangle/\mathfrak{m}$ (this is a finite extension of $\mathbf{Q}_p$). Thus there is a map $\mathrm{Spm}\,\mathbf{Q}_p\langle T\rangle \to \mathrm{Cont}\,\mathbf{Q}_p\langle T\rangle$. But the latter set contains many more points. For instance, we can let $\alpha \in \mathbf{C}_p$ be a transcendental element with $|\alpha| \leq 1$, and define a continuous valuation on $\mathbf{Q}_p\langle T\rangle$ by $f \mapsto |f(\alpha)|$. This is going to address one of the problems in classical rigid geometry brought up in Example 1.3.1.

Addressing the other problem brought up in that example, we can also define an element $x^{-} \in \mathrm{Cont}\,\mathbf{Q}_p\langle T\rangle$ as follows: let $\Gamma = \mathbf{R}_{>0} \times \gamma^{\mathbf{Z}}$, where the order is determined by the relations $a < \gamma < 1$ for all real $a < 1$. (If you like, Γ can be embedded as a subgroup of $\mathbf{R}_{>0} \times \mathbf{R}_{>0}$ by $a\gamma^n \mapsto (a, 1/2^n)$.) Now define x^{-} by

$$\sum_{n=0}^{\infty} a_n T^n \mapsto \sup_{n \geq 0} |a_n| \gamma^n.$$

Thus x^{-} "thinks" that T is infinitesimally smaller than one: we have $|T(x^{-})| = \gamma < 1$, but $|T(x^{-})| > |a|$ for all $a \in \mathbf{Q}_p$ with $|a| < 1$. The point x^{-} prevents us from disconnecting $\operatorname{Cont}\mathbf{Q}_p\langle T\rangle$ by the disjoint open sets $\cup_{n\geq 1}\{|T^n(x)|<|p|\}$ and $\{|T(x)|=1\}$, because neither of these contains x^-!

However, this example suggests that we have more points in $\operatorname{Cont}\mathbf{Q}_p\langle T\rangle$ than we bargained for. There is also a point x^+ with the same definition, except that γ is now infinitesimally greater than 1. Morally, whatever the closed adic disc is, it should not contain any points which think that T is greater than 1, and so we need to modify our model spaces a little.

1.6. Integral subrings.

DEFINITION 1.6.1. Let A be a Huber ring. A subring $A^+ \subset A$ is a *ring of integral elements* if it is open and integrally closed and $A^+ \subset A^\circ$. A *Huber pair*[3] is a pair (A, A^+), where A is Huber and $A^+ \subset A$ is a ring of integral elements. Given a Huber pair, we let $\operatorname{Spa}(A, A^+) \subset \operatorname{Cont}(A)$ be the subset (with its induced topology) of continuous valuations x for which $|f(x)| \leq 1$ for all $f \in A^+$. We will sometimes write $\operatorname{Spa} A$ for $\operatorname{Spa}(A, A^\circ)$.

We remark that $\operatorname{Spa}(A, A^+)$ is always quasi-compact.

Thus the closed adic disc should be $\operatorname{Spa}(A, A^+)$, where $A = \mathbf{Q}_p\langle T\rangle$ and $A^+ = A^\circ = \mathbf{Z}_p\langle T\rangle$. But one could also define an integral subring

$$A^{++} = \left\{\sum_{n=0}^\infty a_n T^n \in A^+ \;\middle|\; |a_n| < 1 \text{ for all } n \geq 1\right\}.$$

We have $A^{++} \subset A^+$, and so $\operatorname{Spa}(A, A^+) \subset \operatorname{Spa}(A, A^{++})$. In fact the complement of $\operatorname{Spa}(A, A^+)$ in $\operatorname{Spa}(A, A^{++})$ is the single point x^+ from our discussion above. Furthermore, if we embed $\operatorname{Spa}(A, A^+)$ into an adic closed disc of larger radius, then it will be an *open* subset of the larger disc, and its closure will be $\operatorname{Spa}(A, A^{++})$.

1.7. The classification of points in the adic unit disc. Suppose C is a nonarchimedean field which is algebraically closed, and suppose that $\alpha \mapsto |\alpha|$ is an absolute value inducing the topology on C. We review here the classification of points in $X = \operatorname{Spa}(C\langle T\rangle, C^\circ\langle T\rangle)$ as in [**Sch12**]. The points of X are divided into five types; we warn that this division into five types breaks down for other adic spaces. Generally, one may work with adic spaces without consciously knowing what each point looks like.

- Points of Type 1 correspond to elements $\alpha \in C$ with $|\alpha| \leq 1$. The corresponding continuous valuation is $f \mapsto |f(\alpha)|$.
- Points of Type 2 and 3, also called Gauss points, correspond to closed discs $D = D(\alpha, r)$. Here $\alpha \in C$ has $|\alpha| \leq 1$, $0 < r \leq 1$ is a real number, and $D = \left\{\beta \in C \;\middle|\; |\alpha - \beta| \leq r\right\}$. The corresponding valuation is

$$f \mapsto \sup_{\beta \in D} |f(\beta)|.$$

  Explicitly, if we expand f as a series in $T-\alpha$, say $f(T) = \sum_{n=0}^\infty a_n(T-\alpha)^n$, then this works out to be $\sup_n |a_n| r^n$.

  If r belongs to $|C|$, then the point is of Type 2; otherwise it is of Type 3.

---

[3]Called an *affinoid algebra* in [**Hub94**].

- Points of Type 4 appear because of the strange phenomenon that C may not be *spherically complete*. That is, there may be a descending sequence of closed discs $D_1 \supset D_2 \supset \cdots$ with empty intersection. (For instance, this occurs when $C = \mathbf{C}_p$.) The corresponding continuous valuation is $f \mapsto \inf_i \sup_{\beta \in D_i} |f(\beta)|$.
- Points of Type 5 have rank 2. For each $\alpha \in C$ with $|\alpha| \leq 1$, each $0 < r \leq 1$, and each sign $\pm$ (excluding the positive sign if $r = 1$), we let $\Gamma = \mathbf{R}_{>0} \times \gamma^{\mathbf{Z}}$ be the ordered abelian group generated by $\mathbf{R}_{>0}$ and an element γ which is infinitesimally less than or greater than r, depending on the sign. The corresponding continuous valuation is

$$\sum_{n=0}^{\infty} a_n (T - \alpha)^n \mapsto \sup_n |a_n| \gamma^n.$$

If C has value group $\mathbf{R}_{>0}$, then there are no points of Type 3. If C is spherically complete, then there are no points of Type 4 either: every descending sequence of closed discs has an intersection which is either itself a closed disc or a single point.

The only non-closed points in X are the Type 2 points, which correspond to discs D: the closure of such a point contains all Type 5 labeled with a triple $(\alpha, r, \pm)$, where $D = D(\alpha, r)$.

1.8. The structure presheaf, and the definition of an adic space. In the construction of affine schemes, one starts with a ring A, defines the topological space $X = \operatorname{Spec} A$, and then defines the structure sheaf $\mathcal{O}_X$ this way: there is a basis of open sets of the form $U_f = \{x | f(x) \neq 0\}$ for $f \in A$, and one puts $\mathcal{O}_X(U_f) = A[1/f]$; it is easy enough to check that there is a unique sheaf of rings $\mathcal{O}_X$ with this property. (Here we use the notational convention that if x corresponds to a prime ideal $\mathfrak{p} \subset A$, then $f(x)$ is the image of x in the residue field of $\mathfrak{p}$.) The idea behind this definition is that U_f should be an affine scheme in its own right, namely $\operatorname{Spec} A[1/f]$. The key observation here is that $\operatorname{Spec} A[1/f] \to \operatorname{Spec} A$ is an open immersion with image U_f, and is universal for this property in the sense that for any A-algebra B, the map $\operatorname{Spec} B \to \operatorname{Spec} A$ factors through U_f if and only if $A \to B$ factors through $A[1/f]$.

It is somewhat more subtle to define $\mathcal{O}_X$ for $X = \operatorname{Spa}(A, A^+)$, where (A, A^+) is a Huber pair. We single out a class of open sets called rational subsets.

DEFINITION 1.8.1. Let $s_1, \ldots, s_n \in A$ and let $T_1, \ldots, T_n \subset A$ be finite subsets such that $T_i A \subset A$ is open for all i. We define a subset

$$U\left(\left\{\frac{T_i}{s_i}\right\}\right) = U\left(\frac{T_1}{s_1}, \ldots, \frac{T_n}{s_n}\right) = \left\{x \in X \;\middle|\; |t_i(x)| \leq |s_i(x)| \neq 0,\ \text{for all}\ t_i \in T_i\right\}.$$

This is open because it is an intersection of a finite collection of the sort of opens which generate the topology on X. Subsets of this form are called *rational subsets*.

Note that a finite intersection of rational subsets is again rational, just by concatenating the data that define the individual rational subsets.

The following theorem shows that rational subsets are themselves adic spectra.

THEOREM 1.8.2 ([**Hub94**], Proposition 1.3]). *Let $U \subset \operatorname{Spa}(A, A^+)$ be a rational subset. Then there exists a complete Huber pair $(A, A^+) \to (\mathcal{O}_X(U), \mathcal{O}_X^+(U))$ such that the map $\operatorname{Spa}(\mathcal{O}_X(U), \mathcal{O}_X^+(U)) \to \operatorname{Spa}(A, A^+)$ factors over U, and is final* *among such maps. Moreover, this map is a homeomorphism onto U. In particular, U is quasi-compact.*

DEFINITION 1.8.3. Define a presheaf $\mathcal{O}_X$ of topological rings on $\mathrm{Spa}(A, A^+)$: If $U \subset X$ is rational, $\mathcal{O}_X(U)$ is as in the theorem. On a general open $W \subset X$, we define
$$\mathcal{O}_X(W) = \varprojlim_{U \subset W \text{ rational}} \mathcal{O}_X(U).$$
We defines $\mathcal{O}_X^+$ analogously. If $\mathcal{O}_X$ is a sheaf, we call (A, A^+) a *sheafy* Huber pair.

PROPOSITION 1.8.4. *For all $U \subset X = \mathrm{Spa}(A, A^+)$,*
$$\mathcal{O}_X^+(U) = \left\{ f \in \mathcal{O}_X(U) \,\bigg|\, |f(x)| \leq 1, \text{ for all } x \in U \right\}.$$
In particular, $\mathcal{O}_X^+$ is a sheaf if $\mathcal{O}_X$ is. If (A, A^+) is complete, then $\mathcal{O}_X(X) = A$ and $\mathcal{O}_X^+(X) = A^+$.

Let (A, A^+) be a sheafy Huber pair, and let $X = \mathrm{Spa}\, A$. Then $(X, \mathcal{O}_X)$ is a locally ringed topological space, and $\mathcal{O}_X$ is a sheaf of topological rings. The locally ringed space $(X, \mathcal{O}_X)$ comes equipped with some extra data: for each $x \in \mathrm{Spa}\, A$, we have a continuous valuation $|\cdot|_x$ on the local ring $\mathcal{O}_{X,x}$. (Note that $\mathcal{O}_X^+$ can be recovered from the data of these valuations, by Proposition 1.8.4.)

We can now define the category of adic spaces.

DEFINITION 1.8.5. An *adic space* is a triple $(X, \mathcal{O}_X, \{|\cdot|_x\}_{x \in X})$, where $(X, \mathcal{O}_X)$ is locally ringed topological space, $\mathcal{O}_X$ is a sheaf of complete topological rings, and for each $x \in X$, $|\cdot|_x$ is a continuous valuation on $\mathcal{O}_{X,x}$. We require that locally on X, this is the triple associated to $\mathrm{Spa}(A, A^+)$, where (A, A^+) is a sheafy Huber pair. A morphism between adic spaces is a morphism between locally ringed topological spaces, which is compatible with the topology on $\mathcal{O}_X$ and with the given valuations $|\cdot|_{x \in X}$, in the evident manner.

Of course one wants some criteria for determining whether a given Huber pair is sheafy.

THEOREM 1.8.6 ([**Hub94**]). *A Huber pair (A, A^+) is sheafy in the following situations.*
(1) *The ring A is discrete. Thus, there is a functor from schemes to adic spaces, which sends $\mathrm{Spec}\, A$ to $\mathrm{Spa}(A, A)$.*
(2) *The ring A is finitely generated (as an algebra) over a noetherian ring of definition. Thus, there is a functor from noetherian formal schemes to adic spaces, which sends $\mathrm{Spf}\, A$ to $\mathrm{Spa}(A, A)$.*
(3) *The ring A is Tate and strongly noetherian, which means that the rings*
$$A\langle X_1, \ldots, X_n \rangle = \left\{ \sum_{\underline{i}=(i_1,\ldots,i_n) \geq 0} a_{\underline{i}} T^{\underline{i}} \,\bigg|\, a_{\underline{i}} \in A,\ a_{\underline{i}} \to 0 \right\}$$
are noetherian for all $n \geq 0$. Thus there is a functor from rigid spaces over a nonarchimedean field K to adic spaces over $\mathrm{Spa}\, K$, which sends $\mathrm{Spm}\, A$ to $\mathrm{Spa}(A, A^\circ)$ for an affinoid K-algebra A.

EXAMPLE 1.8.7 (The adic closed disc over $\mathbf{Q}_p$). Let $A = \mathbf{Q}_p \langle T \rangle$, and let $A^+ = A^\circ = \mathbf{Z}_p \langle T \rangle$. Then $\mathrm{Spa}(A, A^+)$ is the adic closed disc over $\mathbf{Q}_p$.

EXAMPLE 1.8.8 (The adic open disc over $\mathbf{Q}_p$). Let $A = \mathbf{Z}_p[\![T]\!]$. Since A is its own ring of definition and is noetherian, (A, A) is sheafy and $\mathrm{Spa}(A, A)$ is an adic space. We have a morphism $\mathrm{Spa}(A, A) \to \mathrm{Spa}(\mathbf{Z}_p, \mathbf{Z}_p)$. The latter is a two-point space, with generic point $\eta = \mathrm{Spa}(\mathbf{Q}_p, \mathbf{Z}_p)$. The generic fiber of $\mathrm{Spa}(A, A)$ is $\mathrm{Spa}(A, A)_\eta$, the preimage of η. It is worthwhile to study this space in detail.

Let $x \in \mathrm{Spa}(A, A)_\eta$. We have $|p(x)| \neq 0$. We also know that since p and T are topologically nilpotent in A, we have $|T(x)|^n \to 0$ as $n \to \infty$. Therefore, there exists an $n \geq 0$ with $|T^n(x)| \leq |p(x)|$. This means that x lies in the rational subset $U(T^n/p)$. From this we see that the increasing sequence of rational subsets $U(T^n/p)$ covers $\mathrm{Spa}(A, A)_\eta$. Since this covering has no finite subcovering, we can conclude that $\mathrm{Spa}(A, A)_\eta$ is not quasi-compact.

EXAMPLE 1.8.9 (The adic affine line over $\mathbf{Q}_p$). Let D be the adic closed disc over $\mathbf{Q}_p$. We let $\mathbf{A}^1_{\mathbf{Q}_p} = \varinjlim D$, where the colimit is taken over the transition map $T \mapsto pT$. Put another way, $\mathbf{A}^1_{\mathbf{Q}_p}$ is the ascending union of closed discs of unbounded radius. Then $\mathbf{A}^1_{\mathbf{Q}_p}$ is not quasi-compact. As we remarked earlier, the closure of the unit disc $D \subset \mathbf{A}^1_{\mathbf{Q}_p}$ is $\mathrm{Spa}(A, A^{++})$ for a strict subring $A^{++} \subset A^\circ$.

EXAMPLE 1.8.10 (The projective line over $\mathbf{Q}_p$). Let D be the adic closed disc over $\mathbf{Q}_p$. The projective line $\mathbf{P}^1_{\mathbf{Q}_p}$ is obtained by gluing together two copies of D along the map $T \mapsto T^{-1}$ on the "circle" $\{|T| = 1\}$. Then $\mathbf{P}^1_{\mathbf{Q}_p}$ contains $\mathbf{A}^1_{\mathbf{Q}_p}$ as an open subspace; the complement is a single point.

1.9. Partially proper adic spaces. Given an adic space X, one can consider its *functor of points*: whenever (R, R^+) is a complete sheafy Huber pair, we define $X(R, R^+)$ to be the set of morphisms from $\mathrm{Spa}(R, R^+)$ to X. We also have the relative version of this functor: If X is fibered over a base space S, then we may consider the relative functor of points on the category of morphisms $\mathrm{Spa}(R, R^+) \to S$, which sends such an object to the set of S-morphisms $\mathrm{Spa}(R, R^+) \to X$. Since every adic space is covered by affinoid spaces, an adic space is determined by its functor of points.

Let us compute the functor of points for the examples in the previous section.

EXAMPLE 1.9.1. Let (R, R^+) be a complete sheafy Huber pair over $(\mathbf{Q}_p, \mathbf{Z}_p)$.

(1) Let D be the closed unit disc over $\mathbf{Q}_p$. Then
$$D(R, R^+) = \mathrm{Hom}(\mathbf{Z}_p\langle T\rangle, R^+) \cong R^+$$
(via $f \mapsto f(T)$). (The Hom here and below is in the category of topological $\mathbf{Z}_p$-algebras.)

(2) Let D° be the open unit disc over $\mathbf{Q}_p$. Then
$$D^\circ(R, R^+) = \mathrm{Hom}(\mathbf{Z}_p[\![T]\!], R^+) \cong R^{\circ\circ}$$
is the set of topologically nilpotent elements of R, again via $f \mapsto f(T)$. Now, *a priori* the image is $R^{\circ\circ} \cap R^+$. However, the fact that R^+ is open and integrally closed means that if $a \in R^{\circ\circ}$, then $a^n \in R^+$ for n large enough, and thus $a \in R^+$. Thus, $R^{\circ\circ} \subset R^+$.

(3) Let $\mathbf{A}^1_{\mathbf{Q}_p}$ be the adic affine line over $\mathbf{Q}_p$. Then
$$\mathbf{A}^1_{\mathbf{Q}_p}(R, R^+) = R.$$

If $\overline{D}$ is the closure of D in $\mathbf{A}^1_{\mathbf{Q}_p}$, then

$$\overline{D}(R,R^+) = \mathrm{Hom}(A^{++}, R^+) = \left\{ a \in R \;\Big|\; pa^n \in R^\circ \text{ for all } n \geq 1 \right\}.$$

Again, *a priori* the condition on a is that $pa^n \in R^+$ for all $n \geq 1$. But if $pa^n \in R^\circ$ for all $n \geq 1$, then also $(pa^n)^2 = p(pa^{2n}) \in pR^\circ \subset R^+$, so $pa^n \in R^+$ as well.

(4) Let $\mathbf{P}^1_{\mathbf{Q}_p}$ be the adic projective line over $\mathbf{Q}_p$. Then $\mathbf{P}^1_{\mathbf{Q}_p}(R, R^+)$ is the set of projective rank 1 quotients of R^2.

DEFINITION 1.9.2. Let X be an adic space. We say X is *partially proper* if it is quasi-separated[4] and if for every sheafy Huber pair (R, R^+) and every morphism $\mathrm{Spa}(R, R^\circ) \to X$, there exists a unique morphism $\mathrm{Spa}(R, R^+) \to X$ making the diagram commute:

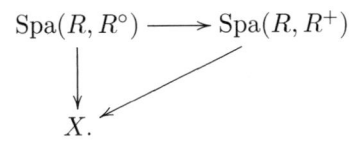

$$\begin{array}{ccc} \mathrm{Spa}(R, R^\circ) & \longrightarrow & \mathrm{Spa}(R, R^+) \\ \downarrow & \swarrow & \\ X. & & \end{array}$$

Thus if X is partially proper, $X(R, R^+) = X(R, R^\circ)$ only depends on R. Finally, X is *proper* if it is quasi-compact and partially proper.

There is a relative definition of partial properness for a morphism $X \to S$, which we leave to the reader to work out. Note that the definition of partial properness is similar to the valuative criteria for properness and separatedness for schemes. There is also a definition of properness involving universally closed morphisms, cf. [**Hub96**].

Intuitively, a space is partially proper when it has no boundary. In the examples above, D°, $\overline{D}$, $\mathbf{A}^1_{\mathbf{Q}_p}$ and $\mathbf{P}^1_{\mathbf{Q}_p}$ are partially proper, but of these only $\overline{D}$ and $\mathbf{P}^1_{\mathbf{Q}_p}$ are proper. Then D is not partially proper, as its functor of points really depends on R^+.

2. Perfectoid fields

We are now going to take a sudden change of direction to talk about perfectoid fields. The idea is that perfectoid fields are the one-point perfectoid spaces, so they are rather a prerequisite to study perfectoid spaces in general. Besides, perfectoid fields have an interesting history, even if the name and formal definition did not appear until [**Sch12**] and [**KL15**].

A class of perfectoid fields plays a crucial role in Tate's study of p-divisible groups [**Tat67**]. Let K be the fraction field of a mixed-characteristic discrete valuation ring with perfect residue field of characteristic p (*e.g.*, a finite extension of $\mathbf{Q}_p$). Tate considered a tower of Galois extensions K_n/K satisfying the conditions (a) $\mathrm{Gal}(K_n/K) \cong (\mathbf{Z}/p^n\mathbf{Z})^h$ for some $h \geq 1$ and (b) K_n/K is totally ramified. (For Tate, such a tower came by adjoining the torsion in a p-divisible group.) Let $K_\infty = \cup_n K_n$ and let $\widehat{K}_\infty$ be its completion.

---

[4] A topological space is quasi-separated if the intersection of any two quasi-compact open subsets of X is again quasi-compact. If (A, A^+) is a Huber pair, then $\mathrm{Spa}(A, A^+)$ is quasi-separated.

Let C be the completion of an algebraic closure of K. Tate proved some basic facts about the cohomology of C as a $\mathrm{Gal}(\overline{K}/K)$-module, using K_∞ as an intermediary. (The ultimate goal was to prove a p-adic Hodge decomposition for p-divisible groups and abelian varieties.) Along the way he proved a curious fact: if L/K_∞ is a finite extension, then the ideal of K_∞° generated by traces of elements of L° contains the maximal ideal $\mathfrak{m}_{K_\infty}$ of K_∞°. (Thus it is either $\mathfrak{m}_{K_\infty}$ or else it is all of K_∞°.) Now, if L were instead a finite extension of K, then this ideal of traces is related to the different ideal of L/K, and measures the ramification: the bigger the ideal, the less ramified L/K is. Tate's result is that any finite extension of K_∞ is *almost unramified*, or put another way, the corresponding extension of K_∞° is *almost étale*.

The next work along these lines comes from Fontaine and Wintenberger [**FW79**]. They considered a more general infinite algebraic extension K_∞/K which is highly ramified, in the technical sense that $G_K^u G_{K_\infty} \subset G_K$ is open for all $u \geq -1$, where G_K^u is a higher ramification group. Such extensions are called *arithmetically profinite* (APF). For instance, if K_∞/K is a totally ramified Galois extension with $\mathrm{Gal}(K_\infty/K)$ a p-adic Lie group, then K_∞/K is APF. To such an extension, Fontaine and Wintenberger attached a nonarchimedean field X, the *field of norms*, whose multiplicative monoid is the inverse limit $\varprojlim K_n$, where the transition maps in the limit are norms. The field X has characteristic p; in fact it is a Laurent series field over the residue field of K. Rather surprisingly, we have an isomorphism of Galois groups $\mathrm{Gal}(\overline{X}/X) \cong \mathrm{Gal}(\overline{K}/K_\infty)$. This isomorphism is fundamental to the classification of p-adic Galois representations via (ϕ, Γ)-modules (see [**Ked15**] for a discussion of these) and the proof of the p-adic local Langlands correspondence for $\mathrm{GL}_2(\mathbf{Q}_p)$ [**Col10**].

The themes of almost étale extensions and passage to characteristic p are the hallmarks of perfectoid fields, which we now define.

DEFINITION 2.0.1. A nonarchimedean field K of residue characteristic p is a *perfectoid field* if (a) its value group is nondiscrete, and (b) the pth power Frobenius map on K°/p is surjective.

EXAMPLE 2.0.2.
(1) The basic examples of perfectoid fields are the completions of $\mathbf{Q}_p(\mu_{p^\infty})$ and $\mathbf{Q}_p(p^{1/p^\infty})$. The completion of any strictly APF extension is perfectoid.
(2) One source of APF extensions (and therefore perfectoid fields) comes from *p-divisible formal group laws*. Let E be a local field of characteristic 0 with residue characteristic p and uniformizer π. Recall that a 1-dimensional formal group law over $\mathcal{O}_E$ is a power series $\mathcal{F}(X, Y) = X + Y +$ higher order terms in $\mathcal{O}_E[\![X, Y]\!]$ which satisfies the axioms of an abelian group. Iterating $\mathcal{F}$ on itself p times produces a power series $[p]_\mathcal{F}(T)$. If $[p]_\mathcal{F}(T)$ modulo π is nonzero, then $\mathcal{F}$ is p-divisible; in that case $[p]_\mathcal{F}(T) \mod \pi = g(T^{p^h})$ for some power series g and some maximal h, called the height of $\mathcal{F}$. The set of roots $\mathcal{F}[p^n]$ of $[p^n]_\mathcal{F}$ is isomorphic to $(\mathbf{Z}/p^n\mathbf{Z})^h$. Let $E_\infty = E(\mathcal{F}[p^\infty])$ be the field obtained by adjoining all p-power torsion points to E. The extension E_∞/E is APF, and therefore the completion of E_∞ is perfectoid.
(3) If a nonarchimedean field has characteristic p, then it is perfectoid if and only if it is perfect. A basic example is $k(\!(t^{1/p^\infty})\!)$, where $k/\mathbf{F}_p$ is a perfect

field: this is defined to be the completion of the perfection of $k((t))$. This example is rather fundamental: if K is a perfectoid field of characteristic p and residue field k, then K contains $k((t^{1/p^\infty}))$, where t is any element of K with $0 < |t| < 1$.

2.1. Tilting. Let K be a perfectoid field with absolute value $|\ |$. We let $K^\circ = \{|x| \leq 1\}$ be its ring of integers.

We define
$$K^\flat = \varprojlim K,$$
where the transition map is $x \mapsto x^p$. Thus, elements of $K^\flat$ are sequences $(a_0, a_1, \dots)$ of elements of K with $a_n^p = a_{n-1}$ for all $n \geq 1$. (If K has characteristic p, then trivially $K^\flat \cong K$; this operation is only interesting in characteristic 0.) *A priori* $K^\flat$ is a topological multiplicative monoid. We define an addition law on $K^\flat$ by the rule $(a_n) + (b_n) = (c_n)$, where

(2.1.1) $$c_n = \lim_{m \to \infty} (a_{m+n} + b_{m+n})^{p^m}.$$

It it easy to check that the limit exists (here we use the fact that K is complete). It can be verified directly that $K^\flat$ is a field, but the easiest route is to pass to the quotient K°/p. The reduction map $K^\circ \to K^\circ/p$ induces a map of topological multiplicative monoids
$$\varprojlim_{x \mapsto x^p} K^\circ \to \varprojlim_{x \mapsto x^p} K^\circ/p.$$
Now one observes that this map is an isomorphism; the inverse sends a sequence $(a_n \bmod p)$ to (b_n), where
$$b_n = \lim_{m \to \infty} a_{m+n}^{p^m}.$$
(The limit does not depend on the choice of lift of a_n.) Therefore $\varprojlim K^\circ$ inherits the structure of a ring, with addition law as in (2.1.1); its fraction field is $K^\flat$. Let $f \mapsto f^\sharp$ denote the projection map $K^\flat \to K$ which sends (a_n) to a_0. We define an absolute value on $K^\flat$ by $|f| = |f^\sharp|$. One checks that this is a nontrivial nonarchimedean absolute value inducing the topology on $K^\flat$, and that $K^\flat$ is complete with respect to it. Finally, the very definition of $K^\flat$ shows that it is perfect of characteristic p. Therefore $K^\flat$ is a perfectoid field of characteristic p; it is called the *tilt* of K.

The perfectoid field $K^\flat$ contains a pseudo-uniformizer ϖ with $|\varpi| = |p|$. An important observation is that $K^{\flat\circ} \cong \varprojlim_{x \mapsto x^p} K^\circ/p$, and that
$$K^{\flat\circ}/\varpi \cong K^\circ/p.$$

EXAMPLE 2.1.1.
(1) Let $K = \mathbf{Q}_p(p^{1/p^\infty})^\wedge$. Then $K^\flat$ contains the element $t = (p, p^{1/p}, \dots)$ with $|t| = |p|$. Thus t is a pseudo-uniformizer of $K^\flat$, and since $K^\flat$ is perfectoid, $K^\flat$ contains $\mathbf{F}_p((t^{1/p^\infty}))$ (as remarked in Example 2.0.2). In fact $K^\flat = \mathbf{F}_p((t^{1/p^\infty}))$. To see this, observe that $K^\circ/p = \mathbf{Z}_p[p^{1/p^\infty}]/p \cong \mathbf{F}_p[t^{1/p^\infty}]/t$, and apply $\varprojlim$ along $x \mapsto x^p$ to both sides.
(2) If $K = \mathbf{Q}_p(\mu_{p^\infty})^\wedge$, then $K^\flat$ (considered as the fraction field of $\varprojlim K^\circ/p$) contains the element $t = (1 - \zeta_p, 1 - \zeta_{p^2}, \dots)$, and then once again

$K^\flat = \mathbf{F}_p((t^{1/p^\infty}))$. In fact if K is the completion of any APF extension of a p-adic field (see Example 2.0.2), then $K^\flat \cong k((t^{1/p^\infty}))$, where k is the residue field of K.

2.2. The tilting equivalence for perfectoid fields. For a perfectoid field K of characteristic 0, the structures of K and $K^\flat$ seem quite different: of course their characteristics are different, and even though there is a multiplicative map $K^\flat \to K$ ($f \mapsto f^\sharp$), this is far from being surjective in general. Nonetheless we will encounter a family of theorems known as *tilting equivalences* which relate the arithmetic of a perfectoid object and its tilt. The most basic tilting equivalence concerns the Galois groups of perfectoid fields.

THEOREM 2.2.1. *Let K be a perfectoid field of characteristic 0. Then for any finite extension L/K (necessarily separable), L is also a perfectoid field, and $L^\flat/K^\flat$ is a finite extension of the same degree as L/K. The categories of finite extensions of K and $K^\flat$ are equivalent, via $L \mapsto L^\flat$. Consequently there is an isomorphism $\mathrm{Gal}(\overline{K}/K) \cong \mathrm{Gal}(\overline{K}^\flat/K^\flat)$.*

EXAMPLE 2.2.2. Theorem 2.2.1 allows us to describe the tilt of the perfectoid field $\mathbf{C}_p = \overline{\mathbf{Q}}_p^\wedge$. Since $\mathbf{C}_p$ is the completion of the algebraic closure of the perfectoid field $K = \mathbf{Q}_p(p^{1/p^\infty})^\wedge$, $\mathbf{C}_p^\flat$ is the completion of the algebraic closure of $K^\flat \cong \mathbf{F}_p((t^{1/p^\infty}))$.

There is an explicit inverse to $L \mapsto L^\flat$ which merits discussion. Since we want to move from characteristic p to characteristic 0, it is not surprising that Witt vectors appear. Recall that for a perfect ring R of characteristic p, we have the ring of Witt vectors $W(R)$, which is characterized by the following properties: $W(R)$ is p-adically complete and p-torsion free, and $W(R)/pW(R) \cong R$. This is a ring which is separated and complete for the p-adic topology; there is a surjective morphism $W(R) \to R$ which admits a multiplicative (not additive) section $R \to W(R)$, written $x \mapsto [x]$. The ring $W(R)$ has the following universal property: For a p-adically complete, p-torsion free ring S and a map of multiplicative monoids $R \to S$ for which the composition $R \to S \to S/p$ is a ring homomorphism, there exists a unique continuous ring homomorphism $W(R) \to S$ such that the diagram

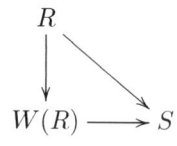

$$\begin{array}{ccc} & R & \\ \swarrow & & \searrow \\ W(R) & \longrightarrow & S \end{array}$$

commutes. Elements of $W(R)$ may be written uniquely as formal power series $[x_0] + [x_1]p + [x_n]p^2 + \ldots$.

In the context of Theorem 2.2.1, we have the perfect ring $K^{\flat\circ}$, the p-adically complete p-torsion free ring K°, and the ring homomorphism $K^{\flat\circ} \to K^\circ/p$, which factors through a map of multiplicative monoids $K^{\flat\circ} \to K^\circ$, namely $f \mapsto f^\sharp$. Therefore by the universal property of Witt vectors, there exists a unique continuous ring homomorphism $\theta \colon W(K^{\flat\circ}) \to K^\circ$ satisfying $\theta([f]) = f^\sharp$. Since p is invertible in K, the map θ extends to a homomorphism of $\mathbf{Q}_p$-algebras $W(K^{\flat\circ})[1/p] \to K$, which we continue to call θ.

LEMMA 2.2.3. *The homomorphism $\theta\colon W(K^{\flat\circ})[1/p] \to K$ is surjective. Its kernel is a principal ideal, generated by an element of the form $[\varpi] + \alpha p$, where $\varpi \in K^\flat$ is a pseudo-uniformizer and $\alpha \in W(K^{\flat\circ})$ is a unit.*

We can now describe the inverse to the tilting functor $L \mapsto L^\flat$ in Theorem 2.2.1. Suppose that $M/K^\flat$ is a finite extension. Then M° is perfect, and $W(M^\circ)$ is an algebra over $W(K^{\flat\circ})$. We put

$$M^\sharp = W(M^\circ) \otimes_{W(K^{\flat\circ}),\theta} K.$$

Then $M^\sharp$ is a perfectoid field, and there is a multiplicative map $M \to M^\sharp$ given by $f \mapsto f^\sharp = [f] \otimes 1$. There is an isomorphism $M \cong M^{\sharp\flat}$ given by $f \mapsto (f^\sharp, (f^{1/p})^\sharp, \ldots)$.

2.3. Untilts of a perfectoid field of characteristic p.

Let K be a perfectoid field of characteristic p. Does there always exists a characteristic 0 perfectoid field whose tilt is K, and if so, can one describe the set of such "untilts"? Certainly an untilt is not unique in general: In Example 2.1.1 we saw that there at least two distinct perfectoid fields whose tilts are isomorphic to $\mathbf{F}_p((t^{1/p^\infty}))$.

DEFINITION 2.3.1. An *untilt* of K is a pair $(K^\sharp, \iota)$, where $K^\sharp$ is a perfectoid field and $\iota\colon K \xrightarrow{\sim} K^{\sharp\flat}$ is an isomorphism.

We remark that our definition includes K as an untilt of itself, since after all $K^\flat = K$.

Given an untilt $(K^\sharp, \iota)$, the multiplicative map $K^\circ \xrightarrow{\iota} K^{\sharp\flat\circ} \xrightarrow{\sharp} K^{\sharp\circ}$ induces a surjective ring homomorphism

$$\begin{aligned} \theta_{K^\sharp}\colon W(K^\circ) &\to K^{\sharp\circ} \\ \sum_{n=0}^\infty [f_n]p^n &\mapsto \sum_{n=0}^\infty f_n^\sharp p^n. \end{aligned}$$

Then $\ker \theta_{K^\sharp}$ is an ideal which is *primitive of degree 1*: this means that I is generated by an element of the form $\sum_{n \geq 0} [f_n]p^n$, where f_0 is topologically nilpotent and $f_1 \in K^\circ$ is a unit.

THEOREM 2.3.2. *The map $I \mapsto (W(K^\circ)/I)[1/p]$ is a bijection between the set of primitive ideals of $W(K^\circ)$ of degree 1, and the set of isomorphism classes of untilts of K.*

Note that $I = (p)$ is the unique ideal which produces the trivial untilt K.

Theorem 2.3.2 suggests that untilts of K of characteristic 0 are parametrized by some kind of geometric object which is related to $W(K^\circ)$. An approximation to this object might be $\operatorname{MaxSpec} W(K^\circ)[1/p[\varpi]]$, where ϖ is a pseudo-uniformizer of K. After all, every characteristic 0 untilt $K^\sharp$ of K induces a surjective ring homomorphism $\theta_{K^\sharp}\colon W(K^\circ)[1/p] \to K^\sharp$ for which $\theta_{K^\sharp}([\varpi]) = \varpi^\sharp$ is a pseudo-uniformizer of $K^\sharp$ (and is therefore nonzero); thus $\ker \theta_{K^\sharp}$ determines a maximal ideal of $W(K^\circ)[1/p[\varpi]]$. However, $\operatorname{MaxSpec} W(K^\circ)[1/p[\varpi]]$ isn't a rigid-analytic space, as $W(K^\circ)[1/p[\varpi]]$ isn't an affinoid algebra.

The approach of Fargues and Fontaine requires looking at $W(K^\circ)$ as a ring equipped with its $([\varpi], p)$-adic topology. (This is called the *weak topology* in [**FF11**].) This makes $W(K^\circ)$ into a Huber ring (with itself as ring of definition), and so we may make the following definition.

DEFINITION 2.3.3 (The adic Fargues–Fontaine curves $\mathcal{Y}_K$ and $\mathcal{X}_K$). Let
$$\mathcal{Y}_K = \operatorname{Spa} W(K^\circ) \setminus \{|p[\varpi]| = 0\},$$
where ϖ is a pseudo-uniformizer of K. The Frobenius automorphism on K° induces a properly discontinuous automorphism $\phi \colon \mathcal{Y}_K \to \mathcal{Y}_K$; we let
$$\mathcal{X}_K = \mathcal{Y}_K / \phi^{\mathbf{Z}}.$$

We claim that $\mathcal{Y}_K$ is covered by rational subsets of the form
$$U\left(\frac{\{p,[\varpi^a]\}}{[\varpi^a]}, \frac{\{p,[\varpi^b]\}}{p}\right) = \left\{|[\varpi^b]| \leq |p| \leq |[\varpi^a]|\right\} \subset \operatorname{Spa} W(K^\circ)$$
as a and b (with $a \leq b$) range through $\mathbf{Z}[1/p]_{>0}$. Indeed, suppose that $x \in \operatorname{Spa} W(K^\circ)$ satisfies $|p[\varpi]|(x) \neq 0$. Since $[\varpi]$ is topologically nilpotent and $|p(x)| \neq 0$, there exists a $b > 0$ with $\left|[\varpi]^b(x)\right| \leq |p(x)|$. Similarly, there exists an $a > 0$ with $|p(x)| \leq |[\varpi^a](x)|$.

For an interval $I = [a,b] \subset (0,\infty)$ with endpoints lying in $\mathbf{Z}[1/p]_{>0}$, let $\mathcal{Y}_{K,I}$ be the rational subset defined above, and let $B_{K,I} = H^0(\mathcal{Y}_{K,I}, \mathcal{O}_{\mathcal{Y}_K})$. Finally, let
$$B_K = H^0(\mathcal{Y}_K, \mathcal{O}_{\mathcal{Y}_K}) = \varprojlim_I B_{K,I}.$$

These rings can be defined in terms of a family of norms on the ring $W(K^\circ)[1/p[\varpi]]$. For $r > 0$, let
$$\left|\sum_{n \in \mathbf{Z}} [x_n] p^n\right|_r = \max\left\{p^{-n} |x_n|^r\right\}.$$
For the interval $I = [a,b]$, the ring $B_{K,I}$ is the completion of $W(K^\circ)[1/p[\varpi]]$ with respect to the norm $\max\{|\ |_a, |\ |_b\}$, and B_K is the Fréchet completion of $W(K^\circ)[1/p[\varpi]]$ with respect to the family of norms $|\ |_r$.

THEOREM 2.3.4 ([**Ked16**]). *The Huber ring $B_{K,I}$ is strongly noetherian. Thus $\mathcal{Y}_K$ and $\mathcal{X}_K$ are adic spaces.*

THEOREM 2.3.5 ([**FF11**], Corollary 2.5.4). *Suppose that $K = C$ is algebraically closed. There is a bijection between the set of isomorphism classes of closed maximal ideals of B_C and the set of isomorphism classes of characteristic 0 untilts of C, given by $I \mapsto B_C/I$.*

This means that there is an embedding of the set of isomorphism classes of characteristic 0 untilts of C into the set of closed points of $\mathcal{Y}_C$ (although this is far from being surjective). For a characteristic 0 untilt $C^\sharp$ with corresponding ideal I, the homomorphism $\theta_{C^\sharp} \colon W(C^\circ) \to C^{\sharp\circ}$ extends to a surjection $\theta_{C^\sharp} \colon B_C \to C^\sharp$ with kernel I.

2.4. Explicit parametrization of untilts by a formal $\mathbf{Q}_p$-vector space. Theorems 2.3.2 and 2.3.5 do not give particularly explicit parametrizations for the set of untilts of a perfectoid field K. The problem is that, even though it is easy to exhibit elements of $W(K^\circ)$ which generate primitive ideals of degree 1, it is not easy to decide whether two such elements generate the same ideal.

We offer now a different perspective. Assume that $K = C$ is an algebraically closed perfectoid field of characteristic p; we want to classify untilts of C. Suppose that $(C^\sharp, \iota \colon C \to C^{\sharp\flat})$ is an untilt of C of characteristic 0. By Theorem 2.2.1, the field $C^\sharp$ is also algebraically closed. Therefore it contains a compatible system of primitive pth power roots of unity: $1, \zeta_p, \zeta_{p^2}, \ldots$. Let $\varepsilon = \iota^{-1}(1, \zeta_p, \zeta_{p^2}, \ldots) \in C$. The idea is that the element $\varepsilon \in C$ is an invariant of the untilt $C^\sharp$. Now, this element isn't quite well-defined, because there is an ambiguity in the choice of system of roots of unity.

Before resolving this ambiguity, we introduce some notation. Let $H = \widehat{\mathbf{G}}_m$ be the formal multiplicative group over $\mathbf{Z}_p$. This is the completion of $\mathbf{G}_{m,\mathbf{Z}_p}$ along the origin of $\mathbf{G}_{m,\mathbf{F}_p}$. It is perhaps easiest to think of H as a functor from complete adic $\mathbf{Z}_p$-algebras to $\mathbf{Z}_p$-modules, which sends R to the abelian group $1 + R^{\circ\circ}$ under multiplication. This group gets its $\mathbf{Z}_p$-module structure this way: for $a \in \mathbf{Z}_p$, the action of a sends x to x^a (defined using power series). The underlying formal scheme of H is isomorphic to $\mathrm{Spf}\, \mathbf{Z}_p[\![T]\!]$. We also define the *universal cover*

$$\widetilde{H} = \varprojlim_{x \mapsto x^p} H,$$

so that if R is an adic $\mathbf{Z}_p$-algebra, $\widetilde{H}(R)$ is the $\mathbf{Q}_p$-*vector space* $\varprojlim_{x \mapsto x^p}(1 + R^{\circ\circ})$. There is a reduction map

(2.4.1) $$\widetilde{H}(R) \to \widetilde{H}(R/p),$$

which one checks is an isomorphism, rather along the lines of the proof that $K^{\flat\circ} \cong \varprojlim_{x \mapsto x^p} K^\circ/p$ for a perfectoid field K. Consequently,

$$\widetilde{H}(R) \cong \widetilde{H}(R/p) \cong \varprojlim_{x \mapsto x^p} R^{\circ\circ}/p \cong \varprojlim_{x \mapsto x^p} R^{\circ\circ},$$

so that $\widetilde{H}$ is representable by the formal scheme $\mathrm{Spf}\, \mathbf{Z}_p[\![T^{1/p^\infty}]\!]$. Thus $\widetilde{H}$ is a $\mathbf{Q}_p$-vector space object in the category of formal schemes, which is to say, a *formal $\mathbf{Q}_p$-vector space*. Whenever K is a perfectoid field, $\widetilde{H}(K^\circ) \cong \widetilde{H}(K^{\flat\circ}) \cong H(K^{\flat\circ})$ (the last isomorphism holds because $K^\flat$ is perfect).

Given a characteristic 0 untilt $C^\sharp$ of C, we obtain a nonzero element $\varepsilon \in \widetilde{H}(C^\circ)$ defined as the image of $(1, \zeta_p, \zeta_{p^2}, \ldots)$ under $\widetilde{H}(C^{\sharp\circ}) \cong \widetilde{H}(C^\circ)$. This element is well-defined up to translation by an element of $\mathbf{Z}_p^\times$. Note that $\theta_{C^\sharp}([\varepsilon^{1/p^n}]) = \zeta_{p^n}$ for all $n \geq 0$; therefore the element

(2.4.2) $$\xi = \frac{[\varepsilon] - 1}{[\varepsilon^{1/p}] - 1} = [1] + [\varepsilon^{1/p}] + \cdots + [\varepsilon^{(p-1)/p}]$$

lies in the kernel of $\theta_{C^\sharp}$. One checks that the ideal (ξ) is primitive of degree 1, and therefore $C^\sharp$ corresponds to the ideal (ξ) under the bijection in Theorem 2.3.2.

On the other hand, we could start with a nonzero element $\varepsilon \in \widetilde{H}(C^\circ)$, form ξ as above, and from this construct the untilt $C^\sharp = W(C^\circ)[1/p]/(\xi)$. In fact:

THEOREM 2.4.1 (see [**FF11**, Proposition 3.4] and [**FF11**, Remarque 3.6]). *The map $C^\sharp \mapsto \varepsilon$ gives a bijection between equivalence classes (respectively, Frobenius-equivalence classes) of characteristic 0 untilts of C and $(\widetilde{H}(C^\circ) \setminus \{0\})/\mathbf{Z}_p^\times$ (respectively, $(\widetilde{H}(C^\circ) \setminus \{0\})/\mathbf{Q}_p^\times$).*

The $\mathbf{Q}_p$-vector space $\widetilde{H}(C^\circ)$ is rather interesting. On the one hand it is huge: it certainly has uncountable dimension. To get a handle on it, we first choose a characteristic 0 untilt $C^\sharp$ of C, so that $\widetilde{H}(C^\circ) \cong \widetilde{H}(C^{\sharp\circ})$. We have a *logarithm map*

$\log\colon H(C^{\sharp\circ}) \to C^{\sharp}$, defined by the usual formula
$$\log x = \sum_{n=1}^{\infty} (-1)^{n-1} \frac{(x-1)^n}{n}.$$
The logarithm map is a $\mathbf{Z}_p$-module homomorphism, which sits in an exact sequence
$$(2.4.3) \qquad 0 \to \mu_{p^\infty}(C^{\sharp}) \to H(C^{\sharp\circ}) \to C^{\sharp} \to 0,$$
where $\mu_{p^\infty}(C^{\sharp}) = H[p^\infty](C^{\sharp\circ})$ is the group of pth power roots of 1 in $C^{\sharp}$. Let us check that the logarithm map is surjective. If $x \in C^{\sharp}$, there exists an n large enough so that $p^n x$ is in the region of convergence of the exponential map; then $z = \exp(p^n x) \in H(C^{\sharp\circ})$ satisfies $\log(z) = p^n x$, so that $\log(z^{1/p^n}) = x$ for any p^nth root z^{1/p^n} of z in $C^{\sharp\circ}$.

Taking inverse limits along multiplication by p in (2.4.3) gives an exact sequence of $\mathbf{Q}_p$-vector spaces:
$$(2.4.4) \qquad 0 \to VH(C^{\sharp}) \to \widetilde{H}(C^{\sharp\circ}) \to C^{\sharp} \to 0,$$
where $VH = \varprojlim_p H[p^\infty](C^{\sharp})$; note that VH is a $\mathbf{Q}_p$-vector space of dimension 1, spanned by a compatible system of primitive pth power roots of 1.

The exact sequence in (2.4.4) sheds some light onto the structure of the $\mathbf{Q}_p$-vector space $\widetilde{H}(C^\circ)$. Once a characteristic 0 untilt $C^{\sharp}$ is chosen, together with a system of pth power roots of 1 in $C^{\sharp}$, there is a "presentation" of $\widetilde{H}(C^\circ)$ as an extension of $C^{\sharp}$ by $\mathbf{Q}_p$.

2.5. The schematic Fargues-Fontaine curve.
We give here another interpretation of the exact sequence (2.4.4). Given an element $\varepsilon \in \widetilde{H}(C^{\sharp\circ})$, we define its logarithm
$$(2.5.1) \qquad t = \log[\varepsilon] = \sum_{n=1}^{\infty} (-1)^{n-1} \frac{([\varepsilon]-1)^n}{n} \in B_C = H^0(\mathcal{Y}_C, \mathcal{O}_{\mathcal{Y}_C}).$$
One has to check here that the sum converges in the Fréchet topology on B_C, but this is just a matter of checking that $|[\varepsilon] - 1|_r < 1$ for all $0 < r < \infty$. Then formally we have
$$\phi(t) = \log \phi([\varepsilon]) = \log[\varepsilon^p] = p\log[\varepsilon] = pt,$$
and so t lies in the $\mathbf{Q}_p$-vector space $B_C^{\phi=p}$ consisting of elements that exhibit this behavior. The element t also has the property that $\theta_{C^{\sharp}}(t) = 0$, since $\theta_{C^{\sharp}}([\varepsilon]) = 1$.

In general we can take any element $\alpha \in \widetilde{H}(C^\circ)$ and produce $\log[\alpha] \in B_C^{\phi=p}$. We have the following commutative diagram, in which the first row is (2.4.4):

$$\begin{array}{ccccccccc}
0 & \longrightarrow & VH(C^{\sharp}) & \longrightarrow & \widetilde{H}(C^{\sharp\circ}) & \longrightarrow & C^{\sharp} & \longrightarrow & 0 \\
& & \downarrow & & {\scriptstyle \log[\cdot]}\downarrow & & \|\downarrow & & \\
0 & \longrightarrow & \mathbf{Q}_p t & \longrightarrow & B_C^{\phi=p} & \xrightarrow{\theta_{C^{\sharp}}} & C^{\sharp} & \longrightarrow & 0.
\end{array}$$

THEOREM 2.5.1 ([**FF11**]). *The map $\varepsilon \mapsto \log[\varepsilon]$ defines an isomorphism of $\mathbf{Q}_p$-vector spaces $\widetilde{H}(C^\circ) \cong B_C^{\phi=p}$. Furthermore, for each $t \in B_C^{\phi=p} \backslash \{0\}$, there is a unique Frobenius-equivalence class of characteristic 0 untilts $C^{\sharp}$ such that $\theta_{C^{\sharp}}(t) = 0$. Therefore there is a bijection between the set of Frobenius-equivalence classes of characteristic 0 untilts of $C^{\sharp}$ and the set $(B_C^{\phi=p} \backslash \{0\})/\mathbf{Q}_p^\times$.*

Recall that $\mathcal{Y}_C$ is the adic space which is (informally) supposed to parametrize equivalence classes of characteristic 0 untilts of C, and $\mathcal{X}_C = \mathcal{Y}_C/\phi^{\mathbf{Z}}$ parametrizes Frobenius-equivalence classes of such untilts. A key insight of [**FF11**] is that $\mathcal{X}_C$ resembles a proper smooth analytic curve, and so should be the analytification of an algebraic curve, just as the Tate curve $\mathbf{G}_{\mathrm{m}}/q^{\mathbf{Z}}$ is the analytification of an elliptic curve over a p-adic field K. In this context, the usual thing to do is to find an ample line bundle $\mathcal{L}$ on $\mathcal{X}_C$, and then define

$$X_C = \operatorname{Proj} \bigoplus_{n \geq 0} H^0(\mathcal{X}_C, \mathcal{L}^{\otimes n}).$$

In the case of $\mathbf{G}_{\mathrm{m}}/q^{\mathbf{Z}}$, the line bundle is $\mathcal{O}(P)$, where P is the origin of $\mathbf{G}_{\mathrm{m}}/q^{\mathbf{Z}}$; the graded ring in the above construction is $K[x,y,z]/f(x,y,z)$, where f is a cubic whose coefficients depend on q according to the usual formulas.

For the Fargues-Fontaine curve, the requisite line bundle $\mathcal{L}$ on $\mathcal{X}_C$ should pull back to a line bundle on $\mathcal{Y}_C$ which is ϕ-equivariant. And so we define a free line bundle $\mathcal{O}_{\mathcal{Y}_C} e$, with the ϕ-equivariance defined by $\phi(e) = p^{-1} e$. This $\mathcal{O}_{\mathcal{Y}_C} e$ descends to a line bundle on $\mathcal{X}_C$, which we call $\mathcal{O}_{\mathcal{X}_C}(1)$. For $n \in \mathbf{Z}$ we define $\mathcal{O}_{\mathcal{X}_C}(n) = \mathcal{O}_{\mathcal{X}_C}^{\otimes n}$ (with the usual convention regarding negative n).

The algebraic Fargues-Fontaine curve is defined by declaring $\mathcal{O}_{\mathcal{X}_C}(1)$ to be very ample. Note that

$$H^0(\mathcal{X}_C, \mathcal{O}_{\mathcal{X}_C}(n)) \cong H^0(\mathcal{Y}_C, \mathcal{O}_{\mathcal{Y}_C} e^{\otimes n})^{\phi=1} \cong B_C^{\phi=p^n}$$

DEFINITION 2.5.2 (The schematic Fargues-Fontaine curve). Define $X_C = \operatorname{Proj} P$, where

$$P = \bigoplus_{d \geq 0} P_d, \text{ where } P_d = B_C^{\phi=p^d}.$$

THEOREM 2.5.3.
(1) The "ring of constants" $H^0(X_C, \mathcal{O}_{X_C}) = P_0 = B_C^{\phi=1}$ is exactly $\mathbf{Q}_p$.
(2) The graded ring P is factorial: the irreducible homogenous elements are exactly the nonzero elements of P_1, and for every $d \geq 1$, a nonzero element of P_d admits a factorization into irreducibles in P_1, unique up to units.
(3) As a result, X_C is an integral Noetherian scheme of dimension 1, which admits a cover by spectra of Dedekind rings (in fact PIDs).

In these respects X_C resembles nothing so much as the projective line $\mathbf{P}_C^1 = \operatorname{Proj} C[S,T]$, where $C[S,T]$ is graded by total degree. But unlike $\mathbf{P}_C^1$, the scheme X_C is not finitely generated over any field.

Since X_C is an integral Noetherian scheme of dimension 1, it is the union of its generic point together with its set of closed points $|X_C|$. In light of Theorem 2.5.3, it is easy to describe the closed points: they correspond to nonzero homogenous prime ideals of P (other than the irrelevant ideal); since every homogenous element of P factors as a product of elements of P_1, every such ideal is generated by a nonzero element of P_1. Since $P^\times = \mathbf{Q}_p$, we find that $|X_C|$ is in bijection with $(P_1 \setminus \{0\})/\mathbf{Q}_p^\times$. Summing up our investigations of untilts of $C^\sharp$ gives the following theorem.

THEOREM 2.5.4. *Let C be an algebraically closed perfectoid field of characteristic p. The following sets are in bijection:*
- *Frobenius-equivalence classes of characteristic 0 untilts of C,*

- $(\widetilde{H}(C^\circ)\setminus\{0\})/\mathbf{Q}_p^\times$,
- *closed points of the scheme X_C.*

2.6. Universal covers of other p-divisible groups. What are the $\mathbf{Q}_p$-vector spaces $P_d = B_C^{\phi=p^d}$ for $d \geq 2$? It is easy enough to exhibit elements of P_d; for $x \in C^{\circ\circ}$ the element

$$\sum_{n \in \mathbf{Z}} \frac{[x^{p^n}]}{p^{dn}}$$

belongs to P_d. However, it is probably not the case that all elements of P_d admit such a presentation, nor is it clear that such a presentation is unique.

The situation is better for the $\mathbf{Q}_p$-vector space $B_C^{\phi^h=p}$, where $h \geq 1$. As in the case $h = 1$, this is isomorphic to the universal cover of a p-divisible formal group. Let $H_{1/h}/\mathbf{Z}_p$ be the 1-dimensional formal group whose logarithm is

$$\log_{H_{1/h}}(T) = \sum_{n=1}^\infty T^{p^{hn}}/p^n.$$

This means that the underlying formal scheme of $H_{1/h}$ is Spf $\mathbf{Z}_p[\![T]\!]$, and the addition law $+_{H_{1/h}}$ is determined by the relation

$$\log_{H_{1/h}}(X +_{H_{1/h}} Y) = \log_{H_{1/h}}(X) + \log_{H_{1/h}}(Y)$$

as power series in $\mathbf{Q}_p[\![X,Y]\!]$. Then $H_{1/h} \otimes_{\mathbf{Z}_p} \mathbf{F}_p$ has height h; in fact $[p]_{H_{1/h}}(T) \equiv T^{p^h} \pmod{p}$ (See [**Haz12**] for proofs of these assertions. The formal group $H_{1/h}$ is an example of a *p-typical* formal group.) We remark that if $\mathbf{Q}_{p^h}/\mathbf{Q}_p$ is the unramified extension of degree h, and if $\mathbf{Z}_{p^h}$ is the ring of integers in $\mathbf{Q}_{p^h}$, then $H_{1/h} \otimes_{\mathbf{Z}_p} \mathbf{Z}_{p^h}$ admits endomorphisms by $\mathbf{Z}_{p^h}$. In fact $H_{1/h} \otimes_{\mathbf{Z}_p} \mathbf{Z}_{p^h}$ is a Lubin-Tate formal $\mathbf{Z}_{p^h}$-module in the sense of [**LT65**].

Let

$$\widetilde{H}_{1/h} = \varprojlim_{x \mapsto [p]_{H_{1/h}}(x)} H_{1/h},$$

a priori as a functor from adic $\mathbf{Z}_p$-algebras to $\mathbf{Q}_p$-vector spaces. Just as with H_1, one uses the congruence between $[p]_{H_{1/h}}$ and a power of Frobenius to show that for any adic $\mathbf{Z}_p$-algebra R, we have isomorphisms

$$\widetilde{H}_{1/h}(R) \cong \widetilde{H}_{1/h}(R/p) \cong \varprojlim_{x \mapsto x^p} R^{\circ\circ}.$$

Applied to $R = W(C^\circ)$, the first isomorphism has inverse

$$\widetilde{H}_{1/h}(C^\circ) \to \widetilde{H}_{1/h}(W(C^\circ))$$
$$(x_n) \mapsto (y_n),$$

where

$$y_n = \lim_{m \to \infty} p^m [x_{m+n}].$$

This isomorphism respects the action of Frobenius ϕ on either side, and therefore the identity $\phi^h = p$ holds in End $\widetilde{H}_{1/h}(W(C^\circ))$, since it holds in End $\widetilde{H}_{1/h}(C^\circ)$. Given an element $(x_n) \in \widetilde{H}_{1/h}(W(C^\circ))$, its logarithm $\log_{\widetilde{H}_{1/h}}(x_0)$ lies in $B_C^{\phi^h=p}$.

THEOREM 2.6.1 ([**FF11**, Proposition 3.4.5]). *The map $(x_n) \mapsto \log_{\widetilde{H}_{1/h}}([x_0])$ gives an isomorphism $\widetilde{H}_{1/h}(C^\circ) \overset{\sim}{\to} B_C^{\phi^h=p}$.*

We can be quite explicit about this isomorphism. There is a commutative diagram

$$\widetilde{H}_{1/h}(C^\circ) \xrightarrow{\sim} \widetilde{H}_{1/h}(W(C^\circ)) \xrightarrow{(x_n) \mapsto \log_{H_{1/h}}(x_0)} B_C^{\phi^h = p}$$
$$\cong \downarrow$$
$$C^{\circ\circ}$$

in which all maps are isomorphisms; the diagonal map is

$$x \mapsto \lim_{m \to \infty} p^m \log_{H_{1/h}}[x^{1/p^m}] = \sum_{n \in \mathbf{Z}} \frac{[x^{p^{hn}}]}{p^n}.$$

Note that the latter expression visibly lies in $B_C^{\phi^h = p}$.

Theorem 2.6.1 generalizes to p-divisible groups of arbitrary height $h \geq 1$ and dimension $d \geq 0$, whenever $0 \leq d/h \leq 1$. The universal cover of such a formal group parametrizes $B_C^{\phi^h = p^d}$.

2.7. Interpretation in terms of vector bundles on X. A major theorem in [**FF11**] is the classification of vector bundles on the Fargues-Fontaine curve X. This classification is in terms of isocrystals.

DEFINITION 2.7.1. Let k be a perfect field of characteristic $p > 0$, and let $K = W(k)[1/p]$. Let $\phi \colon K \to K$ be the Frobenius automorphism. An *isocrystal* over k is a finite-dimensional K-vector space N together with an isomorphism $\phi_N \colon \phi^* N \to N$.

These form an abelian tensor category. When k is algebraically closed, the category of isocrystals over k is well understood. It is a semisimple category, with one irreducible object $N = N_{d/h}$ for each pair (d, h), where $d \in \mathbf{Z}$ and $h \geq 1$ are relatively prime. The underlying K-vector space of N has basis $e, \phi_N(e), \ldots, \phi_N^{h-1}(e)$, and $\phi_N^h(e) = p^d e$. Morphisms between the simple objects go as follows: There are no nonzero morphisms between distinct $N_{d/h}$s, and the endomorphism algebra of $N_{d/h}$ is a central division algebra over K of rank h^2, with invariant $d/h \in \mathbf{Q}/\mathbf{Z}$.

Given an isocrystal N over k, and an algebraically closed perfectoid field C of characteristic p with residue field k, we can define the graded P-module

$$\widetilde{N} = \bigoplus_{d \geq 0} (B_C \otimes_{W(k)} N)^{\phi = p^d}.$$

Let $\mathcal{E}(N)$ be the corresponding $\mathcal{O}_X$-module. Then $\mathcal{E}(N)$ is a vector bundle of rank $\dim N$. For a relatively prime pair (d, h) with $d \geq 0$ and $h \geq 1$, we let $\mathcal{O}_{X_C}(d/h) = \mathcal{E}(N_{-d/h})$. Then $H^0(X_C, \mathcal{O}_X(d/h)) \cong B_C^{\phi^h = p^d}$.

THEOREM 2.7.2 ([**FF11**]). *Let C be an algebraically closed perfectoid field of characteristic $p > 0$. Every vector bundle on X_C is isomorphic to $\mathcal{E}(N)$ for an isocrystal N, which is unique up to isomorphism.*

It must be emphasized that the functor $N \mapsto \mathcal{E}(N)$ is far from being an equivalence of categories, as it is not full. Each nonzero element of $B^{\phi=p}$ gives a morphism $\mathcal{O}_X \to \mathcal{O}_X(1)$ which does not arise from a map of isocrystals. However if $N = N_{d/h}$ as above, then $\operatorname{End} N \to \operatorname{End} \mathcal{E}(N)$ is an isomorphism.

In the last subsection we saw that if $0 \leq d/h \leq 1$, then there is a p-divisible group $\overline{H} = \overline{H}_{d/h}/\mathbf{F}_p$ of height h and dimension d, and a natural isomorphism $\overline{H}(C^\circ) \cong H^0(X_C, \mathcal{O}_{X_C}(d/h))$. Let $C^\sharp$ be a characteristic 0 untilt of C, and let H be a lift of $\overline{H}$ to $C^{\sharp\circ}$. (The question of the existence of such lifts is addressed in [**Mes72**, Chapter IV]. As a special case, p-divisible groups can always be lifted from C°/p to C°.) Then there is an exact sequence of $\mathbf{Z}_p$-modules

$$0 \to H[p^\infty](C^{\sharp\circ}) \to H(C^{\sharp\circ}) \overset{\log_H}{\to} \operatorname{Lie} H \otimes_{C^{\sharp\circ}} C^\sharp \to 0,$$

Taking an inverse limit along multiplication by p (this is right-exact because $H[p^\infty](C^{\sharp\circ})$ is p-divisible) gives an exact sequence of $\mathbf{Q}_p$-vector spaces

(2.7.1) $$0 \to VH(C^{\sharp\circ}) \to \widetilde{H}(C^{\sharp\circ}) \overset{\log_H}{\to} \operatorname{Lie} H \otimes_{C^{\sharp\circ}} C^\sharp \to 0.$$

Note that the middle term, which is naturally isomorphic to $\overline{H}(C^\circ)$, does not depend on the lift H. Also note that this exact sequence presents a very large $\mathbf{Q}_p$-vector space as an extension of a finite-dimensional $C^\sharp$-vector space by a finite-dimensional $\mathbf{Q}_p$-vector space; this is an instance of the theory of *Banach–Colmez spaces*, which we will investigate systematically in the last lecture.

Let $x \in |X_C|$ be the closed point corresponding to the Frobenius equivalence class of $C^\sharp$ under Theorem 2.5.4. The exact sequence in (2.7.1) can be reinterpreted as the global sections of the following exact sequence of $\mathcal{O}_{X_C}$-modules:

$$0 \to \mathcal{O}_{X_C} \otimes_{\mathbf{Q}_p} VH \to \mathcal{O}_{X_C}(d/h) \to i_* \operatorname{Lie} H \otimes C^\sharp \to 0,$$

where i is the inclusion of $x = \operatorname{Spec} C^\sharp$ into X_C.

We mention in passing that [**FF11**] deduces the following theorem from Theorem 2.7.2:

THEOREM 2.7.3. *The curve X_C is geometrically simply connected over $\mathbf{Q}_p$. That is, any finite étale cover of X_C is isomorphic to $X_C \times_{\operatorname{Spec} \mathbf{Q}_p} \operatorname{Spec} A$ for an étale $\mathbf{Q}_p$-algebra A. Thus, the étale fundamental group of the scheme X_C is $\operatorname{Gal}(\overline{\mathbf{Q}}_p/\mathbf{Q}_p)$.*

There are versions of Theorems 2.7.2 and 2.7.3 for the adic curve $\mathcal{X}_C$, owing to the equivalence of categories between coherent sheaves on X_C and $\mathcal{X}_C$ [**Far15**].

3. Perfectoid spaces and diamonds

3.1. Definitions.

DEFINITION 3.1.1. Let A be a Huber ring. A Huber ring A is *perfectoid* if the following conditions hold:
  (1) A is Tate, meaning it contains a pseudo-uniformizer (a topologically nilpotent unit),
  (2) A is uniform, meaning that $A^\circ \subset A$ is bounded,
  (3) A contains a pseudo-uniformizer ϖ such that $\varpi^p | p$ in A°, and such that the pth power map $A^\circ/\varpi \to A^\circ/\varpi^p$ is an isomorphism.

REMARK 3.1.2. In the definition above it is always possible to choose a pseudo-uniformizer ϖ which contains a compatible system of pth power roots.

THEOREM 3.1.3. *Let (A, A^+) be a Huber pair, with A perfectoid. Then (A, A^+) is sheafy, so that $X = \operatorname{Spa}(A, A^+)$ is an adic space. Furthermore, $\mathcal{O}_X(U)$ is a perfectoid ring for every rational subset $U \subset X$.*

Theorem 3.1.3 shows that adic spaces $\mathrm{Spa}(R, R^+)$ with R perfectoid can serve as model spaces for the category of perfectoid spaces:

DEFINITION 3.1.4. A *perfectoid space* is an adic space that may be covered by affinoids of the form $\mathrm{Spa}(A, A^+)$, where A is perfectoid.

EXAMPLE 3.1.5.
- If K is a perfectoid field and $K^+ \subset K$ is a ring of integral elements, then $\mathrm{Spa}(K, K^+)$ is a perfectoid space.
- (The perfectoid closed disc.) Let K be a perfectoid field. Let $A = K\langle T^{1/p^\infty}\rangle$; this is the completion of the polynomial algebra $K[T^{1/p^\infty}]$. Then A is a perfectoid ring, and $\mathrm{Spa}(A, A^\circ)$ is a perfectoid space.
- (The perfectoid open disc.) This time let $A = K^\circ[\![T^{1/p^\infty}]\!]$, the completion of $K^\circ[T^{1/p^\infty}]$ with respect to the (ϖ, T)-adic topology (here ϖ is a pseudo-uniformizer of K). Then A is not a perfectoid ring, because it is not Tate. It is not even clear that (A, A) is sheafy (although this is probably true). Nonetheless, the generic fiber of $\mathrm{Spa}\, A$ over $\mathrm{Spa}\, K^\circ$ is perfectoid: it is covered by the affinoids $\mathrm{Spa}(A_n, A_n^\circ)$, where $A_n = K\langle (T/\varpi^{1/p^n})^{1/p^\infty}\rangle$.
- Let k be a perfect field of characteristic p with its discrete topology. Let $A = k[\![T_1^{1/p^\infty}, \ldots, T_n^{1/p^\infty}]\!]$; this is defined as the $(T_1, \ldots, T_n)$-adic completion of $k[T_1^{1/p^\infty}, \ldots, T_n^{1/p^\infty}]$. Then A is not a perfectoid ring (it is not Tate), but the analytic locus in $\mathrm{Spa}\, A$ is perfectoid. This is the complement in $\mathrm{Spa}\, A$ of the single non-analytic point satisfying $|T_i| = 0$ for $i = 1, \ldots, n$. Note that if $n > 1$, this perfectoid space does not live over any particular perfectoid field.
- (Some totally disconnected perfectoid spaces.) Let K be a perfectoid field and let S be a profinite set. Let $A = \mathrm{Cont}(S, K)$ be the ring of continuous maps $S \to K$. Give A the structure of a Banach K-algebra under the sup norm; we have $A^\circ = \mathrm{Cont}(S, K^\circ)$. Then $\mathrm{Spa}(A, A^\circ)$ is a perfectoid space whose underlying topological space is S. The construction globalizes to the case that S is only locally profinite. If K is understood, we write $\underline{S}$ for the resulting perfectoid space.

The tilting operation we discussed in 2.1 extends to perfectoid spaces. For a perfectoid ring A with pseudo-uniformizer ϖ as in Remark 3.1.2, we define its tilt by
$$A^\flat = \left(\varprojlim_{x \mapsto x^p} A^\circ/\varpi \right) [1/\varpi^\flat],$$
where $\varpi^\flat = (\varpi, \varpi^{1/p}, \ldots)$. Then $A^\flat$ is a perfectoid ring of characteristic p.

We gather here some results from [**Sch12**] (which assumes a fixed perfectoid field of scalars, but the proofs carry over in general).

THEOREM 3.1.6. *Let A be a perfectoid ring.*
(1) *There is a homeomorphism of topological monoids:*
$$A^\flat \cong \varprojlim_{x \mapsto x^p} A.$$
If $f \in A^\flat$ corresponds to the sequence (f_n) with $f_n \in A$, define $f^\sharp = f_0$.
(2) *There is a bijection $A^+ \mapsto A^{\flat+} = \varprojlim_{x \mapsto x^p} A^+/\varpi$ between rings of integral elements of A and $A^\flat$.*

(3) *Given a ring of integral elements $A^+ \subset A$, there is a homeomorphism*

$$\operatorname{Spa}(A, A^+) \xrightarrow{\sim} \operatorname{Spa}(A^\flat, A^{\flat+})$$
$$x \mapsto x^\flat$$

where $x^\flat$ is defined by $|f(x^\flat)| = |f^\sharp(x)|$ for $f \in A^\flat$. This homeomorphism identifies rational subsets on either side.

(4) *The categories of perfectoid algebras over A and $A^\flat$ are equivalent, via $B \mapsto B^\flat$.*

(5) *Let B be a finite étale A-algebra, so that B becomes a topological ring. Then B is also perfectoid. The categories of finite étale algebras over A and $A^\flat$ are equivalent, via $B \mapsto B^\flat$.*

One way to construct perfectoid spaces comes from universal covers of p-divisible groups, which we discussed in (2.6). Let k be a perfect field of characteristic p, and let H be a p-divisible group over k. We have the universal cover $\widetilde{H} = \varprojlim_p H$, which we may consider as a functor from k-algebras to $\mathbf{Q}_p$-vector spaces. For now let us assume that H is connected, so that H is representable by $\operatorname{Spf} k[\![T_1, \ldots, T_d]\!]$, where $d = \dim H$; then $\widetilde{H}$ is representable by a formal scheme $\operatorname{Spf} k[\![T_1^{1/p^\infty}, \ldots, T_d^{1/p^\infty}]\!]$. (This follows from two facts: multplication by p in H factors through Frobenius, and a sufficiently high power of Frobenius on H factors through multiplication by p.)

Let $\widetilde{H}^{\mathrm{ad}}$ be the corresponding adic space. Then $\widetilde{H}^{\mathrm{ad}}$ isn't quite a perfectoid space (it isn't analytic), but the punctured version $\widetilde{H}^{\mathrm{ad}} \backslash \{0\}$ is a perfectoid space, as in Example 3.1.5. If we want to create a perfectoid space version of $\widetilde{H}$ without puncturing it, we can introduce a separate perfectoid field K/k, and define $\widetilde{H}_K$ as the adic generic fiber of $\widetilde{H} \times_{\operatorname{Spec} k} \operatorname{Spf} K^\circ$. Then $\widetilde{H}_K$ is a $\mathbf{Q}_p$-vector space object in the category of perfectoid spaces over K.

A similar object exists in characteristic 0. Suppose now that K is a perfectoid field of characteristic 0 whose residue field contains k. Then the ring homomorphism $K^\circ/p \to k$ admits a canonical section, namely $k \to K^{\flat\circ} \to K^{\flat\circ}/p \cong K^\circ/p$. We may define $\widetilde{H}_K$ as the perfectoid space over K whose tilt is $\widetilde{H}_{K^\flat}/K^\flat$. Then if G is any lift of $H \otimes_k K^\circ/p$ to K°, then we have the following functorial interpretation of $\widetilde{H}_K$: it is the sheafification of the functor $R \mapsto \widetilde{G}(R^\circ)$ on perfectoid K-algebras R. Note that this does not depend on the choice of lift G.

In fact, the requirement that H be formal is just a red herring; there is a functor $H \mapsto \widetilde{H}_K$ from the whole category of p-divisible groups over k to the category of perfectoid spaces with $\mathbf{Q}_p$-vector space structure. For instance if $H = \mathbf{Q}_p/\mathbf{Z}_p$ is the constant p-divisible group, then $\widetilde{H} = \underline{\mathbf{Q}}_p$ is the constant $\mathbf{Q}_p$-vector space.

Finally, if we allow K to be any nonarchimedean field with residue field containing k, then $\widetilde{H}_K$ will be a *pre-perfectoid space*, meaning that it becomes perfectoid after extending scalars from K to any perfectoid field.

3.2. Untilts of perfectoid spaces in characteristic p, and a motivation for diamonds.

Let X be a perfectoid space lying over $\operatorname{Spa} \mathbf{F}_p$. As we did with perfectoid fields, we can investigate the set of equivalence classes of untilts of X. What we would like is a *moduli space M* lying over $\operatorname{Spa} \mathbf{F}_p$, for which there is a

natural bijection between the following sets:
- Morphisms $X \to M$, and
- Equivalence classes of characteristic 0 untilts $X^\sharp \to \operatorname{Spa} \mathbf{Q}_p$.

This object M will ultimately be called $\operatorname{Spd} \mathbf{Q}_p$, where the "d" stands for *diamond*; it lives in a category of diamonds, which contains the category of perfectoid spaces as a full subcategory.

In the special case $X = \operatorname{Spa} C$ for a perfectoid field C of characteristic p, Theorem 2.5.4 gave the following parametrizations:

(1) Equivalence classes of untilts correspond to primitive ideals $I \subset W(C^\circ)$ of degree 1, via $C^\sharp \mapsto \ker \theta_{C^\sharp}$.
(2) Frobenius-equivalence classes of characteristic 0 untilts correspond to closed points on the Fargues-Fontaine curve X constructed from C as in (2.5); the inverse map sends a point to its residue field.
(3) Frobenius-equivalence classes of characteristic 0 untilts correspond to elements of the quotient $(\widetilde{H}(C^\circ)\setminus\{0\})/\mathbf{Q}_p^\times$, where $\widetilde{H}$ is the universal cover of the formal multiplicative group as in (2.4).

The parametrization described in (1) relativizes quite easily. Suppose R is a perfectoid $\mathbf{F}_p$-algebra with pseudo-uniformizer ϖ. Then we have the Witt ring $W(R^\circ)$, equipped with its $(p, [\varpi])$-adic topology. A *primitive ideal of degree 1* in $W(R^\circ)$ is a principal ideal generated by an element of the form $\xi = \sum_{n=0}^\infty [x_n]p^n$, where $x_0 \in R$ is topologically nilpotent and $x_1 \in R^\circ$ is a unit.

THEOREM 3.2.1 ([**Fon13**]). *Ideals $I \subset W(R^\circ)$ which are primitive of degree 1 are in bijection with isomorphism classes of untilts of R, via $I \mapsto (W(R^\circ)/I)[1/p]$.*

As in the case with perfectoid fields, however, this does not give us much in the way of defining the object $\operatorname{Spd} \mathbf{Q}_p$; it is not easy to tell whether two such ideals are the same, given their generators.

We turn now to (2). It is easy to define a relative Fargues–Fontaine curve: given a perfectoid ring $R/\mathbf{F}_p$, first define the relative adic curve

$$\mathcal{Y}_R = \operatorname{Spa} W(R^\circ) \setminus \{|p[\varpi]| = 0\}$$

and the ring $B_R = H^0(\mathcal{Y}_R, \mathcal{O}_{\mathcal{Y}_R})$. Then B_R has an action of Frobenius ϕ, and we define the *relative schematic Fargues-Fontaine curve* as

$$X_R = \operatorname{Proj} \bigoplus_{d \geq 0} B_R^{\phi = p^d}.$$

However, when R is not a field, we cannot expect X_R to have any nice properties (*e.g.* it may not be Noetherian). Nor should we expect that closed points of X_R parametrize Frobenius-equivalence classes of characteristic 0 untilts; after all, the residue field of such a point is a field, whereas an untilt $R^\sharp$ very well may not be.

Perhaps (3) has more promise. In the case that $R = C$ is an algebraically closed field of characteristic p, Theorem 2.5.4 says that isomorphism classes of characteristic 0 untilts of C are in bijection with $(\widetilde{H}(C^\circ)\setminus\{0\})/\mathbf{Z}_p^\times$, where H is the formal multiplicative group over $\mathbf{F}_p$. Recall the construction: if $C^\sharp$ is a characteristic 0 untilt, we choose a compatible system $(1, \zeta_p, \zeta_{p^2}, \dots)$ of primitive pth power roots of 1 in $C^\sharp$, which determines a nonzero element of $\widetilde{H}(C^{\sharp\circ}) \cong \widetilde{H}(C^\circ)$, well-defined up to multiplication by an element of $\mathbf{Z}_p^\times$.

Let $\mathbf{Q}_p^{\mathrm{cycl}}$ be the completion of $\mathbf{Q}_p(\mu_{p^\infty})$. Then $\mathrm{Gal}(\mathbf{Q}_p(\mu_{p^\infty})/\mathbf{Q}_p) \cong \mathbf{Z}_p^\times$ acts continuously on $\mathbf{Q}_p^{\mathrm{cycl}}$. Finally, let $\widetilde{H}^{\mathrm{ad}} \setminus \{0\}$ be the punctured adic space attached to the formal scheme $\widetilde{H}$.

LEMMA 3.2.2. *There is an isomorphism $\widetilde{H}^{\mathrm{ad}} \setminus \{0\} \cong \mathrm{Spa}\, \mathbf{Q}_p^{\mathrm{cycl},\flat}$ which is $\mathbf{Z}_p^\times$-equivariant.*

PROOF. Since $\widetilde{H} \cong \mathrm{Spf}\, \mathbf{F}_p[\![t^{1/p^\infty}]\!]$, we have $\widetilde{H}^{\mathrm{ad}} \setminus \{0\} \cong \mathrm{Spa}\, \mathbf{F}_p(\!(t^{1/p^\infty})\!)$. we have already identified the latter with $\mathbf{Q}_p^{\mathrm{cycl},\flat}$ in Example 2.1.1, so one only needs to check that the $\mathbf{Z}_p^\times$-action is preserved. □

(There is a generalization of this lemma to Lubin–Tate extensions of any local field [**Wei17**, Proposition 3.5.3].)

Therefore, we can restate our parametrization of untilts of C as follows:
$$(3.2.1)$$
$$\{\text{Char. 0 untilts of } C\} \cong \mathrm{Hom}_{\mathrm{cont}}(\mathbf{Q}_p^{\mathrm{cycl},\flat}, C)/\mathbf{Z}_p^\times = \mathrm{Hom}(\mathrm{Spa}\, C, \mathrm{Spa}\, \mathbf{Q}_p^{\mathrm{cycl},\flat})/\mathbf{Z}_p^\times.$$

We could also have derived this directly: if $C^\sharp$ is a characteristic 0 untilt of C, then there exists an embedding $\mathbf{Q}_p^{\mathrm{cycl}} \hookrightarrow C^\sharp$ which is well-defined up to the action of $\mathbf{Z}_p^\times$; tilting this gives $\mathbf{Q}_p^{\mathrm{cycl},\flat} \hookrightarrow C$.

The bijections in (3.2.1) suggest that $\mathrm{Spd}\, \mathbf{Q}_p$ should be the quotient

$$\text{``}(\mathrm{Spa}\, \mathbf{Q}_p^{\mathrm{cycl},\flat})/\mathbf{Z}_p^\times\text{.''}$$

But this quotient doesn't exist in the category of adic spaces. The subfield of $\mathbf{Q}_p^{\mathrm{cycl},\flat}$ fixed by $\mathbf{Z}_p^\times$ is just $\mathbf{F}_p$.

We would like to formulate a generalization of (3.2.1) for general perfectoid rings $R/\mathbf{F}_p$. We begin with the case that $R = K$ is a perfectoid field which is not algebraically closed. Let $K^\sharp/\mathbf{Q}_p$ be an untilt. Then $K^\sharp$ might not contain all pth power roots of unity. For each $n \geq 1$, the field $K_n^\sharp := K^\sharp(\mu_{p^n})$ is a perfectoid field, whose tilt K_n is a finite Galois extension of K. Let $\widehat{K}_\infty^\sharp$ be the completion of $\cup_{n \geq 1} K_n^\sharp$; then $\widehat{K}_\infty^\sharp$ is perfectoid. Let $K_\infty = \widehat{K}_\infty^{\sharp\flat}$. Let $G = \mathrm{Gal}(K^\sharp(\mu_{p^\infty})/K^\sharp)$; then G acts continuously on K_∞. If we choose a compatible sequence of pth power roots of 1 in $\widehat{K}_\infty^\sharp$, we obtain a nonzero element $\varepsilon \in \widetilde{H}(\widehat{K}_\infty^{\sharp\circ}) \cong \widetilde{H}(\widehat{K}_\infty^{\sharp\flat\circ}) = \widetilde{H}(K_\infty^\circ)$. Since G acts on ε through the cyclotomic character, the class of ε in $\widetilde{H}(K_\infty^\circ)/\mathbf{Z}_p^\times$ is G-invariant.

Thus, given an untilt $K^\sharp/\mathbf{Q}_p$, there exists a perfectoid field K_∞/K, equal to the completion of a Galois extension with group G, together with a class $\varepsilon \in \mathrm{Hom}(\mathrm{Spa}\, K_\infty, \mathrm{Spa}\, \mathbf{Q}_p^{\mathrm{cycl},\flat})/\mathbf{Z}_p^\times$ which is G-invariant. Conversely, if we are given such data, the class ε determines a characteristic 0 untilt $K_\infty^\sharp$ of K_∞ together with a continuous action of G; then $K^\sharp := (K_\infty^\sharp)^G$ is a characteristic 0 untilt of K.

It may happen that two data of the form (K_∞, ε) give rise to the same untilt. The proper way to sort this out is in the language of sheaves on the *pro-étale site*, in which $\mathrm{Spa}\, K_\infty \to \mathrm{Spa}\, K$ is considered a covering.

3.3. The pro-étale topology.

The extension of fields K_∞/K appearing in the previous section was the completion of the union of a tower of finite separable (that is, étale) extensions of K. Such an extension K_∞/K is said to be *pro-étale*. The definition works in families as follows.

DEFINITION 3.3.1. A morphism $f\colon X \to Y$ of perfectoid spaces is *pro-étale* if locally on X it is of the form $\mathrm{Spa}(A_\infty, A_\infty^+) \to \mathrm{Spa}(A, A^+)$, where A and A_∞ are perfectoid rings, and
$$(A_\infty, A_\infty^+) = \left[\varinjlim (A_i, A_i^+)\right]^\wedge$$
is a filtered colimit of pairs (A_i, A_i^+), such that $\mathrm{Spa}(A_i, A_i^+) \to \mathrm{Spa}(A, A^+)$ is étale.

(The notion of an étale morphism between analytic affinoid adic spaces appears in [**Sch12**, Definition 7.1].)

EXAMPLE 3.3.2. Let K be a perfectoid field, and let S be a profinite set; we have the perfectoid space $\underline{S}$ as in Example 3.1.5. Then $\underline{S} \to \mathrm{Spa}\, K$ is pro-étale. If $K = C$ is algebraically closed and $X \to \mathrm{Spa}\, C$ is pro-étale, then $X = \underline{S}$ for a locally profinite set S.

EXAMPLE 3.3.3. Somewhat counterintuitively, the inclusion of a Zariski-closed subset is pro-étale. For instance, let K be a perfectoid field, let ϖ be a pseudo-uniformizer of K, and let $Y = \mathrm{Spa}\, K\langle T^{1/p^\infty}\rangle$. For $n = 1, 2, \ldots$, let $Y_n \subset Y$ be the rational subset $\{|T| \leq |\varpi|^n\}$. Then "evaluation at 0" induces an isomorphism $\left[\varinjlim \mathcal{O}_Y(Y_n)\right]^\wedge \to K$, so that the inclusion-at-0 map $\mathrm{Spa}\, K \to Y$ is pro-étale.

DEFINITION 3.3.4. Consider the category Pfd of perfectoid spaces of characteristic p. We endow this with the structure of a site by declaring that a collection of morphisms $\{f_i\colon X_i \to X\}$ is a covering (a *pro-étale cover*) if the f_i are pro-étale, and if for all quasi-compact open $U \subset X$, there exists a finite subset $I_U \subset I$, and a quasi-compact open $U_i \subset X_i$ for $i \in I_U$, such that $U = \cup_{i \in I_U} f_i(U_i)$.

If K is eitehr a discrete perfect field (such as $\overline{\mathbf{F}}_p$) or a perfectoid field of characteristic p, we write Pfd_K for the category of perfectoid spaces lying over $\mathrm{Spa}\, K$, endowed with the topology obtained by restriction from Pfd.

REMARK 3.3.5. The finiteness condition in Definition 3.3.4 excludes certain "pointwise" morphisms from being pro-étale covers. For instance if Y is the perfectoid unit disc, we can consider the inclusion $f_x\colon \mathrm{Spa}(K_x, K_x^+) \to Y$ for each point $x \in |Y|$; this is pro-étale by similar reasoning as in Example 3.3.3, but we don't want $\{f_x\}_{x \in |Y|}$ to be a pro-étale covering.

REMARK 3.3.6. The notions of a pro-étale morphism of schemes and of a pro-étale site appear in [**BS15**], where they were used to define a pro-étale fundamental group of a scheme, and also to give the "morally correct" definition of the ℓ-adic cohomology group $H^i(X, \mathbf{Q}_\ell)$ for a scheme X.

It now makes sense to talk about a sheaf on Pfd: this is a presheaf on Pfd (that is, a contravariant set-valued functor) which satisfies the sheaf axioms with respect to the pro-étale topology. If X is a perfectoid space of characteristic p, we have the representable presheaf h_X defined by $h_X(Y) = \mathrm{Hom}(Y, X)$.

PROPOSITION 3.3.7 ([**SW**, Proposition 8.2.7]). *The presheaf h_X is a sheaf.*

If $\mathcal{F}$ is a sheaf on Pfd, and if X is an object of Pfd, then a morphism $h_X \to \mathcal{F}$ is the same thing as a section of $\mathcal{F}(X)$. Note that the functor $X \mapsto h_X$ exhibits Pfd as a full subcategory of the category of sheaves on Pfd.

DEFINITION 3.3.8.
(1) A morphism $\mathcal{F} \to \mathcal{G}$ of sheaves on Pfd is *pro-étale* if for all perfectoid spaces X and maps $h_X \to \mathcal{G}$, the pullback $h_X \times_{\mathcal{G}} \mathcal{F}$ is representable by a perfectoid space Y, and the morphism $Y \to X$ (corresponding to $h_Y = h_X \times_{\mathcal{G}} \mathcal{F} \to h_X$) is pro-étale.
(2) Let $\mathcal{F}$ be a sheaf on Pfd. A *pro-étale equivalence relation* is a monomorphism of sheaves $\mathcal{R} \hookrightarrow \mathcal{F} \times \mathcal{F}$, such that each projection $\mathcal{R} \to \mathcal{F}$ is pro-étale, and such that for all objects S of Pfd, the image of the map $\mathcal{R}(S) \to \mathcal{F}(S) \times \mathcal{F}(S)$ is an equivalence relation on $\mathcal{F}(S)$.
(3) A *diamond* is a sheaf $\mathcal{F}$ on Pfd which is the quotient of a perfectoid space by a pro-étale equivalence relation. That is, there exists a perfectoid space X and a pro-étale equivalence relation $\mathcal{R} \to h_X \times h_X$ such that
$$\mathcal{R} \rightrightarrows h_X \to \mathcal{F}$$
is a coequalizer diagram in the category of sheaves on Pfd.
(4) If X is a perfectoid space (of whatever characteristic), let $X^{\diamond}$ be the representable sheaf $h_{X^{\flat}}$; this is a diamond. In the case $X = \mathrm{Spa}(A, A^+)$ is affinoid perfectoid, we also write $\mathrm{Spd}(A, A^+)$ for $X^{\diamond}$.
(5) A diamond X is *partially proper* if it satisfies the criterion appearing in Definition 1.9.2: for a perfectoid Huber pair (R, R^+), we have $X(R, R^{\circ}) \xrightarrow{\sim} X(R, R^+)$ only depends on R. If X is partially proper we write $X(R) = X(R, R^{\circ})$.

REMARK 3.3.9. The definition of diamonds given above is meant to mimic the notion of an *algebraic space*, which is the quotient of a scheme by an étale equivalence relation. The category of algebraic spaces is a mild generalization of the category of schemes. Some algebraic spaces arise as quotients of schemes by finite groups. Suppose that X is a scheme and G is a finite group acting on X. Assume that the action is *free* in the sense that for all nontrivial $g \in G$ and all $x \in X$ fixed by g, the action of g on the residue field of x is nontrivial. Then the quotient X/G is an algebraic space [**Sta14**, Tag 02Z2]; it is the quotient of X by the étale equivalence relation $G \times X \to X \times X$, $(g, x) \mapsto (x, g(x))$. (The freeness condition is necessary for this morphism to be a monomorphism.) Algebraic spaces are not to be confused with the larger category of algebraic stacks, which include stacky quotients $[X/G]$ for arbitrary actions of G on X.

3.4. The diamond $\mathrm{Spd}\, \mathbf{Q}_p$.
Let us recall that we seek an object like "$(\mathrm{Spa}\, \mathbf{Q}_p^{\mathrm{cycl},\flat})/\mathbf{Z}_p^{\times}$" which parametrizes characteristic 0 untilts of a perfectoid space of characteristic p. Now that we have the category of diamonds, we may make the following *ad hoc* definition.

DEFINITION 3.4.1. We define $\mathrm{Spd}\, \mathbf{Q}_p = (\mathrm{Spd}\, \mathbf{Q}_p^{\mathrm{cycl},\flat})/\underline{\mathbf{Z}}_p^{\times}$. That is, $\mathrm{Spd}\, \mathbf{Q}_p$ is the coequalizer of

(3.4.1) $$\underline{\mathbf{Z}}_p^{\times} \times \mathrm{Spd}\, \mathbf{Q}_p^{\mathrm{cycl}} \rightrightarrows \mathrm{Spd}\, \mathbf{Q}_p^{\mathrm{cycl}},$$

where one map is the projection and the other is the action.

Thus $\mathrm{Spd}\, \mathbf{Q}_p$ is the sheafification of the presheaf on Pfd which assigns to an object S the set $\mathrm{Hom}(S, \mathrm{Spa}\, \mathbf{Q}_p^{\mathrm{cycl},\flat})/\mathbf{Z}_p^{\times}$.

LEMMA 3.4.2. $\mathrm{Spd}\, \mathbf{Q}_p$ *is a partially proper diamond.*

PROOF. Each of the maps $\mathbf{Z}_p^\times \times \operatorname{Spd}\mathbf{Q}_p^{\text{cycl}} \to \operatorname{Spd}\mathbf{Q}_p^{\text{cycl}}$ is pro-étale (see Example 3.3.2). One must show that $\mathbf{Z}_p^\times \times \operatorname{Spd}\mathbf{Q}_p^{\text{cycl}} \to \operatorname{Spd}\mathbf{Q}_p^{\text{cycl}} \times \operatorname{Spd}\mathbf{Q}_p^{\text{cycl}}$ is a monomorphism, which ultimately boils down to the fact that the map $\mathbf{Z}_p^\times \to \operatorname{Aut}\mathbf{Q}_p^{\text{cycl},\flat}$ is injective. From there it is formal that (3.4.1) is a pro-étale equivalence relation, and thus that $\operatorname{Spd}\mathbf{Q}_p$ is a diamond. The partial properness of $\operatorname{Spd}\mathbf{Q}_p$ follows from that of $\operatorname{Spa}\mathbf{Q}_p^{\text{cycl},\flat}$. □

If S is an object of Pfd, then to give an element of $(\operatorname{Spd}\mathbf{Q}_p)(S)$ is to give a pro-étale cover $\widetilde{S} \to S$ and an element of the set $\operatorname{Hom}(\widetilde{S}, \operatorname{Spa}\mathbf{Q}_p^{\text{cycl},\flat})/\mathbf{Z}_p^\times$ which comes equipped with a descent datum along $\widetilde{S} \to S$. In the case $S = \operatorname{Spa}K$ for a perfectoid field $K/\mathbf{F}_p$, one way to do this would be to give a perfectoid field $\widetilde{K}/K$, equal to the completion of a Galois extension of K with group G, and an element of $\operatorname{Hom}_{\text{cont}}(\mathbf{Q}_p^{\text{cycl},\flat}, \widetilde{K})/\mathbf{Z}_p^\times$ which is G-invariant. We have already seen that such data gives an untilt of K. More generally, we have the following theorem.

THEOREM 3.4.3 ([**SW**, Theorem 3.4.5]). *Let X be a perfectoid space of characteristic p. Then the set of isomorphism classes of untilts of X to characteristic 0 is naturally in bijection with $(\operatorname{Spd}\mathbf{Q}_p)(X)$. In other words, there is an equivalence of categories between perfectoid spaces over $\mathbf{Q}_p$, and the category of perfectoid spaces X of characteristic p together with a "structure morphism" $X^\diamond \to \operatorname{Spd}\mathbf{Q}_p$.*

3.5. The functor $X \mapsto X^\diamond$. The construction of $\operatorname{Spd}\mathbf{Q}_p$ from $\operatorname{Spa}\mathbf{Q}_p$ is a special case of a general phenomenon.

DEFINITION 3.5.1. Let X be an analytic adic space on which p is topologically nilpotent (that is, X is fibered over $\operatorname{Spa}\mathbf{Z}_p$). Let $X^\diamond$ be the functor on Pfd which sends an object S to the set of isomorphism classes of pairs $(S^\sharp \to X, \iota)$, where $S^\sharp$ is a perfectoid space and $\iota\colon S^{\sharp\flat} \xrightarrow{\sim} S$ is an isomorphism.

Thus $X^\diamond$ classifies "untilts to X". If X itself is a perfectoid space and S is a test object in Perf, then the tilting equivalence in Theorem 3.1.6(3) shows that morphisms $S \mapsto X^\flat$ are in bijection with untilts $S^\sharp \to X$. Thus $X^\diamond$ agrees with the notation introduced in Definition 3.3.8, namely $X^\diamond = h_{X^\flat}$. Finally, Theorem 3.4.3 shows that $\operatorname{Spd}\mathbf{Q}_p = (\operatorname{Spa}\mathbf{Q}_p)^\diamond$.

If $X = \operatorname{Spa}(R, R^+)$ is affinoid, we may write $\operatorname{Spd}(R, R^+)$ (or just $\operatorname{Spd}R$, if $R^+ = R^\circ$) to mean $X^\diamond$.

THEOREM 3.5.2 ([**SW**, Theorem 10.1.3]). *The functor $X^\diamond$ is a diamond.*

The idea behind this, which appears in [**Fal02**] and [**Col02**], is that if $X = \operatorname{Spa}R$ for a Tate Huber $\mathbf{Z}_p$-algebra R, then there exists a tower of finite étale R-algebras R_i, such that $\widetilde{R} = [\varinjlim R_i]^\wedge$ is a perfectoid ring. Let $\widetilde{X} = \operatorname{Spa}\widetilde{R}$; then
$$\widetilde{X}^\diamond \times_{X^\diamond} \widetilde{X}^\diamond \rightrightarrows \widetilde{X}^\diamond \to X^\diamond$$
presents $X^\diamond$ as a quotient of a perfectoid space by a pro-étale equivalence relation.

EXAMPLE 3.5.3. Let K be a perfectoid field of characteristic 0, and let $R = K\langle T^{\pm 1}\rangle$. Then $\widetilde{R} = K\langle T^{\pm 1/p^\infty}\rangle$ is pro-étale over R. If K contains all pth power roots of 1, then $\widetilde{R}/R$ is even a $\mathbf{Z}_p$-torsor.

EXAMPLE 3.5.4. Let K be a perfectoid field of characteristic 0, and let ϖ be a pseudo-uniformizer which divides p in K°. Let $R = K\langle T\rangle$. This time, adjoining pth roots of T produces ramification at the origin in $\operatorname{Spa} R$ (and everywhere if K has characteristic p!), so that $K\langle T^{1/p}\rangle$ will not be finite étale over R. Instead one can adjoin a root of an Artin–Schreier polynomial, such as $U^p - \varpi U = T$, to produce a finite étale R-algebra R_1 for which T is a pth power in R_1°/ϖ. Iteration of this process produces the desired $\widetilde{R}$.

Thus we have a well-defined functor $X \to X^\diamond$ from analytic adic spaces over $\operatorname{Spa} \mathbf{Z}_p$ to diamonds. One might wonder whether this functor is fully faithful, which would allow us to view analytic adic spaces over $\mathbf{Z}_p$ as a subcategory of the category of diamonnds. This cannot be true as stated, since $\operatorname{Spa} \mathbf{Q}_p^{\operatorname{cycl}}$ and $\operatorname{Spa} \mathbf{Q}_p^{\operatorname{cycl},\flat}$ are non-isomorphic adic spaces, while $(\operatorname{Spd} \mathbf{Q}_p^{\operatorname{cycl}})^\diamond \cong (\operatorname{Spd} \mathbf{Q}_p^{\operatorname{cycl},\flat})^\diamond$. But if we fix a nonarchimedean scalar field K, we may instead consider the functor $X \mapsto X^\diamond$ from analytic adic spaces over $\operatorname{Spa} K$ to diamonds over $\operatorname{Spd} K$. This also fails to be fully faithful, as shown by the following example.

EXAMPLE 3.5.5. Let X be the cuspidal cubic $y^2 = x^3$, considered as an adic space over $\mathbf{Q}_p$. Let $X' \to X$ be the usual desingularization of X. That is, X' is the affine line in one variable t, and $X' \to X$ is $t \mapsto (t^2, t^3)$. We claim that $(X')^\diamond \to X^\diamond$ is an isomorphism. This is equivalent to the claim that $X'(R) \to X(R)$ is a bijection for every perfectoid $\mathbf{Q}_p$-algebra R. We leave injectivity as an exercise to the reader (hint: R is reduced). Surjectivity is a little subtle; we refer to the reader to [**KL**, Theorem 3.7.4] for details.

A ring is R *seminormal* $t \mapsto (t^2, t^3)$ is a bijection from R onto the set of pairs $(x, y) \in R^2$ satisfying $y^2 = x^3$. A rigid-analytic space X over a nonarchimedean field K is seminormal if locally it is $\operatorname{Spa}(A, A^+)$, where A is a seminormal ring. The following theorem states that Example 3.5.5 is essentially the only obstruction to $X \mapsto X^\diamond$ being fully faithful.

THEOREM 3.5.6 ([**SW**, Proposition 10.2.4]). *For a nonarchimedean field K of characteristic 0, the functor $X \mapsto X^\diamond$ from seminormal rigid-analytic spaces over K onto diamonds over $\operatorname{Spd} K$ is fully faithful.*

3.6. A diamond version of the Fargues-Fontaine curve. Let C be an algebraically closed perfectoid field of characteristic $p > 0$, and let ϖ be a pseudo-uniformizer of C. Recall the adic space
$$\mathcal{Y}_C = \operatorname{Spa} W(C^\circ) \setminus \{|p[\varpi]| = 0\}.$$
Since $\mathcal{Y}_C$ is a analytic adic space over $\operatorname{Spa} \mathbf{Q}_p$, Theorem 3.5.2 indicates that $\mathcal{Y}_C^\diamond$ makes sense and is a diamond.

PROPOSITION 3.6.1 (The diamond formula). $\mathcal{Y}_C^\diamond \cong \operatorname{Spd} C \times \operatorname{Spd} \mathbf{Q}_p$

PROOF. (Sketch.) The isomorphism says that for a perfectoid ring R in chararacteristic p, the following categories are equivalent:
(1) Pairs consisting of an untilt $R^\sharp/\mathbf{Q}_p$ of R and a continuous homomorphism $C \to R$, and
(2) Pairs consisting of an untilt $R^\sharp$ of R and a morphism $\operatorname{Spa} R^\sharp \to \mathcal{Y}_C$ (whose existence means that $R^\sharp/\mathbf{Q}_p$).

(Both sides are partially proper, so there is no need to discuss rings of integral elements.) We now describe the equivalence assuming an untilt $R^\sharp/\mathbf{Q}_p$: A continuous homomorphism $C \to R$ induces a homomorphism $\theta_C \colon W(C^\circ) \to R^\sharp$, in which the images of p and $[\varpi]$ are invertible; then θ_C induces a morphism $\operatorname{Spa} R^\sharp \to \mathcal{Y}_C$. Conversely if $\operatorname{Spa} R^\sharp \to \mathcal{Y}_C$ is given, we get a homomorphism $W(C^\circ) \to R^{\sharp\circ}$, in which the images of p and $[\varpi]$ are invertible in $R^\sharp$. This induces $C^\circ \to R^{\sharp\circ}/p$. Take the inverse limit under Frobenius to get $C^\circ \to R^\circ$, and then invert ϖ to get $C \to R$. $\square$

As for the adic Fargues–Fontaine curve $\mathcal{X}_C$, we have the diamond formula
$$\mathcal{X}_C^\diamond \cong (\operatorname{Spd} C \times \operatorname{Spd} \mathbf{Q}_p)/(\phi \times \operatorname{id}),$$
where ϕ is the Frobenius automorphism of C. Recall from Theorem 2.7.3 (or rather the adic version of this theorem) that the étale fundamental group of $\mathcal{X}_C$ is $\operatorname{Gal}(\overline{\mathbf{Q}}_p/\mathbf{Q}_p)$. The notion of an étale morphism exists for diamonds, and for an analytic adic space Y, there is an equivalence of sites $Y_{\text{ét}} \xrightarrow{\sim} Y^\diamond_{\text{ét}}$, via the diamond functor. Therefore:

(3.6.1) $\qquad \pi_1((\operatorname{Spa} C \times \operatorname{Spd} \mathbf{Q}_p)/(\phi \times \operatorname{id})) \cong \operatorname{Gal}(\overline{\mathbf{Q}}_p/\mathbf{Q}_p).$

Scholze observed that this formula resembles a theorem of Drinfeld [**Dri80**]. Suppose U and V are two algebraic curves (not necessarily projective) over a common algebraically closed field of characteristic p. The Künneth formula $\pi_1(U \times V) \cong \pi_1(U) \times \pi_1(V)$ fails (the left side is much larger), but it can be salvaged by means of a "partial Frobenius". There is a group $\pi_1((U \times V)/(\phi \times \operatorname{id}))$ classifying finite étale covers of $U \times V$ equipped with an automorphism lying over $\phi \times \operatorname{id}$. Drinfeld's theorem is that $\pi_1((U \times V)/(\phi \times \operatorname{id})) \cong \pi_1(U) \times \pi_1(V)$. The goal of [**Dri80**] (and its successor [**Laf02**]) was to establish the Langlands correspondence for GL_n over a function field, using moduli spaces of *shtukas*. Scholze's goal as laid out in [**SW**] is to define a moduli space of *mixed-characteristic local shtukas* to establish a local Langlands correspondence for p-adic groups.

There are yet other versions of the diamond formula. Let H/C° be the multiplicative formal group. We have seen that the universal cover $\widetilde{H}$ is a formal $\mathbf{Q}_p$-vector space, whose adic generic fiber $\widetilde{H}_C$ is a $\mathbf{Q}_p$-vector space object in the category of perfectoid spaces. The underlying perfectoid space of $\widetilde{H}_C$ is the perfectoid open unit disc. Let $\widetilde{H}_C^* = \widetilde{H}_C \setminus \{0\}$. Then $\widetilde{H}_C^*$ admits an action of $\mathbf{Q}_p^\times$.

PROPOSITION 3.6.2 ([**Wei17**]). *There is an isomorphism of diamonds*
$$\widetilde{H}_C^{*\diamond}/\mathbf{Q}_p^\times \cong (\operatorname{Spd} C \times \operatorname{Spd} \mathbf{Q}_p)/(\operatorname{id} \times \phi).$$
The étale fundamental group of $\widetilde{H}_C^{*\diamond}/\mathbf{Q}_p^\times$ is isomorphic to $\operatorname{Gal}(\overline{\mathbf{Q}}_p/\mathbf{Q}_p)$.

PROOF. By Lemma 3.2.2, $\widetilde{H}_C^*$ is isomorphic to $\operatorname{Spa} C \times \operatorname{Spa} \mathbf{Q}_p^{\text{cycl},\flat}$, where the $\mathbf{Z}_p^\times$ action on $\widetilde{H}_C^*$ becomes the Galois action on $\operatorname{Spa} \mathbf{Q}_p^{\text{cycl},\flat}$. Therefore $\widetilde{H}_C^{*\diamond}/\mathbf{Z}_p^\times$ is isomorphic to $\operatorname{Spd} C \times (\operatorname{Spd} \mathbf{Q}_p^{\text{cycl},\flat}/\mathbf{Z}_p^\times) \cong \operatorname{Spd} C \times \operatorname{Spd} \mathbf{Q}_p$. One can also check that the action of p on $\widetilde{H}_C^*$ corresponds to the action of Frobenius on $\operatorname{Spa} \mathbf{Q}_p^{\text{cycl},\flat}$, which gives the claimed isomorphism.

The statement about the étale fundamental group looks like (3.6.1), but the partial Frobenius is on the wrong side. No matter: the composition of two partial Frobenii is the absolute Frobenius, which is an equivalence on the étale site of any diamond. $\square$

REMARK 3.6.3. There is a generalization of the above proposition which concerns a finite extension $E/\mathbf{Q}_p$. One has to replace H with the Lubin-Tate formal

$\mathcal{O}_E$-module H_E. Then the diamond $Z_E = \widetilde{H}_{E,C}^{*\diamond}/E^\times$ classifies untilts of a perfectoid C-algebra to a perfectoid E-algebra, up to Frobenius.

REMARK 3.6.4. The diamond $(\operatorname{Spd} C \times \operatorname{Spd} \mathbf{Q}_p)/(\operatorname{id} \times \phi)$ is called the *mirror curve* by Fargues, who identifies it as the moduli space of divisors of degree 1 on X_C.

4. Banach-Colmez spaces

4.1. Definition and first examples. So far we have considered objects belonging to a progression of categories: rigid spaces over a nonarchimedean field of residue characteristic p, analytic adic spaces over $\operatorname{Spa} \mathbf{Z}_p$, perfectoid spaces, and finally diamonds, which (in the limited sense of Theorem 3.5.6) generalize all three. This lecture will introduce some examples of diamonds which carry the structure of $\mathbf{Q}_p$-vector spaces. Throughout, we fix an algebraically closed perfectoid field $C/\mathbf{Q}_p$ with residue field k.

EXAMPLE 4.1.1. The following are two examples of sheaves of $\mathbf{Q}_p$-vector spaces on Pfd_C.
  (1) If V is a finite-dimensional $\mathbf{Q}_p$-vector space, we have the constant sheaf $\underline{V}$. If (R, R^+) is a perfectoid Huber pair over (C, C°), then $\underline{V}(R, R^+)$ is the $\mathbf{Q}_p$-vector space of continuous maps $|\operatorname{Spa}(R, R^+)| \to V$. If $\operatorname{Spa}(R, R^+)$ is connected, then $\underline{V}(R, R^+) = V$.
  (2) The additive group $\mathbf{G}_a$ may be considered as a sheaf of $\mathbf{Q}_p$-vector spaces on Pfd_C, by $\mathbf{G}_a(R, R^+) = R$. For a finite-dimensional C-vector space W, the sheaf $W \otimes_C \mathbf{G}_a$ is $(R, R^+) \mapsto W \otimes_C R$.

Both sorts of examples are diamonds arising from analytic adic spaces over C.

DEFINITION 4.1.2. The category of *Banach-Colmez spaces* over C is the smallest abelian subcategory of the category of sheaves of $\mathbf{Q}_p$-vector spaces on Pfd_C which contains the objects $\underline{V}$ and $W \otimes_{\mathbf{Q}_p} \mathbf{G}_a$ from Example 4.1.1 and which is closed under extensions.

An equivalent category was introduced by Colmez [**Col02**] without reference to perfectoid spaces; the definition above appears in [**Bra**], where it shown that the two definitions are equivalent. The "Banach" half of the name refers to Colmez' definition, in which the objects are functors taking values in the category of $\mathbf{Q}_p$-Banach spaces.

EXAMPLE 4.1.3. Let H_0 be a p-divisible group over k, and let H be a lift of $H_0 \otimes_k C^\circ/p$ to C°. We have seen that the universal cover $\widetilde{H} = \varprojlim_p H$ does not depend on the choice of lift H, and that the generic fiber $\widetilde{H}_C$ is a $\mathbf{Q}_p$-vector space object in the category of perfectoid spaces over C. The logarithm map on H induces an exact sequence of sheaves of $\mathbf{Q}_p$-vector spaces on Pfd_C, as in (2.7.1):

$$0 \to \underline{VH} \to \widetilde{H}_C \to \operatorname{Lie} H \otimes_{C^\circ} \mathbf{G}_a \to 0.$$

One has to check exactness on the right. This is a matter of showing that, for any perfectoid C-algebra R and any $v \in \operatorname{Lie} H \otimes_{C^\circ} R$, that there exists a pro-étale R'/R and a sequence $(x_0, x_1, \cdots) \in \widetilde{H}(R^\circ)$ with $\log_H(x_0) = v$. After replacing v with $p^n v$ for $n \gg 0$, we may assume that $\exp_H(v)$ converges to $x_0 \in H(R^\circ)$. The pro-étale extension R' is then obtained by adjoining all pth power division points of x_0 to R.

Since $\widetilde{H}_C$ is an extension of $(\operatorname{Lie} H)[1/p] \otimes_C \mathbf{G}_a$ by $\underline{VH}$, it is a Banach-Colmez space, which happens to be representable by a perfectoid space.

In (2.7) we saw a connection between $\widetilde{H}_C$ and vector bundes on the Fargues-Fontaine curve $X_{C^\flat}$. Let $D(H)$ be the (contravariant) Dieudonné module of H, so that $D(H)$ is a free finite-rank $W(k)$-module equipped with actions of Frobenius and Verschiebung. Then $N := \operatorname{Hom}_{W(k)}(D(H), W(k)[1/p])$ is an isocrystal, all of whose slopes lie in the range $[0,1]$. Let $\mathcal{E}(N)$ be the associated vector bundle. For a perfectoid C-algebra R, we have the relative Fargues–Fontaine curve $X_{R^\flat}$ constructed in (3.2), which lies over $X_{C^\flat}$.

PROPOSITION 4.1.4. *Let R be a perfectoid C-algebra. There is an isomorphism $\widetilde{H}(R^\circ) \overset{\sim}{\to} H^0(X_{R^\flat}, \mathcal{E}(N))$.*

PROOF. (Sketch.) The left-hand side is $\widetilde{H}(R^\circ) \cong \widetilde{H}(R^{\flat\circ})$ and right-hand side is $(B_{R^\flat} \otimes_{W(k)} N)^{\phi=1}$. (Thus both sides only depend on the tilt R°.) $R^{\flat\circ}$ is a perfect ring; by [**SW13**, Theorem 4.1.4], the covariant crystalline Dieudonné module functor on p-divisible groups over $R^{\flat\circ}$ up to isogeny is fully faithful. Applied to morphisms $(\mathbf{Q}_p/\mathbf{Z}_p)_{R^{\flat\circ}} \to H_{R^{\flat\circ}}$, that result gives an isomorphism $\widetilde{H}(R^{\flat\circ}) \overset{\sim}{\to} (B_{\operatorname{cris}}(R^{\flat\circ}) \otimes_{W(k)} N)^{\phi=1}$, where $B_{\operatorname{cris}}(R^{\flat\circ})$ is the crystalline period ring. A hint as to why $(B_{R^\flat} \otimes_{W(k)} N)^{\phi=1} \cong (B_{\operatorname{cris}}(R^{\flat\circ}) \otimes_{W(k)} N)^{\phi=1}$ is [**FF11**, Corollaire 1.10.13], although strictly speaking that result only applies to the case that R is a field. □

EXAMPLE 4.1.5. Let $C'/\mathbf{Q}_p$ be an untilt of $C^\flat$, not necessarily equal to C itself. We define a sheaf $\mathbf{G}'_a$ on Pfd_C by sending a perfectoid C-algebra R to the untilt of $R^\flat$ over C'. That is, $\mathbf{G}'_a(R) = W(R^\flat) \otimes_{W(C^{\flat\circ})} C'$. We claim that $\mathbf{G}'_a$ is a Banach-Colmez space. To see this, let H be the formal multiplicative group over C°. Theorem 2.5.1 produces two nonzero elements $t, t' \in \widetilde{H}(C^{\flat\circ}) \cong \widetilde{H}(C^\circ)$, well-defined up to multiplication by $\mathbf{Q}_p^\times$, corresponding to the untilts C and C' of $C^\flat$, respectively. There are now two intersecting exact sequences of sheaves of $\mathbf{Q}_p$-vector spaces on Pfd_C:

$$\begin{array}{ccccccccc}
& & & & 0 & & & & \\
& & & & \downarrow & & & & \\
& & & & \underline{\mathbf{Q}}_p t' & & & & \\
& & & & \downarrow & & & & \\
0 & \longrightarrow & \underline{\mathbf{Q}}_p t & \longrightarrow & \widetilde{H} & \longrightarrow & \mathbf{G}_a & \longrightarrow & 0. \\
& & & & \downarrow & & & & \\
& & & & \mathbf{G}'_a & & & & \\
& & & & \downarrow & & & & \\
& & & & 0 & & & &
\end{array}$$

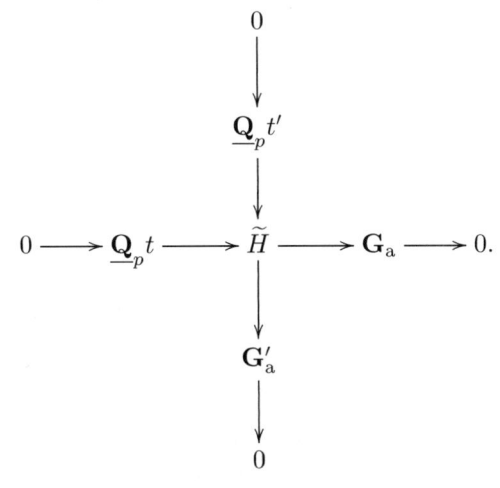

Thus $\mathbf{G}'_a$ is the quotient of a Banach–Colmez space, and so must be one itself.

4.2. Banach-Colmez spaces of slope > 1.
Now suppose N is a general isocrystal over k, which doesn't necessarily arise from a p-divisible group. We may consider the functor $H^0(\mathcal{E}(N))$ on Pfd_C, which sends a perfectoid C-algebra R to the $\mathbf{Q}_p$-vector space $H^0(X_{R^\flat}, \mathcal{E}(N))$. It suffices to consider the case $\mathcal{E}(N) = \mathcal{O}_X(\lambda)$ for $\lambda \in \mathbf{Q}$, because a general $\mathcal{E}(N)$ is isomorphic to a direct sum of these. If $\lambda < 0$ then $H^0(X, \mathcal{O}_X(\lambda)) = 0$, and if $\lambda \in [0, 1]$, then Proposition 4.1.4 shows that $H^0(\mathcal{O}_X(\lambda)) \cong \widetilde{H}_\lambda$ is an absolute perfectoid space.

What if $\lambda > 1$? For instance, if $\lambda = 2$, then $H^0(X_{R^\flat}, \mathcal{O}_X(2)) = B_{R^\flat}^{\phi = p^2}$. For brevity's sake, let $B^{\phi = p^2} = H^0(\mathcal{O}_X(2))$. Let C' be an untilt of $C^\flat$ which is not Frobenius-equivalent to C. As in Example 4.1.5, the two untilts C and C' correspond to $\mathbf{Q}_p$-linearly independent elements $t, t' \in B_{C^\flat}^{\phi = p} \cong H^0(X_{C^\flat}, \mathcal{O}_X(1))$.

PROPOSITION 4.2.1. *There is an exact sequence of sheaves of $\mathbf{Q}_p$-vector spaces on Pfd_C:*

$$0 \to \underline{\mathbf{Q}}_p \to B^{\phi = p} \times B^{\phi = p} \to B^{\phi = p^2} \to 0$$
$$t \mapsto (t, t')$$
$$(x, x') \mapsto xt' - x't$$

PROOF. First we check that the map $B^{\phi = p} \times B^{\phi = p} \to B^{\phi = p^2}$ is surjective. Let R be a perfectoid C-algebra, and let $s \in B_{R^\flat}^{\phi = p^2}$. Let R' be the untilt of $R^\flat$ over C'. We have two ring homomorphisms $\theta \colon B_{R^\flat} \to R$ and $\theta' \colon B_{R^\flat} \to R'$, such that $\ker \theta \cap B_{R^\flat}^{\phi = p} = \mathbf{Q}_p t$ and $\ker \theta' \cap B_{R^\flat}^{\phi = p} = \mathbf{Q}_p t'$. Thus $\theta(t')$ and $\theta'(t)$ are both nonzero. We have the elements $\theta(t')^{-1}\theta(s) \in R$ and $\theta'(t)^{-1}\theta'(s) \in R'$. After replacing R with a pro-étale extension, we can find elements $x, x' \in B_{R^\flat}^{\phi = p}$ such that $\theta(x) = \theta(t')^{-1}\theta(s)$ and $\theta'(x') = -\theta'(t)^{-1}\theta'(s)$. Then the element

$$\alpha = xt' - x't - s \in B_{R'}^{\phi = p^2}$$

has the property that $\theta(\alpha) = 0$ and $\theta'(\alpha) = 0$. This implies that $\alpha = att'$ for some $a \in H^0(X_{R^\flat}, \mathcal{O}_X) = \underline{\mathbf{Q}}_p(R)$; this shows that $s = (x - at)t' - x't$ lies in the image of $B^{\phi = p} \times B^{\phi = p}$ as required. □

As a corollary, we find that $B^{\phi = p^2}$ is a Banach-Colmez space, and also a diamond. Indeed, Proposition 4.2.1 gives a presentation of $B^{\phi = p^2}$ as a quotient of a perfectoid space by a pro-étale equivalence relation. More generally, if N is an isocrystal over k, then $H^0(\mathcal{E}(N))$ is a Banach-Colmez space and a diamond.

4.3. The de Rham period ring.
We begin with a definition from p-adic Hodge theory.

DEFINITION 4.3.1. Let R be a perfectoid ring. The *de Rham period ring* $B_{\mathrm{dR}}^+(R)$ is the completion of $W(R^{\flat\circ})[1/p]$ with respect to the kernel of $\theta_R \colon W(R^{\flat\circ})[1/p] \to R$.

If $R = C$ is an algebraically closed perfectoid field, then $B_{\mathrm{dR}}^+(C)$ is a discrete valuation ring with uniformizer ξ_C, residue field C and fraction field $B_{\mathrm{dR}}(C)$. These objects were constructed by Fontaine. They appear in the context of p-adic p-adic Galois representations, particularly in the comparison isomorphism linking étale and de Rham cohomology of a variety over a p-adic field [**Fal89**]. They also appear in the study of the Fargues-Fontaine curve: the untilt C of $C^\flat$ determines a closed point $\infty \in X_{C^\flat}$, and $B_{\mathrm{dR}}^+(C)$ is the completed local ring $\widehat{\mathcal{O}}_{X_{C^\flat},\infty}$.

DEFINITION 4.3.2. For $n \geq 1$, let $B_{\mathrm{dR}}^+/\mathrm{Fil}^n$ be the sheaf on Pfd_C which assigns to (R, R^+) the $\mathbf{Q}_p$-vector space $B_{\mathrm{dR}}^+(R)/(\ker \theta_R)^n$.

THEOREM 4.3.3. $B_{\mathrm{dR}}^+/\mathrm{Fil}^n$ *is a Banach–Colmez space and a diamond.*

Note that $B_{\mathrm{dR}}^+/\mathrm{Fil}^1 = \mathbf{G}_\mathrm{a}$, because for a perfectoid C-algebra R, we have $B_{\mathrm{dR}}(R)^+/(\ker \theta_R) = R$.

PROOF. We sketch the proof for $B_{\mathrm{dR}}/\mathrm{Fil}^2$; the general case works by induction. Consider the complex of sheaves of $\mathbf{Q}_p$-vector spaces on Pfd_C:

(4.3.1) $\quad\quad\quad 0 \to \mathrm{Fil}^1/\mathrm{Fil}^2 \to B_{\mathrm{dR}}^+/\mathrm{Fil}^2 \to B_{\mathrm{dR}}^+/\mathrm{Fil}^1 \to 0.$

We already observed that $B_{\mathrm{dR}}^+/\mathrm{Fil}^1 = \mathbf{G}_\mathrm{a}$. As for $\mathrm{Fil}^1/\mathrm{Fil}^2$, we claim that it is $\mathbf{G}_\mathrm{a}(1) = \mathbf{G}_\mathrm{a} \otimes_{\mathbf{Q}_p} \mathbf{Q}_p(1)$. We construct an isomorphism $\mathrm{Fil}^1/\mathrm{Fil}^2 \to \mathbf{G}_\mathrm{a}$ over $\mathrm{Spd}\,C$. Form the element t as in (2.5.1), and consider it as an element of $B_{\mathrm{dR}}^+(C)$. Then t generates the kernel of $B_{\mathrm{dR}}^+(R) \to R$ for any perfectoid C-algebra R. This shows that $\mathrm{Fil}^1/\mathrm{Fil}^2 \cong \mathbf{G}_\mathrm{a}$, and therefore that $B_{\mathrm{dR}}^+/\mathrm{Fil}^2$ is a Banach-Colmez space.

Given a perfectoid C-algebra R, a section of $\mathrm{Fil}^1/\mathrm{Fil}^2$ over R consists of a pro-étale cover $\mathrm{Spa}\,\widetilde{R} \to \mathrm{Spa}\,R$ and an element $\alpha \in tB_{\mathrm{dR}}(\widetilde{R})^+$, together with a descent datum through $\mathrm{Spa}\,\widetilde{R} \to \mathrm{Spa}\,R$ for the image of α modulo t^2. Our morphism sends this section to $\theta_{\widetilde{R}}(\alpha/t)$, which (because of the descent datum) lies in R. (We leave it to the reader to construct the morphism in the opposite direction.) Note that $\mathrm{Gal}(\mathbf{Q}_p^\mathrm{cycl}/\mathbf{Q}_p)$ acts on t via the cyclotomic character; this is what we need to descend the morphism through $\mathbf{Q}_p^\mathrm{cycl}/\mathbf{Q}_p$.

Now we claim that the complex in (4.3.1) locally splits. Let H be the formal multiplicative group over C, and let $\widetilde{H}_C$ be the generic fiber of its universal cover. Then $\widetilde{H}_C$ is a perfectoid space, and the logarithm map $\widetilde{H}_C \to \mathbf{G}_\mathrm{a}$ is a pro-étale cover. Define a morphism $\widetilde{H}_C \to B_{\mathrm{dR}}^+$ by $(x_0, x_1, \dots) \mapsto \log[(x_i)]$. Then the following diagram commutes:

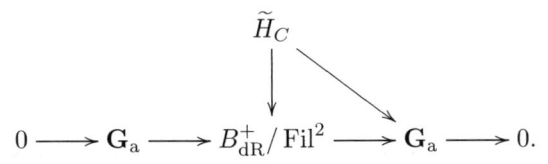

$$\begin{array}{ccccccccc} & & & & \widetilde{H}_C & & & & \\ & & & \swarrow & \downarrow & \searrow & & & \\ 0 & \longrightarrow & \mathbf{G}_\mathrm{a} & \longrightarrow & B_{\mathrm{dR}}^+/\mathrm{Fil}^2 & \longrightarrow & \mathbf{G}_\mathrm{a} & \longrightarrow & 0. \end{array}$$

We can now give a presentation of $B_{\mathrm{dR}}^+/\mathrm{Fil}^2$: it is the quotient of $\mathbf{G}_\mathrm{a} \times \widetilde{H}_C$ by the pro-étale equivalence relation of "having the same image in $B_{\mathrm{dR}}^+/\mathrm{Fil}^2$". $\square$

In general, $B_{\mathrm{dR}}^+/\mathrm{Fil}^i$ is a Banach-Colmez space admitting an i-step filtration, where the quotients are isomorphic to $\mathbf{G}_\mathrm{a}$.

As with the Banach-Colmez spaces of the previous section, $B_{\mathrm{dR}}^+/\mathrm{Fil}^n$ is the space of global sections of a (Zariski) sheaf on the Fargues-Fontaine curve $X_{C^\flat}$. The untilt C of $C^\flat$ determines a closed point $\infty \in X_{C^\flat}$. The completion of $X_{C^\flat}$ at ∞ is $\mathrm{Spec}\,B_{\mathrm{dR}}^+(C)$. Let $i_\infty \colon \mathrm{Spec}\,B_{\mathrm{dR}}^+(C) \to X_{C^\flat}$ be the corresponding morphism. Then $\mathcal{F} := i_{\infty *}(B_{\mathrm{dR}}^+/(\ker \theta_C)^n)$ is a coherent sheaf on $X_{C^\flat}$ supported at ∞. Proposition 4.3.3 says that $H^0(\mathcal{F})$ (meaning the sheaf $R \mapsto H^0(X_{R^\flat}, \mathcal{F})$) is a Banach-Colmez space and a diamond.

Our examples show a strong connection between Banach–Colmez spaces and coherent sheaves on the Fargues-Fontaine curve. Indeed, for any coherent sheaf $\mathcal{F}$ on $X_{C^\flat}$, the sheaf $R \mapsto H^0(X_{R^\flat}, \mathcal{F})$ is a Banach-Colmez space. The complete story is a theorem of Le Bras [**Bra**], which gives an equivalence between the category $\mathcal{BC}(C)$ of Banach–Colmez spaces relative to C and the core of a certain t-structure on the derived category of coherent sheaves on $X_{C^\flat}$. Every object of $\mathcal{BC}(C)$ is isomorphic to $H^0(\mathcal{F}^+) \oplus H^1(\mathcal{F}^-)$, where $\mathcal{F}^+$ and $\mathcal{F}^-$ are coherent $\mathcal{O}_{X_{C^\flat}}$-modules. (An example of type $H^1(\mathcal{F}^-)$ is discussed in Project 5.5.) As a corollary, $\mathcal{BC}(C)$ only depends on the tilt $C^\flat$. Finally, every Banach-Colmez space is a diamond.

4.4. A survey of the diamond landscape.

In these lectures we have introduced a hierarchy of nonarchimedean analytic spaces: rigid spaces, adic spaces, perfectoid spaces, and diamonds. We have highlighted the role of $\mathbf{Q}_p$-vector space objects in each category. In the last two sections, we studied $\mathbf{Q}_p$-vector space diamonds arising as global sections of sheaves on the Fargues-Fontaine curve (vector bundles and torsion sheaves, respectively).

Since we presented these objects without much context, you have a right to wonder about motivation. Why do we care that certain sheaves on Pfd are diamonds? And why are these particular objects so important?

Fundamentals of diamond geometry. The device of étale cohomology allows us to apply our intuitions about algebraic topology to schemes. To wit, if X is a scheme, there is a notion of an étale site $X_{\text{ét}}$, whose objects are étale morphisms over X; these can be used to define the ℓ-adic cohomology groups $H^i(X_{\text{ét}}, \mathbf{Q}_\ell)$. If in addition X is a smooth projective variety over an algebraically closed field k, and ℓ is invertible in k, then the $H^i(X_{\text{ét}}, \mathbf{Q}_\ell)$ have some nice properties: they are zero outside of the range $i = 0, 1, \ldots, 2\dim X$, they satisfy Poincaré duality, there is a Lefschetz fixed-point formula one can apply to endomorphisms of X, and so on.

Underpinning these properties is a framework of results concerning different kinds of morphisms (finite type, étale, proper, smooth, etc.) and their effects on étale sheaves. For instance, we have a notion of a smooth morphism of schemes $f \colon X \to Y$, which is meant to mimic the same notion for manifolds, and which can be checked using a Jacobian criterion. The Poincaré duality theorem mentioned above is a special case of a relative version: there is an isomorphism $f^!\mathcal{F} \cong f^*\mathcal{F}[2d](d)$, valid whenever f is a smooth morphism of relative dimension d, and $\mathcal{F}$ is an étale sheaf of $(\mathbf{Z}/n\mathbf{Z})$-modules on Y, where n is invertible on Y.

Many of these fundamentals are carried over into the world of rigid and adic spaces in [**Hub96**]. Huber defines the important classes of morphisms of adic spaces (finite type, étale, proper, smooth, etc.), and proves theorems (base change theorems, Poincaré duality) about how they interact with étale cohomology.

Perfectoid spaces seem at first glance to be immune to this sort of treatment. For instance, let K be a perfectoid field, and let $\widetilde{D} = \operatorname{Spa} K\langle T^{1/p^\infty}\rangle$ be the perfectoid closed disc from Example 3.1.5. The ring $K\langle T^{1/p^\infty}\rangle$ isn't finitely generated over K, nor is it even topologically finitely generated, so already we run into problems if we wish to think of $\widetilde{D}$ as being "finite type" over K. The situation seems even worse if one tries to define smooth morphisms of perfectoid spaces using a

Jacobian criterion. (If f belongs to $A = K\langle T^{1/p^\infty}\rangle$, then "$df/dT$", naïvely defined, may fail to lie in A, so this is certainly not the right way to proceed.)

Nonetheless, Scholze [**Sch17**] has defined a notion of *cohomological smoothness* for a morphism of diamonds (relative to a prime ℓ distinct from the residue characteristic), which essentially says that relative Poincaré duality holds. In this sense, $\widetilde{D} \to \operatorname{Spa} K$ is cohomologically smooth, as are the Banach–Colmez spaces $\widetilde{H}$ and $B_{\mathrm{dR}}^+/\mathrm{Fil}^n$ (over the base $\operatorname{Spd} C$). Recent work of Fargues-Scholze [**FS**] even gives a Jacobian criterion of sorts to determine whether a morphism of diamonds is cohomologically smooth.

Moduli spaces of mixed-characteristic local shtukas. Let $C/\mathbf{Q}_p$ be an algebraically closed perfectoid field with residue field k, and let H_0 be a p-divisible group over k. Recall from the discussion in (2.7) that we have an exact sequence of $\mathbf{Q}_p$-vector spaces

$$0 \to VH(C^\circ) \to \widetilde{H}(C^\circ) \to \operatorname{Lie} H \otimes C \to 0,$$

which can be interpreted as an exact sequence of $\mathcal{O}_{X_{C^\flat}}$-modules:

$$0 \to \mathcal{O}_{X_{C^\flat}} \otimes_{\mathbf{Q}_p} VH \to \mathcal{E}(H_0) \to i_* \operatorname{Lie} H \otimes C \to 0,$$

where $\mathcal{E}(H_0)$ is the vector bundle corresponding to (the isocrystal corresponding to) H_0, and i is the morphism $\operatorname{Spec} B_{\mathrm{dR}}^+(C) \to X_{C^\flat}$.

Define a (partially proper) sheaf $\mathcal{M}_{H_0}$ on Pfd_C as follows. For a perfectoid C-algebra R, we define $\mathcal{M}_{H_0}(R)$ to be the set of injective morphisms $s\colon \mathcal{O}_{X_{R^\flat}}^h \to \mathcal{E}(H_0)$ of $\mathcal{O}_{X_{R^\flat}}$-modules, whose cokernel is a sheaf of the form $i_* W$, where W is a projective R-module. (We have used the same letter i to denote the morphism $\operatorname{Spec} B_{\mathrm{dR}}^+(R) \to X_{R^\flat}$.)

Results in [**SW13**] show that $\mathcal{M}_{H_0}$ is a perfectoid space, and that it is isomorphic to the moduli space of deformations H of H_0 together with a $\mathbf{Q}_p$-basis for VH. The space $\mathcal{M}_{H_0}$ admits commuting actions of the groups $J = \operatorname{Aut}^0 H_0$ (automorphisms up to isogeny; this acts on $\mathcal{E}(H_0)$) and $\mathrm{GL}_h(\mathbf{Q}_p)$ (which acts on $\mathcal{O}_{X_{R^\flat}}^h$). The cohomology groups $H_c^i(\mathcal{M}_{H_0,\mathbf{C}_p}, \overline{\mathbf{Q}}_\ell)$ admit an action of $\mathrm{GL}_h(\mathbf{Q}_p) \times J \times W_{\mathbf{Q}_p}$, where $W_{\mathbf{Q}_p}$ is the Weil group. In the case that H_0 is basic, the *Kottwitz conjectures* predict that these cohomology groups realize Langlands functoriality. In the special case that H_0 is connected of dimension 1, the space $\mathcal{M}_{\overline{H}}$ is called a Lubin-Tate space (at infinite level). The Kottwitz conjectures are known to be true in for Lubin-Tate space [**HT01**].

The introduction of diamonds allows us to generalize the situation considerably. Fix an integer $h \geq 1$ and an isocrystal b of rank h. Write $\mathcal{E}_b$ for the corresponding vector bunde on X_C. Fix an h-tuple of integers $\mu = (a_1, a_2, \ldots, a_h)$ with $a_1 \geq \cdots \geq a_h \geq 0$. Such a μ determines a class of modules over a discrete valuation ring (A, M), namely those of the form $\bigoplus_{i=1}^h A/M^{a_i}$. The set of such μ forms a partially ordered set.

DEFINITION 4.4.1 (The space of infinite-level local shtukas with one leg [**SW**]). Let $\mathcal{M}_{b,\mu}$ be the (partially proper) functor on Pfd_C which assigns to R the set of exact sequences

$$0 \to \mathcal{O}_{X_{R^\flat}}^h \to \mathcal{E}_{b,R^\flat} \to i_* W \to 0,$$

where W is a $B_{\mathrm{dR}}^+(R)$-module quotient of $i^*\mathcal{E}_{b,R^\flat}$ which (at every geometric point of $\mathrm{Spa}\,R$) is of type $\leq \mu$.

One refers to the exact sequence above as a *modification* of $\mathcal{E}_b$ of type $\leq \mu$ which produces the trivial vector bundle.

When b is an isocrystal with slopes in $[0,1]$ and μ is *minuscule* (meaning $a_i \leq 1$ for all i), we recover the moduli space $\mathcal{M}_{\overline{H}_0}$ as above, so long as a certain compatibility is satisfied between b and μ.

The name "shtuka" recalls Drinfeld's constructions for a smooth projective curve over a finite field [**Dri80**]. Drinfeld defined a space of rank 2 shtukas and studied the cohomology of this space, and in doing so proved the Langlands conjectures for GL_2 over a function field. This was generalized to GL_n by L. Lafforgue [**Laf02**]. (There is a strong but highly non-obvious analogy between the two sorts of shtukas.)

THEOREM 4.4.2 ([**SW**]). *The sheaf $\mathcal{M}_{b,\mu}$ is a diamond.*

The idea is that $\mathcal{M}_{b,\mu}$ admits a pro-étale morphism to the space of possible Ws, which is a kind of flag variety; one wants to show that this latter space is a diamond. For this it helps to know that $B_{\mathrm{dR}}^+/\mathrm{Fil}^i$ is a diamond, which is Theorem 4.3.3. (More details are supplied by the lecture notes of Kedlaya in this series.)

It therefore makes sense to consider the étale cohomology of the $\mathcal{M}_{b,\mu}$, and to pose generalizations of the Kottwitz conjecture for it. The construction of the $\mathcal{M}_{b,\mu}$ answers a question of Rapoport–Viehmann about the existence of "local Shimura varieties" [**RV14**].

A geometric Langlands program for p-adic fields. Let X be a smooth projective curve over a finite field k, with function field K. The set

$$\prod_{x \in |X|} \mathrm{GL}_n(K_x^\circ) \backslash \mathrm{GL}_n(\mathbf{A}_K) / \mathrm{GL}_n(K)$$

has two interpretations: (1) it classifies the set of isomorphism classes of rank n vector bundles on X, and (2) functions on this set are automorphic forms on K of level 1. Now, automorphic forms on K of level 1 which are Hecke eigenforms are supposed to correspond to n-dimensional Galois representations of K which are unramified everywhere, which is to say, rank n local systems on X.

The idea behind *geometric Langlands* is to geometrize the above statement, along the lines of the function-sheaf correspondence of Grothendieck. The set $\prod_{x \in |X|} \mathrm{GL}_n(K_x^\circ) \backslash \mathrm{GL}_n(\mathbf{A}_K) / \mathrm{GL}_n(K)$ is the set of k-points of the stack Bun_n which classifies vector bundles of rank n. Instead of considering functions on this set, we consider $\overline{\mathbf{Q}}_\ell$-sheaves on Bun_n.

The Hecke operators from the usual theory get geometrized as well. The stack Bun_n admits Hecke correspondences indexed by n-tuples $\mu = (a_1, \ldots, a_n)$, with $a_1 \geq \cdots \geq a_n$. For each such μ, there is a diagram of stacks

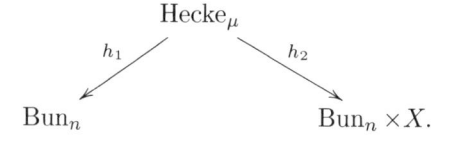

$$\begin{array}{ccc} & \mathrm{Hecke}_\mu & \\ {}^{h_1}\swarrow & & \searrow^{h_2} \\ \mathrm{Bun}_n & & \mathrm{Bun}_n \times X. \end{array}$$

Here Hecke$_\mu$ classifies pairs of rank n vector bundles $\mathcal{E}_1$ and $\mathcal{E}_2$, together with a modification of $\mathcal{E}_2$ at a point $P \in X$ which produces $\mathcal{E}_1$; the morphisms h_1 and h_2 take such a datum to $\mathcal{E}_1$ and $(\mathcal{E}_2, P)$, respectively. The Hecke operator $\mathcal{H}_\mu$ inputs a sheaf on Bun$_n$ and outputs a sheaf on Bun$_n \times X$. In the case that μ is minuscule (meaning all a_i are 0 or 1), then $\mathcal{H}_\mu(\mathcal{F})) = (h_2)_! h_1^* \mathcal{F}$.

THEOREM 4.4.3 ([**FGV02**]). *For every irreducible and everywhere unramified ℓ-adic representation $\phi \colon \mathrm{Gal}(K^s/K) \to \mathrm{GL}_n(\overline{\mathbf{Q}}_\ell)$, there exists a nonzero perverse sheaf $\mathcal{F}_\phi$ on* Bun$_n$, *which is a Hecke eigensheaf with respect to ϕ in the following sense: for all μ, $\mathcal{H}_\mu(\mathcal{F}) \cong \mathcal{F} \boxtimes (r_\mu \circ \phi)$, where r_μ is the algebraic representation of* GL_n *with highest weight μ.*

There is a marvelous suite of conjectures due to Fargues [**Far**] which replaces X with the Fargues–Fontaine curve in the above discussion. In this context we define the stack Bun$_n$ as the sheaf on Pfd which assigns to a perfectoid $\mathbf{F}_p$-algebra R the groupoid of rank n vector bundles on X_R.

THEOREM 4.4.4 ([**FS**]). *The sheaf* Bun$_n$ *is a smooth Artin stack in the category of perfectoid spaces: it admits a smooth surjective morphism from a smooth diamond.*

As before, the stack Bun$_n$ admits Hecke correspondences. For each μ, there is a corresponding Hecke operator H_μ which inputs a sheaf on Bun$_n$ and outputs a sheaf on Bun$_n \times \mathrm{Spd}\,\mathbf{Q}_p$. Part of Fargues' conjecture is the following.

CONJECTURE 4.4.5. *Let $\phi \colon W_{\mathbf{Q}_p} \to \mathrm{GL}_n(\overline{\mathbf{Q}}_\ell)$ be an irreducible ℓ-adic representation. There exists a nonzero perverse sheaf $\mathcal{F}_\phi$ on* Bun$_n$ *such that for all μ we have $H_\mu(\mathcal{F}_\phi) \cong \mathcal{F}_\phi \otimes (r_\mu \circ \phi)$.*

There is a connection between the Hecke operators H_μ and spaces of shtukas $\mathcal{M}_{b,\mu}$, and in fact the full statement Fargues' conjecture implies the generalized Kottwitz conjecture for $\mathcal{M}_{b,\mu}$ in the case that b is basic.

5. Projects

5.1. Basic examples of adic spaces.

(1) Classify points in $\mathrm{Spa}\,\mathbf{Q}_p\langle T\rangle$; describe the set-theoretic fibers of $\mathrm{Spa}\,\mathbf{C}_p\langle T\rangle \to \mathrm{Spa}\,\mathbf{Q}_p\langle T\rangle$.
(2) Classify points in $\mathrm{Spa}\,W(C^\circ)$, where C is an algebraically closed perfectoid field of characteristic $p > 0$.

5.2. Perfectoid fields.

(1) Let $K = \mathbf{Q}_2(2^{1/2^\infty})^\wedge$. Identify $K^\flat$ with $\mathbf{F}_2((t^{1/2^\infty}))$, where t corresponds to the sequence $(2, 2^{1/2}, \dots)$. Let $L = K(\sqrt{-1})$, so that L/K has degree 2. Thus L is perfectoid. Identify $L^\flat$ as a separable extension of $\mathbf{F}_2((t^{1/p^\infty}))$. Repeat for all other quadratic field extensions of K.
(2) Let K be a perfectoid field with residue field k. Show that $K^\flat \cong k((t^{1/p^\infty}))$ if and only if the following criterion holds: K admits no proper perfectoid subfields with the same residue field and value group.

5.3. Some commutative algebra.

(1) Let K be a perfectoid field. Describe the group of units in $K\langle T^{1/p^\infty}\rangle$.
(2) Let C be an algebraically closed perfectoid field of characteric p, and let $f \in C\langle T^{1/p^\infty}\rangle$ be a non-unit. Let $D = \{|x| \leq 1\} \subset C$. Does there always exist $\alpha \in D$ with $f(\alpha) = 0$? Is the set of zeros of f finite? Profinite? Which subsets of D are zero sets of such f?
(3) Is there a generalization of the preceding exercise in characteristic 0?
(4) Continuing this theme, let C be an algebraically closed perfectoid field of characteristic p, and let $f_1, \ldots, f_m \in A = C\langle T_1^{1/p^\infty}, \ldots, T_n^{1/p^\infty}\rangle$ be elements which do not generate the unit ideal. Does there exist a common zero of the f_i in D^n? (This is something like a perfectoid Nullstellensatz statement.)

5.4. Closed subsets of adic spaces.
For a scheme X, a closed subset $T \subset X$ is (rather by definition) Zariski closed: it is the zero locus of an ideal sheaf in $\mathcal{O}_X$. There is a scheme, the *reduced induced subscheme* Z, and a closed immersion $Z \to X$ whose set-theoretic image is T. This property is universal: for a reduced scheme Y, a morphism $f\colon Y \to X$ has $f(Y) \subset T$ (set-theoretically) if and only if f factors as $Y \to Z \to X$.

It is quite different with adic spaces. One difference is that closed subsets are not necessarily Zariski-closed.

(1) Consider $\mathbf{Q}_p$ as a closed subset of the underlying topological space of $\mathbf{A}^1$, considered as an adic space over $\mathbf{Q}_p$. Show that $\mathbf{Q}_p$ is not Zariski closed.
(2) Nonetheless, show that there exists a reduced adic space Z and a morphism $Z \to \mathbf{A}^1$, which is a monomorphism and has image $\mathbf{Q}_p$, and which satisfies a universal property.

Let $H/\overline{\mathbf{F}}_p$ be a formal p-divisible group of height 2 and dimension 1. Its universal cover $\widetilde{H}$ lifts to a formal $\mathbf{Q}_p$-vector space over $\breve{\mathbf{Z}}_p = W(\overline{\mathbf{F}}_p)$; let $\widetilde{H}_{\breve{\mathbf{Q}}_p}$ be its generic fiber. Then $\widetilde{H}_{\breve{\mathbf{Q}}_p}$ is a preperfectoid space. Let $M(H)$ be the Dieudonné module of H; this is a free $\breve{\mathbf{Z}}_p$-module of rank 2. There is a quasi-logarithm map of adic spaces

$$\mathrm{qlog}_H\colon \widetilde{H}_{\breve{\mathbf{Q}}_p} \to M(H) \otimes_{\breve{\mathbf{Z}}_p} \mathbf{G}_a \cong \mathbf{G}_a^2$$

which respects the $\mathbf{Q}_p$-vector space structure on either side. We describe it as a natural transformation between functors from $\mathrm{Pfd}_{\breve{\mathbf{Q}}_p}$ to $\mathbf{Q}_p$-vector spaces. Let R be a perfectoid $\breve{\mathbf{Q}}_p$-algebra. We have an isomorphism $\widetilde{H}_{\breve{\mathbf{Q}}_p}(R) = \widetilde{H}(R^\circ) \cong (B(R^\flat) \otimes_{\breve{\mathbf{Z}}_p} M(H))^{\phi=1}$. Then $\mathrm{qlog}_H(R)$ is the composition of this map with $\theta_R \otimes 1\colon B(R^\flat) \otimes_{\breve{\mathbf{Z}}_p} M(H) \to R \otimes_{\breve{\mathbf{Z}}_p} M(H)$.

(3) Prove that qlog is a monomorphism.
(4) Let Z be the image of qlog_H, considered as a subset of the underlying topological space of $\mathbf{G}_a^2$. Show that Z is closed and generalizing.
(5) Show that the residue fields of nonzero points of Z are never finite extensions of $\breve{\mathbf{Q}}_p$. That is, the image of qlog_H contains no "classical points" other than the origin.
(6) Show that if Y is a perfectoid space over $\mathrm{Spa}\,\breve{\mathbf{Q}}_p$ and $f\colon Y \to \mathbf{G}_a^2$ has set-theoretic image contained in Z, then f factors through qlog_H.

Thus we have a closed subset of the adic space $\mathbf{G}_a^2$ which (considered as a subfunctor on the category of perfectoid spaces) is representable by a preperfectoid space. In fact, it is a theorem of Scholze [**Sch17**] that any closed generalizing subset of a diamond, when considered as a subfunctor on the category of perfectoid spaces, is itself a diamond.

5.5. Computations with Banach-Colmez spaces. Recall our discussion of Banach-Colmez spaces, which are sheaves of $\mathbf{Q}_p$-algebras on the category of perfectoid spaces. There are two projects here. The first has to do with some "ineffective" Banach-Colmez spaces. Fix an algebraically closed perfectoid fied C of characteristic 0.

(1) We begin with the space $H^1(\mathcal{O}_X(-1))$, which inputs a perfectoid C-algebra R and outputs the $\mathbf{Q}_p$-vector space $H^1(X_R, \mathcal{O}_X(-1))$. Show that there is an isomorphism of sheaves of $\mathbf{Q}_p$-vector spaces on Pfd_C:
$$H^1(\mathcal{O}_X(-1)) \cong \mathbf{G}_a/\underline{\mathbf{Q}}_p.$$

(2) The sheaf $H^1(\mathcal{O}_X(-1))$ parametrizes extension classes
$$0 \to \mathcal{O}_X(-1) \to \mathcal{E} \to \mathcal{O}_X \to 0,$$
or (after twisting by $\mathcal{O}_X(1)$) extension classes
$$0 \to \mathcal{O}_X \to \mathcal{E} \to \mathcal{O}_X(1) \to 0.$$
Show that if this latter extension is nonsplit, then there exists an isomorphism $\mathcal{E} \cong \mathcal{O}_X(1/2)$. Recall that global sections of $\mathcal{O}_X(1/2)$ are representable by a formal scheme $\widetilde{H}$, where $H/\bar{\mathbf{F}}_p$ is a formal p-divisible group of dimension 1 and height 2. Let us abbreviate $\widetilde{H}_C^* = \widetilde{H}_C \setminus \{0\}$; this is a perfectoid space over C. Show that there is an isomorphism
$$H^1(\mathcal{O}_X(-1)) \setminus \{0\} \cong (\widetilde{H}_C^* \times \underline{\mathbf{Q}}_p(1)^*)/D^\times,$$
where $D = \mathrm{Aut}^0 H$ is the nonsplit quaternion algebra over $\mathbf{Q}_p$, where $\mathbf{Q}_p(1)^* = \mathbf{Q}_p(1) \setminus \{0\}$, and where $D^\times$ acts on $\mathbf{Q}_p(1)^*$ through the reduced norm map.

(3) Let $\Omega = \mathbf{G}_a \backslash \underline{\mathbf{Q}}_p$. Combining the previous two exercises gives an isomorphism $\Omega/\underline{\mathbf{Q}}_p \cong (\widetilde{H}_C^* \times \underline{\mathbf{Q}}_p(1)^*)/D^\times$. This isomorphism means there is a diamond M carrying an action of $\mathbf{Q}_p \times D^\times$, whose quotient by $D^\times$ is Ω, and whose quotient by $\mathbf{Q}_p$ is $\widetilde{H}_C^* \times \mathbf{Q}_p(1)^*$. Show that M (with this action) is isomorphic to the Lubin-Tate tower for $\mathrm{GL}_2(\mathbf{Q}_p)$.

(4) Is there a similar story for $H^1(X, \mathcal{O}_X(\lambda))$ for other negative values of $\lambda \in \mathbf{Q}$?

The other project is due to David Hansen. Let $M \to \mathrm{Spd}\, C$ be the infinite-level Lubin–Tate tower for $\mathrm{GL}_2(\mathbf{Q}_p)$. Then M can be interpreted as the space of "mixed-characteristic shtukas" of a certain type. To wit, M is the sheafification of the presheaf which assigns to a perfectoid C-algebra R, the set of exact sequences of the form
$$0 \to \mathcal{O}_{X_{R^\flat}}^2 \to \mathcal{O}_{X_{R^\flat}}(1/2) \to i_*W \to 0,$$
where $i\colon \mathrm{Spec}\, B_{\mathrm{dR}}(R) \to X_{R^\flat}$ is the usual morphism, and W is a rank 1 projective quotient of $i^*\mathcal{O}_{X_{R^\flat}}(1/2)$. Then M admits an action of the product group $\mathrm{GL}_2(\mathbf{Q}_p) \times D^\times$, where $D = \mathrm{Aut}_{\mathcal{O}_X} \mathcal{O}_X(1/2)$ is the nonsplit quaternion algebra over $\mathbf{Q}_p$.

Here is a different space of shtukas, which we'll call N: it is the sheafification of the presheaf which assigns to a perfectoid C-algebra R the set of exact sequences of the form
$$0 \to \mathcal{O}_{X_{R^\flat}}^2 \to \mathcal{O}_{X_{R^\flat}}(1)^2 \to i_*V \to 0,$$
where this time V is a projective $B_{\mathrm{dR}}^+(R^\flat)/(\ker\theta_R)^2$-module of rank 1. Then N admits an action of $\mathrm{GL}_2(\mathbf{Q}_p) \times \mathrm{GL}_2(\mathbf{Q}_p)$.

(5) Show that N is a perfectoid space.
(6) Show that, in the category of diamonds admitting an action of $\mathrm{GL}_2(\mathbf{Q}_p) \times \mathrm{GL}_2(\mathbf{Q}_p)$, the sheaf N is isomorphic to the quotient $(M \times M)/D^\times$, where the action of $D^\times$ is the diagonal one.
(7) Are there other isomorphisms of these type, for different spaces of shtukas?

References

[Ber91] Pierre Berthelot, *Cohomologie rigide et cohomologie rigide à supports propres*, Prépublication de l'université de Rennes 1, 1991.

[Bra] Arthur-César Le Bras, *Espaces de Banach-Colmez et faisceaux cohérents sur la courbe de Fargues-Fontaine* (French, with English and French summaries), Duke Math. J. **167** (2018), no. 18, 3455–3532, DOI 10.1215/00127094-2018-0034. MR3881201

[BS15] Bhargav Bhatt and Peter Scholze, *The pro-étale topology for schemes* (English, with English and French summaries), Astérisque **369** (2015), 99–201. MR3379634

[Col02] Pierre Colmez, *Espaces de Banach de dimension finie* (French, with English and French summaries), J. Inst. Math. Jussieu **1** (2002), no. 3, 331–439, DOI 10.1017/S1474748002000099. MR1956055

[Col10] Pierre Colmez, *Représentations de $\mathrm{GL}_2(\mathbf{Q}_p)$ et (ϕ, Γ)-modules* (French, with English and French summaries), Astérisque **330** (2010), 281–509. MR2642409

[Dri80] V. G. Drinfel'd, *Langlands' conjecture for $\mathrm{GL}(2)$ over functional fields*, Proceedings of the International Congress of Mathematicians (Helsinki, 1978), Acad. Sci. Fennica, Helsinki, 1980, pp. 565–574. MR562656

[Fal89] Gerd Faltings, *Crystalline cohomology and p-adic Galois-representations*, Algebraic analysis, geometry, and number theory (Baltimore, MD, 1988), Johns Hopkins Univ. Press, Baltimore, MD, 1989, pp. 25–80. MR1463696

[Fal02] Gerd Faltings, *Almost étale extensions*, Astérisque **279** (2002), 185–270. Cohomologies p-adiques et applications arithmétiques, II. MR1922831

[Far] Laurent Fargues, *Geometrization of the local Langlands correspondence: an overview*, https://arxiv.org/pdf/1602.00999.pdf.

[Far15] Laurent Fargues, *Quelques résultats et conjectures concernant la courbe* (French, with English and French summaries), Astérisque **369** (2015), 325–374. MR3379639

[FF11] Laurent Fargues and Jean-Marc Fontaine, *Courbes et fibrés vectoriels en théorie de Hodge p-adique*, preprint, 2011.

[FGV02] E. Frenkel, D. Gaitsgory, and K. Vilonen, *On the geometric Langlands conjecture*, J. Amer. Math. Soc. **15** (2002), no. 2, 367–417, DOI 10.1090/S0894-0347-01-00388-5. MR1887638

[Fon13] Jean-Marc Fontaine, *Perfectoïdes, presque pureté et monodromie-poids (d'après Peter Scholze)* (French, with French summary), Astérisque **352** (2013), Exp. No. 1057, x, 509–534. Séminaire Bourbaki. Vol. 2011/2012. Exposés 1043–1058. MR3087355

[FS] Laurent Fargues and Peter Scholze, *Geometrization of the local Langlands correspondence*, in preparation.

[FW79] Jean-Marc Fontaine and Jean-Pierre Wintenberger, *Extensions algébrique et corps des normes des extensions APF des corps locaux* (French, with English summary), C. R. Acad. Sci. Paris Sér. A-B **288** (1979), no. 8, A441–A444. MR527692

[Haz12] Michiel Hazewinkel, *Formal groups and applications*, AMS Chelsea Publishing, Providence, RI, 2012. Corrected reprint of the 1978 original. MR2987372

[HT01] Michael Harris and Richard Taylor, *The geometry and cohomology of some simple Shimura varieties*, Annals of Mathematics Studies, vol. 151, Princeton University Press, Princeton, NJ, 2001. With an appendix by Vladimir G. Berkovich. MR1876802

[Hub94] R. Huber, *A generalization of formal schemes and rigid analytic varieties*, Math. Z. **217** (1994), no. 4, 513–551, DOI 10.1007/BF02571959. MR1306024

[Hub96] Roland Huber, *Étale cohomology of rigid analytic varieties and adic spaces*, Aspects of Mathematics, E30, Friedr. Vieweg & Sohn, Braunschweig, 1996. MR1734903

[Ked15] Kiran S. Kedlaya, *New methods for (Γ, φ)-modules*, Res. Math. Sci. **2** (2015), Art. 20, 31, DOI 10.1186/s40687-015-0031-z. MR3412585

[Ked16] Kiran S. Kedlaya, *Noetherian properties of Fargues-Fontaine curves*, Int. Math. Res. Not. IMRN **8** (2016), 2544–2567, DOI 10.1093/imrn/rnv227. MR3519123

[KL] Kiran S. Kedlaya and Ruochuan Liu, *Relative p-adic Hodge theory: foundations* (English, with English and French summaries), Astérisque **371** (2015), 239. MR3379653

[KL15] Kiran S. Kedlaya and Ruochuan Liu, *Relative p-adic Hodge theory: foundations* (English, with English and French summaries), Astérisque **371** (2015), 239. MR3379653

[Laf02] Laurent Lafforgue, *Chtoucas de Drinfeld et correspondance de Langlands* (French, with English and French summaries), Invent. Math. **147** (2002), no. 1, 1–241, DOI 10.1007/s002220100174. MR1875184

[LT65] Jonathan Lubin and John Tate, *Formal complex multiplication in local fields*, Ann. of Math. (2) **81** (1965), 380–387, DOI 10.2307/1970622. MR0172878

[Mes72] William Messing, *The crystals associated to Barsotti-Tate groups: with applications to abelian schemes*, Lecture Notes in Mathematics, Vol. 264, Springer-Verlag, Berlin-New York, 1972. MR0347836

[RV14] Michael Rapoport and Eva Viehmann, *Towards a theory of local Shimura varieties*, Münster J. Math. **7** (2014), no. 1, 273–326. MR3271247

[Sch12] Peter Scholze, *Perfectoid spaces*, Publ. Math. Inst. Hautes Études Sci. **116** (2012), 245–313, DOI 10.1007/s10240-012-0042-x. MR3090258

[Sch17] Peter Scholze, *The étale cohomology of diamonds*, ARGOS Seminar in Bonn, 2017.

[Sta14] The Stacks Project Authors, *Stacks Project*, http://stacks.math.columbia.edu, 2014.

[SW] Peter Scholze and Jared Weinstein, *Berkeley lectures on p-adic geometry*, Book manuscript, 2018.

[SW13] Peter Scholze and Jared Weinstein, *Moduli of p-divisible groups*, Camb. J. Math. **1** (2013), no. 2, 145–237, DOI 10.4310/CJM.2013.v1.n2.a1. MR3272049

[Tat67] J. T. Tate, *p-divisible groups*, Proc. Conf. Local Fields (Driebergen, 1966), Springer, Berlin, 1967, pp. 158–183. MR0231827

[Wei17] Jared Weinstein, $\mathrm{Gal}(\overline{\mathbf{Q}}_p/\mathbf{Q}_p)$ *as a geometric fundamental group*, Int. Math. Res. Not. IMRN **10** (2017), 2964–2997, DOI 10.1093/imrn/rnw072. MR3658130

Sheaves, stacks, and shtukas

Kiran S. Kedlaya

These are extended and slightly[1] revised notes from a four-lecture series at the 2017 Arizona Winter School on the topic of perfectoid spaces. The appendix describes the proposed student projects (including contributions from David Hansen and Sean Howe). See the table of contents below for a list of topics covered.

These notes were deliberately written to include *much more* material than could possibly be presented in four one-hour lectures. Some of this material is original, including the definition of an analytic Huber ring, the extension of various basic results from Tate rings to analytic rings, and the general form of Drinfeld's lemma for diamonds.

The reader should be warned that these lectures were interleaved with the other lectures at the AWS; specifically, each lecture of ours followed the corresponding lecture of Weinstein [**177**]. In particular, the reader who has never seen adic spaces before should turn with [**177**, Lecture 1] before continuing here.

These notes have benefited tremendously from detailed feedback from a number of people, including Bastian Haase, David Hansen, Shizhang Li, Ruochuan Liu, Peter Scholze, and Alex Youcis, and from a detailed review of the post-AWS revisions by Sean Howe. In addition, special thanks are due to the participants of the UCSD winter 2017 reading seminar on perfectoid spaces: Annie Carter, Zonglin Jiang, Jake Postema, Daniel Smith, Claus Sorensen, Xin Tong, and Peter Wear.

The author was supported by NSF (grants DMS-1501214, DMS-1802161), UCSD (Warschawski Professorship), and IAS (Visiting Professorship 2018–2019).

CONVENTION 0.0.1. Throughout these lecture notes, the following conventions are in force unless specifically overridden.

- All rings are commutative and unital.
- A *complete* topological space is required to be Hausdorff (as usual).
- All Huber rings and pairs we consider are complete. (This is not the convention used in [**177**, Lecture 1].)
- A *nonarchimedean field* is a field which is complete with respect to a nontrivial nonarchimedean absolute value.

---

[1] One significant revision is the treatment of pseudocoherent modules in Lecture 1, necessitated by Scholze's discovery of Example 1.4.8. The presentation of §2.4 has also been modified to account for Theorem 2.9.12.

1. Sheaves on analytic adic spaces

We begin by picking up where the first lecture of Weinstein [**177**, Lecture 1], on the adic spectrum associated to a Huber pair, leaves off. We collect the basic facts we need about the structure sheaf, vector bundles, and coherent sheaves on the adic spectrum. The approach is in some sense motivated by the analogy between the theories of "varieties" (here meaning schemes locally of finite type over a field) and of general schemes. In our version of this analogy, the building blocks of the finite-type case are affinoid algebras over a nonarchimedean field (with which we assume some familiarity, e.g., at the level of [**23**] or [**72**]), and we are trying to extend to more general Huber rings in order to capture examples that are very much not of finite type (notably perfectoid rings). However, this passage does not go quite as smoothly as in the theory of schemes, so some care is required to assemble a theory that is both expansive enough to include perfectoid rings and robust enough to allow us to assert the general theorems we will need.

In order to streamline the exposition, we have opted to state most of the key theorems first without proof (see §1.2–1.4). We then follow with discussion of the overall strategy of proof of these theorems (see §1.6), and finally treat the technical details of the proofs (see §1.7–1.9). Along the way, we include some technical subsections that can be skimmed or skipped on first reading: one on the open mapping theorem (§1.1), one on Banach rings (§1.5), one on the étale topology (§1.10), and one on preadic spaces (§1.11).

HYPOTHESIS 1.0.1. *Throughout §1, let (A, A^+) be a fixed Huber pair (with A complete, as per our conventions) and put $X := \mathrm{Spa}(A, A^+)$. Unless otherwise specified, we assume also that A is analytic (see Definition 1.1.2); however, there is little harm done if the reader prefers to assume in addition that A is Tate (see Definition 1.1.2 and Remark 1.1.5).*

1.1. Analytic rings and the open mapping theorem.
We begin with a brief technical discussion, which can mostly be skipped on first reading. This has to do with the fact that Huber's theory of adic spaces includes the theory of formal schemes as a subcase, but we are primarily interested in the complementary subcase.

REMARK 1.1.1. In any Huber ring, the set of units is open: if x is a unit and y is sufficiently close to x, then $x^{-1}(x-y)$ is topologically nilpotent and its powers sum to an inverse of $x^{-1}y$. This implies that any maximal ideal is closed.

This observation is often used in conjunction with [**95**, Proposition 3.6(i)]: if $A \neq 0$, then $X \neq \emptyset$. For a derivation of this result, see Corollary 1.5.18.

DEFINITION 1.1.2. Recall that the Huber ring A is said to be *Tate* (or sometimes *microbial*) if it contains a topologically nilpotent unit (occasionally called a *microbe* by analogy with terminology used in real algebraic geometry [**52**]; more commonly a *pseudouniformizer*). For example, if A is an algebra over a nonarchimedean field, then A is Tate; this includes the case where A contains a field F on which the induced topology is discrete (e.g., when A is of characteristic p), as then for any pseudouniformizer $\varpi \in A$ we have $F((\varpi)) \subseteq A$. However, it is possible for A to be Tate without being an algebra over any nonarchimedean field; see [**113**, Example 2.16].

More generally, we say that A is *analytic* if its topologically nilpotent elements generate the trivial ideal in A; Example 1.5.7 separates the Tate and analytic conditions. The term *analytic* is not standard (the corresponding term in [**74**, §0.B.1.(c)] is *extremal*), but is motivated by Lemma 1.1.3 below. By convention, the zero ring is both Tate and analytic.

We say that a Huber pair (A, A^+) is *Tate* (resp. *analytic*) if A is Tate (resp. analytic).

LEMMA 1.1.3. *The following conditions on a general Huber pair (A, A^+) are equivalent.*
  (a) *The ring A is analytic.*
  (b) *Any ideal of definition in any ring of definition generates the unit ideal in A.*
  (c) *Every open ideal of A is trivial.*
  (d) *For every nontrivial ideal I of A, the quotient topology on A/I is not discrete.*
  (e) *The only discrete topological A-module is the zero module.*
  (f) *The set X contains no point on whose residue field the induced valuation is trivial.*

PROOF. We start with some easy implications:
- (b) implies (a) (any ideal of definition consists of topologically nilpotent elements);
- (b) and (c) are equivalent (any ideal of definition is open, and any open ideal contains an ideal of definition);
- (c) and (d) are equivalent (trivially);
- (e) implies (d) (trivially).

We next check that (a) implies (b). Suppose that A is analytic, A_0 is a ring of definition, and I is an ideal of definition. For any topologically nilpotent elements $x_1, \ldots, x_n \in A$ which generate the unit ideal, for any sufficiently large m the elements $x_1^m, \ldots, x_n^m$ belong to I and still generate the unit ideal in A.

At this point, we have the equivalence among (a)–(d). To add (e), we need only check that (c) implies (e), which we achieve by checking the contrapositive. Let M be a nonzero discrete topological A-module, and choose any nonzero $m \in M$. The map $A \to M$, $a \mapsto am$ is continuous; its kernel is a nontrivial open ideal of A.

We next check that (a) implies (f). If A is analytic, then for each $v \in X$, we can find a topologically nilpotent element $x \in A$ with $v(x) \neq 0$. We must then have $0 < v(x) < 1$, so the induced valuation on the residue field is nontrivial.

We finally check that (f) implies (d), by establishing the contrapositive. Let I be a nontrivial ideal of A such that A/I is discrete for the quotient topology. Then the trivial valuation on the residue field of any maximal ideal of A/I gives rise to a point of X on whose residue field the induced valuation is trivial. $\square$

COROLLARY 1.1.4. *If (A, A^+) is an analytic Huber pair, then $\mathrm{Spa}(A, A^+) \to \mathrm{Spa}(A^+, A^+)$ is injective. (We will show later that it is also a homeomorphism onto its image; see Lemma 1.6.5.)*

PROOF. For $v \in \mathrm{Spa}(A, A^+)$, by Lemma 1.1.3 there exists a topologically nilpotent element x of A such that $0 < v(x) < 1$. For $w \in \mathrm{Spa}(A, A^+)$ agreeing with v on A^+, for any $y, z \in A$, any sufficiently large positive integer n has the property that $x^n y, x^n z \in A^+$; it follows that the order relations in the pairs

$$(v(y), v(z)), (v(x^n y), v(x^n z)), (w(x^n y), w(x^n z)), (w(y), w(z))$$

all coincide, yielding $v = w$. $\square$

REMARK 1.1.5. Lemma 1.1.3 shows that a Huber pair (A, A^+) is analytic if and only if $\mathrm{Spa}(A, A^+)$ is analytic in the sense of Huber; that is, if (A, A^+) is analytic, then $\mathrm{Spa}(A, A^+)$ is covered by rational subspaces (see Definition 1.2.1) which are the adic spectra of Tate rings. Consequently, from the point of view of adic spaces, escalating the level of generality of Huber pairs from Tate to analytic does not create any new geometric objects. However, it does improve various statements about acyclicity of sheaves, as in the rest of this lecture.

EXERCISE 1.1.6. Let A be a Huber ring. If there exists a finite, faithfully flat morphism $A \to B$ such that B is Tate (resp. analytic) under its natural topology as an A-module (see Definition 1.1.11), then A is Tate (resp. analytic).

EXERCISE 1.1.7. A (continuous) morphism $f: A \to B$ of general Huber rings is *adic* if one can choose rings of definition A_0, B_0 of A, B and an ideal of definition A such that $f(A_0) \subseteq B_0$ and $f(I)B_0$ is an ideal of definition of B_0. Prove that this condition is always satisfied when A is analytic.

From now on, assume (unless otherwise indicated) that A is analytic. In the classical theory of Banach spaces, the *open mapping theorem* of Banach plays a fundamental role in showing that topological properties are often controlled by algebraic properties. The same theorem is available in the nonarchimedean setting for analytic rings.

DEFINITION 1.1.8. A morphism of topological abelian groups is *strict* if the subspace and quotient topologies on its image coincide. For a surjective morphism, this is equivalent to the map being open.

THEOREM 1.1.9 (Open mapping theorem). *Let $f: M \to N$ be a continuous morphism of topological A-modules which are first-countable (i.e., 0 admits a countable neighborhood basis) and complete (which implies Hausdorff). If f is surjective, then f is open. (Note that A itself is first-countable.)*

PROOF. As in the archimedean case, this comes down to an application of Baire's theorem that every complete metric space is a Baire space (i.e., the union of countably many nowhere dense subsets is never open). The case where A is a nonarchimedean field can be treated in parallel with the archimedean case, as in Bourbaki [**25**, I.3.3, Théorème 1]; see also [**155**, Proposition 8.6]. It was observed by Huber [**96**, Lemma 2.4(i)] that the argument carries over to the case where A is Tate; this was made explicit by Henkel [**93**]. The analytic case is similar; see Problem A.3.1. □

REMARK 1.1.10. Theorem 1.1.9 is in fact a characterization of analytic Huber rings: if A is not analytic, there exists a morphism $f: M \to N$ of complete first-countable topological A-modules which is continuous but not open. For example, let I be a nontrivial open ideal and take M, N to be two copies of $\prod_{n \in \mathbb{Z}} (A/I)$ equipped with the discrete topology and the product topology, respectively. (Thanks to Zonglin Jiang for this example.)

Before stating an immediate corollary of Theorem 1.1.9, we need a definition.

DEFINITION 1.1.11. Let M be a finitely generated A-module. For any A-linear surjective morphism $F \to M$ where F is a finite free A-module, we may form the quotient topology of M; the resulting topology does not depend on the choice. (It suffices to compare with a second surjection $F \oplus F' \to M$ by factoring the map $F' \to M$ through $F \oplus F'$.) This topology is called the *natural topology* on M.

If A is noetherian, then M is always complete for its natural topology (see Corollary 1.1.15 below). In general, M need not be complete for the natural topology, but the only way for completeness to fail is for M to fail to be Hausdorff. Namely, if M is Hausdorff, then $\ker(F \to M)$ is closed, so quotienting by it gives a complete A-module.

Even if M is complete for its topology, that does not mean that its image under a morphism of finitely generated A-modules must be complete (unless A is noetherian). For example, for $f \in A$, it can happen that $\times f : A \to A$ is injective but its image is not closed; see Remark 1.8.3.

COROLLARY 1.1.12. *Suppose that A is analytic. Let M be a finitely generated A-module. If M admits the structure of a complete first-countable topological A-module for some topology, then that topology must be the natural topology.*

PROOF. Apply Theorem 1.1.9 to an A-linear surjection $F \to M$ with F finite free. □

Let us now see some examples of this theorem in action. The following argument is essentially [**24**, Proposition 3.7.2/1] or [**72**, Lemma 1.2.3].

LEMMA 1.1.13. *Let M be a finitely generated A-module which is complete for the natural topology. Then any dense A-submodule of M equals M itself. (This argument does not require A to be analytic, but the following corollary does.)*

PROOF. We may lift the problem to the case where M is free on the basis $\mathbf{e}_1, \ldots, \mathbf{e}_n$. Let N be a dense submodule of M; we may then choose $\mathbf{e}'_1, \ldots, \mathbf{e}'_n \in N$ such that $\mathbf{e}'_j = \sum_i B_{ij} \mathbf{e}_i$ with B_{ij} being topologically nilpotent if $i \neq j$ and $B_{ii} - 1$ being topologically nilpotent if $i = j$. Then the matrix B is invertible (its determinant equals 1 plus a topological nilpotent), so $N = M$. □

COROLLARY 1.1.14. *Let M be a finitely generated A-module which is complete for the natural topology. Then any A-submodule of M whose closure is finitely generated is itself closed.*

PROOF. Let N be an A-submodule whose closure $\widehat{N}$ is finitely generated. By Corollary 1.1.12, the subspace topology on $\widehat{N}$ coincides with the natural topology, so Lemma 1.1.13 may be applied to see that $N = \widehat{N}$. □

COROLLARY 1.1.15. *The following statements hold.*
  (a) *If A is noetherian, then every finitely generated A-module is complete for the natural topology, and every submodule of such a module is closed.*
  (b) *Conversely, if every ideal of A is closed, then A is noetherian.*

PROOF. Suppose first that A is noetherian. For M a finitely generated A-module and $F \to M$ an A-linear surjection with F finite free, Corollary 1.1.14 implies that $\ker(F \to M)$ is closed, so M is complete. Applying Corollary 1.1.14 again shows that every submodule of M is closed, yielding (a).

Conversely, suppose that every ideal of A is closed. To prove (b), we will obtain a contradiction under the hypothesis that there exists an ascending chain of ideals $I_1 \subseteq I_2 \subseteq \cdots$ which does not stabilize, by showing that the union I of the chain is not closed. In fact this already follows from Baire's theorem, but we give a more elementary argument below.

Since A is analytic, we can find some finite set $x_1, \ldots, x_n$ of topologically nilpotent units which generate the unit ideal in A. For each m, choose an element $y_m \in I_m - I_{m-1}$. We can then choose an index $i_m \in \{1, \ldots, n\}$ such that $x_{i_m}^j y_m \notin I_{m-1}$ for all positive integers j.

Let $V_1, V_2, \ldots$ be a cofinal sequence of neighborhoods of 0 in A. We now choose positive integers $j_1, j_2, \ldots$ and open subgroups U_m of A subject to the following conditions (by choosing j_m sufficiently large and U_m sufficiently small given the choice of $j_1, \ldots, j_{m-1}$ and $U_1, \ldots, U_{m-1}$).

(a) For each positive integer m, $x_{i_m}^{j_m} y_m \in V_m \cap \bigcap_{m' < m} U_{m'}$.
(b) For each positive integer m, $\overline{U}_m \subseteq \bigcap_{m' < m} U_{m'}$.
(c) For each positive integer m, $(x_{i_m}^{j_m} y_m + U_m) \cap I_{m-1} = 0.$.

Then $\sum_{m=1}^{\infty} x_{i_m}^{j_m} y_m$ converges to a limit y which is in the closure of I by (a), but not in I by (b) (for each m we have $y \in I_{m-1} + x_{i_m}^{j_m} + \overline{U}_m$ and hence $y \notin I_{m-1}$), a contradiction. □

As a concrete example of what happens when A is not noetherian, we offer the following exercise.

DEFINITION 1.1.16. For A a Huber ring, let $A\langle T\rangle$ be the completion of $A[T]$ for the topology with a neighborhood basis given by $U[T] = \{\sum_{n=0}^{\infty} a_n T^n : a_n \in U \text{ for all } n\}$ as U runs over neighborhoods of 0 in A. We may similarly define $A\langle T_1, \ldots, T_m\rangle$, or even the analogue with infinitely many variables. When the topology on A is induced by a norm, this can be interpreted in terms of a Gauss[2] norm; see Definition 1.5.3.

EXERCISE 1.1.17. Let p be a prime. Let A be the quotient of the infinite Tate algebra $\mathbb{Q}_p\langle T, U_1, V_1, U_2, V_2, \ldots\rangle$ by the closure of the ideal $(TU_1 - pV_1, TU_2 - p^2 V_2, \ldots)$.

(a) Show that A is uniform (see Definition 1.2.12).
(b) Show that T is not a zero-divisor in A.
(c) Show that the ideal TA is not closed in A.

The following argument can be found in [96, II.1], [97, Lemma 1.7.6].

LEMMA 1.1.18. *Let M be an A-module which is the cokernel of a strict morphism between finite projective A-modules. Equivalently by Theorem 1.1.9, M is finitely presented and complete for the natural topology.*

(a) *Let $M\langle T\rangle$ be the set of formal sums $\sum_{n=0}^{\infty} x_n T^n$ with $x_n \in M$ forming a null sequence. Then the natural map $M \otimes_A A\langle T\rangle \to M\langle T\rangle$ is an isomorphism.*
(b) *Let $M\langle T^{\pm}\rangle$ be the set of formal sums $\sum_{n \in \mathbb{Z}} x_n T^n$ with $x_n \in M$ forming a null sequence in each direction. Then the natural map $M \otimes_A A\langle T^{\pm}\rangle \to M\langle T^{\pm}\rangle$ is an isomorphism.*

PROOF. We treat only (a), since (b) is similar. If M is finitely generated and complete for the natural topology, then it is apparent that $M \otimes_A A\langle T\rangle \to M\langle T\rangle$ is surjective. Suppose now that as in the statement of the lemma, M is the cokernel

---

[2] Correctly spelled "Gauß", but I'll stick to the customary English transliteration.

of a strict morphism $F_1 \to F_0$ between finite projective A-modules. Put $N := \ker(F_0 \to M)$; then N is finitely generated and complete for the natural topology. We thus have a commutative diagram

$$\begin{array}{ccccccc}
N \otimes_A A\langle T\rangle & \longrightarrow & F_0 \otimes_A A\langle T\rangle & \longrightarrow & M \otimes_A A\langle T\rangle & \longrightarrow & 0 \\
\downarrow & & \downarrow & & \downarrow & & \\
0 \longrightarrow N\langle T\rangle & \longrightarrow & F\langle T\rangle & \longrightarrow & M\langle T\rangle & \longrightarrow & 0
\end{array}$$

with exact rows in which the middle vertical arrow is an isomorphism and both vertical arrows are surjective. By the five lemma, it follows that the right vertical arrow is injective. $\square$

LEMMA 1.1.19. *Suppose that A is noetherian.*
(a) *The homomorphism $A \to A\langle T\rangle$ is flat.*
(b) *If $A\langle T\rangle$ is also noetherian, then $A[T] \to A\langle T\rangle$ is also flat.*

PROOF. Let $0 \to M \to N \to P \to 0$ be an exact sequence of finite A-modules; by Corollary 1.1.15, it is also a strict exact sequence for the natural topologies. Consequently, the exact sequence

$$0 \to M\langle T\rangle \to N\langle T\rangle \to P\langle T\rangle \to 0$$

is the base extension of the previous sequence from A to $A\langle T\rangle$. This proves (a).

Suppose now that $A\langle T\rangle$ is noetherian. To prove (b), by [**166**, Tag 00MP] it suffices to check that for every prime ideal $\mathfrak{p}$ of A, the map $A[T]\otimes_A \kappa(\mathfrak{p}) \to A\langle T\rangle \otimes_A \kappa(\mathfrak{p})$ is flat. Since $A\langle T\rangle$ is noetherian (and analytic because A is), Corollary 1.1.15 implies that $\mathfrak{p}A\langle T\rangle$ is a closed ideal; we may thus identify $\mathfrak{p}A\langle T\rangle$ with the subset $\mathfrak{p}\langle T\rangle$ of $A\langle T\rangle$ (again as in Lemma 1.1.18. In particular, as a module over the principal ideal domain $A[T] \otimes_A \kappa(\mathfrak{p}) = \kappa(\mathfrak{p})[T]$, $A\langle T\rangle \otimes_A \kappa(\mathfrak{p}) = A\langle T\rangle/\mathfrak{p}\langle T\rangle$ is torsion-free and hence flat. $\square$

1.2. The structure sheaf. We continue with the definition and analysis of the structure presheaf. As in the theory of affine schemes, we have in mind a formula for certain distinguished open subsets, in this case the rational subspaces; the shape of the general definition is meant to enforce this formula. However, we will almost immediately hit a serious difficulty which echoes throughout the entire theory.

We recall some facts about rational subsets of X from the previous lecture [**177**, Lecture 1].

DEFINITION 1.2.1. A *rational subspace* of X is one of the form

$$X\left(\frac{f_1,\ldots,f_n}{g}\right) := \{v \in X : v(f_i) \leq v(g) \neq 0 \quad (i = 1,\ldots,n)\}$$

where $f_1,\ldots,f_n, g \in A$ are some elements which generate an open ideal in A; such subspaces form a neighborhood basis in X. Since we are assuming that A is analytic, by Lemma 1.1.3 any open ideal is in fact the trivial ideal; in particular, we may rewrite the previous formula as

(1.2.1.1) $$X\left(\frac{f_1,\ldots,f_n}{g}\right) := \{v \in X : v(f_i) \leq v(g) \quad (i = 1,\ldots,n)\}.$$

There is a morphism $(A, A^+) \to (B, B^+)$ of (complete) Huber pairs which is initial among morphisms for which $\mathrm{Spa}(B, B^+)$ maps into $X\left(\frac{f_1,\ldots,f_n}{g}\right)$; this morphism induces a map $\mathrm{Spa}(B, B^+) \cong X\left(\frac{f_1,\ldots,f_n}{g}\right)$ which not only is a homeomorphism, but matches up rational subspace of $\mathrm{Spa}(B, B^+)$ with rational subspaces of X contained in $X\left(\frac{f_1,\ldots,f_n}{g}\right)$. We call any such morphism "the" *rational localization* corresponding to $X\left(\frac{f_1,\ldots,f_n}{g}\right)$, using the definite article since the morphism is unique up to unique isomorphism.

Thanks to (1.2.1.1), the ring B in the pair (B, B^+) may be identified explicitly as the quotient of $A\langle T_1,\ldots,T_n\rangle$ by the closure of the ideal $(gT_1 - f_1,\ldots,gT_n - f_n)$; we denote this ring by $A\left\langle\frac{f_1,\ldots,f_n}{g}\right\rangle$. (We will see later that when the structure presheaf on X is a sheaf, it is not necessary to take the closure; see Theorem 1.2.7.) The ring B^+ may be identified as the integral closure of the image of $A^+\langle T_1,\ldots,T_n\rangle$ in B; we denote this ring by $A^+\left\langle\frac{f_1,\ldots,f_n}{g}\right\rangle$.

EXERCISE 1.2.2. Given $f_1,\ldots,f_n, g \in A$ which generate the unit ideal, there exists a neighborhood W of 0 in A such that any $f'_1,\ldots,f'_n, g' \in A$ satisfying $f'_1 - f_1,\ldots,f'_n - f_n, g' - g \in W$ generate the unit ideal and define the same rational subspace as do $f_1,\ldots,f_n, g$. (See [**156**, Remark 2.8], [**117**, Remark 2.4.7].)

DEFINITION 1.2.3. Define the *structure presheaf* $\mathcal{O}$ on X as follows: for $U \subseteq X$ open, let $\mathcal{O}(U)$ be the inverse limit of B over all rational localizations $(A, A^+) \to (B, B^+)$ with $\mathrm{Spa}(B, B^+) \subseteq U$. In particular, if $U = \mathrm{Spa}(B, B^+)$ then $\mathcal{O}(U) = B$.

Let $\mathcal{O}^+$ be the subpresheaf of $\mathcal{O}$ defined as follows: for $U \subseteq X$ open, let $\mathcal{O}(U)$ be the inverse limit of B^+ over all rational localizations $(A, A^+) \to (B, B^+)$ with $\mathrm{Spa}(B, B^+) \subseteq U$. Equivalently,

$$\mathcal{O}^+(U) = \{f \in \mathcal{O}(U) : v(f) \leq 1 \text{ for all } v \in U\}.$$

In particular, if $U = \mathrm{Spa}(B, B^+)$ then $\mathcal{O}(U) = B^+$.

REMARK 1.2.4. For any open subset U of X, the ring $\mathcal{O}(U)$ is complete for the inverse limit topology, but in general it is not a Huber ring. A typical example is the open unit disc inside the closed unit disc, which is *Fréchet complete* with respect to the supremum norms over all of the closed discs around the origin of radii less than 1. (This ring cannot be Huber because the topologically nilpotent elements do not form an open set.)

REMARK 1.2.5. For each $x \in X$, the stalk $\mathcal{O}_{X,x}$ is a direct limit of complete rings, and hence is a henselian local ring; in particular, the categories of finite étale algebras over $\mathcal{O}_{X,x}$ and over its residue field are equivalent. Compare [**117**, Lemma 2.4.17].

REMARK 1.2.6. In order to follow the theory of affine schemes, one would next expect to prove that the presheaf $\mathcal{O}$ is a sheaf. This is indeed true when A is an affinoid algebra over a nonarchimedean field, as this follows (after a small formal argument; see Lemma 1.6.3) from Tate's acyclicity theorem in rigid analytic geometry [**169**, Theorem 8.2], [**24**, Theorem 8.2.1/1].

Unfortunately, there exist examples where $\mathcal{O}$ is not a sheaf. This remains true if we assume that A is Tate, as shown by an example of Huber [**96**, §1]; or even if we assume that A is Tate and uniform, as shown by examples of Buzzard–Verberkmoes [**27**, Proposition 18] and Mihara [**138**, Theorem 3.15].

Since $\mathcal{O}$ can fail to be a sheaf, we must track of the distinction between its sections on an open subset U, denoted $\mathcal{O}(U)$, and the locally-defined sections of $\mathcal{O}$ on U (i.e., the sections of the sheafification of $\mathcal{O}$ on U). We write $H^0(U, \mathcal{O})$ for the latter.

A conceptual explanation for the previous examples is given by the following result.

THEOREM 1.2.7 (original). *Suppose that $\mathcal{O}$ is a sheaf. Then for any $f_1, \ldots, f_n, g \in A$ which generate the unit ideal, the ideal $(gT_1 - f_1, \ldots, gT_n - f_n)$ in $A\langle T_1, \ldots, T_n \rangle$ is closed.*

PROOF OF THEOREM 1.2.7. Let $(A, A^+) \to (B, B^+)$ be the rational localization defined by the parameters $f_1, \ldots, f_n$; then the kernel of the map $A\langle T_1, \ldots, T_n \rangle \to B$ taking T_i to f_i/g is the closure of the ideal in question. By Corollary 1.1.14, it thus suffices to check that this kernel is finitely generated; this will follow from Lemma 1.9.23. □

In light of the previous remarks, we are forced to introduced and study the following definition.

DEFINITION 1.2.8. We say that (A, A^+) is *sheafy* if $\mathcal{O}$ is a sheaf. Although it is not immediately obvious from the definition, we will see shortly that this property depends only on A, not on A^+ (Remark 1.6.9).

DEFINITION 1.2.9. When (A, A^+) is sheafy, we may equip X in a natural way with the structure of a *locally v-ringed space*, i.e., a locally ringed space in which the stalk of the structure sheaf at each point is equipped with a distinguished valuation (with morphisms required to correctly pull back these valuations). By considering locally v-ringed spaces which are locally of this form, we obtain Huber's notion of an *analytic adic space*.

As explained in [**177**, Lecture 1], Huber's theory also allows the use of rings A which are not analytic; this for example allows ordinary schemes and formal schemes to be treated as adic spaces. In addition, Huber shows that a Huber ring A which need not be analytic, but which admits a noetherian ring of definition, is sheafy [**96**, Theorem 2.5]. However, allowing nonanalytic Huber rings creates some extra complications which are not pertinent to the examples we have in mind (with a small number of exceptions), e.g., the distinction between continuous and adic morphisms (see Exercise 1.1.7). For expository treatments of adic spaces without the analytic restriction, see [**32**] or [**175**].

We will establish sheafiness for two primary classes of Huber rings. The first includes the class of affinoid algebras.

DEFINITION 1.2.10. The ring A is *strongly noetherian* if for every nonnegative integer n, the ring $A\langle T_1, \ldots, T_n \rangle$ is noetherian; note that this property passes to rational localizations. For example, if A is an affinoid algebra over a nonarchimedean field K, then A is strongly noetherian: this reduces to the fact that $K\langle T_1, \ldots, T_n \rangle$ is noetherian, for which see [**169**, Theorem 4.5] or [**24**, Theorem 5.2.6/1].

When A is Tate, the following result is due to Huber [**96**, Theorem 2.5]. The general case incorporates an observation of Gabber to treat the case where A is analytic but not Tate; see §1.7 for the proof.

THEOREM 1.2.11 (Huber plus Gabber's method). *If A is strongly noetherian, then A is sheafy.*

The second class of sheafy rings we consider includes the class of perfectoid rings. (We will mostly handle the condition of uniformity via the corresponding definition for Banach rings; see Definition 1.5.11 and Exercise 1.5.13.)

DEFINITION 1.2.12. Recall that A is said to be *uniform* if the ring of power-bounded elements of A is a bounded subset. (A subset S of A is *bounded* if for every neighborhood U of 0 in A, there exists a neighborhood V of 0 in A such that $S \cdot V \subseteq U$. If A is topologized using a norm, this corresponds to boundedness in the usual sense.) For example, if K is a nonarchimedean field, then $K\langle T\rangle/(T^2)$ is not uniform because the K-line spanned by T is unbounded, but consists of nilpotent and hence power-bounded elements; by the same token, any uniform (analytic) Huber ring is reduced, and conversely for affinoid algebras (see Remark 1.2.16).

The pair (A, A^+) is *stably uniform* if for every rational localization $(A, A^+) \to (B, B^+)$, the ring B is uniform. Again, this depends only on A, not on A^+: one may quantify over all rational localizations by running over finite sequences $f_1, \ldots, f_n, g$ of parameters which generate the unit ideal, rather than over rational subspaces; and in this formulation A^+ does not appear. (What is affected by the choice of A^+ is whether or not two different sets of parameters define the *same* rational subspace.)

The case of the following result where A is Tate is due to Buzzard–Verberkmoes [**27**, Theorem 7] and independently Mihara [**138**, Theorem 4.9] (see also [**117**, Theorem 2.8.10]). The general case is again obtained by modifying the argument slightly using Gabber's method; see again §1.7 for the proof.

THEOREM 1.2.13 (Buzzard–Verberkmoes, Mihara plus Gabber's method). *If A is stably uniform, then (A, A^+) is sheafy.*

REMARK 1.2.14. If A is uniform, then the natural map from A to the ring $H^0(X, \mathcal{O})$ of global sections of $\mathcal{O}$ is automatically injective (Remark 1.5.25); if A is stably uniform, then the analogous map for any rational subspace is also injective. The content of Theorem 1.2.13 is to show that these maps are all surjective.

Let us now discuss the previous two definitions in more detail.

REMARK 1.2.15. Unfortunately, it is rather difficult to exhibit examples of strongly noetherian Banach rings, in part because there is no general analogue of the Hilbert basis theorem: if A is noetherian and Tate, it does not follow that $A\langle T\rangle$ is noetherian. See Remark 1.2.17 for further discussion.

For K a discretely valued field, one can build another class of strongly noetherian rings by considering *semiaffinoid algebras*, i.e., the quotients of rings of the form
$$\mathfrak{o}_K[\![T_1, \ldots, T_m]\!]\langle U_1, \ldots, U_n\rangle \otimes_{\mathfrak{o}_K} K;$$
these rings, and the *uniformly rigid spaces* associated to them, have been studied by Kappen [**100**]. Beware that the identification of rigid spaces with certain adic spaces does not extend to uniformly rigid spaces, and as a result certain phenomena do not exhibit the same behavior.

A third class of strongly noetherian rings will arise from studying Fargues–Fontaine curves in a subsequent lecture. See Remark 3.1.10.

REMARK 1.2.16. Every reduced affinoid algebra over a nonarchimedean field is stably uniform; this follows from the facts that any reduced affinoid algebra is uniform [**24**, Theorem 6.2.4/1], [**72**, Theorem 3.4.9] and any rational localization of a reduced affinoid algebra is again reduced [**24**, Corollary 7.3.2/10], [**117**, Lemma 2.5.9]. However, this argument does not apply to reduced strongly noetherian rings; see Remark 1.2.17 for further discussion.

Additionally, every perfectoid ring is stably uniform; this is because any rational localization is again perfectoid. These examples are genuinely separate from the strongly noetherian case, because a perfectoid ring is noetherian if and only if it is a finite direct product of perfectoid fields (Corollary 2.9.3).

REMARK 1.2.17. One can construct examples where (the underlying ring of) A is a field but is not uniform; a particularly interesting example has been given by Fujiwara–Gabber–Kato [**73**, §8.3]. Note that any such A is a nondiscrete topological ring; hence A is Tate but cannot be a nonarchimedean field. The underlying ring of A is obviously noetherian and reduced.

The aforementioned example has the additional property that $A\langle T\rangle$ is not noetherian; it thus witnesses the failure of the Hilbert basis theorem for Huber rings (Remark 1.2.15). By contrast, if one could find such a field which is strongly noetherian, it would provide an example of a reduced, strongly noetherian, Tate ring which is not even uniform, let alone stably uniform (Remark 1.2.16).

It is not straightforward to check that a given uniform Huber ring A is stably uniform. Most known examples which are not strongly noetherian are derived from perfectoid algebras (to be introduced in the next lecture) using the following observation. (See Exercise 2.5.8 for an exception.)

LEMMA 1.2.18. *Suppose that there exist a stably uniform Huber ring B and a continuous A-linear morphism $A \to B$ which splits in the category of topological A-modules. Then A is stably uniform.*

PROOF. The existence of the splitting implies that $A \to B$ is strict, so A is uniform. Moreover, the existence of the splitting is preserved by taking the completed tensor product over A with a rational localization. It follows that A is stably uniform. □

REMARK 1.2.19. Rings satisfying the hypothesis of Lemma 1.2.18 with B being a perfectoid ring (as in Corollary 2.5.5 below) are called *sousperfectoid* rings in [**86**], where their basic properties are studied in some detail. This refines the concept of a *preperfectoid* ring considered in [**161**].

The following question is taken from [**117**, Remark 2.8.11].

PROBLEM 1.2.20. Is it possible for A to be uniform and sheafy without being stably uniform?

REMARK 1.2.21. At this point, it is natural to ask whether the inclusion functor from sheafy Huber rings to arbitrary Huber rings admits a spectrum-preserving left adjoint. This would be clear if $H^0(X, \mathcal{O})$ were guaranteed to be a sheafy Huber ring; however, it is not even clear that it is complete, due to the implicit direct limit in the definition of $H^0(X, \mathcal{O})$. By contrast, if X admits a single covering by the spectra of sheafy rings, then the subspace topology gives $H^0(X, \mathcal{O})$ the structure of a Huber ring, and it turns out (but not trivially) that this ring is sheafy; see Theorem 1.2.22.

Another approach to working around the failure of sheafiness for general Huber rings is to use techniques from the theory of algebraic stacks. For this approach, see §1.11.

For the proof of the following result, see §1.9.

THEOREM 1.2.22 (original). *Suppose that there exists a finite covering $\mathfrak{V}$ of X by rational subspaces such that $\mathcal{O}|_V$ is a sheaf for each $V \in \mathfrak{V}$. Put*
$$\tilde{A} := H^0(X, \mathcal{O}), \qquad \tilde{A}^+ := H^0(X, \mathcal{O}^+);$$
note that these rings constitute a Huber pair for the subspace topology on $\tilde{A}$.

(a) *The map $A \to \tilde{A}$ induces a homeomorphism $\mathrm{Spa}(\tilde{A}, \tilde{A}^+) \cong \mathrm{Spa}(A, A^+)$ of topological spaces such that rational subspaces pull back to rational subspaces (but possibly not conversely) and on each $V \in \mathfrak{V}$, the structure presheaf pulls back to the structure presheaf.*

(b) *The ring $\tilde{A}$ is sheafy.*

In particular, by Theorem 1.3.4, $\mathcal{O}$ is acyclic.

REMARK 1.2.23. In Theorem 1.2.22, it is obvious that if $\mathcal{O}(V)$ is stably uniform for each $V \in \mathfrak{V}$, then so is $\tilde{A}$. The analogous statement for the strongly noetherian property is true but much less obvious; see Corollary 1.4.19. See Problem 1.2.25 for a related problem.

REMARK 1.2.24. One of the examples of Buzzard–Verberkmoes [27, Proposition 13] is a construction in which there exists a finite covering $\mathfrak{V}$ of X by rational subspaces such that $\mathcal{O}(V)$ is a perfectoid (and hence stably uniform and sheafy) Huber ring for each $V \in \mathfrak{V}$, so Theorem 1.2.22 applies, but the map $A \to H^0(X, \mathcal{O})$ is not injective. (In this example, one has $A^+ = A^\circ$.) See Remark 2.5.11 for further discussion.

Another example of Buzzard–Verberkmoes [27, Proposition 16] is a construction in which $A \to H^0(X, \mathcal{O})$ is injective but not surjective. However, in this example, we do not know whether A is uniform (injectivity is instead established using Corollary 1.5.24), or whether $H^0(X, \mathcal{O})$ is a Huber ring (because the construction does not immediately yield local sheafiness).

PROBLEM 1.2.25. Suppose that A is uniform and that every open subset of X can be written as the union of rational subspaces $\mathrm{Spa}(B, B^+)$ for which B is uniform. Does it follow that A is stably uniform?

1.3. Cohomology of sheaves. Recall that Tate's acyclicity theorem asserts more than the fact that $\mathcal{O}$ is a sheaf: it also asserts the vanishing of higher cohomology of $\mathcal{O}$ on rational subspaces, and makes similar assertions for the presheaves associated to finitely generated A-modules. We turn next to generalizing these statements to more general Huber rings.

DEFINITION 1.3.1. We say that a sheaf $\mathcal{F}$ on X is *acyclic* if $H^i(U, \mathcal{F}) = 0$ for every rational subspace U of $\mathcal{F}$ and every positive integer i.

DEFINITION 1.3.2. For any A-module M, let $\tilde{M}$ be the presheaf on X such that for $U \subseteq X$ open, $\tilde{M}(U)$ is the inverse limit of $M \otimes_A B$ over all rational localizations $(A, A^+) \to (B, B^+)$ with $\mathrm{Spa}(B, B^+) \subseteq U$. In particular, if $U = \mathrm{Spa}(B, B^+)$ then $\tilde{M}(U) = M \otimes_A B$.

REMARK 1.3.3. Beware that the definition of $\tilde{M}$ uses the ordinary tensor product, and makes no reference to any topology on M. However, if M is finitely generated and both M and its base extension are complete for the natural topology (Definition 1.1.11), then the ordinary tensor product coincides with the completed tensor product. Note that the condition on completeness of the base extension cannot be omitted; see Exercise 1.4.7.

In the Tate case, the following result is due to Kedlaya–Liu [**117**, Theorem 2.4.23]; this again generalizes results of Tate and Huber for affinoid algebras and strongly noetherian rings, respectively. For the proof, see §1.8.

THEOREM 1.3.4 (Kedlaya–Liu plus Gabber's method). *If A is sheafy, then for any finite projective A-module M, the presheaf $\tilde{M}$ is an acyclic sheaf.*

REMARK 1.3.5. One serious impediment to extending Theorem 1.3.4 to more general modules is that it is not known that rational localization maps are flat. This is true in rigid analytic geometry [**169**, Lemma 8.6], [**24**, Corollary 7.3.2/6]; the same result extends to strongly noetherian Tate rings, as shown by Huber [**96**, II.1], [**97**, Lemma 1.7.6]. It is not at all clear whether flatness should hold in general; however, one can prove a weaker result which nonetheless implies all of the previously asserted flatness results, and is useful in applications. See Theorem 1.4.14.

REMARK 1.3.6. By contrast with the situation for schemes (e.g., see [**166**, Tag 01XE]), in rigid analytic geometry there is no cohomological criterion for detecting affinoid spaces among quasicompact rigid spaces. To wit, let X be a quasicompact rigid analytic space over a field K (identified with its corresponding adic space) for which $H^i(X, \mathcal{F}) = 0$ for every coherent sheaf $\mathcal{F}$ on X and every $i > 0$ (that is, X is a *quasicompact Stein space* over K). It can then happen that X is not an affinoid space over K; examples have been given by Q. Liu [**134**, **135**]. (The example in [**134**] is not normal, and its normalization is an affinoid space. The example in [**135**] is normal and even smooth over K, because it is a subset of the affine plane.)

For X a quasicompact Stein space over K, by Remark 1.2.21, $\mathcal{O}(X)$ is a Huber ring; by [**135**, Proposition 1.2(a)], $\mathcal{O}(X)$ is noetherian. Since the product of X over K with any affinoid space is again a quasicompact Stein space [**135**, Proposition 4.1], $\mathcal{O}(X)$ is even strongly noetherian, and hence sheafy by Theorem 1.2.11. Put $A := \mathcal{O}(X)$, $A^+ := \mathcal{O}^+(X)$; there is a natural map $X \to \mathrm{Spa}(A, A^+)$ of adic spaces, which by [**135**, Proposition 1.3] induces a bijection of the rigid analytic points of X with the maximal ideals of A. By [**135**, §3, Théorème 1], there exists a covering of $\mathrm{Spa}(A, A^+)$ by rational subspaces whose pullbacks to X are affinoid spaces over K; by applying [**24**, Proposition 7.3.3/5] to each of these subspaces, we deduce that $X \to \mathrm{Spa}(A, A^+)$ is an isomorphism. That is, X is the adic affinoid space associated to the Huber pair (A, A^+), but the ring A is not *topologically of finite type* (or *tft* for short) over K. In the example of [**135**, §4, Théorème 4], this is witnessed by the fact that the reduction $A^\circ/A^{\circ\circ}$ is not noetherian; if A were an affinoid algebra over K, this reduction would instead be of finite type over the residue field of K [**24**, Corollary 6.3.4/3].

The upshot of this discussion is that Liu's counterexamples *do not* rule out the existence of a cohomological criterion for detecting adic affinoid spaces among quasicompact analytic adic spaces. See Problem 1.3.7 below.

PROBLEM 1.3.7. Let X be an analytic adic space satisfying the following conditions.

- The space X is *separated*: X is covered by a family of open subspaces among which any nontrivial finite intersection is an adic affinoid space.
- The space X is *holomorphically separable*: the natural map
$$X \to \mathrm{Spa}(\mathcal{O}(X), \mathcal{O}^+(X))$$
is injective.
- For each $i > 0$, the group $H^i(X, \mathcal{O})$ vanishes.

Does it then follow that $X \cong \mathrm{Spa}(\mathcal{O}(X), \mathcal{O}^+(X))$, and hence that X is an adic affinoid space?

The conditions on X correspond to the definition of an *S-space* in [135]; when X is a rigid analytic space, these conditions are equivalent to X being a quasicompact Stein space [135, §4, Théorème 2]. In particular, Remark 1.3.6 gives an affirmative answer to this question when X is a rigid analytic space; we expect that the methods of [135] can be adapted to the general case.

1.4. Vector bundles and pseudocoherent sheaves. To further continue the analogy with affine schemes, one would now like to define coherent sheaves (or pseudocoherent sheaves, in the absence of noetherian hypotheses) and verify that they are precisely the sheaves arising from pseudocoherent modules. In rigid analytic geometry, this is a theorem of Kiehl [121, Theorem 1.2]; however, here we are hampered by a lack of flatness (Remark 1.3.5). Before remedying this in a way that leads to a full generalization of Kiehl's result, let us consider separately the case of vector bundles.

DEFINITION 1.4.1. A *vector bundle* on X is a sheaf $\mathcal{F}$ of $\mathcal{O}$-modules on X which is locally of the form $\mathcal{O}^{\oplus n}$ for some positive integer n. In other words, there exists a finite covering $\{U_i\}_{i=1}^n$ of X by rational subspaces such that for each i, $M_i := \mathcal{F}(U_i)$ is a finite free $\mathcal{O}(U_i)$-module and the canonical morphism $\tilde{M}_i \to \mathcal{F}|_{U_i}$ of sheaves of $\mathcal{O}|_{U_i}$-modules is an isomorphism. Let **Vec**$_X$ denote the category of vector bundles on X.

Let **FPMod**$_A$ denote the category of finite projective A-modules. For $M \in$ **FPMod**$_A$, M is already locally free over $\mathrm{Spec}(A)$, and so $\tilde{M} \in$ **Vec**$_X$; moreover, the resulting functor **FPMod**$_A \to$ **Vec**$_X$ taking M to $\tilde{M}$ is exact. (Note that this exactness must be derived from the flatness of finite projective modules because rational localizations are not known to be flat; see Remark 1.3.5.)

When A is Tate, the following result is due to Kedlaya–Liu [117, Theorem 2.7.7]; again, the Tate hypothesis can be removed using Gabber's method. See §1.9 for the proof.

THEOREM 1.4.2 (Kedlaya–Liu plus Gabber's method). *If A is sheafy, then the functor* **FPMod**$_A \to$ **Vec**$_X$ *taking M to $\tilde{M}$ is an equivalence of categories, with quasi-inverse taking $\mathcal{F}$ to $\mathcal{F}(X)$. In particular, by Theorem 1.3.4, every sheaf in* **Vec**$_X$ *is acyclic.*

REMARK 1.4.3. If one restricts attention to finite étale A-algebras and finite étale $\mathcal{O}_X$-modules, then the functor $M \mapsto \tilde{M}$ is an equivalence of categories even if A is not sheafy. See for example [117, Theorem 2.6.9] in the case where A is Tate.

REMARK 1.4.4. Theorem 1.4.2 may be reformulated as the statement that the functor $\mathbf{Vec}_{\mathrm{Spec}(A)} \to \mathbf{Vec}_X$ given by pullback along the canonical morphism $X \to \mathrm{Spec}(A)$ of locally ringed spaces (coming from the adjunction property of affine schemes) is an equivalence of categories. It also implies that $\mathbf{Vec}_X$ depends only on A, not on A^+. (The same will be true for $\mathbf{PCoh}_X$ by Theorem 1.4.18.)

We now turn to more general (but still finitely generated) modules; here we give a streamlined presentation of material from [118].

DEFINITION 1.4.5. An A-module M is *pseudocoherent* if it admits a projective resolution (possibly of infinite length) by finite projective A-modules (which may even be taken to be free modules); when A is noetherian, this is equivalent to A being finitely generated. (The term *pseudocoherent* appears to have originated in SGA 6 [14, Exposé I], and is used systematically in the paper of Thomason–Trobaugh [171].)

We say that M is *stably pseudocoherent* if M is pseudocoherent and, for every rational localization $(A, A^+) \to (B, B^+)$, $M \otimes_A B$ is complete for its natural topology as a B-module. Write $\mathbf{PCoh}_A$ for the category of stably pseudocoherent A-modules; if A is strongly noetherian, by Corollary 1.1.15 this is exactly the category of finitely generated A-modules. Otherwise, it is rather difficult to directly exhibit elements of $\mathbf{PCoh}_A$ other than elements of $\mathbf{FPMod}_A$; we can exhibit some elements indirectly using Theorem 1.4.20.

REMARK 1.4.6. If $f \in A$ is not a zero-divisor and fA is not closed in A (which can occur; see Exercise 1.1.17 or Remark 1.8.3), then A/fA is pseudocoherent but not complete for the natural topology, and hence not an object of $\mathbf{PCoh}_A$.

Unfortunately, even if fA is closed in A, it is possible for there to exist a rational localization $(A, A^+) \to (B, B^+)$ for which fB is not a closed ideal; see Example 1.4.8. In this case, A/fA is pseudocoherent and complete for the natural topology, yet not stably pseudocoherent.

EXERCISE 1.4.7. Set notation as in Exercise 1.1.17.
(a) Show that A is uniform. I do not know whether A is stably uniform.
(b) Show that the natural map $\mathbb{Q}_p\langle T\rangle \to A$ is flat. Recall that $\mathbb{Q}_p\langle T\rangle$ is a principal ideal domain (see [72, Theorem 2.2.9] or [108, Proposition 8.3.2]), so this amounts to checking that no nonzero element of $\mathbb{Q}_p\langle T\rangle$ maps to a zero-divisor in A (the case of T itself having been checked in Exercise 1.1.17).
(c) Conclude that if R is a noetherian Huber ring, $R \to A$ is a flat homomorphism of Huber rings with A analytic, and M is a finitely generated R-module, then $M \otimes_R A$ is pseudocoherent but not necessarily complete for the natural topology. (Take $R := \mathbb{Q}_p\langle T\rangle$, $M := R/TR$.)

EXAMPLE 1.4.8 (Scholze, private communication). We construct an example of a rational localization $(A, A^+) \to (B, B^+)$ and an element $f \in A$ for which f is not a zero-divisor of A or B, and fA is closed in A but fB is not closed in B. See Remark 1.4.6 for context and [118, Example 2.4.2] for more details.

Let K be any perfectoid field. Fix an integer $N \geq 3$ coprime to p, so that the (compactified) modular curve $X(N)_K$ exists as a scheme (rather than a Deligne-Mumford stack), and hence as an adic space. Let V_0 be the ordinary locus of $X(N)_K$, including all cuspidal discs. Let U_0 be an affinoid strict neighborhood of V_0 in

$X(N)_K$; by making U_0 sufficiently small, we may ensure that $\mathcal{O}(U_0)$ contains a lift h of the Hasse invariant and that V_0 is the rational subspace of U_0 of the form $|h| \leq 1$. (See [**30**] for more context on the p-adic geometry of modular curves.)

Let U_1, V_1 be the respective inverse images of U_0, V_0 in the infinite-level modular curve $X(Np^\infty)_K$. The Hodge-Tate period map [**28**] induces a morphism $\psi: U_1 \to \mathbf{P}^1_K$ such that $\overline{V}_1 = \psi^{-1}(\mathbf{P}^1(\mathbb{Q}_p))$. (More precisely, $\psi^{-1}(\mathbf{P}^1(\mathbb{Q}_p))$ consists of V_1 plus some rank-2 points corresponding to generically ordinary families of elliptic curves with supersingular specialization). Let $D \subseteq \mathbf{P}^1_K$ be the closed unit disc, and put

$$U := \psi^{-1}(D) \subseteq U_1, \qquad V = U \times_{U_1} V_1.$$

The space $U = \mathrm{Spa}(A, A^+)$ is an affinoid perfectoid space, so A is sheafy (Corollary 2.5.4). The morphism ψ from U to the closed unit disc defines an element $f \in A$. Let Z be the Shilov boundary of U; it is the inverse image of the Shilov boundary of U_0, and hence consists entirely of points not in V. In particular, f vanishes nowhere on Z; since Z is compact, f is bounded below on Z, and so multiplication by f defines a strict inclusion on A.

The space $V = \mathrm{Spa}(B, B^+)$ is again the rational subspace of U defined by the Hasse invariant; however, ψ carries V to the Cantor set $D(\mathbb{Q}_p) \cong \mathbb{Z}_p$. In particular, any continuous function from $\mathbb{Z}_p$ to K pulls back to an element of B, the identity function corresponding to f itself. Moreover, any continuous function $g: \mathbb{Z}_p \to K$ with $g(0) = 0$ corresponds to an element of the closure of fB; however, g only corresponds to an element of fB if and only if $g(x)/x$ is bounded on $\mathbb{Z}_p \setminus \{0\}$. Consequently, fB is not closed.

REMARK 1.4.9. An easy fact about pseudocoherent A-modules is the "two out of three" property: in a short exact sequence

(1.4.9.1) $$0 \to M_1 \to M \to M_2 \to 0$$

of A-modules, if any two of M, M_1, M_2 are pseudocoherent, then so is the third.

The "two out of three" property is not true as stated for $\mathbf{PCoh}_A$: if $M_1, M \in \mathbf{PCoh}_A$, then M_2 need not be complete for its natural topology (as in Definition 1.1.11, this can occur for $M_1 = M = A$). However, suppose that $M_2 \in \mathbf{PCoh}_A$. In this case, if M_1 is complete for the natural topology, then so is M_2 (see Exercise 1.4.10); while if M is complete for the natural topology, then M_1 is Hausdorff for the subspace topology and hence also for its natural topology. Using the fact that $M \in \mathbf{PCoh}_A$ implies that (1.4.9.1) remains exact upon tensoring with a rational localization (see Corollary 1.4.15), we conclude that if $M_2 \in \mathbf{PCoh}_A$, then $M \in \mathbf{PCoh}_A$ if and only if $M_1 \in \mathbf{PCoh}_A$.

EXERCISE 1.4.10. Let

$$0 \to M_1 \to M \to M_2 \to 0$$

be an exact sequence of topological A-modules in which M_1, M_2 are complete and M is finitely generated over some Huber ring B over A. Then M is complete for its natural topology as a B-module. (Hint: Choose a B-linear surjection $F \to M$ and apply the open mapping theorem to the composition $F \to M \to M_2$ as a morphism of topological A-modules. This implies that the surjection $M \to M_2$ has a bounded set-theoretic section; using this section, separate the problem of summing a null sequence in M to analogous problems in M_1 and M_2.)

REMARK 1.4.11. Note that a pseudocoherent module is not guaranteed to have a *finite* projective resolution by finite projective modules, even over a noetherian ring; this is the stronger property of being of *finite projective dimension*. For example, for any field k, over the local ring $k[T]/(T^2)$, the residue field is a module which is pseudocoherent but not of finite projective dimension. More generally, every pseudocoherent module over a noetherian local ring is of finite projective dimension if and only if the ring is regular [**166**, Tag 0AFS]. (Modules of finite projective dimension are sometimes called *perfect modules*, as in [**166**, Tag 0656], since they are the ones whose associated singleton complexes are perfect.)

REMARK 1.4.12. It is not hard to show that a tensor product of pseudocoherent modules is pseudocoherent. However, we do not know whether $\mathbf{PCoh}_A$ is stable under formation of tensor products, due to the completeness requirement.

REMARK 1.4.13. It is possible to characterize stably pseudocoherent modules without testing the completeness condition for all rational localizations. For an example of this, see Remark 1.6.17.

When the ring A is Tate, the following result is due to Kedlaya–Liu [**118**, Theorem 2.4.15]. See again §1.9 for the proof.

THEOREM 1.4.14 (Kedlaya–Liu plus Gabber's method). *If A is sheafy, then for any rational localization $(A, A^+) \to (B, B^+)$, base extension from A to B defines an exact functor $\mathbf{PCoh}_A \to \mathbf{PCoh}_B$. In particular, if A is strongly noetherian, then $A \to B$ is flat (because every finitely generated module belongs to $\mathbf{PCoh}_A$ by Corollary* 1.1.15*).*

COROLLARY 1.4.15. *Suppose that A is sheafy. For any rational localization $(A, A^+) \to (B, B^+)$ and any $M \in \mathbf{PCoh}_A$, we have $\mathrm{Tor}_1^A(M, B) = 0$.*

PROOF. By Remark 1.4.9, if we write M as F/N for some finite free A-module F, then $N \in \mathbf{PCoh}_A$. The claim thus follows from Theorem 1.4.14. □

Theorem 1.4.14 makes it possible to consider sheaves constructed from pseudocoherent modules, starting with the following statement which in the Tate case is [**118**, Theorem 2.5.1]. See again §1.9 for the proof.

THEOREM 1.4.16 (Kedlaya–Liu plus Gabber's method). *If A is sheafy, then for any $M \in \mathbf{PCoh}_A$, the presheaf $\tilde{M}$ is an acyclic sheaf.*

DEFINITION 1.4.17. A *pseudocoherent sheaf* on X is a sheaf $\mathcal{F}$ of $\mathcal{O}$-modules on X which is locally of the form $\tilde{M}$ for M a stably pseudocoherent module. In other words, there exists a finite covering $\{U_i\}_{i=1}^n$ of X by rational subspaces such that for each i, $M_i := \mathcal{F}(U_i) \in \mathbf{PCoh}_{\mathcal{O}(U_i)}$ and the canonical morphism $\tilde{M}_i \to \mathcal{F}|_{U_i}$ of sheaves of $\mathcal{O}|_{U_i}$-modules is an isomorphism. Let $\mathbf{PCoh}_X$ denote the category of pseudocoherent sheaves on X; by Theorem 1.4.14, the functor $\mathbf{PCoh}_A \to \mathbf{PCoh}_X$ taking M to $\tilde{M}$ is exact.

In case A is strongly noetherian, we refer to pseudocoherent sheaves also as *coherent sheaves*, and denote the category of them also by $\mathbf{Coh}_X$.

When A is Tate, the following result is due[3] to Kedlaya–Liu [**118**, Theorem 2.5.5]. Somewhat surprisingly, the strongly noetherian case cannot be found[4] in Huber's work. See again §1.9 for the proof.

THEOREM 1.4.18 (Kedlaya–Liu plus Gabber's method). *If A is sheafy, then the functor $\mathbf{PCoh}_A \to \mathbf{PCoh}_X$ taking M to $\tilde{M}$ is an exact (by Theorem 1.4.14) equivalence of categories, with quasi-inverse taking $\mathcal{F}$ to $\mathcal{F}(X)$. In particular, by Theorem 1.4.16, every sheaf in $\mathbf{PCoh}_X$ is acyclic.*

COROLLARY 1.4.19. *In Theorem 1.2.22, if $\mathcal{O}(V)$ is strongly noetherian for each $V \in \mathfrak{V}$, then so is $\tilde{A}$.*

PROOF. It suffices to check that $\tilde{A}$ is noetherian, as we may then apply the same logic to the pullback coverings of the spectra of $A\langle T_1, \ldots, T_n\rangle$ for all n. We may further assume that $\tilde{A} = A$.

Let I be any ideal of A and put $M := A/IA$. For $V \in \mathfrak{V}$, $\mathcal{O}(V)$ is strongly noetherian and so $M \otimes_A \mathcal{O}(V) \in \mathbf{PCoh}_{\mathcal{O}(V)}$; this means that $\tilde{M} \in \mathbf{PCoh}_X$. By Theorem 1.4.16 and Theorem 1.4.18, we have $M = H^0(X, \tilde{M}) \in \mathbf{PCoh}_A$. Hence I is finitely generated; since I was arbitrary, it follows that A is noetherian. □

Using similar methods, we obtain the following result. See again §1.9 for the proof.

THEOREM 1.4.20 (original). *Suppose that A is sheafy. Let I be a closed ideal of A which is pseudocoherent as an A-module. Then A/I is sheafy if and only if $A/I \in \mathbf{PCoh}_A$.*

REMARK 1.4.21. For example, if A and $A\langle T\rangle$ are both sheafy, then Theorem 1.4.20 implies that $A \in \mathbf{PCoh}_{A\langle T\rangle}$ and hence, by Remark 1.4.9, that $TA\langle T\rangle \in \mathbf{PCoh}_{A\langle T\rangle}$. It is not at all obvious how to give a direct proof of this assertion.

REMARK 1.4.22. In algebraic geometry, one knows that the theory of quasicoherent sheaves on affine schemes is "the same" whether one uses the Zariski topology or the étale topology, in that both the category of sheaves and their cohomology groups are the same. Roughly speaking, the same is true for adic spaces, but one needs to be careful about technical hypotheses. See §1.10 for a detailed discussion.

At this point, we have completed the statements of the main results of this lecture. The notes continue with some technical tools needed for the proofs; the reader impatient to get to the main ideas of the proofs is advised to skip ahead to §1.6 at this point, coming back as needed later.

1.5. Huber versus Banach rings. Although most of our discussion will be in terms of Huber rings, which play the starring role in the study of rigid analytic spaces and adic spaces, it is sometimes useful to translate certain statements into the parallel language of Banach rings, which underlie the theory of Berkovich spaces. We explain briefly how these two points of view interact, as in [**117, 118**] where most of the local theory is described in terms of Banach rings. A key application

---

[3] A very general result along these lines is included as part of the overall development of adic spaces using formal geometry in the book of Fujiwara–Kato [**74**, §I.7.2]. However, we have not checked whether the general result actually implies Theorem 1.4.18 as written.

[4] In [**97**], one finds a citation to a work in preparation by Huber called "Coherent sheaves on adic spaces". We have not seen any version of this paper.

will be to show that certain multiplication maps on Tate algebras are strict; see Lemma 1.5.26.

DEFINITION 1.5.1. By a *Banach ring* (more precisely, a *nonarchimedean commutative Banach ring*), we will mean a ring B equipped with a function $|\bullet| : B \to \mathbb{R}_{\geq 0}$ satisfying the following conditions.

(a) On the additive group of B, $|\bullet|$ is a norm (i.e., a nonarchimedean absolute value, so that $|x - y| \leq \max\{|x|, |y|\}$ for all $x, y \in B$) with respect to which B is complete.
(b) The norm on B is *submultiplicative*: for all $x, y \in B$, we have $|xy| \leq |x||y|$.

A ring homomorphism $f : B \to B'$ of Banach rings is *bounded* if there exists $c \geq 0$ such that $|f(x)|' \leq c|x|$ for all $x \in B$; the minimum such c is called the *operator norm* of f.

We view Banach rings as a category with the morphisms being the bounded ring homomorphisms. In particular, if two norms on the same ring differ by a bounded multiplicative factor on either side, then they define isomorphic Banach rings.

As for Huber rings, we say that a Banach ring B is *analytic* if its topologically nilpotent elements of B generate the unit ideal. (In [117], the corresponding condition is for B to be *free of trivial spectrum*.)

REMARK 1.5.2. In condition (b) of Definition 1.5.1, one could instead insist that there exist some constant $c > 0$ such that for all $x, y \in B$, we have $|xy| \leq c|x||y|$. However, this adds no essential generality, as replacing $|x|$ with the operator norm of $y \mapsto xy$ gives an isomorphic Banach ring which does satisfy (b).

In the category of Banach rings, we have the following analogue of Tate algebras.

DEFINITION 1.5.3. For B a Banach ring and $\rho > 0$, let $B\langle \frac{T}{\rho} \rangle$ be the completion of $B[T]$ for the *weighted Gauss norm*
$$\left| \sum_{n=0}^{\infty} x_n T^n \right|_\rho = \max\{|x_n| \rho^n\}.$$
For $\rho = 1$, this coincides with the usual Tate algebra $B\langle T \rangle$.

If we define the associated graded ring
$$\operatorname{Gr} B := \bigoplus_{r > 0} \frac{\{x \in B : |x| \leq r\}}{\{x \in B : |x| < r\}},$$
then $\operatorname{Gr} B\langle \frac{T}{\rho} \rangle$ is the graded ring $(\operatorname{Gr} B)[\overline{T}]$ with $\overline{T}$ placed in degree ρ. One consequence of this (which generalizes the usual Gauss's lemma) is that if the norm on B is multiplicative, then $\operatorname{Gr} B$ is an integral domain, as then is $(\operatorname{Gr} B)[\overline{T}]$, so the weighted Gauss norm on $B\langle \frac{T}{\rho} \rangle$ is multiplicative. (See Lemma 1.8.1 for another use of the graded ring construction.)

REMARK 1.5.4. As usual, let A be an analytic Huber ring. Choose a ring of definition A_0 of A (which must be open in A) and an ideal of definition I (which must be finitely generated). Using these choices, we can *promote* A to a Banach ring as follows: for $x \in A$, let $|x|$ be the infimum of e^{-n} over all integers n for which $xI^m \subseteq I^{m+n}$ for all nonnegative integers m for which $m + n \geq 0$.

Note that this works even if A is not analytic; in particular, any Huber ring is metrizable and in particular first-countable. However, even analytic Huber rings need not be second-countable; consider for example $\mathbb{Q}_p \langle T_1, T_2, \dots \rangle$.

REMARK 1.5.5. In the other direction, starting with a Banach ring B, it is not immediately obvious that its underlying topological ring is a Huber ring; the difficulty is to find a finitely generated ideal of definition. However, this is always possible if B is analytic: if $x_1, \ldots, x_n$ are topologically nilpotent elements of B generating the unit ideal, then for any ring of definition A_0, for any sufficiently large m the elements $x_1^m, \ldots, x_n^m$ of A belong to A_0 and generate an ideal of definition.

EXAMPLE 1.5.6. The infinite polynomial ring $\mathbb{Q}[T_1, T_2, \ldots]$ admits a submultiplicative norm where for $x \neq 0$, $|x| = e^{-n}$ where n is the largest integer such that $x \in (T_1, T_2, \ldots)^n$. Let B be the Banach ring obtained by taking the completion with respect to this norm. Then the underlying ring of B is not a Huber ring.

As an example of viewing a Banach ring as a Huber ring, we give an example of a Huber ring which is analytic but not Tate. Of course this example can be described perfectly well without Banach rings, but we find the presentation using norms a bit more succinct.

EXAMPLE 1.5.7. Equip $\mathbb{Z}$ with the trivial norm, choose any $\rho > 1$, and equip
$$ A := \mathbb{Z}\left\langle \frac{a}{\rho}, \frac{b}{\rho}, \frac{x}{\rho^{-1}}, \frac{y}{\rho^{-1}} \right\rangle / (ax + by - 1) $$
with the quotient norm; this is an analytic Banach ring (because x and y are topologically nilpotent), so it may be viewed as a Huber ring. If we view A as a filtered ring, the associated graded ring is $\mathbb{Z}[a, b, x, y]/(ax+by-1)$ with a, b placed in degree -1 and x, y placed in degree $+1$. Since the graded ring is an integral domain and its only units are ± 1, it follows that the norm on A is multiplicative and every unit in A has norm 1. In particular, A is not Tate.

In order to explain the extent to which passage between Huber and Banach rings can be made functorial, we need to introduce the notion of a Banach module.

REMARK 1.5.8. Let B be an analytic Banach ring and let M be a complete metrizable topological B-module. Then M may be equipped with the structure of a *Banach module* over B, i.e., a module complete with respect to a norm $|\bullet|_M$ satisfying
(1.5.8.1) $$ |bm| \leq |b|\,|m| \qquad (b \in B, m \in M). $$
Namely, if one chooses a bounded (in the sense of Definition 1.2.12) open neighborhood M_0 of 0 in M, one can define a surjective morphism $N \to M$ of topological B-modules by taking N to be the completed direct sum of B^{M_0} for the supremum norm, then mapping the generator of N corresponding to $m \in M_0$ to $m \in M$. By Theorem 1.1.9, this morphism is a strict surjection, so the quotient norm from N defines the desired topology on M.

In case B is a nonarchimedean field equipped with a multiplicative norm, one can say more: for $b \in B$ nonzero, we also have
$$ |m| \leq |bm|\,\left|b^{-1}\right| = |bm|\,|b|^{-1}, $$
which upgrades (1.5.8.1) to an equality
(1.5.8.2) $$ |bm| = |b|\,|m| \qquad (b \in B, m \in M). $$

REMARK 1.5.9. Let B be an analytic Banach ring, and let M_1, M_2 be Banach modules over B. Let $f: M_1 \to M_2$ be a morphism of B-modules. We say that f

is *bounded* if there exists $c > 0$ such that $|f(m)| \leq c|m|$ for all $m \in M_1$. If f is bounded, then evidently f is continuous.

Conversely, if B is a nonarchimedean field equipped with a multiplicative norm and f is continuous, then f is bounded. To see this, let u be a topologically nilpotent unit of B, so that $0 < |u| < 1$. If f fails to be bounded, then for every positive integer n there exists an element $x_n \in M_1$ such that $|f(x_n)| > 2^n |x_n|$. Choose an integer m_n such that $|u|^{m_n} |x_n| \in [|u|, 1]$. By (1.5.8.2), we have
$$|u^{m_n} x_n| \leq 1, \qquad |u^{m_n} f(x_n)| > 2^n |u|,$$
so $\{u^{m_n} x_n\}$ is a bounded sequence in M_1 whose image in M_2 is not bounded. This contradicts the continuity of f.

For general B, it is not the case that f being continuous implies that f is bounded. For example, it is possible for the same Huber ring A to arise as the underlying ring of two Banach rings B_1, B_2 in such a way that the composition $B_1 \to A \to B_2$ is not a bounded morphism of B_1-modules. To make a concrete example, choose any $\rho \in (0, 1)$, let k be a field equipped with the trivial norm, and define
$$B_1 := k\left\langle x, \frac{y}{\rho}, \frac{y^{-1}}{\rho^{-1}}, T, U \right\rangle / (y - Tx, x^2 - Uy),$$
$$B_2 := k\left\langle \frac{x}{\rho}, \frac{x^{-1}}{\rho^{-1}}, y, T, U \right\rangle / (y - Tx, x^2 - Uy).$$
Then there is a k-linear homeomorphism $f : B_1 \to B_2$ identifying x and y on both sides, but in B_1 we have $|y^{-n}| = \rho^{-n}$ while in B_2 we have $|y^{-n}| = \rho^{-2n}$; hence f is not bounded.

Nonetheless, from the previous discussion plus Remark 1.5.8, we see that for general B, the forgetful functor from Banach modules over B to complete metrizable topological modules over B admits a (highly noncanonical) left adjoint; namely, for each module M in the target category, choose a bounded open neighborhood M_0 of 0 in M and run the construction of Remark 1.5.8. To check that this functor indeed turns continuous morphisms into bounded morphisms, we proceed in two steps.

- First, we check that if M_0' is another bounded open neighborhood of 0 in M, then the identity map on M is bounded for the norms defined by M_0 and M_0'. To this end, choose topologically nilpotent units $x_1, \ldots, x_n \in B$ generating the unit ideal; then for any sufficiently large integer m, we have $x_i^m M_0 \subseteq M_0'$ for $i = 1, \ldots, n$. Choose elements $b_1, \ldots, b_n \in B$ such that $x_1^m b_1 + \cdots + x_n^m b_n = 1$; if $x \in M_0$, then $x_i^{-m} x \in M_0'$ and so the norm of x with respect to M_0' is at most $\max_i \{|x_i^{-m} b_i|\}$, and similarly in the opposite direction.
- Second, we note that if $f : M_1 \to M_2$ is continuous, then we may choose open bounded neighborhoods $M_{1,0}, M_{2,0}$ of 0 in M_1, M_2 in such a way that $f(M_{1,0}) \subseteq M_{2,0}$. We thus get a bounded morphism for the corresponding norms, and hence for any other choice of norms (by the previous point). The adjoint property follows similarly.

REMARK 1.5.10. Let B be an analytic Banach ring, and let $B \to A$ be a (continuous) morphism of Huber rings. We may then take $M = A$ in Remark 1.5.8; this amounts to promoting A to a Banach ring using an ideal of definition extended from B. By Remark 1.5.9, the map $B \to A$ is bounded; this remains true if we replace the norm on A with its associated operator norm as per Remark 1.5.2, since the norm topology does not change. To summarize, the forgetful functor from Banach rings over B to Huber rings over B admits a left adjoint, and is even an equivalence of categories if B is itself a Banach algebra over a nonarchimedean field (by Remark 1.5.9).

We now continue to introduce basic structures associated to Banach rings, keeping an eye on the relationship with Huber rings.

DEFINITION 1.5.11. For B a Banach ring, the *spectral seminorm* on B is the function $|\bullet|_{\mathrm{sp}} : B \to \mathbb{R}_{\geq 0}$ given by

$$|x|_{\mathrm{sp}} = \lim_{n \to \infty} |x^n|^{1/n} \qquad (x \in B).$$

(Using submultiplicativity, it is an elementary exercise in real analysis to show that the limit exists.) In general, the spectral seminorm is not a norm; for example, it maps all nilpotent elements to 0. The spectral seminorm need not be multiplicative, but it is *power-multiplicative*: for any $x \in B$ and any positive integer n, $|x^n|_{\mathrm{sp}} = |x|_{\mathrm{sp}}^n$.

Even if the spectral seminorm is a norm, it need not define the same topology as the original norm. This does however hold if B satisfies the equivalent conditions of Exercise 1.5.13 below; in this case, we say that B is *uniform*. If B is uniform, then it is reduced.

EXERCISE 1.5.12. Let B be a Banach ring. Then $x \in B$ is topologically nilpotent if and only if $|x|_{\mathrm{sp}} < 1$.

EXERCISE 1.5.13. For any integer $m > 1$, the following conditions on a Banach ring B are equivalent.

(a) There exists $c > 0$ such that $|x|_{\mathrm{sp}} \geq c|x|$ for all $x \in B$.
(b) There exists $c > 0$ such that $|x^m| \geq c|x|^m$ for all $x \in B$.

If the underlying topological ring of B is a Huber ring, then these conditions also imply the following equivalent conditions.

(c) The spectral seminorm defines the same topology as the original norm (and in particular is a norm).
(d) The underlying Huber ring of B is uniform (in the sense of Definition 1.2.12).

In addition, if B is analytic and is the promotion of a Huber ring as per Remark 1.5.4 (e.g., because it is an algebra over a nonarchimedean field), then all four conditions are equivalent. More precisely, if B is analytic, then (d) still implies the restricted form of (a) where x is limited to some bounded neighborhood of 0 in B, the constant c depending on this neighborhood; assuming that B is the promotion of a Huber ring, one may extend (a) to larger x.

Beware that it is falsely claimed in [**117**, Definition 2.8.1] that if B is Tate, then (a)–(d) are equivalent. For a counterexample, start with a nonarchimedean field K with norm $|\bullet|$, then take B to be a copy of K equipped with the new norm

$$x \mapsto \begin{cases} |x| & \text{if } |x| \leq 1 \\ |x|(1 + \log|x|) & \text{if } |x| > 1. \end{cases}$$

DEFINITION 1.5.14. For B a Banach ring, let $\mathcal{M}(B)$ denote the *Gelfand spectrum*[5] of B, which as a set consists of the multiplicative seminorms on B which are bounded by the given norm. Under the evaluation topology (i.e., the subspace topology from the product topology on $\mathbb{R}^B$), $\mathcal{M}(B)$ is compact.

REMARK 1.5.15. For (A, A^+) a Huber ring with A promoted to a Banach ring B as per Remark 1.5.4, there is a natural map $\mathcal{M}(B) \to \mathrm{Spa}(A, A^+)$ obtained by viewing a multiplicative seminorm as a valuation; however, this map is not continuous. If A is analytic, this map is a section of a continuous morphism $\mathrm{Spa}(A, A^+) \to \mathcal{M}(B)$ which takes a valuation v to the bounded multiplicative seminorm defining the topology on the residue field (whose underlying valuation is the maximal generization of v). This map identifies $\mathcal{M}(B)$ with the *maximal Hausdorff quotient* of $\mathrm{Spa}(A, A^+)$.

The following is [13, Theorem 1.2.1], with essentially the same proof.

LEMMA 1.5.16. *For B a Banach ring, $B = 0$ if and only if $\mathcal{M}(B) = \emptyset$.*

PROOF. The content is that if $B \neq 0$, then $\mathcal{M}(B) \neq \emptyset$. Note that for any maximal ideal $\mathfrak{m}$ of B, $\mathfrak{m}$ is closed (see Remark 1.1.1) and $\mathcal{M}(B/\mathfrak{m})$ may be identified with a subset of $\mathcal{M}(B)$; we may thus assume that B is a field. (This does not by itself imply that B is complete for a multiplicative norm; see Remark 1.2.17.)

By Zorn's lemma, we may construct a minimal bounded seminorm β on B; it will suffice to check that β is multiplicative. Note that β must already be power-multiplicative, or else we could replace it with its spectral seminorm and violate minimality.

We can now finish in (at least) two different ways.

- Here is the approach taken in [13, Theorem 1.2.2]. Suppose $x \in B$ is nonzero. For any $\rho < \beta(x)$, the map $B \to B\langle \frac{T}{\rho}\rangle/(T-x)$ must be zero: otherwise, we could restrict the quotient norm on the target to get a seminorm on B bounded by β and strictly smaller at x, contradicting minimality. Since B is nonzero, the map can only be zero if the target is the zero ring, or equivalently if $T - x = -x(1 - x^{-1}T)$ has an inverse in $B\langle \frac{T}{\rho}\rangle$, or equivalently if the unique inverse $\sum_{n=0}^{\infty} -x^{-n-1}T^n$ in $B[\![T]\!]$ converges in $B\langle \frac{T}{\rho}\rangle$. That is, we must have
$$\lim_{n \to \infty} \beta(x^{-n})\rho^n = 0 \qquad \text{for all } \rho \in (0, \beta(x)),$$
and in particular $\beta(x^{-n}) < \rho^{-n}$ for n sufficiently large. By power-multiplicativity, this implies that $\beta(x^{-1}) \leq \beta(x)^{-1}$. For all $y \in B$, we now have
$$\beta(xy) \leq \beta(x)\beta(y) \leq \beta(x^{-1})^{-1}\beta(y) \leq \beta(xy);$$
hence β is multiplicative, as needed.
- Another approach (suggested by Zonglin Jiang) is to use Exercise 1.5.17 below to show that for any nonzero $x \in B$, the formula
$$\beta'(y) = \lim_{n \to \infty} \frac{\beta(x^n y)}{\beta(x^n)}$$

---

[5] This definition is usually attributed to Berkovich [13], who established many key properties. However, one finds isolated references to it in earlier literature, notably in the work of Guennebaud [84]. We thank Ofer Gabber for bringing this fact to our attention.

defines a power-multiplicative seminorm β' on B. For all $y \in B$, we have $\beta'(y) \leq \beta(y)$ and $\beta(xy) = \beta(x)\beta(y)$; by minimality, we must have $\beta = \beta'$, proving multiplicativity.

Using either approach, the proof is complete. $\square$

EXERCISE 1.5.17. Prove [24, Proposition 1.3.2/2]: for any uniform Banach ring B equipped with its spectral norm and any nonzero $x \in B$, the limit
$$|y|_x := \lim_{n \to \infty} \frac{|yx^n|}{|x^n|}$$
exists and defines a power-multiplicative seminorm $|\bullet|_x$ on B.

We now recover [95, Proposition 3.6(i)].

COROLLARY 1.5.18. *For (A, A^+) a (not necessarily analytic) Huber pair with $A \neq 0$, we have* $\mathrm{Spa}(A, A^+) \neq \emptyset$.

PROOF. Promote A to a Banach ring as per Remark 1.5.4, then apply Lemma 1.5.16. $\square$

COROLLARY 1.5.19. *For B a nonzero Banach ring, an ideal I of B is trivial if and only if for each $\beta \in \mathcal{M}(B)$, there exists $x \in I$ with $\beta(x) > 0$. In particular, an element x of B is invertible if and only if $\beta(x) \neq 0$ for all $\beta \in \mathcal{M}(B)$.*

PROOF. By Remark 1.1.1, we may assume that I is closed. If I is trivial, then obviously $x = 1$ satisfies $\beta(x) > 0$ for all $\beta \in \mathcal{M}(B)$. Otherwise, B/I is a nonzero Banach ring (since I is now closed), $\mathcal{M}(B/I)$ is nonempty by Lemma 1.5.16, and any element of $\mathcal{M}(B/I)$ restricts to an element $\beta \in \mathcal{M}(B)$ with $\beta(x) = 0$ for all $x \in I$. $\square$

COROLLARY 1.5.20. *For (A, A^+) a (not necessarily analytic) Huber pair, an ideal I of A is trivial if and only if for each $v \in \mathrm{Spa}(A, A^+)$, there exists $x \in I$ with $v(x) > 0$. In particular, $x \in A$ is invertible if and only if $v(x) > 0$ for all $v \in \mathrm{Spa}(A, A^+)$.*

PROOF. Promote A to a Banach ring as per Remark 1.5.4, then apply Corollary 1.5.19. $\square$

The following is an analogue of the maximum modulus principle in rigid analytic geometry [24, Proposition 6.2.1/4]. The proof is taken from [13, Theorem 1.3.1].

LEMMA 1.5.21. *For B a Banach ring, the spectral seminorm of B equals the supremum over $\mathcal{M}(B)$.*

PROOF. In one direction, it is obvious that any multiplicative seminorm bounded by the original norm is also bounded by the spectral seminorm. In the other direction, we must check that if $x \in B$, $\rho > 0$ satisfy $\beta(x) < \rho$ for all $\beta \in \mathcal{M}(B)$, then $|x|_{\mathrm{sp}} < \rho$. The input condition implies that $1 - xT$ vanishes nowhere on the spectrum of $B\langle \frac{T}{\rho^{-1}} \rangle$, hence is invertible by Corollary 1.5.19. In the larger ring $B[\![T]\!]$, the inverse of $1 - xT$ equals $1 + xT + x^2T^2 + \cdots$; the fact that this belongs to $B\langle \frac{T}{\rho^{-1}} \rangle$ implies that $|x|_{\mathrm{sp}} < \rho$ as in the proof of Lemma 1.5.16. $\square$

COROLLARY 1.5.22. *Let B be a Banach ring. If the* uniform completion *of B (i.e., the separated completion of B with respect to the spectral seminorm) is zero, then so is B itself.*

PROOF. By Lemma 1.5.21 the given condition implies that $\mathcal{M}(B) = \emptyset$, at which point Lemma 1.5.16 implies that $B = 0$. □

COROLLARY 1.5.23. *For B a Banach ring, an element $x \in B$ is topologically nilpotent if and only if $\beta(x) < 1$ for all $\beta \in \mathcal{M}(B)$.*

PROOF. This is immediate from Exercise 1.5.12, Lemma 1.5.21, and the compactness of $\mathcal{M}(B)$. □

This immediately yields the following corollary of Lemma 1.5.21; see also [**27**, Lemma 5] for a purely topological proof.

COROLLARY 1.5.24. *Let (A, A^+) be a (not necessarily analytic) Huber pair. Then the kernel of $A \to H^0(X, \mathcal{O})$ contains only topologically nilpotent elements.*

PROOF. Promote A to a Banach ring as per Remark 1.5.4. Then any $x \in A$ mapping to zero in $H^0(X, \mathcal{O})$ satisfies $\alpha(x) = 0$ for all $\alpha \in \mathcal{M}(A)$; by Corollary 1.5.23, x is topologically nilpotent. □

REMARK 1.5.25. Lemma 1.5.21 implies that for any covering $\mathfrak{V}$ of X, the map $A \to \bigoplus_{V \in \mathfrak{V}} \mathcal{O}(V)$ is an isometry for the spectral seminorms on all terms. In particular, if A is uniform, then $A \to \mathcal{O}(X)$ is injective; by Remark 1.5.4, the same statement holds for Huber rings.

This can be used to prove the following key lemma (compare [**27**, Lemma 3]).

LEMMA 1.5.26. *Suppose that A is uniform. Choose $x = \sum_{n=0}^{\infty} x_n T^n \in A\langle T \rangle$ such that the x_n generate the unit ideal. Then multiplication by x defines a strict inclusion $A\langle T \rangle \to A\langle T \rangle$. (The analogous statement for $A\langle T^{\pm} \rangle$ also holds, with an analogous proof.)*

PROOF. Using Remark 1.5.4, we may reduce to considering the analogous problem where A is a uniform Banach ring. For $\alpha \in \mathcal{M}(A)$, write $\tilde{\alpha} \in \mathcal{M}(A\langle T \rangle)$ for the Gauss extension; note that the latter is the maximal seminorm on $A\langle T \rangle$ restricting to β on $\mathcal{M}(A)$. By Lemma 1.5.21, we may then compute the spectral seminorm on $A\langle T \rangle$ as the supremum of $\tilde{\alpha}$ as α runs over $\mathcal{M}(A)$.

Choose $n \geq 0$ such that $x_0, \ldots, x_n$ generate the unit ideal in A; then the quantity
$$c := \inf_{\alpha \in \mathcal{M}(B)} \{\alpha(x_0), \ldots, \alpha(x_n)\}$$
is positive. For all $y \in A\langle T \rangle$, we have
$$\sup_{\alpha \in \mathcal{M}(A)} \{\tilde{\alpha}(xy)\} = \sup_{\alpha \in \mathcal{M}(A)} \{\tilde{\alpha}(x)\tilde{\alpha}(y)\} \geq c \sup_{\alpha \in \mathcal{M}(A)} \{\tilde{\alpha}(y)\};$$
this proves the claim. □

EXERCISE 1.5.27. Let $\{A_i\}_{i \in I}$ be a filtered direct system of uniform Banach rings, equipped with their spectral norms, and let A be the completed direct limit of the A_i. Prove that every finite projective module on A is the base extension of some finite projective module over some A_i. (See [**118**, Lemma 5.6.8].)

REMARK 1.5.28. It is possible to construct a version of the theory of adic spaces in which Banach rings, rather than Huber rings, form the building blocks; this is the theory of *reified adic spaces* described in [**112**]. The main structural distinction is that valuations do not map into totally arbitrary value groups; rather, each value group must be normalized using a fixed inclusion of the positive real numbers. In the reified theory, many occurrences of Tate or analytic hypotheses can be relaxed, because the normalization of the value group can be used in the place of topologically nilpotent elements. However, the open mapping theorem, being a purely topological statement, provides a stumbling block.

1.6. A strategy of proof: variations on Tate's reduction. We next collect some general observations that will be used to complete the omitted proofs from earlier in the lecture. The reader is again reminded to keep in mind the analogy with affine schemes, as many of the ideas are similar.

To begin, we reduce the sheafy property, and the acyclicity of sheaves, to a statement about sufficiently fine coverings of and by basic open subsets.

DEFINITION 1.6.1. By a *cofinal family of rational coverings*, we will mean a function C assigning to each rational subspace U of X a set of finite coverings of U by rational subspaces which is *cofinal*: every covering of U by open subspaces is refined by some covering in $C(U)$. For example, since U is quasicompact, one obtains a cofinal family of rational coverings by taking $C(U)$ to be all finite coverings by rational subspaces.

DEFINITION 1.6.2. We use the following notation for Čech cohomology groups. For $\mathcal{F}$ a presheaf on X, $U \subseteq X$ open, $\mathfrak{V} = \{V_i\}_{i \in I}$ a covering of U by open subspaces, and j a nonnegative integer, let $\check{C}^j(U, \mathcal{F}; \mathfrak{V})$ be the product of $\mathcal{F}(V_{i_0} \cap \cdots \cap V_{i_j})$ over all distinct $i_0, \ldots, i_j \in I$. Let $d^j : \check{C}^j(U, \mathcal{F}; \mathfrak{V}) \to \check{C}^{j+1}(U, \mathcal{F}; \mathfrak{V})$ be the map given by the formula

$$(s_{i_0,\ldots,i_j})_{i_0,\ldots,i_j \in I} \mapsto \left(\sum_{k=0}^{j+1} (-1)^k s_{i_0,\ldots,\widehat{i_k},\ldots,i_{j+1}} \right)_{i_0,\ldots,i_{j+1} \in I} ;$$

with these differentials, $\check{C}^\bullet(U, \mathcal{F}; \mathfrak{V})$ form a complex whose cohomology groups we denote by $\check{H}^\bullet(U, \mathcal{F}; \mathfrak{V})$.

The following is the same standard argument used to establish the basic properties of the structure sheaf on affine schemes; compare [**166**, Tag 01EW].

LEMMA 1.6.3. *Let C be a cofinal family of rational coverings. Let $\mathcal{F}$ be a presheaf on X with the property that for any open subset U, $\mathcal{F}(U)$ is the inverse limit of $\mathcal{F}(V)$ over all rational subspaces $V \subseteq U$.*

(a) *Suppose that for every rational subspace U of X and every covering $\mathfrak{V} \in C(U)$, the natural map*

$$\mathcal{F}(U) \to \check{H}^0(U, \mathcal{F}; \mathfrak{V})$$

is an isomorphism. Then $\mathcal{F}$ is a sheaf.

(b) *Suppose that $\mathcal{F}$ is a sheaf, and that for every rational subspace U of X and every covering $\mathfrak{V} \in C(U)$, we have $\check{H}^i(U, \mathcal{F}; \mathfrak{V}) = 0$ for all $i > 0$. Then $\mathcal{F}$ is acyclic.*

PROOF. We start with (a). To show that $\mathcal{F}$ is a sheaf, we must check that $\mathcal{F}(U) \to \check{H}^0(U, \mathcal{F}; \mathfrak{V})$ is an isomorphism for every open subspace U of X and every open covering $\mathfrak{V}$ of U. We will check injectivity, then surjectivity; note that each of these assertions follows formally from the case where U is rational.

Suppose that U is rational and that $\mathfrak{V}$ is a covering of U. Let $\mathfrak{V}' \in C(U)$ be a refinement of $\mathfrak{V}$. The map $\mathcal{F}(U) \to \check{H}^0(U, \mathcal{F}; \mathfrak{V}')$ then factors as the map $\mathcal{F}(U) \to \check{H}^0(U, \mathcal{F}; \mathfrak{V})$ followed by the restriction map $\check{H}^0(U, \mathcal{F}; \mathfrak{V}) \to \check{H}^0(U, \mathcal{F}; \mathfrak{V}')$. Since $\mathcal{F}(U) \to \check{H}^0(U, \mathcal{F}; \mathfrak{V}')$ is injective, the map $\mathcal{F}(U) \to \check{H}^0(U, \mathcal{F}; \mathfrak{V})$ is injective for U rational, and hence for arbitrary U.

By the previous paragraph, the map $\check{H}^0(U, \mathcal{F}; \mathfrak{V}) \to \check{H}^0(U, \mathcal{F}; \mathfrak{V}')$ is also injective. Consequently, the surjectivity of $\check{H}^0(U, \mathcal{F}; \mathfrak{V}) \to \check{H}^0(U, \mathcal{F}; \mathfrak{V}')$ implies the surjectivity of $\mathcal{F}(U) \to \check{H}^0(U, \mathcal{F}; \mathfrak{V})$ for U rational, and hence for arbitrary U. This completes the proof of (a).

To establish (b), we will show that $H^i(U, \mathcal{F}) = 0$ for all rational subspaces U and all $i > 0$; we do this by induction on i. Given that $\mathcal{F}$ is a sheaf and that $H^j(U, \mathcal{F}) = 0$ for all rational subspaces U and all $j < i$, a standard spectral sequence argument [24, Corollary 8.1.4/3] produces a canonical morphism $H^i(U, \mathcal{F}) \to \check{H}^i(U, \mathcal{F}; \mathfrak{V})$ for any open covering $\mathfrak{V}$ of U, with the property that if $\mathfrak{V}'$ is a refinement of $\mathfrak{V}$ then the morphism $H^i(U, \mathcal{F}) \to \check{H}^i(U, \mathcal{F}; \mathfrak{V}')$ factors as $H^i(U, \mathcal{F}) \to \check{H}^i(U, \mathcal{F}; \mathfrak{V})$ followed by the natural morphism $\check{H}^i(U, \mathcal{F}; \mathfrak{V}) \to \check{H}^i(U, \mathcal{F}; \mathfrak{V}')$. With this in mind, we may imitate the proof of (a) to conclude. $\square$

In order to maximally exploit this argument, we construct some special finite coverings of rational subspaces, so as to cut down the required number of explicit calculations of Čech cohomology.

DEFINITION 1.6.4. For $f_1, \ldots, f_n \in A$ which generate the unit ideal, the sets
$$X\left(\frac{f_1, \ldots, f_n}{f_i}\right) \qquad (i = 1, \ldots, n)$$
form a covering of X by rational subspaces; this covering is called the *standard rational covering* defined by the parameters $f_1, \ldots, f_n$. A standard rational covering with $n = 2$ will be called a *standard binary rational covering*.

Although we have generically been assuming that A is analytic, the previous definition makes sense without this hypothesis. However, in order for the sets $X\left(\frac{f_1, \ldots, f_n}{f_i}\right)$ to form a covering, the elements $f_1, \ldots, f_n$ must still generate the unit ideal in A, not an arbitrary open ideal (because the condition that $v \in X$ belongs to $X\left(\frac{f_1, \ldots, f_n}{f_i}\right)$ includes the requirement that $v(f_i) \neq 0$).

We record the following statement for use in a subsequent lecture.

LEMMA 1.6.5. *The map $X = \mathrm{Spa}(A, A^+) \to \mathrm{Spa}(A^+, A^+)$ (which by Corollary 1.1.4 is injective) is an open immersion (i.e., a homeomorphism onto an open subset of the target).*

PROOF. Let $x_1, \ldots, x_n \in A^+$ be topologically nilpotent elements which generate the unit ideal in A. The image of X in $\mathrm{Spa}(A^+, A^+)$ can then be written as the union of the open subsets $\mathrm{Spa}(A^+, A^+)(\frac{x_1, \ldots, x_n}{x_i})$ for $i = 1, \ldots, n$; it is thus open. To complete the argument, we need only check that each of these open subsets is itself homeomorphic to the rational subspace $X(\frac{x_1, \ldots, x_n}{x_i})$ of X; that is, we may reduce to the case where A is Tate.

Now suppose that $x \in A^+$ is a topologically nilpotent unit in A. In this case, we must check that for any $f_1, \ldots, f_n, g \in A$ which generate the unit ideal, the rational subspace $X\left(\frac{f_1, \ldots, f_n}{g}\right)$ of X is the pullback of a rational subspace of $\mathrm{Spa}(A^+, A^+)$.

To see this, let m be an integer which is large enough that $x^m f_1, \ldots, x^m f_n, x^m g \in A^+$; these parameters then generate an open ideal in A^+, and so define a rational subspace of $\mathrm{Spa}(A^+, A^+)$ of the desired form. □

DEFINITION 1.6.6. It will be useful to single out some special types of standard binary rational coverings. The covering with parameters $f_1 = f$, $f_2 = 1$ will be called the *simple Laurent covering* defined by f. The covering with parameters $f_1 = f$, $f_2 = 1 - f$ will be called the *simple balanced covering* defined by f; note that the terms in this covering can be rewritten as $X(\frac{1}{f}), X(\frac{1}{1-f})$.

REMARK 1.6.7. The concept of a standard rational covering of X is the closest analogue in this context to a covering of an affine scheme by distinguished open affine subschemes. That is because for R a ring and $f_1, \ldots, f_n, g \in R$ generating the unit ideal, the ring obtained from R by adjoining $f_1/g, \ldots, f_n/g$ is precisely $R[1/g]$; consequently, for $f_1, \ldots, f_n \in R$ generating the unit ideal, the "rational covering" defined by $f_1, \ldots, f_n$ is nothing but the covering of $\mathrm{Spec}\, R$ by $\mathrm{Spec}\, R[1/f_1], \ldots, \mathrm{Spec}\, R[1/f_n]$.

The previous remark suggests the following lemma, due in this form to Huber [96, Lemma 2.6]; see also [24, Lemma 8.2.2/2] for the case where A is an affinoid algebra over a nonarchimedean field, or [117, Lemma 2.4.19(a)] for the case where A is Tate.

LEMMA 1.6.8 (Huber). *Every open covering of a rational subspace of X can be refined by some standard rational covering.*

PROOF. Since every rational subspace of X is itself the spectrum of a Huber pair, it suffices to consider coverings of X itself. Since X is quasicompact, we may start with a finite covering $\mathfrak{V} = \{V_i\}_{i \in I}$ of X by rational subspaces. For $i \in I$, write

$$V_i = X\left(\frac{f_{i1}, \ldots, f_{in_i}}{g_i}\right) \qquad (i \in I)$$

for some $f_{i1}, \ldots, f_{in_i}, g_i$ which generate the unit ideal in A. Let S_0 be the set of products $\prod_{i \in I} s_i$ with $s_i \in \{f_{i1}, \ldots, f_{in_i}, g_i\}$; then S_0 generates the unit ideal.

Let S be the subset of S_0 consisting of those products $\prod_{i \in I} s_i$ where $s_i = g_i$ for at least one $i \in I$. These also generate the unit ideal: to see this, by Corollary 1.5.20 it suffices to check that for each $v \in X$ there exists some $s \in S$ for which $v(s) \neq 0$. To see this, choose an index $i \in I$ for which $v \in V_i$, put $s_i = g_i$, and for each $j \neq i$ choose $s_j \in \{f_{j1}, \ldots, f_{jn_j}, g_j\}$ to maximize $v(s_j)$. Since $f_{j1}, \ldots, f_{jn_j}, g_j$ generate the unit ideal, we must have $v(s_j) \neq 0$; it follows that $v(s) \neq 0$.

We may thus form the standard covering by S. This refines the original covering: if $s \in S$ with $s_i = g_i$, then $X(\frac{S}{s}) \subseteq V_i$. □

REMARK 1.6.9. Using Lemma 1.6.8, we may see that the property of (A, A^+) being sheafy depends only on A, not on A^+: both the collection of standard rational coverings, and the assertions of the sheaf axiom for these coverings, depend only on A.

REMARK 1.6.10. For A not necessarily analytic, a rational subspace of X is defined by parameters which generate an open ideal of A, rather than the trivial ideal. Recall that these two conditions coincide if and only if A is analytic (Lemma 1.1.3). For this reason, the proof of Lemma 1.6.8 does not extend to the case where A is not analytic.

To see just how different the nonanalytic case is, we consider the following example. The ring $\mathbf{A}_{\inf}$ to be introduced later exhibits similar behavior; see Remark 3.1.10.

EXAMPLE 1.6.11. Let k be a field, equip $A := k[\![x,y]\!]$ with the (x,y)-adic topology, and put $A^+ := A$. The space X then contains a unique valuation v with $v(x) = v(y) = 0$. The only rational subspace of X containing v is X itself: for $f_1, \ldots, f_n, g \in A$ generating an open ideal, we have $v \in X\left(\frac{f_1,\ldots,f_n}{g}\right)$ if and only if $v(g) \neq 0$, which forces g to be a unit. Thus in this case the conclusion of Lemma 1.6.8 does turn out to be correct: any covering of X is refined by the trivial covering of X by itself, whereas any proper rational subspace of X is the spectrum of an analytic ring and so is subject to Lemma 1.6.8.

Continuing the analogy with affine schemes, we may further reduce the cofinal family consisting of the standard rational coverings by considering compositions of special coverings. The following argument is due to Gabber–Ramero (taken from [**75**]).

LEMMA 1.6.12 (Gabber–Ramero). *Every open covering of a rational subspace of X can be refined by some composition of standard binary rational coverings.*

PROOF. Again, we need only consider coverings of X itself. By Lemma 1.6.8, there is no harm in starting with the standard rational covering defined by some $f_1, \ldots, f_n \in A$ generating the unit ideal. We induct on the smallest value of m for which some m-element subset of $\{f_1, \ldots, f_n\}$ generates the unit ideal; there is nothing to check unless $m \geq 3$. Without loss of generality, we may assume that $f_1, \ldots, f_m$ generate the unit ideal, and choose $g_1, \ldots, g_m \in A$ for which $f_1 g_1 + \cdots + f_m g_m = 1$. Now define

$$h = \sum_{i=1}^{\lfloor m/2 \rfloor} f_i g_i, \qquad h' = \sum_{i=\lfloor m/2 \rfloor + 1}^{m} f_i g_i$$

and form the standard binary rational covering generated by h, h'. On each of $X\langle \frac{h}{h'}\rangle$ and $X\langle \frac{h'}{h}\rangle$, the unit ideal is generated by a subset of $f_1, \ldots, f_m$ of size at most $\lceil m/2 \rceil \leq m-1$; we may thus apply the induction hypothesis to conclude. □

The following refinement of Lemma 1.6.12 will be useful for checking flatness.

LEMMA 1.6.13. *Every open covering of a rational subspace of X can be refined by some composition of coverings, each of which is either a simple Laurent covering or a simple balanced covering.*

PROOF. By Lemma 1.6.12, it suffices to prove the claim for the standard binary rational covering of X defined by some $f, g \in A$ which generate the unit ideal. Choose $a, b \in A$ with $af + bg = 1$. We then may refine the original covering by taking the simple balanced covering defined by af, and forming the simple Laurent coverings of $X(\frac{1}{af}), X(\frac{1}{bg})$ defined by the respective parameters $g/f, f/g$. □

REMARK 1.6.14. In the Tate case, one can do even better than Lemma 1.6.13: it is only necessary to use simple Laurent coverings. This was shown for affinoid algebras in [**24**, Lemma 8.2.2/3, Lemma 8.2.2/4] and in general in [**96**, Theorem 2.5, (II.1)(iv)], [**117**, Lemma 2.4.19]. In light of Lemma 1.6.12, one may see this simply

by checking that for any $f, g \in A$ generating the unit ideal, the simple balanced covering defined by f is refined by a composition of simple Laurent coverings. To this end, choose a topologically nilpotent unit $x \in X$; then for any sufficiently large integer n, we have
$$\max\{v(f/x^n), v(g/x^n)\} \geq 1 \qquad (v \in X).$$
From the ensuing equality (and its symmetric counterpart)
$$X\left(\frac{f}{g}\right) = X\left(\frac{f/x^n}{1}\right)\left(\frac{1}{g/x^n}\right) \cup X\left(\frac{1}{f/x^n}\right)\left(\frac{g/f}{1}\right),$$
we see that the original covering is refined by a suitable composition of simple Laurent coverings.

Let us now sketch how we will use the preceding lemmas to carry out the various proofs that we are still due to provide.

REMARK 1.6.15. To show that some particular A is sheafy (as in Theorem 1.2.11 or Theorem 1.2.13), by Lemma 1.6.3 it suffices to check the isomorphism $\mathcal{O}(U) \cong \check{H}^0(U, \mathcal{O}; \mathfrak{V})$ for every rational subspace U and every finite covering $\mathfrak{V}$ by rational subspaces in some cofinal family. By Lemma 1.6.12, we may take this collection to be the compositions of standard binary rational coverings. Checking the isomorphism for such coverings immediately reduces to checking for a single standard binary rational covering; by Lemma 1.6.13 it is even sufficient to check for a simple Laurent covering and a simple balanced covering. That is, we must show that for every rational localization $(A, A^+) \to (B, B^+)$ and every pair $f, g \in B$ with $g \in \{1, 1-f\}$, the sequence

(1.6.15.1) $$0 \to B \to B\left\langle\frac{f}{g}\right\rangle \oplus B\left\langle\frac{g}{f}\right\rangle \to B\left\langle\frac{f}{g}, \frac{g}{f}\right\rangle \to 0$$

is exact at the left and middle.

In the same vein, if $\mathcal{O}$ is a sheaf, then the above sequence is known to be exact at the left and middle; to show that $\mathcal{O}$ is acyclic, it suffices to check exactness at the right. Similarly, to prove that $\tilde{M}$ is an acyclic sheaf for some given A-module M (as in Theorem 1.3.4 and Theorem 1.4.16), it suffices to check that tensoring (1.6.15.1) over B with $M \otimes_A B$ gives another exact sequence.

REMARK 1.6.16. Given that A is sheafy and $\tilde{M}$ is acyclic for every finite projective A-module M, to show that every vector bundle on X arises from some finite projective A-module (as in Theorem 1.4.2), by Lemma 1.6.13 it suffices to consider a bundle which is specified by modules on each term of a composition of simple Laurent coverings and simple balanced coverings. It then suffices to check that for every rational localization $(A, A^+) \to (B, B^+)$ and every pair $f, g \in B$ with $g \in \{1, 1-f\}$, the functor

$$\mathbf{FPMod}_B \to \mathbf{FPMod}_{B\langle\frac{f}{g}\rangle} \times_{\mathbf{FPMod}_{B\langle\frac{f}{g}, \frac{g}{f}\rangle}} \mathbf{FPMod}_{B\langle\frac{g}{f}\rangle}$$

is an equivalence of categories. (Note that since we are only considering a covering by two open sets, there is no need to impose a cocycle condition on the objects on the right-hand side.) Similarly, given that $\tilde{M}$ is acyclic for every stably pseudocoherent A-module M, to show that every pseudocoherent sheaf on X arises from some stably pseudocoherent A-module (as in Theorem 1.4.18), we must check that for B, f, g as above, the functor

$$\mathbf{PCoh}_B \to \mathbf{PCoh}_{B\langle \frac{f}{g} \rangle} \times_{\mathbf{PCoh}_{B\langle \frac{f}{g}, \frac{g}{f} \rangle}} \mathbf{PCoh}_{B\langle \frac{g}{f} \rangle}$$

is an equivalence of categories. However, in order to even have such a functor, we must first establish the preservation of stably pseudocoherent modules under base extension along rational localizations; see Remark 1.6.17.

REMARK 1.6.17. Given that A is sheafy and acyclic, to show that base extension along a rational localization preserves the category of stably pseudocoherent modules (as in Theorem 1.4.14) and is exact, we must proceed somewhat indirectly. Namely, we initially only consider rational localizations which occur in compositions of simple Laurent and simple balanced coverings, which for convenience we temporarily refer to as *nice* localizations. Let $\mathbf{PCoh}_A^{\mathrm{nice}}$ denote the category of pseudocoherent A-modules such that for any nice localization $(A, A^+) \to (B, B^+)$, $M \otimes_A B$ is complete for its natural topology as a B-module; this category contains $\mathbf{PCoh}_A$, and by the end of the proof we will see that the two categories coincide.

We first verify that for $(A, A^+) \to (B, B^+)$ a nice localization, base extension defines an exact functor $\mathbf{PCoh}_A^{\mathrm{nice}} \to \mathbf{PCoh}_B^{\mathrm{nice}}$. This reduces to a calculation for each term in a simple Laurent covering or a simple balanced covering; note that the Laurent case implies the balanced case because

$$X\left(\frac{f}{1-f}\right) = X\left(\frac{1}{1-f}\right), \qquad X\left(\frac{1-f}{f}\right) = X\left(\frac{1}{f}\right).$$

Next, let $U = \mathrm{Spa}(B, B^+)$ be a not necessarily nice rational subspace. Let

$$0 \to N \to F \to M \to 0$$

be an exact sequence in $\mathbf{PCoh}_A^{\mathrm{nice}}$ in which F is finite free. By Lemma 1.6.13, we can find a covering $\mathfrak{V}$ of U by rational subspaces which are nice in X (and hence in U). By the previous paragraph, for each $\mathrm{Spa}(C, C^+) \in \mathfrak{V}$, tensoring over A with C converts the previous exact sequence into an exact sequence in $\mathbf{PCoh}_C^{\mathrm{nice}}$. Glueing as in Remark 1.6.16 then yields an exact sequence

$$0 \to N' \to F' \to M' \to 0$$

in $\mathbf{PCoh}_B^{\mathrm{nice}}$ fitting into a natural diagram

$$\begin{array}{ccccccccc} & N \otimes_A B & \to & F \otimes_A B & \to & M \otimes_A B & \to & 0 \\ & \downarrow & & \downarrow & & \downarrow & & \\ 0 & \to & N' & \to & F' & \to & M' & \to & 0 \end{array}$$

in which the middle vertical arrow is an isomorphism, so the right vertical arrow must be surjective. By repeating the argument with N playing the role previously played by M, we see that the left vertical arrow must also be surjective. By the five lemma, the right vertical arrow must be surjective; again, repeating the argument shows that the left vertical arrow is surjective. In other words, the original exact sequence remains exact, and the terms remain complete for the natural topology, after tensoring over A with B. We thus conclude that base extension defines an exact functor $\mathbf{PCoh}_A^{\mathrm{nice}} \to \mathbf{PCoh}_B^{\mathrm{nice}}$, which in turn implies that $\mathbf{PCoh}_A^{\mathrm{nice}} = \mathbf{PCoh}_A$.

These arguments are summarized in [117, Proposition 2.4.20] as follows.

LEMMA 1.6.18. *Let $\mathcal{P}_{an}$ be the set of pairs $(U, \mathfrak{V})$ where U is a rational subspace and $\mathfrak{V}$ is a finite covering of U by rational subspaces. Suppose that $\mathcal{P} \subseteq \mathcal{P}_{an}$ satisfies the following conditions.*

(i) *Locality: if $(U, \mathfrak{V})$ admits a refinement in $\mathcal{P}$, then $(U, \mathfrak{V}) \in \mathcal{P}$.*
(ii) *Transitivity: Any composition of coverings in $\mathcal{P}$ is in $\mathcal{P}$.*
(iii) *Every standard binary rational covering is in $\mathcal{P}$.*

Then $\mathcal{P} = \mathcal{P}_{an}$.

REMARK 1.6.19. While glueing theorems fit neatly into the framework of Lemma 1.6.18, writing the proofs of sheafiness and acyclicity theorems in this language still requires an argument in the style of Lemma 1.6.3. See [**117**, Proposition 2.4.21] for a presentation of this form.

1.7. Proofs: sheafiness. We now use the formalism we have set up to establish sheafiness when A is strongly noetherian (Theorem 1.2.11) or stably uniform (Theorem 1.2.13).

HYPOTHESIS 1.7.1. Throughout §1.7, let $(A, A^+) \to (B, B^+)$ be a rational localization, and let $f, g \in B$ be elements which generate the unit ideal. We will use frequently and without comment the fact that $B\langle \frac{f}{g} \rangle$ is the quotient of $B\langle T \rangle$ by the closure of the ideal $(f - gT)$ (and the same with f and g reversed), and similarly $B\langle \frac{f}{g}, \frac{g}{f} \rangle$ is the quotient of $B\langle T^{\pm} \rangle$ by the closure of the ideal $(f - gT)$.

LEMMA 1.7.2. *With notation as in Hypothesis* 1.7.1, *suppose that for each of the pairs*

$$(R, x) = (B\langle T \rangle, f - gT), (B\langle T^{-1} \rangle, g - fT^{-1}), (B\langle T^{\pm} \rangle, f - gT),$$

the ideal xR is closed. Then the sequence (1.6.15.1) *is exact at the middle.*

PROOF. By hypothesis, we have a commutative diagram
(1.7.2.1)

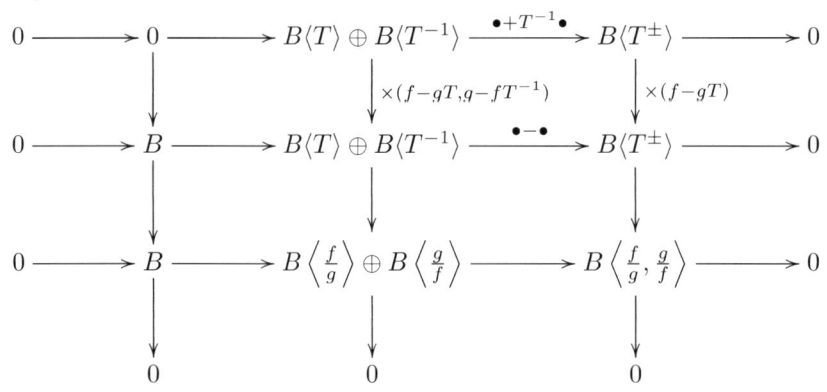

$$\begin{array}{ccccccccc}
0 & \to & 0 & \to & B\langle T \rangle \oplus B\langle T^{-1} \rangle & \xrightarrow{\bullet + T^{-1} \bullet} & B\langle T^{\pm} \rangle & \to & 0 \\
& & \downarrow & & \downarrow{\scriptstyle \times (f-gT, g-fT^{-1})} & & \downarrow{\scriptstyle \times (f-gT)} & & \\
0 & \to & B & \to & B\langle T \rangle \oplus B\langle T^{-1} \rangle & \xrightarrow{\bullet - \bullet} & B\langle T^{\pm} \rangle & \to & 0 \\
& & \downarrow & & \downarrow & & \downarrow & & \\
0 & \to & B & \to & B\langle \frac{f}{g} \rangle \oplus B\langle \frac{g}{f} \rangle & \to & B\langle \frac{f}{g}, \frac{g}{f} \rangle & \to & 0 \\
& & \downarrow & & \downarrow & & \downarrow & & \\
& & 0 & & 0 & & 0 & &
\end{array}$$

in which all three columns, and the first two rows, are exact. By diagram-chasing, or applying the snake lemma to the first two rows, we deduce the claim. □

At this point, it is easy to finish the proof of sheafiness in the stably uniform case.

LEMMA 1.7.3. *With notation as in Hypothesis* 1.7.1, *if B is uniform, then* (1.6.15.1) *is exact at the left and middle.*

PROOF. From Lemma 1.5.26, we see that on one hand, the criterion of Lemma 1.7.2 applies, so (1.6.15.1) is exact in the middle; on the other hand, the middle and right columns in (1.7.2.1) may be augmented to short exact sequences, so (1.6.15.1) is exact at the left. (To obtain exactness at the left, one can also invoke Remark 1.5.25.) □

PROOF OF THEOREM 1.2.13. By Remark 1.6.15, this reduces immediately to Lemma 1.7.3. □

In the strongly noetherian case, exactness at the middle is no issue, but we must do a bit of work to check exactness at the left. Here we must essentially give Huber's proof that rational localization maps are flat in the strongly noetherian case [**96**, II.1], [**97**, Lemma 1.7.6]. We warn the reader that [**72**, Lemma 4.2.5] is somewhat sketchy on this point.

LEMMA 1.7.4. *Suppose that A is strongly noetherian. With notation as in Hypothesis 1.7.1, the maps $B \to B\langle\frac{f}{g}\rangle, B \to B\langle\frac{g}{f}\rangle$ are flat.*

PROOF. By symmetry, we need only check the first claim. By Lemma 1.1.19, the map $B[T] \to B\langle T\rangle$ is flat; we thus obtain a flat map

$$B = \frac{B[T]}{(f-gT)} \to \frac{B\langle T\rangle}{(f-gT)} = B\left\langle\frac{f}{g}\right\rangle,$$

using Corollary 1.1.15 to make the last identification. □

LEMMA 1.7.5. *With notation as in Hypothesis 1.7.1, if the map $B \to B\langle\frac{f}{g}\rangle \oplus B\langle\frac{g}{f}\rangle$ is flat, then it is faithfully flat.*

PROOF. By [**166**, Tag 00HQ], it suffices to show that the image of the map

$$\operatorname{Spec}\left(B\left\langle\frac{f}{g}\right\rangle \oplus B\left\langle\frac{g}{f}\right\rangle\right) \to \operatorname{Spec}(B)$$

includes every maximal ideal $\mathfrak{m}$ of B. To see this, note that since $\mathfrak{m}$ is necessarily closed (see Remark 1.1.1), $B/\mathfrak{m}$ is again a nonzero Huber ring and so has nonzero spectrum (by Lemma 1.5.16); we can thus choose $v \in \operatorname{Spa}(B, B^+)$ containing $\mathfrak{m}$ in its kernel. The point v must appear in the spectra of one of $B\langle\frac{f}{g}\rangle$ or $B\langle\frac{g}{f}\rangle$; taking the kernel of the resulting valuation gives a prime ideal of the corresponding ring which contracts to $\mathfrak{m}$. □

LEMMA 1.7.6. *Suppose that A is strongly noetherian. With notation as in Hypothesis 1.7.1, (1.6.15.1) is exact at the left and middle.*

PROOF. By Corollary 1.1.15, every ideal of $B\langle T\rangle$ is closed; by Lemma 1.7.2, this means that (1.6.15.1) is exact at the middle. To show exactness at the left, note that the map in question is flat by Lemma 1.7.4, and hence faithfully flat by Lemma 1.7.5. □

PROOF OF THEOREM 1.2.11. By Remark 1.6.15, this reduces immediately to Lemma 1.7.6. □

The following is proved in [**86**] assuming that A is Tate, but the argument generalizes easily.

EXERCISE 1.7.7. Suppose that A is uniform; $f, g \in A$ generate the unit ideal; and $A\langle \frac{f}{g}, \frac{g}{f} \rangle$ is also uniform. Prove that $A\langle \frac{f}{g} \rangle$, $A\langle \frac{g}{f} \rangle$ are uniform. (Hint: promote A to a Banach ring. Given $x \in A\langle \frac{f}{g} \rangle$ such that x^2 is "small", produce $y \in A$ such that y is "small" in $A\langle \frac{g}{f} \rangle$ and $x - y$ is "small" in $A\langle \frac{f}{g} \rangle$. Then bound $\alpha(y)$ for an arbitrary $\alpha \in \mathcal{M}(A)$.)

1.8. Proofs: acyclicity. We next turn to acyclicity of sheaves associated to finite projective A-modules (Theorem 1.3.4). Throughout §1.8, continue to set notation as in Hypothesis 1.7.1.

LEMMA 1.8.1. *With notation as in Hypothesis 1.7.1, the map $B\langle \frac{f}{g} \rangle \oplus B\langle \frac{g}{f} \rangle \to B\langle \frac{f}{g}, \frac{g}{f} \rangle$ is strict surjective, i.e., the sequence (1.6.15.1) is always strict exact at the right.*

PROOF. In the commutative diagram

$$\begin{array}{ccc} B\langle T \rangle \oplus B\langle T^{-1} \rangle & \longrightarrow & B\langle T^{\pm} \rangle \\ \downarrow & & \downarrow \\ B\langle \frac{f}{g} \rangle \oplus B\langle \frac{g}{f} \rangle & \longrightarrow & B\langle \frac{f}{g}, \frac{g}{f} \rangle \end{array}$$

both vertical arrows and the top horizontal arrow are strict surjections; this yields the claim. □

PROOF OF THEOREM 1.3.4. Since M is a direct summand of a finite free A-module, we may assume without loss of generality that $M = A$. By Remark 1.6.15, this reduces immediately to Lemma 1.8.1. □

To obtain acyclicity for $\tilde{M}$ for more general M, we must study the sequence (1.6.15.1) a bit more closely. A key step is the following converse of sorts to Lemma 1.7.2.

LEMMA 1.8.2. *With notation as in Hypothesis 1.7.1, suppose that (1.6.15.1) is exact at the left and middle (e.g., because A is sheafy). Then multiplication by $f - gT$ defines injective maps $B\langle T \rangle \to B\langle T \rangle, B\langle T^{\pm} \rangle \to B\langle T^{\pm} \rangle$ with closed image.*

PROOF. We treat the case of $B\langle T \rangle$, the case of $B\langle T^{\pm} \rangle$ being similar. We first argue that we may check the claim after replacing B with each of $B\langle \frac{f}{g} \rangle$, $B\langle \frac{g}{f} \rangle$, $B\langle \frac{f}{g}, \frac{g}{f} \rangle$. Given these cases, for $x \in B\langle T \rangle$, we may recover x from $x(f - gT)$ by doing so in each of $B\langle \frac{f}{g}, T \rangle$ and $B\langle \frac{g}{f}, T \rangle$ and noting that the answers must agree in $B\langle \frac{f}{g}, \frac{g}{f}, T \rangle$. Since (1.6.15.1) is strict exact (by our hypothesis plus Lemma 1.8.1 and Theorem 1.1.9), it follows that the map $x(f - gT) \mapsto x$ is continuous, as desired.

Now promote B to a Banach ring as per Remark 1.5.4, and let $\operatorname{Gr} B$ denote the associated graded ring (Definition 1.5.3), so that $\operatorname{Gr} B\langle T \rangle = (\operatorname{Gr} B)[\overline{T}]$ with $\overline{T}$ placed in degree 1. By our initial reduction, we may assume that either $g = 1$ and $|f| \leq 1$, or $f = 1$ and $|g| \leq 1$. (Namely, when passing from B to $B\langle \frac{f}{g} \rangle$ or $B\langle \frac{g}{f} \rangle$, we replace f, g with $\frac{f}{g}, 1$ or with $1, \frac{g}{f}$.) Consequently, we may assume that the image of $g - fT$ in $\operatorname{Gr} B\langle T \rangle$ has the form $\overline{x} = \overline{x}_0 + \overline{x}_1 \overline{T}$ with $1 \in \{\overline{x}_0, \overline{x}_1\}$. It follows easily from this (by examining the effect of multiplication by $\overline{x}$ on constant and leading coefficients; see also Exercise 1.8.5) that $\overline{x}$ is not a zero-divisor in $\operatorname{Gr} B\langle T \rangle$. This in turn implies that multiplication by $g - fT$ defines an isometric (hence strict) inclusion of $B\langle T \rangle$ into itself, thus proving the claim. □

REMARK 1.8.3. Lemma 1.8.2 provides a key special case of Theorem 1.2.7: if either of the ideals $(T - f)$ or $(1 - fT)$ in $A\langle T \rangle$ is not closed for some $f \in A$, then A is not sheafy. This is precisely the mechanism of Mihara's example [**138**, Proposition 3.14]; we have not checked whether the same criterion applies directly to the example of Buzzard–Verberkmoes.

REMARK 1.8.4. At this point, it is tempting to try to emulate the proof of [**72**, Lemma 4.2.5] to show that the sequence (1.6.15.1), if it is exact, also splits in the category of topological B-modules. However, this is not true in general; the proof of [**72**, Lemma 4.2.5] is correspondingly incorrect, although the result stated there does turn out to be correct for other reasons. A related phenomenon is that for R a ring and $f \in R$, in general the exact sequence

$$0 \to R \to R\left[\frac{1}{f}\right] \oplus R\left[\frac{1}{1-f}\right] \to R\left[\frac{1}{f}, \frac{1}{1-f}\right] \to 0$$

does not necessarily split in the category of A-modules.

We mention a purely algebraic fact related to the proof of Lemma 1.8.2.

EXERCISE 1.8.5. Let R be a ring. Let $f \in R[T_1, \ldots, T_n]$ be a polynomial whose coefficients generate the unit ideal in R. Prove that f is not a zero-divisor in $R[T_1, \ldots, T_n]$.

1.9. Proofs: vector bundles and pseudocoherent sheaves. We finally establish the local nature of sheafiness (Theorem 1.2.22), the base change property for pseudocoherent modules (Theorem 1.4.14), the acyclicity of sheaves associated to pseudocoherent modules (Theorem 1.4.16), the glueing theorems for vector bundles (Theorem 1.4.2) and pseudocoherent sheaves (Theorem 1.4.18), and corollaries (Theorem 1.2.7 and Theorem 1.4.20). Throughout §1.9, continue to set notation as in Hypothesis 1.7.1; in order to address Theorem 1.2.22, we refrain from globally assuming that A is sheafy.

In order to treat the two glueing theorems in parallel, we start with the base change and acyclicity arguments. In the pseudocoherent case, we will temporarily work with a formally different definition before reconciling it with the original one.

DEFINITION 1.9.1. As in Remark 1.6.17, we say that a rational localization $(A, A^+) \to (B, B^+)$ is *nice* if it occurs in a covering of $\mathrm{Spa}(A, A^+)$ given by a composition of simple Laurent and balanced coverings. Equivalently, B is obtained from A by repeatedly passing from A to $A\langle \frac{f}{g} \rangle$ or $A\langle \frac{g}{f} \rangle$ where $g \in \{1, 1-f\}$. We say that a rational subspace of X is *nice* if it corresponds to a nice rational localization.

Let $\mathbf{PCoh}_A^{\text{nice}}$ be the category of pseudocoherent A-modules M such that for every nice rational localization $(A, A^+) \to (B, B^+)$, $M \otimes_A B$ is complete for its natural topology as a B-module.

REMARK 1.9.2. We will use in a couple of places the fact that the analogue of the sequence (1.6.15.1) with B, f, g replaced by $B\langle \frac{f}{g} \rangle, \frac{f}{g}, 1$ is the sequence

$$0 \to B\left\langle \frac{f}{g} \right\rangle \to B\left\langle \frac{f}{g} \right\rangle \oplus B\left\langle \frac{f}{g}, \frac{g}{f} \right\rangle \to B\left\langle \frac{f}{g}, \frac{g}{f} \right\rangle \to 0,$$

which is trivially exact.

LEMMA 1.9.3. *With notation as in Hypothesis 1.7.1, suppose that* $g \in \{1, 1-f\}$ *and* (1.6.15.1) *is exact (e.g., because A is sheafy). Let M be a B-module which is finitely presented and complete for the natural topology. Then* $\operatorname{Tor}_1^B(M, B\langle \frac{g}{f} \rangle) = 0$.

PROOF. As in Remark 1.6.17, we may use the equality $B\left\langle \frac{1-f}{f} \right\rangle = B\left\langle \frac{1}{f} \right\rangle$ to reduce to the case $g = 1$. By Lemma 1.1.18, we may identify $M \otimes_B B\langle T \rangle$ with $M\langle T \rangle$. By Lemma 1.8.2, we may identify $B\langle \frac{1}{f} \rangle$ with $B\langle T \rangle/(1-fT)$. Now let $M[\![T]\!]$ be the set of formal sums $\sum_{n=0}^\infty x_n T^n$ with $x_n \in M$ with no convergence condition on the sequence $\{x_n\}$; on $M[\![T]\!]$, multiplication by $1 - fT$ has an inverse given by multiplication by $1 + fT + f^2T^2 + \cdots$. It follows that multiplication by $1 - fT$ on $M\langle T \rangle$ is injective, so $\operatorname{Tor}_1^B(M, B\langle \frac{1}{f} \rangle) = 0$ as desired. □

LEMMA 1.9.4. *With notation as in Hypothesis 1.7.1, suppose that* $g \in \{1, 1-f\}$ *and* (1.6.15.1) *is exact. Let M be a finitely presented B-module such that for any nice rational localization* $(B, B^+) \to (C, C^+)$, $M \otimes_B C$ *is complete for its natural topology as a C-module. Then* $\operatorname{Tor}_1^B(M, B\langle \frac{f}{g} \rangle) = \operatorname{Tor}_1^B(M, B\langle \frac{f}{g}, \frac{g}{f} \rangle) = 0$.

PROOF. Let $0 \to N \to F \to M \to 0$ be a short exact sequence of B-modules with F finite free. (Note that even if $M \in \mathbf{PCoh}_B$, we cannot yet invoke Remark 1.4.9 to assert that $N \in \mathbf{PCoh}_B$.) By Lemma 1.9.3, the sequence

$$0 \to N \otimes_B B\left\langle \frac{g}{f} \right\rangle \to F \otimes_B B\left\langle \frac{g}{f} \right\rangle \to M \otimes_B B\left\langle \frac{g}{f} \right\rangle \to 0$$

is exact. The module $M \otimes_B B\langle \frac{g}{f} \rangle$ over $B\langle \frac{g}{f} \rangle$ is again finitely presented and (by hypothesis) complete for its natural topology. In light of Remark 1.9.2, we may apply Lemma 1.9.3 a second time (replacing f, g with g, f if $g = 1 - f$, or with $1, f^{-1}$ if $g = 1$) to see that

$$0 \to N \otimes_B B\left\langle \frac{f}{g}, \frac{g}{f} \right\rangle \to F \otimes_B B\left\langle \frac{f}{g}, \frac{g}{f} \right\rangle \to M \otimes_B B\left\langle \frac{f}{g}, \frac{g}{f} \right\rangle \to 0$$

is exact. This implies the equality $\operatorname{Tor}_1^B(M, B\langle \frac{f}{g}, \frac{g}{f} \rangle) = 0$; we may thus tensor (1.6.15.1) with M to deduce that $\operatorname{Tor}_1^B(M, B\langle \frac{f}{g} \rangle) = 0$. □

COROLLARY 1.9.5. *Suppose that A is sheafy. Let M be a finitely presented A-module such that for any nice rational localization* $(A, A^+) \to (B, B^+)$, $M \otimes_A B$ *is complete for the natural topology. Then for any nice rational localization* $(A, A^+) \to (B, B^+)$, *we have* $\operatorname{Tor}_1^A(M, B) = 0$.

PROOF. This is immediate from Lemma 1.9.3 and Lemma 1.9.4. □

As a partial result, we now have a variant form of Theorem 1.4.14.

COROLLARY 1.9.6. *Suppose that A is sheafy. Then for every (not necessarily nice) rational localization* $(A, A^+) \to (B, B^+)$ *and every nice rational localization* $(B, B^+) \to (C, C^+)$, *base extension defines an exact functor* $\mathbf{PCoh}_B^{\mathrm{nice}} \to \mathbf{PCoh}_C^{\mathrm{nice}}$.

PROOF. This is immediate from Corollary 1.9.5. □

LEMMA 1.9.7. *With notation as in Hypothesis 1.7.1, suppose that* $g \in \{1, 1-f\}$ *and* (1.6.15.1) *is exact. Let M be a finitely presented B-module such that for any nice rational localization* $(B, B^+) \to (C, C^+)$, $M \otimes_B C$ *is complete for its natural topology as a C-module. Then tensoring* (1.6.15.1) *over* B *with* M *yields another exact sequence.*

PROOF. This is immediate from Lemma 1.9.4. □

This gives us a variant form of Theorem 1.4.16.

COROLLARY 1.9.8. *Suppose that* A *is sheafy and choose* $M \in \mathbf{PCoh}_A^{\mathrm{nice}}$. *Then* $\tilde{M}$ *is acyclic on nice rational subspaces; that is, for any nice rational subspace* $U := \mathrm{Spa}(B, B^+)$ *of* X,
$$H^0(U, \tilde{M}) = M \otimes_A B, \qquad H^i(U, \tilde{M}) = 0 \quad (i > 0).$$

PROOF. By Remark 1.6.15, this follows from Lemma 1.9.7. □

We now begin to work on glueing. The following lemma is essentially [**117**, Lemma 2.7.2]; compare [**72**, Lemma 4.5.3].

LEMMA 1.9.9. *With notation as in Hypothesis* 1.7.1, *there exists a neighborhood* U *of* 0 *in* $B\langle \frac{f}{g}, \frac{g}{f} \rangle$ *such that for every positive integer* n, *every matrix* $V \in \mathrm{GL}_n(B\langle \frac{f}{g}, \frac{g}{f} \rangle)$ *for which* $V - 1$ *has entries in* U *can be factored as* $V_1 \cdot V_2$ *with* $V_1 \in \mathrm{GL}_n(B\langle \frac{f}{g} \rangle)$, $V_2 \in \mathrm{GL}_n(B\langle \frac{g}{f} \rangle)$.

PROOF. This is a direct consequence of Lemma 1.8.1 via a contraction mapping argument. □

The following lemma combines [**117**, Lemma 1.3.8, Lemma 2.7.4], together with some minor modifications to work around the fact that we are not limiting ourselves to simple Laurent coverings. (The relevant special feature of a Laurent covering is that the map $B\langle \frac{1}{f} \rangle \to B\langle f, \frac{1}{f} \rangle$ has dense image.)

LEMMA 1.9.10. *With notation as in Hypothesis* 1.7.1, *let* M_1, M_2, M_{12} *be finitely generated modules over* $B\langle \frac{f}{g} \rangle$, $B\langle \frac{g}{f} \rangle$, $B\langle \frac{f}{g}, \frac{g}{f} \rangle$, *respectively, and let* $\psi_1 : M_1 \otimes_{B\langle \frac{f}{g} \rangle} B\langle \frac{f}{g}, \frac{g}{f} \rangle \to M_{12}$, $\psi_2 : M_2 \otimes_{B\langle \frac{g}{f} \rangle} B\langle \frac{f}{g}, \frac{g}{f} \rangle \to M_{12}$ *be isomorphisms.*

(a) *The map* $\psi : M_1 \oplus M_2 \to M_{12}$ *taking* $(\mathbf{v}, \mathbf{w})$ *to* $\psi_1(\mathbf{v}) - \psi_2(\mathbf{w})$ *is strict surjective.*

(b) *For* $M = \ker(\psi)$, *the induced maps*
$$M \otimes_B B\left\langle \frac{f}{g} \right\rangle \to M_1, \qquad M \otimes_B B\left\langle \frac{g}{f} \right\rangle \to M_2$$
are strict surjective.

PROOF. Let $\mathbf{v}_1, \ldots, \mathbf{v}_n$ and $\mathbf{w}_1, \ldots, \mathbf{w}_n$ be generating sets of M_1 and M_2, respectively, of the same cardinality. We may then choose $n \times n$ matrices V and W over $B\langle \frac{f}{g}, \frac{g}{f} \rangle$ such that $\psi_2(\mathbf{w}_j) = \sum_i V_{ij} \psi_1(\mathbf{v}_i)$ and $\psi_1(\mathbf{v}_j) = \sum_i W_{ij} \psi_2(\mathbf{w}_i)$.

Choose U as in Lemma 1.9.9. Since $B\langle \frac{f}{g} \rangle[f^{-1}]$ is dense in $B\langle \frac{f}{g}, \frac{g}{f} \rangle$, we can choose a nonnegative integer m and an $n \times n$ matrix W' over $B\langle \frac{g}{f} \rangle$ so that $V(f^{-m}W' - W)$ has entries in U. We may thus write $1 + V(f^{-m}W' - W) = X_1 X_2^{-1}$ with $X_1 \in \mathrm{GL}_n(B\langle \frac{f}{g} \rangle)$, $X_2 \in \mathrm{GL}_n(B\langle \frac{g}{f} \rangle)$.

We now define elements $\mathbf{x}_j \in M_1 \oplus M_2$ by the formula
$$\mathbf{x}_j = (\mathbf{x}_{j,1}, \mathbf{x}_{j,2}) = \left(\sum_i f^m (X_1)_{ij} \mathbf{v}_i, \sum_i (W' X_2)_{ij} \mathbf{w}_i \right) \qquad (j = 1, \ldots, n).$$

Then
$$\psi_1(\mathbf{x}_{j,1}) - \psi_2(\mathbf{x}_{j,2}) = \sum_i (f^m X_1 - VW'X_2)_{ij}\psi_1(\mathbf{v}_i)$$
$$= \sum_i f^m((1-VW)X_2)_{ij}\psi_1(\mathbf{v}_i) = 0,$$
so $\mathbf{x}_j \in M$. Since $X_1 \in \mathrm{GL}_n(B\langle \frac{f}{g}\rangle)$, we deduce that the map $M \otimes_B B\langle \frac{f}{g}\rangle \to M_1$ induces a strict surjection onto $f^m M_1$.

The induced map $M \otimes_B B\langle \frac{f}{g}, \frac{g}{f}\rangle \to M_{12}$ is strict surjective (because f is invertible in $B\langle \frac{f}{g}, \frac{g}{f}\rangle$), so using Lemma 1.8.1 we obtain a strict surjection $M \otimes_B (B\langle \frac{f}{g}\rangle \oplus B\langle \frac{g}{f}\rangle) \to M_{12}$. Since this map factors through ψ, we obtain (a).

For each $\mathbf{v} \in M_2$, $\psi_2(\mathbf{v})$ lifts to $M \otimes_B (B\langle \frac{f}{g}\rangle \oplus B\langle \frac{g}{f}\rangle)$ as above, so we can find $\mathbf{w}_1 \in M_1, \mathbf{w}_2 \in M_2$ in the images of the base extension maps from M with $\psi_1(\mathbf{w}_1) - \psi_2(\mathbf{w}_2) = \psi_2(\mathbf{v})$. Then $(\mathbf{w}_1, \mathbf{v} + \mathbf{w}_2) \in M$, so both $\mathbf{w}_2$ and $\mathbf{v} + \mathbf{w}_2$ are elements of M_2 in the image of the base extension map. This proves that $M \otimes_B B\langle \frac{g}{f}\rangle \to M_2$ is strict surjective; we may reverse the roles of f and g to deduce (b). □

LEMMA 1.9.11. *With notation as in Lemma 1.9.10, suppose that (1.6.15.1) is exact and that M_1, M_2, M_{12} are stably pseudocoherent modules over their respective base rings. Then M is a pseudocoherent B-module.*

PROOF. By Lemma 1.9.10, we can choose a finite free B-module F and a (not necessarily surjective) B-linear map $F \to M$ such that for F_1, F_2, F_{12} the respective base extensions of F, the induced maps
$$F_1 \to M_1, \qquad F_2 \to M_2, \qquad F_{12} \to M_{12}$$
are surjective. Let N_1, N_2, N_{12} be the kernels of these maps and put $N = \ker(N_1 \oplus N_2 \to N_{12})$; by applying the snake lemma to the second and third columns in the diagram

(1.9.11.1)
$$\begin{array}{ccccccccc}
 & & 0 & & 0 & & 0 & & \\
 & & \downarrow & & \downarrow & & \downarrow & & \\
0 & \to & N & \to & N_1 \oplus N_2 & \to & N_{12} & \dashrightarrow & 0 \\
 & & \downarrow & & \downarrow & & \downarrow & & \\
0 & \to & F & \to & F_1 \oplus F_2 & \to & F_{12} & \to & 0 \\
 & & \downarrow & & \downarrow & & \downarrow & & \\
0 & \to & M & \to & M_1 \oplus M_2 & \to & M_{12} & \to & 0 \\
 & & \downarrow & & \downarrow & & \downarrow & & \\
 & & 0 & & 0 & & 0 & &
\end{array}$$

we obtain the first column and its exactness (minus the dashed arrows). In particular, $N = \ker(F \to M)$.

By Remark 1.9.2 and Lemma 1.9.4, we have isomorphisms
$$N_1 \otimes_{B\langle \frac{f}{g}\rangle} B\left\langle \frac{f}{g}, \frac{g}{f}\right\rangle \cong N_{12}, \qquad N_2 \otimes_{B\langle \frac{g}{f}\rangle} B\left\langle \frac{f}{g}, \frac{g}{f}\right\rangle \cong N_{12}.$$

By the "two out of three" property of pseudocoherent modules (as in Remark 1.4.9), the modules N_1, N_2, N_{12} again form an instance of the desired result; hence any general statement we can make about M, M_1, M_2, M_{12} also applies to N, N_1, N_2, N_{12}. This means that we may apply Lemma 1.9.10 to see that the dashed arrows in the previous diagram are surjective. Also, in the diagram

$$\begin{array}{ccccccc} N \otimes_B B\left\langle\frac{f}{g}\right\rangle & \longrightarrow & F \otimes_B B\left\langle\frac{f}{g}\right\rangle & \longrightarrow & M \otimes_B B\left\langle\frac{f}{g}\right\rangle & \longrightarrow & 0 \\ \downarrow & & \downarrow & & \downarrow & & \\ 0 \longrightarrow N_1 & & \longrightarrow F_1 & & \longrightarrow M_1 & \longrightarrow & 0 \end{array}$$

with exact rows, we know from Lemma 1.9.10 that *both* outside vertical arrows are surjective. Since the middle arrow is an isomorphism, we may apply the five lemma to obtain injectivity of the right vertical arrow; this (and a similar argument with M_1 replaced with M_2) yields that the maps

(1.9.11.2) $$M \otimes_B B\left\langle\frac{f}{g}\right\rangle \to M_1, \qquad M \otimes_B B\left\langle\frac{g}{f}\right\rangle \to M_2$$

are isomorphisms.

To prove that M is pseudocoherent, it will suffice to prove that for every positive integer m, M admits a projective resolution in which the last m terms are finite projective modules. This holds for $m = 1$ by Lemma 1.9.10 as above; however, given the claim for any given m, it applies not only to M but also to N, which formally implies the claim about M for $m+1$. We may thus conclude by induction on m. $\square$

LEMMA 1.9.12. *With notation as in Hypothesis 1.7.1, the image of the map* $\operatorname{Spec}(B\langle\frac{f}{g}\rangle \oplus B\langle\frac{g}{f}\rangle) \to \operatorname{Spec}(B)$ *includes all maximal ideals.*

PROOF. By Remark 1.1.1 and Corollary 1.5.18, every maximal ideal $\mathfrak{m}$ of B occurs as the kernel of some valuation v. That valuation extends to one of $B\langle\frac{f}{g}\rangle$ or $B\langle\frac{g}{f}\rangle$, and the kernel of that extension is a prime ideal contracting to $\mathfrak{m}$. $\square$

LEMMA 1.9.13. *With notation as in Hypothesis 1.7.1, suppose that A is sheafy.*

(a) *There is an exact functor*

$$\mathbf{FPMod}_{B\langle\frac{f}{g}\rangle} \times_{\mathbf{FPMod}_{B\langle\frac{f}{g},\frac{g}{f}\rangle}} \mathbf{FPMod}_{B\langle\frac{g}{f}\rangle} \to \mathbf{FPMod}_B$$

given by taking equalizers. Moreover, the composition of this functor with the base extension functor in the opposite direction is naturally isomorphic to the identity.

(b) *There is an exact, fully faithful functor*

$$\mathbf{PCoh}_{B\langle\frac{f}{g}\rangle}^{\mathrm{nice}} \times_{\mathbf{PCoh}_{B\langle\frac{f}{g},\frac{g}{f}\rangle}^{\mathrm{nice}}} \mathbf{PCoh}_{B\langle\frac{g}{f}\rangle}^{\mathrm{nice}} \to \mathbf{PCoh}_B^{\mathrm{nice}}$$

given by taking equalizers. Moreover, the composition of this functor with the base extension functor in the opposite direction is well-defined (that is, for $M \in \mathbf{PCoh}_B^{\mathrm{nice}}$ in the essential image, we have $M \otimes_B B\langle*\rangle \in \mathbf{PCoh}_{B\langle*\rangle}^{\mathrm{nice}}$) and naturally isomorphic to the identity.

PROOF. To prove (b), set notation as in Lemma 1.9.11 and its proof; we know already that M is pseudocoherent, and must prove that for any nice rational localization $(B, B^+) \to (C, C^+)$, $M \otimes_B C$ is complete for the natural topology. To this end, perform a base extension on the diagram (1.9.11.1) to get a new diagram

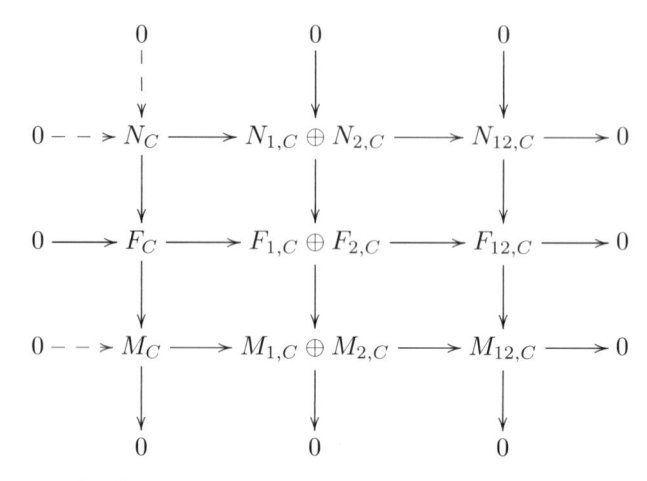

$$\begin{array}{ccccccccc}
 & & 0 & & 0 & & 0 & & \\
 & & \vdots & & \downarrow & & \downarrow & & \\
0 & \dashrightarrow & N_C & \to & N_{1,C} \oplus N_{2,C} & \to & N_{12,C} & \to & 0 \\
 & & \downarrow & & \downarrow & & \downarrow & & \\
0 & \to & F_C & \to & F_{1,C} \oplus F_{2,C} & \to & F_{12,C} & \to & 0 \\
 & & \downarrow & & \downarrow & & \downarrow & & \\
0 & \dashrightarrow & M_C & \to & M_{1,C} \oplus M_{2,C} & \to & M_{12,C} & \to & 0 \\
 & & \downarrow & & \downarrow & & \downarrow & & \\
 & & 0 & & 0 & & 0 & &
\end{array}$$

with exact rows and columns excluding the dashed arrows; here we are using sheafiness in the second row and Corollary 1.9.6 in the second and third column. By diagram chasing, we obtain exactness in the third row; consequently, $M_C = M \otimes_B C$ is the kernel of a strict surjective morphism, and so is complete for the subspace topology and hence (by Corollary 1.1.12) for the natural topology.

To deduce (a), we may first apply (b) to obtain a module $M \in \mathbf{PCoh}_B^{\mathrm{nice}}$. To check that $M \in \mathbf{FPMod}_B$, note that M is finitely presented and $M_{\mathfrak{m}}$ is a finite free $B_{\mathfrak{m}}$-module for every maximal ideal $\mathfrak{m}$ of B (by Lemma 1.9.12); by [**166**, Tag 00NX], M is a finite projective B-module. □

For vector bundles, this glueing is all we need.

PROOF OF THEOREM 1.4.2. By Remark 1.6.16, this reduces immediately to the statement that with notation as in Hypothesis 1.7.1,

$$\mathbf{FPMod}_B \to \mathbf{FPMod}_{B\langle\frac{f}{g}\rangle} \times_{\mathbf{FPMod}_{B\langle\frac{f}{g},\frac{g}{f}\rangle}} \mathbf{FPMod}_{B\langle\frac{g}{f}\rangle}$$

is an exact equivalence of categories. This functor is fully faithful by Theorem 1.3.4, exact trivially, and essentially surjective by Lemma 1.9.13(a). □

For pseudocoherent sheaves, we must work harder to get rid of the distinction between $\mathbf{PCoh}_*$ and $\mathbf{PCoh}_*^{\mathrm{nice}}$.

LEMMA 1.9.14. *The nice rational subspaces form a neighborhood basis of X.*

PROOF. Since this is a local assertion on X, we may assume at once that A is Tate. In this case, we argue as in Remark 1.6.14: given a rational subspace $U = X(\frac{f_1,\ldots,f_n}{g})$, if we choose a topologically nilpotent x of A, then for every sufficiently large positive integer m we have $U \subseteq X(\frac{1}{x^{-m}g})$. We may thus reduce to the case where g is invertible in A, at which point U becomes a nice rational subspace: it is the intersection of $X(\frac{f_i g^{-1}}{1})$ for $i = 1, \ldots, n$. □

DEFINITION 1.9.15. Suppose that A is sheafy. For $M \in \mathbf{PCoh}_A^{\mathrm{nice}}$, let $\tilde{M}^{\mathrm{nice}}$ denote the sheafification of the presheaf $\tilde{M}$. By Corollary 1.9.8, the values of $\tilde{M}$ and $\tilde{M}^{\mathrm{nice}}$ coincide on any nice rational subspace. Let $\mathbf{PCoh}_X^{\mathrm{nice}}$ denote the category of sheaves which are locally of the form $\tilde{M}^{\mathrm{nice}}$; there is an obvious functor $\mathbf{PCoh}_A^{\mathrm{nice}} \to \mathbf{PCoh}_X^{\mathrm{nice}}$ taking M to $\tilde{M}^{\mathrm{nice}}$. For any (not necessarily nice) rational subspace $U = \mathrm{Spa}(B, B^+)$ of X, Lemma 1.9.14 implies that U admits a neighborhood basis consisting of nice rational subspaces of X; consequently, restriction of sheaves defines a functor $\mathbf{PCoh}_X^{\mathrm{nice}} \to \mathbf{PCoh}_Y^{\mathrm{nice}}$.

We obtain a "nice" analogue of Theorem 1.4.18.

LEMMA 1.9.16. *If A is sheafy, then the functor $\mathbf{PCoh}_A^{\mathrm{nice}} \to \mathbf{PCoh}_X^{\mathrm{nice}}$ taking M to $\tilde{M}^{\mathrm{nice}}$ is an exact equivalence of categories, with quasi-inverse taking $\mathcal{F}$ to $\mathcal{F}(X)$.*

PROOF. By Remark 1.6.16, this reduces immediately to the statement that with notation as in Hypothesis 1.7.1 with $g \in \{1, 1-f\}$, there is an exact equivalence of categories

$$(1.9.16.1) \qquad \mathbf{PCoh}_B^{\mathrm{nice}} \to \mathbf{PCoh}_{B\langle \frac{f}{g} \rangle}^{\mathrm{nice}} \times_{\mathbf{PCoh}_{B\langle \frac{f}{g}, \frac{g}{f} \rangle}^{\mathrm{nice}}} \mathbf{PCoh}_{B\langle \frac{g}{f} \rangle}^{\mathrm{nice}}.$$

To begin with, by Corollary 1.9.6, we have exact base extension functors from $\mathbf{PCoh}_B^{\mathrm{nice}}$ to each of $\mathbf{PCoh}_{B\langle \frac{f}{g} \rangle}^{\mathrm{nice}}$, $\mathbf{PCoh}_{B\langle \frac{g}{f} \rangle}^{\mathrm{nice}}$, $\mathbf{PCoh}_{B\langle \frac{f}{g}, \frac{g}{f} \rangle}^{\mathrm{nice}}$. This yields an exact functor as in (1.9.16.1); this functor is fully faithful by Lemma 1.9.7 and essentially surjective by Lemma 1.9.13(b). □

LEMMA 1.9.17. *If A is sheafy, then for any (not necessarily nice) rational localization $(A, A^+) \to (B, B^+)$, base extension from A to B defines an exact functor $\mathbf{PCoh}_A^{\mathrm{nice}} \to \mathbf{PCoh}_B^{\mathrm{nice}}$.*

PROOF. Put $U := \mathrm{Spa}(B, B^+)$, choose $M \in \mathbf{PCoh}_A^{\mathrm{nice}}$, choose a surjective morphism $F \to M$ with F a finite free A-module, and put $N := \ker(F \to M)$; note that $N \in \mathbf{PCoh}_A^{\mathrm{nice}}$, so as in the proof of Lemma 1.9.13 every statement we prove about M will automatically hold for N also. By Lemma 1.9.16, the exact sequence

$$0 \to N \to F \to M \to 0$$

corresponds to an exact sequence of sheaves on X

$$0 \to \tilde{N}^{\mathrm{nice}} \to \tilde{F}^{\mathrm{nice}} \to \tilde{M}^{\mathrm{nice}} \to 0$$

which we may then restrict to U. By Corollary 1.9.8, taking sections over U yields an exact sequence of B-modules; that is, in the commuting diagram

$$\begin{array}{ccccccc}
N \otimes_A B & \longrightarrow & F \otimes_A B & \longrightarrow & M \otimes_A B & \longrightarrow & 0 \\
\downarrow & & \downarrow & & \downarrow & & \\
0 \longrightarrow H^0(U, \tilde{N}^{\mathrm{nice}}) & \longrightarrow & H^0(U, \tilde{F}^{\mathrm{nice}}) & \longrightarrow & H^0(U, \tilde{M}^{\mathrm{nice}}) & \longrightarrow & 0
\end{array}$$

the rows are exact. By Theorem 1.3.4, the middle vertical arrow is an isomorphism, so the right vertical arrow is surjective. The same is then true with M replaced by N, that is, the left vertical arrow is surjective; by the five lemma, the right vertical arrow is an isomorphism. That is, the base extension functor can be written as a composite of exact functors by going around the 2-commuting diagram

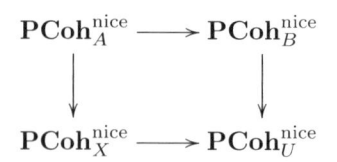

$$\begin{array}{ccc} \mathbf{PCoh}_A^{\mathrm{nice}} & \longrightarrow & \mathbf{PCoh}_B^{\mathrm{nice}} \\ \downarrow & & \downarrow \\ \mathbf{PCoh}_X^{\mathrm{nice}} & \longrightarrow & \mathbf{PCoh}_U^{\mathrm{nice}} \end{array}$$

in which the bottom horizontal arrow is exact, while the vertical arrows are exact equivalences by Lemma 1.9.16. This proves the claim. □

At last, we may now eliminate the "nice" qualifier and establish the results we really want.

COROLLARY 1.9.18. *The inclusion* $\mathbf{PCoh}_A \to \mathbf{PCoh}_A^{\mathrm{nice}}$ *of categories is an equality.*

PROOF. By Lemma 1.9.17, for every $M \in \mathbf{PCoh}_A^{\mathrm{nice}}$ and every rational localization $(A, A^+) \to (B, B^+)$, $M \otimes_A B \in \mathbf{PCoh}_B^{\mathrm{nice}}$; in particular, $M \otimes_A B$ is complete for the natural topology. Since M is already assumed to be pseudocoherent, it is now stably pseudocoherent. □

PROOF OF THEOREM 1.4.14. This follows by combining Lemma 1.9.17 with Corollary 1.9.18. □

PROOF OF THEOREM 1.4.16. We must check that for $M \in \mathbf{PCoh}_A$, $(A, A^+) \to (B, B^+)$ a rational localization, and $U := \mathrm{Spa}(B, B^+)$, we have $H^0(U, \tilde{M}) = M \otimes_A B$ and $H^i(U, \tilde{M}) = 0$ for $i > 0$. By Theorem 1.4.14, we have $M \otimes_A B \in \mathbf{PCoh}_B$. Since $\tilde{M}^{\mathrm{nice}}$ is defined as the sheafification of $\tilde{M}$, by convention it has the same cohomology as $\tilde{M}$. From the proof of Lemma 1.9.17, we see that $H^0(U, \tilde{M}) = M \otimes_A B$. Since $\tilde{M}|_B \in \mathbf{PCoh}_U$, Corollary 1.9.8 implies that $H^i(U, \tilde{M}) = 0$ for $i > 0$. □

PROOF OF THEOREM 1.4.18. By Corollary 1.9.18, $\mathbf{PCoh}_A = \mathbf{PCoh}_A^{\mathrm{nice}}$ and $\mathbf{PCoh}_X = \mathbf{PCoh}_X^{\mathrm{nice}}$. Consequently, the desired result follows immediately from Lemma 1.9.16. □

REMARK 1.9.19. One source of inspiration for our glueing arguments is the Beauville–Laszlo theorem [11], [166, Tag 0BNI], [15], which asserts (among other things) that if R is a ring, $f \in R$ is not a zero-divisor, and $\widehat{R}$ is the f-adic completion of R, then the functor

$$\mathbf{FPMod}_R \to \mathbf{FPMod}_{R_f} \times_{\mathbf{FPMod}_{\widehat{R}_f}} \mathbf{FPMod}_{\widehat{R}}$$

is an equivalence of categories. See [117, Remark 2.7.9] for further explanation of how this result can be derived in the style of the arguments given above.

Another similarity that should be noted is that the Beauville–Laszlo theorem was originally introduced in order to construct and study affine Grassmannians associated to algebraic groups in the context of geometric Langlands. The glueing results discussed here are themselves relevant for the construction and study of certain mixed-characteristic analogues of affine Grassmannians, to be introduced in a later lecture (§4.6).

REMARK 1.9.20. One difficulty in applying the previous results is the fact that if A is sheafy, it does not follow that $A\langle T_1, \ldots, T_n\rangle$ is sheafy. It may be useful to focus attention on rings for which this does hold (see Definition A.5.1); here, we instead establish some variants of the previous results incorporating extra series variables.

DEFINITION 1.9.21. For n a nonnegative integer, let $\mathcal{O}\langle T_1, \ldots, T_n\rangle$ be the presheaf on X defined as follows: for $U \subseteq X$ open, let $\mathcal{O}\langle T_1, \ldots, T_n\rangle$ be the inverse limit of $B\langle T_1, \ldots, T_n\rangle$ over all rational localizations $(A, A^+) \to (B, B^+)$ with $\mathrm{Spa}(B, B^+) \subseteq U$. By Theorem 1.3.4 and Lemma 1.6.3, $\mathcal{O}\langle T_1, \ldots, T_n\rangle$ is an acyclic sheaf.

Define the category $\mathbf{PCoh}_A\langle T_1, \ldots, T_n\rangle$ to be the category of pseudocoherent $A\langle T_1, \ldots, T_n\rangle$-modules M such that for each rational localization $(A, A^+) \to (B, B^+)$, $M \otimes_{A\langle T_1, \ldots, T_n\rangle} B\langle T_1, \ldots, T_n\rangle$ is complete for its natural topology as a $B\langle T_1, \ldots, T_n\rangle$-module. (By contrast, the definition of $\mathbf{PCoh}_{A\langle T_1, \ldots, T_n\rangle}$ is quantified over rational localizations of the ring $A\langle T_1, \ldots, T_n\rangle$, and hence is more restrictive.) For such a module, let $\tilde{M}$ denote the associated presheaf on X (whose definition we leave to the reader).

THEOREM 1.9.22. *Suppose that A is sheafy and let n be a nonnegative integer.*
  (a) *For any rational localization $(A, A^+) \to (B, B^+)$, base extension defines an exact functor $\mathbf{PCoh}_A\langle T_1, \ldots, T_n\rangle \to \mathbf{PCoh}_B\langle T_1, \ldots, T_n\rangle$.*
  (b) *For any $M \in \mathbf{PCoh}_A\langle T_1, \ldots, T_n\rangle$, the presheaf $\tilde{M}$ on X is an acyclic sheaf.*
  (c) *The functor $M \mapsto \tilde{M}$ defines an equivalence between $\mathbf{PCoh}_A\langle T_1, \ldots, T_n\rangle$ and the category of sheaves of $\mathcal{O}_X$-modules locally of this form, with a quasi-inverse given by the global sections functor.*

PROOF. This is a straightforward variation on the proofs of Theorem 1.4.14, Theorem 1.4.16, and Theorem 1.4.18. We leave the details to the reader. □

With Theorem 1.9.22 in hand, we may complete the proof of Theorem 1.2.7.

LEMMA 1.9.23. *Suppose that A is sheafy. Then for any rational localization $(A, A^+) \to (B, B^+)$ and any surjective homomorphism $h : A\langle T_1, \ldots, T_n\rangle \to B$ with kernel I, we have $B \in \mathbf{PCoh}_A\langle T_1, \ldots, T_n\rangle$.*

PROOF. Choose parameters $f_1, \ldots, f_m, g$ defining the rational localization. (Note that we are not assuming that $m = n$ or that h is the surjection induced by these parameters.) By Theorem 1.9.22, we may prove the claim locally on X; using the standard rational covering defined by $f_1, \ldots, f_m, g$, we reduce to the situation where one of $f_1, \ldots, f_m, g$ is itself a unit. In this case $(A, A^+) \to (B, B^+)$ factors as a composition of rational localizations, each of which occurs in a simple Laurent covering:
- if g is a unit, then the rational subspace is defined by the conditions $v(f_1/g) \leq 1, \ldots, v(f_n/g) \leq 1$;
- if f_1 is a unit, then after imposing the condition $v(g/f_1) \geq 1$, g becomes a unit and we may continue as in the previous case.

We are thus reduced to checking the claim in case $m = 1$ and $(f_1, g) \in \{(f, 1), (1, f)\}$ for some $f \in A$. In these cases, we first identify B with the quotient of $A\langle U\rangle$ generated by the closure of the ideal I generated by $U - f$ or $1 - fU$, respectively, and check that $B \in \mathbf{PCoh}_A\langle U \rangle$. By Lemma 1.8.2, multiplication by $U - f$ or $1 - fU$ on $A\langle U \rangle$ is a strict inclusion, so B is pseudocoherent as an $A\langle U \rangle$-module. By the same token, for any rational localization $(A, A^+) \to (C, C^+)$, $(B, B^+) \to (B\widehat{\otimes}_A C, (B\widehat{\otimes}_A C)^+)$ is the corresponding localization of (B, B^+), so $B\widehat{\otimes}_A C$ is the quotient of $C\langle U \rangle$ by the principal ideal generated by either $f - U$ or $1 - fU$. In particular, this quotient equals $B \otimes_{A\langle U \rangle} C\langle U \rangle$, so the latter is complete for the natural topology. We deduce that $B \in \mathbf{PCoh}_A\langle U \rangle$.

We finally check the claim about the original h (still assuming that $m = 1$ and $(f_1, g) \in \{(f, 1), (1, f)\}$). Choose lifts $y_1, \ldots, y_n$ of $g(T_1), \ldots, g(T_n)$ to $A\langle U \rangle$. By considering the morphism $A\langle T_1, \ldots, T_n, U \rangle \to A\langle U \rangle$ taking T_i to y_i, we have $A\langle U \rangle \in \mathbf{PCoh}_A\langle T_1, \ldots, T_n, U \rangle$ and hence $B \in \mathbf{PCoh}_A\langle T_1, \ldots, T_n, U \rangle$. Now choose a lift z to $A\langle T_1, \ldots, T_n \rangle$ of the image of U in B; we may then view $A\langle T_1, \ldots, T_n \rangle$ as a quotient of $A\langle T_1, \ldots, T_n, U \rangle$ via the map $U \mapsto z$, and then conclude that $B \in \mathbf{PCoh}_A\langle T_1, \ldots, T_n \rangle$. $\square$

PROOF OF THEOREM 1.2.22. There is no harm in refining the covering $\mathfrak{V}$; by Lemma 1.6.12, we may reduce to the case where $\mathfrak{V}$ is the simple binary rational covering generated by some $f, g \in A$. By hypothesis, the sequence

$$0 \to \tilde{A} \to A\left\langle \frac{f}{g} \right\rangle \oplus A\left\langle \frac{g}{f} \right\rangle \to A\left\langle \frac{f}{g}, \frac{g}{f} \right\rangle \to 0$$

is strict exact, as then is

$$0 \to \tilde{A}\langle T \rangle \to A\left\langle \frac{f}{g}, T \right\rangle \oplus A\left\langle \frac{g}{f}, T \right\rangle \to A\left\langle \frac{f}{g}, \frac{g}{f}, T \right\rangle \to 0.$$

Using Lemma 1.8.2, we obtain a commutative diagram

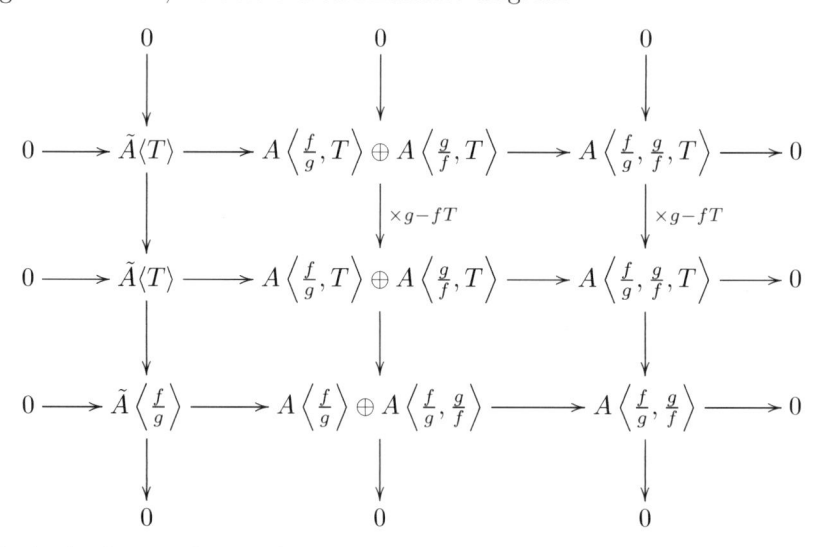

$$\begin{array}{ccccccccc}
& & 0 & & 0 & & 0 & & \\
& & \downarrow & & \downarrow & & \downarrow & & \\
0 & \to & \tilde{A}\langle T \rangle & \to & A\langle \tfrac{f}{g}, T\rangle \oplus A\langle \tfrac{g}{f}, T\rangle & \to & A\langle \tfrac{f}{g}, \tfrac{g}{f}, T\rangle & \to & 0 \\
& & \downarrow & & \downarrow \scriptstyle{\times g - fT} & & \downarrow \scriptstyle{\times g - fT} & & \\
0 & \to & \tilde{A}\langle T \rangle & \to & A\langle \tfrac{f}{g}, T\rangle \oplus A\langle \tfrac{g}{f}, T\rangle & \to & A\langle \tfrac{f}{g}, \tfrac{g}{f}, T\rangle & \to & 0 \\
& & \downarrow & & \downarrow & & \downarrow & & \\
0 & \to & \tilde{A}\langle \tfrac{f}{g}\rangle & \to & A\langle \tfrac{f}{g}\rangle \oplus A\langle \tfrac{f}{g}, \tfrac{g}{f}\rangle & \to & A\langle \tfrac{f}{g}, \tfrac{g}{f}\rangle & \to & 0 \\
& & \downarrow & & \downarrow & & \downarrow & & \\
& & 0 & & 0 & & 0 & &
\end{array}$$

in which the first and second rows are exact, the second and third columns are exact, and the left column is exact at the top and bottom (but not *a priori* in the middle). By diagram chasing, the third row is exact. From this (and the analogous argument with f, g interchanged), we obtain natural isomorphisms

$$A\left\langle \frac{f}{g} \right\rangle \cong \tilde{A}\left\langle \frac{f}{g} \right\rangle, \quad A\left\langle \frac{g}{f} \right\rangle \cong \tilde{A}\left\langle \frac{g}{f} \right\rangle, \quad A\left\langle \frac{f}{g}, \frac{g}{f} \right\rangle \cong \tilde{A}\left\langle \frac{f}{g}, \frac{g}{f} \right\rangle;$$

from this we deduce (a). Note that we cannot say that any rational subspace of $\mathrm{Spa}(\tilde{A}, \tilde{A}^+)$ arises by pullback from $\mathrm{Spa}(A, A^+)$, since such a subspace is defined by parameters in $\tilde{A}$ which we cannot immediately replace with parameters in A.

In light of the previous arguments, to finish the proof of both (a) and (b), it now suffices to check (b) in the case where $\tilde{A} = A$. To this end, let $(A, A^+) \to (B, B^+)$ be the rational localization defined by the parameters $h_1, \ldots, h_n, k \in A$, and identify B with the quotient of $A\langle T_1, \ldots, T_n\rangle$ by the closure of the ideal generated by $kT_1 - h_1, \ldots, kT_n - h_n$. By Lemma 1.9.23, for each nonempty subset $*$ of $\{\frac{f}{g}, \frac{g}{f}\}$, we obtain an exact sequence in $\mathbf{PCoh}_{A\langle * \rangle}\langle T_1, \ldots, T_n\rangle$ of the form

$$(1.9.23.1) \qquad A\langle *, T_1, \ldots, T_n\rangle^n \to A\langle *, T_1, \ldots, T_n\rangle \to B\langle *\rangle \to 0$$

in which the first map takes the generators to $kT_1 - h_1, \ldots, kT_n - h_n$.

Although we do not presently know that either A or $A\langle T_1, \ldots, T_n\rangle$ is sheafy, we do have the exact sequence

$$0 \to A\langle T_1, \ldots, T_n\rangle \to A\left\langle \frac{f}{g}, T_1, \ldots, T_n\right\rangle \oplus A\left\langle \frac{g}{f}, T_1, \ldots, T_n\right\rangle$$
$$\to A\left\langle \frac{f}{g}, \frac{g}{f}, T_1, \ldots, T_n\right\rangle \to 0.$$

We may thus apply Lemma 1.9.11 to establish B is a pseudocoherent $A\langle T_1, \ldots, T_n\rangle$-module, and in particular finitely presented; hence the closure of the ideal of $A\langle T_1, \ldots, T_n\rangle$ generated by $kT_1 - h_1, \ldots, kT_n - h_n$ is finitely generated. By Corollary 1.1.14, the ideal $(kT_1 - h_1, \ldots, kT_n - h_n)$ is itself closed; it follows that

$$B = \ker\left(B\left\langle \frac{f}{g}\right\rangle \oplus B\left\langle \frac{g}{f}\right\rangle \to B\left\langle \frac{f}{g}, \frac{g}{f}\right\rangle\right) = H^0(\mathrm{Spa}(B, B^+), \mathcal{O}),$$

as desired. $\square$

LEMMA 1.9.24. *Let I be a (not necessarily closed) ideal of A. For any $f_1, \ldots, f_n \in A$ which generate the unit ideal in A/I, there exists a rational localization $(A, A^+) \to (B, B^+)$ such that $\mathrm{Spa}(B, B^+)$ contains the zero locus of I on $\mathrm{Spa}(A, A^+)$ and $f_1, \ldots, f_n$ generate the unit ideal in B.*

PROOF. By hypothesis, there exist $b_1, \ldots, b_n \in A$ such that $a_1 b_1 + \cdots + a_n b_n \equiv 1 \pmod{I}$. The rational localization corresponding to the subspace $\{v \in \mathrm{Spa}(A, A^+) : v(a_1 b_1 + \cdots + a_n b_n) \geq 1\}$ has the desired property. $\square$

LEMMA 1.9.25. *Let I be a closed ideal of A such that $A/I \in \mathbf{PCoh}_A$. Let $(A, A^+) \to (B, B^+)$ be a rational localization.*

(a) *Put $\overline{A} := A/I$ and let $\overline{A}^+$ be the integral closure of the image of A^+ in $\overline{A}$. Let $(\overline{A}, \overline{A}^+) \to (\overline{B}, \overline{B}^+)$ be the base extension of the given rational localization. Then $\overline{B} \cong B/IB$.*
(b) *Suppose that $\mathrm{Spa}(B, B^+)$ contains the zero locus of I on $\mathrm{Spa}(A, A^+)$. Then $A/IA \cong B/IB$.*

PROOF. By Theorem 1.4.14, the sequence

$$0 \to I \to A \to A/I \to 0$$

remains exact upon tensoring over A with B. In particular, $IB \cong I \otimes B \in \mathbf{PCoh}_B$ is a closed ideal, so B/IB coincides with the completed tensor product $B \widehat{\otimes}_A A/I$; this proves (a). From (a), (b) is obvious. $\square$

PROOF OF THEOREM 1.4.20. If A/I is sheafy, then Lemma 1.9.23 immediately implies that $A/I \in \mathbf{PCoh}_A$. We thus concentrate on the reverse implication.

Suppose that $A/I \in \mathbf{PCoh}_A$. Put $\overline{A} := A/I$ and let $\overline{A}^+$ be the integral closure of the image of A^+ in $\overline{A}$. By Remark 1.6.15, it suffices to check that for any rational localization $(\overline{A}, \overline{A}^+) \to (\overline{B}, \overline{B}^+)$ and any $\overline{f}, \overline{g} \in \overline{B}$ with $\overline{g} \in \{1, 1 - \overline{f}\}$, the sequence

$$0 \to \overline{B} \to \overline{B}\left\langle \frac{\overline{f}}{\overline{g}} \right\rangle \oplus \overline{B}\left\langle \frac{\overline{g}}{\overline{f}} \right\rangle \to B\left\langle \frac{\overline{f}}{\overline{g}}, \frac{\overline{g}}{\overline{f}} \right\rangle \to 0$$

is exact. By Lemma 1.9.25, there is no harm in replacing (A, A^+) with the rational localization corresponding to a subspace containing the zero locus of I on X. By Lemma 1.9.24, we may thus ensure that some set of parameters in $\overline{A}$ defining the rational localization $(\overline{A}, \overline{A}^+) \to (\overline{B}, \overline{B}^+)$ lift to elements of A which generate the unit ideal. By Lemma 1.9.25 again, there is no harm in replacing (A, A^+) by the corresponding rational localization; that is, we may assume that $(\overline{A}, \overline{A}^+) = (\overline{B}, \overline{B}^+)$.

Lift $\overline{f}, \overline{g}$ to $f, g \in A$ with $g \in \{1, 1 - f\}$. In the diagram

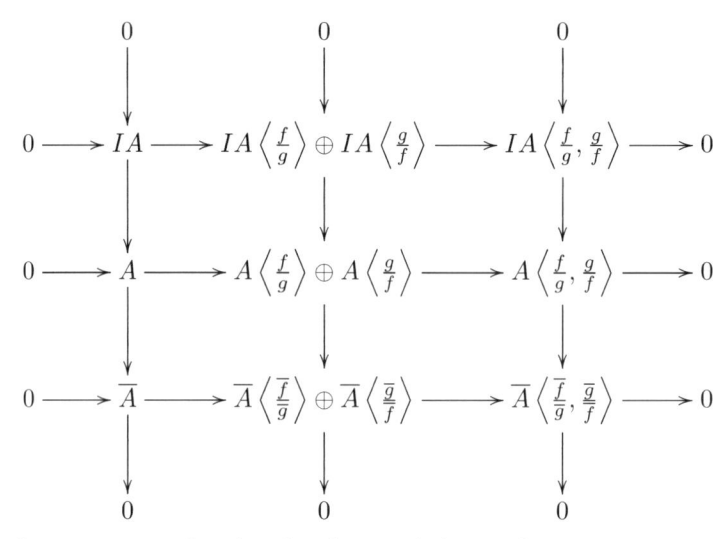

$$\begin{array}{ccccccccc}
& & 0 & & 0 & & 0 & & \\
& & \downarrow & & \downarrow & & \downarrow & & \\
0 & \to & IA & \to & IA\left\langle \frac{f}{g} \right\rangle \oplus IA\left\langle \frac{g}{f} \right\rangle & \to & IA\left\langle \frac{f}{g}, \frac{g}{f} \right\rangle & \to & 0 \\
& & \downarrow & & \downarrow & & \downarrow & & \\
0 & \to & A & \to & A\left\langle \frac{f}{g} \right\rangle \oplus A\left\langle \frac{g}{f} \right\rangle & \to & A\left\langle \frac{f}{g}, \frac{g}{f} \right\rangle & \to & 0 \\
& & \downarrow & & \downarrow & & \downarrow & & \\
0 & \to & \overline{A} & \to & \overline{A}\left\langle \frac{\overline{f}}{\overline{g}} \right\rangle \oplus \overline{A}\left\langle \frac{\overline{g}}{\overline{f}} \right\rangle & \to & \overline{A}\left\langle \frac{\overline{f}}{\overline{g}}, \frac{\overline{g}}{\overline{f}} \right\rangle & \to & 0 \\
& & \downarrow & & \downarrow & & \downarrow & & \\
& & 0 & & 0 & & 0 & &
\end{array}$$

the second row is exact by the sheafiness of A, the first row is exact by Theorem 1.4.14, the first column is exact by definition, and the second and third columns are exact by Lemma 1.9.25. This implies exactness of the third row, as needed. □

EXERCISE 1.9.26. Suppose that A is sheafy. Let $(A, A^+) \to (B, B^+)$ be the rational localization defined by the parameters $f_1, \ldots, f_n, g$. Show that the Koszul complex corresponding to the elements $f_1 - gT_1, \ldots, f_n - gT_n$ in $A\langle T_1, \ldots, T_n\rangle$ is quasi-isomorphic to the singleton complex $A\langle \frac{f_1}{g}, \ldots, \frac{f_n}{g}\rangle$. (Compare the case $n = 1$ with Lemma 1.8.2.)

1.10. Remarks on the étale topology. As promised at the end of §1.4, we include some remarks about the étale topology on X.

LEMMA 1.10.1. *Suppose that A is uniform. Then every finite étale A-algebra is uniform for its natural topology as an A-module.*

PROOF. In the case where A is Tate, this is [**117**, Proposition 2.8.16]. The analytic case can be treated similarly, but can also be handled as follows. Extend

A to a Huber pair (A, A^+), let B be a finite étale A-algebra of constant rank (it suffices to treat this case), and let B^+ be the integral closure of A^+ in B. Since B is finitely generated as an A-module and the unit ideal of A is generated by topologically nilpotent elements, we can find $b_1, \ldots, b_n \in B^+$ which generate B as an A-module. We now see B^+ is contained in the set

$$\{b \in B : \mathrm{Trace}_{B/A}(bb_1), \ldots, \mathrm{Trace}_{B/A}(bb_n) \in A^+\},$$

which is bounded (because the trace pairing is nondegenerate); hence B is uniform. □

REMARK 1.10.2. Let B be a finite étale A-algebra and let B^+ be the integral closure of A^+ in B. If A is strongly noetherian, then so is B, so there are no technical issues with considering the étale site of X. On the other hand, if A is only sheafy, or even stably uniform, then we cannot immediately infer the same about B: we have the sheaf axiom for coverings of $\mathrm{Spa}(B, B^+)$ arising by pullback from X, but any covering that separates points within some fiber of the projection $\mathrm{Spa}(B, B^+) \to X$ will fail to be refined by such a covering. (That said, we do not have a counterexample in mind.)

In light of the previous remark, we make the following hypothesis.

HYPOTHESIS 1.10.3. For the remainder of §1.10, let X_{et} be the site whose morphisms are compositions of rational localizations and finite étale morphisms. (Note that this gives the "right" definition of the étale site for analytic adic spaces, but not for schemes.) Assume that X_{et} admits a basis $\mathcal{B}$ closed under formation of rational localizations and finite étale covers, and consisting of subspaces of the form $\mathrm{Spa}(B, B^+)$ where B is sheafy. For example, this hypothesis is satisfied if A is perfectoid, or even sousperfectoid (see Remark 1.2.19).

For the étale topology, one has the following analogue of Lemma 1.6.18; however, the proof is somewhat less straightforward, and uses a method introduced by de Jong–van der Put [**38**, Proposition 3.2.2].

LEMMA 1.10.4. *Let $\mathcal{P}_{\mathrm{et}}$ be the collection*[6] *of pairs $(U, \mathfrak{V})$ where $U \in \mathcal{B}$ and $\mathfrak{V}$ is a finite covering of U in X_{et} by elements of $\mathcal{B}$. Suppose that $\mathcal{P} \subseteq \mathcal{P}_{\mathrm{et}}$ satisfies the following conditions.*

   (i) *Locality: if $(U, \mathfrak{V})$ admits a refinement in $\mathcal{P}$, then $(U, \mathfrak{V}) \in \mathcal{P}$.*
   (ii) *Transitivity: Any composition of coverings in $\mathcal{P}$ is in $\mathcal{P}$.*
   (iii) *Every standard binary rational covering is in $\mathcal{P}$.*
   (iv) *Every finite étale surjective morphism, viewed as a covering, is in $\mathcal{P}$.*

Then $\mathcal{P} = \mathcal{P}_{\mathrm{et}}$.

PROOF. See [**117**, Proposition 8.2.20]. □

REMARK 1.10.5. Using Lemma 1.10.4, it is straightforward to extend Theorems 1.3.4, and 1.4.2 to X_{et}. One can also extend 1.4.14, 1.4.16, and 1.4.18 to X_{et} provided that one modifies the definition of a stably pseudocoherent module to quantify over étale localizations rather than rational localizations. We leave the details to the reader (or see [**118**, §2.5]).

---

[6]This is not a set in general, but a proper class.

1.11. Preadic spaces.

We end with a remark about how to formally build "spaces" out of Huber pairs even when they are not sheafy. This discussion is taken from [**161**, §2], although our terminology[7] instead follows [**117**, §8.2]. (To keep within our global context, we build only *analytic* preadic spaces here.)

DEFINITION 1.11.1. For $\mathcal{C}$ a category equipped with a Grothendieck topology (so in particular admitting fiber products), a *sheaf* on $\mathcal{C}$ is by definition a contravariant functor F from $\mathcal{C}$ to some target category satisfying the sheaf axiom: for $\{U_i \to X\}_i$ a covering in $\mathcal{C}$, the sequence

$$F(X) \to \prod_i F(U_i) \rightrightarrows \prod_{i,j} F(U_i \times_X U_j)$$

is an equalizer.

We say that $\mathcal{C}$ is *subcanonical* if for each $X \in \mathcal{C}$, the representable functor $h_X : Y \mapsto \mathrm{Hom}_{\mathcal{C}}(Y, X)$ is a sheaf. For example, the Zariski topology on schemes is subcanonical: this reduces immediately to the corresponding statement for affine schemes, which asserts that if R and S are rings and $f_1, \ldots, f_n \in R$ generate the unit ideal in R, then the diagram

$$\mathrm{Hom}(S, R) \to \prod_i \mathrm{Hom}(S, R_{f_i}) \rightrightarrows \prod_{i,j} \mathrm{Hom}(S, R_{f_i f_j})$$

is an equalizer; this follows from the sheaf axiom for the structure sheaf. By the same reasoning, the étale, fppf, and fpqc topologies on the category of schemes are subcanonical.

Similarly, if $\mathcal{C}$ is a subcategory of the category of adic spaces admitting fiber products (e.g., locally noetherian spaces, or perfectoid spaces), the analytic topology on $\mathcal{C}$ (i.e., the site coming from the underlying topology on underlying spaces) is subcanonical. By contrast, the analytic topology on the full category of Huber pairs is not subcanonical.

DEFINITION 1.11.2. Let $\mathcal{C}$ be the opposite category of (analytic but not necessarily sheafy) Huber pairs, equipped with the analytic topology. Let $\mathcal{C}^{\sim}$ be the associated topos. For $(A, A^+) \in \mathcal{C}$, let $\widetilde{\mathrm{Spa}}(A, A^+) \in \mathcal{C}^{\sim}$ be the sheafification of the representable functor on $\mathcal{C}$ defined by (A, A^+).

By an *open immersion* in $\mathcal{C}^{\sim}$, we will mean a morphism $f : \mathcal{F} \to \mathcal{G}$ such that for every $(A, A^+) \in \mathcal{C}$ and every morphism $\widetilde{\mathrm{Spa}}(A, A^+) \to \mathcal{G}$ in $\mathcal{C}^{\sim}$, there is an open subset U of $\mathrm{Spa}(A, A^+)$ such that

$$\mathcal{F} \times_{\mathcal{G}} \widetilde{\mathrm{Spa}}(A, A^+) = \varinjlim_{V \subseteq U, V \text{ rational}} \widetilde{\mathrm{Spa}}(\mathcal{O}_{\mathrm{Spa}(A, A^+)}(V), \mathcal{O}^+_{\mathrm{Spa}(A, A^+)}(V)).$$

A *preadic*[8] space, or more precisely an *analytic preadic space*, is an object $\mathcal{F} \in \mathcal{C}^{\sim}$ such that

$$\mathcal{F} = \varinjlim_{\widetilde{\mathrm{Spa}}(A, A^+) \to \mathcal{F} \text{ open}} \widetilde{\mathrm{Spa}}(A, A^+).$$

The obvious functor from analytic adic spaces to analytic preadic spaces is a full embedding.

---

[7]In [**161**], what we call *preadic spaces* and *adic spaces* are called *adic spaces* and *honest adic spaces*, respectively. We prefer to leave the term *adic spaces* with the original meaning specified by Huber.

[8]Or *pre-adic space* if you prefer to hyphenate prefixes.

A morphism $f : \mathcal{F} \to \mathcal{G}$ is *finite étale* if for every $(A, A^+) \in \mathcal{C}$ and every morphism $\widetilde{\mathrm{Spa}}(A, A^+) \to \mathcal{G}$ in $\mathcal{C}^\sim$, we have $\mathcal{F} \times_\mathcal{G} \widetilde{\mathrm{Spa}}(A, A^+) \cong \widetilde{\mathrm{Spa}}(B, B^+)$ for some finite étale morphism $(A, A^+) \to (B, B^+)$. Using finite étale morphisms and open immersions, we may define the *etale topology* on analytic preadic spaces.

REMARK 1.11.3. Similar sheaf-theoretic considerations give rise to other types of objects that one might think of as analytic analogues of *algebraic spaces* (or more generally *algebraic stacks*). Notably, they underlie the construction of *diamonds*, which are introduced in [177, Lecture 3] and used in our subsequent lectures (see Definition 4.3.1 and beyond).

2. Perfectoid rings and spaces

In this lecture, we define perfectoid rings and spaces, picking up from the discussion of perfectoid fields in [177, Lecture 2]. As for fields, there is a "tilting" construction that converts these spaces into related objects in characteristic p; however, this time the "Galois" component of the tilting correspondence is augmented by a "spatial" component.

As in the first lecture, we state all of the main results first, then return to the proofs. Along the way, we attempt to lay flat some of the tangled history[9] surrounding these results. (See also the introduction of [117] for a further comparison of that paper with [156].)

Hereafter, we fix a prime number p and consider only Huber rings in which p is topologically nilpotent.

2.1. Perfectoid rings and pairs. We now define perfectoid rings and pairs, postponing some proofs for the time being.

DEFINITION 2.1.1. Let (A, A^+) be a uniform analytic Huber pair; by uniformity, A^+ is a ring of definition of A. We say (A, A^+) is *perfectoid* if there exists an ideal of definition $I \subseteq A^+$ such that $p \in I^p$ and $\varphi : A^+/I \to A^+/I^p$ is surjective (but not necessarily injective; see Remark 2.3.16). This turns out to depend only on A (Corollary 2.3.10); we may thus say also that A is a *perfectoid ring*.

EXAMPLE 2.1.2. Recall that a ring A of characteristic p is *perfect* if its absolute Frobenius map φ is a bijection; any such ring is reduced. It is a basic result of abstract algebra that a field is perfect in the sense of Galois theory (i.e., every finite extension is separable) if and only if either it is of characteristic 0, or it is of characteristic p and perfect in the present sense.

Let (A, A^+) be an analytic Huber pair of characteristic p. If A is uniform, then (A, A^+) is perfectoid if and only if A is perfect: if A is perfect, then A^+ must also be perfect because it is integrally closed in A.

In addition, if A is a perfect Huber ring of characteristic p, then A is automatically uniform and hence a perfectoid ring. To see this, note that A can be written as the quotient of a perfect uniform Huber ring, such as the completed perfect closure of a polynomial ring over $\mathbb{F}_p$, then apply Corollary 2.4.5. (A direct argument using the open mapping theorem is also possible.)

---

[9]I am reminded here of a famous quote of David Mumford [140, Preface]: "When I first started doing research in algebraic geometry, I thought the subject attractive... because it was a small, quiet field where a dozen people did not leap on each new idea the minute it became current."

EXAMPLE 2.1.3. Any algebraically closed nonarchimedean field is a perfectoid ring. More generally, recall that a *perfectoid field* is defined as a nonarchimedean field F which is not discretely valued for which $\varphi : \mathfrak{o}_F/(p) \to \mathfrak{o}_F/(p)$ is surjective. If F is a perfectoid field, then $(F, \mathfrak{o}_F)$ is a perfectoid Huber pair; this is obvious in characteristic p (because F is then perfect), and otherwise we may take $I = (\mu)$ for any topologically nilpotent element μ such that μ^p divides p in $\mathfrak{o}_F$ (which exists because F is not discretely valued). Conversely, if A is a perfectoid ring which is a field, then A is a perfectoid field; see Corollary 2.3.11 and Theorem 2.9.1.

EXAMPLE 2.1.4. Let (A, A^+) be any perfectoid Huber pair (e.g., $(F, \mathfrak{o}_F)$ for F a perfectoid field as in Example 2.1.3). Then for every nonnegative integer n,

$$(A\langle T_1^{p^{-\infty}}, \ldots, T_n^{p^{-\infty}}\rangle, A^+\langle T_1^{p^{-\infty}}, \ldots, T_n^{p^{-\infty}}\rangle)$$

and

$$(A\langle T_1^{\pm p^{-\infty}}, \ldots, T_n^{\pm p^{-\infty}}\rangle, A^+\langle T_1^{\pm p^{-\infty}}, \ldots, T_n^{\pm p^{-\infty}}\rangle)$$

are also perfectoid Huber pairs.

REMARK 2.1.5. If (A, A^+) is perfectoid, then $\varphi : A^+/((p)+I) \to A^+/((p)+I^p)$ is surjective for any ideal of definition I as in Definition 2.1.1. By Lemma 2.7.4, the same is true for any ideal of definition I whatsoever. In particular, the criterion of Definition 2.1.1 is satisfied for *every* ideal of definition I for which $p \in I^p$. However, the existence of such an ideal of definition is a genuine condition; for instance, it precludes the case $(\mathbb{Q}_p, \mathbb{Z}_p)$, for which $\varphi : A^+/((p)+I) \to A^+/((p)+I^p)$ is surjective for any ideal of definition I.

Note that the previous paragraph does not immediately imply that $\varphi : A^+/(p) \to A^+/(p)$ is surjective. However, this will follow from the tilting correspondence (Theorem 2.3.9); see Remark 2.3.12.

REMARK 2.1.6. Note that for A a perfectoid ring, the ring $A\langle T_1^{p^{-\infty}}, \ldots, T_n^{p^{-\infty}}\rangle$ of Example 2.1.4 is evidently not noetherian: the ideal $(T_1^{p^{-n}} : n = 0, 1, \ldots)$ is not finitely generated. In fact, perfectoid rings can never be noetherian except in the trivial case where they are finite direct sums of perfectoid fields; see Corollary 2.9.3.

This means that we cannot hope to use noetherian properties to show that perfectoid rings are sheafy. Instead, we will have to show that they are stably uniform, by establishing the preservation of the perfectoid property under rational localizations using the tilting construction (Corollary 2.5.4).

REMARK 2.1.7. For (A, A^+) a perfectoid ring, I an ideal of definition as in Definition 2.1.1, and $(A, A^+) \to (B, B^+)$ a morphism of uniform Huber pairs, the pair (B, B^+) is perfectoid if and only if $\varphi : B^+/IB^+ \to B^+/I^pB^+$ is surjective. This is a consequence of Remark 2.1.5, which allows us to use IB^+ as an ideal of definition to check the perfectoid condition.

We record some historical aspects of the definition of perfectoid fields and rings.

REMARK 2.1.8. The term *perfectoid* was introduced by Scholze [156], but various aspects of the general concept had appeared several times before then. Here we report on these appearances.

The term *perfectoid field* was introduced by Scholze in [156]. A similar definition was given independently[10] by Kedlaya in [110] and incorporated into the work of Kedlaya–Liu [117]; while much of the work on [117] predates the appearance of [156], some terminology from the latter was adapted in the published version of the former. See Remark 2.5.13 for more details. It was subsequently discovered that Matignon–Reversat [137] had introduced the same concept in 1984 as a *hyperperfect field (corps hyperparfait)*, but the importance of this development seems to have gone unnoticed at the time.

Some examples of perfectoid rings which are not fields appear in the hypothesis of the *almost purity theorem* of Faltings [56, Theorem 3.1], [58, Theorem 4]. These examples served as a key motivation for the general construction.

In [31, §5], Colmez introduces the concept of a *sympathetic algebra (algèbre sympathique)*, which in our terminology is a uniform connected Banach algebra A over an algebraically closed perfectoid field in which every element of $1 + A^{\circ\circ}$ has a p-th root; any such ring is perfectoid. He then uses sympathetic algebras to define what are commonly known as *Banach-Colmez spaces*, which are discussed in [177, Lecture 4] and used in our student project.

The term *perfectoid ring* was introduced by Scholze [156] to refer to a perfectoid ring in the present sense over an arbitrary perfectoid field (of characteristic either 0 or p). The concept of a perfectoid ring over $\mathbb{Q}_p$ was introduced independently by Kedlaya–Liu [117], with terminology adapted from [156]; some alternate characterizations of perfectoid rings over $\mathbb{Q}_p$, phrased in terms of Witt vectors, can be found in [34].

On the characteristic p side, Kedlaya–Liu consider *perfect uniform Banach $\mathbb{F}_p$-algebras*, which in the present terminology are exactly the perfectoid rings which are Tate and of characteristic p. Note that (by analogy with Definition 1.1.2) any such ring A is an algebra over the completed perfect closure of $\mathbb{F}_p((\varpi))$ for any pseudouniformizer $\varpi \in A$, and hence is a perfectoid ring in the sense of [156]. Moreover, the word *uniform* is redundant, as any perfect Banach ring A of characteristic p is automatically uniform. (Uniformity at the level of Huber rings follows from Example 2.1.2; since A is necessarily an algebra over a nonarchimedean field, Exercise 1.5.13 then implies uniformity at the level of Banach rings.)

In his Bourbaki seminar on the work of Scholze, Fontaine [68] introduced the concept of a Tate perfectoid ring, phrasing the definition in terms of condition (a) of Corollary 2.6.16. As [68] is primarily a survey of [156], the theory of Tate perfectoid rings is not developed in any detail there; this development was subsequently carried out by Kedlaya–Liu [118].

The definition of perfectoid rings used here, which allows for arbitrary analytic rings, is original to these notes. For examples that separate the various definitions, see the following references herein:

- for a perfectoid ring over $\mathbb{Q}_p$ which is not an algebra over a perfectoid field, see Exercise 2.4.8;
- for a perfectoid ring which is Tate but not a $\mathbb{Q}_p$-algebra, see the proof of Lemma 3.1.3;
- for a perfectoid ring which is analytic but not Tate, see Exercise 2.4.7.

---

[10]The paper [110] was originally written as a supplement to the lecture notes from the 2009 Clay Mathematics Institute summer school on p-adic Hodge theory. As of this writing, those notes remain unpublished.

The massive work-in-progress [75] should ultimately be even more inclusive, in ways we do not attempt to treat here (for instance, it includes some rings which are not analytic).

2.2. Witt vectors. In order to say more about perfectoid rings, we need to recall some basic facts about Witt vectors (compare [110, §1.1]).

DEFINITION 2.2.1. A *strict p-ring* is a p-adically complete (so in particular p-adically separated) ring S which is flat over $\mathbb{Z}_p$ (that is, p is not a zero-divisor) with the property that S/pS is perfect.

EXAMPLE 2.2.2. The ring $S := \mathbb{Z}_p$ is a strict p-ring with $S/pS \cong \mathbb{F}_p$. Similarly, for any finite unramified extension F of $\mathbb{Q}_p$ with residue field $\mathbb{F}_q$, the integral closure S of $\mathbb{Z}_p$ in F is a strict p-ring with $S/pS \cong \mathbb{F}_q$. Similarly, for F the maximal unramified extension of $\mathbb{Q}_p$, the completed integral closure S of $\mathbb{Z}_p$ in F is a strict p-ring with $S/pS \cong \overline{\mathbb{F}}_p$.

EXAMPLE 2.2.3. For n a nonnegative integer, the p-adic completion S of $\mathbb{Z}[T_1^{p^{-\infty}}, \ldots, T_n^{p^{-\infty}}]$ is a strict p-ring with $S/pS \cong \mathbb{F}_p[\overline{T}_1^{p^{-\infty}}, \ldots, \overline{T}_n^{p^{-\infty}}]$.

LEMMA 2.2.4. *For any ring R, any ideal I of R, and any nonnegative integer n, the map $x \mapsto x^{p^n}$ induces a morphism of multiplicative monoids*

$$R/((p)+I) \to R/((p)^{n+1} + (p)^n I + \cdots + (p)I^{p^{n-1}} + I^{p^n}).$$

PROOF. This is an immediate consequence of the p-divisibility of binomial coefficients. □

COROLLARY 2.2.5. *For S a strict p-ring, the map $S \to S/pS$ admits a unique multiplicative section $x \mapsto [x]$, called the* Teichmüller map. *In particular, the element $[x] \in S$ (called the* Teichmüller lift *of x) is the unique lift of x which admits p^n-th roots for all positive integers n.*

COROLLARY 2.2.6. *For S a strict p-ring, every element x of S has a unique representation as a p-adically convergent series $\sum_{n=0}^{\infty} p^n [\overline{x}_n]$ with $\overline{x}_n \in S/pS$. The $\overline{x}_n$ are called the* Teichmüller coordinates *of x.*

LEMMA 2.2.7. *Let S be a strict p-ring. Let S' be any p-adically complete ring. Then every ring homomorphism $\overline{\pi} : S/pS \to S'/pS'$ lifts uniquely to a homomorphism $S \to S'$.*

PROOF. By Lemma 2.2.4, $\overline{\pi}$ lifts uniquely to a multiplicative map $\pi : S/pS \to S'$. One then shows that the formula

$$\sum_{n=0}^{\infty} p^n [\overline{x}_n] \mapsto \sum_{n=0}^{\infty} p^n \pi(\overline{x}_n)$$

defines the desired ring homomorphism, by checking additivity modulo p^m by induction on m. For details, see [110, Lemma 1.1.6]. □

REMARK 2.2.8. By applying Lemma 2.2.7 in the case where S is as in Example 2.2.3, we may see that arithmetic in a strict p-ring can be expressed in terms of certain universal "polynomials" in the Teichmüller coordinates. For example, if one writes

$$[x] + [y] = \sum_{n=0}^{\infty} p^n [z_n],$$

then z_n is given by a certain homogeneous polynomial over $\mathbb{F}_p$ in $x^{p^{-n}}, y^{p^{-n}}$ of degree 1 (for the convention that $\deg(x) = \deg(y) = 1$) divisible by $x^{p^{-n}} y^{p^{-n}}$.

REMARK 2.2.9. A corollary of the previous remark is that if S is a strict p-ring and I is a perfect ideal of S/pS, then the set of $x \in S$ whose Teichmüller coordinates all belong to I is an ideal of S. Note that the quotient by this ideal is again a strict p-ring.

THEOREM 2.2.10 (Witt). *The functor $S \mapsto S/pS$ defines an equivalence of categories between strict p-rings and perfect $\mathbb{F}_p$-algebras.*

PROOF. Full faithfulness follows from Lemma 2.2.7. To check essential surjectivity, we first lift perfect polynomial rings over $\mathbb{F}_p$ in any (possibly infinite) number of variables as in Example 2.2.3, then use Remark 2.2.9 to lift quotients of such rings. That covers all perfect $\mathbb{F}_p$-algebras. $\square$

REMARK 2.2.11. There is no analogue of Theorem 2.2.10 for general $\mathbb{F}_p$-algebras. For example, if R is a field which is not perfect, then the Cohen structure theorem implies that R can be realized as S/pS for some flat p-adically complete $\mathbb{Z}_p$-algebra S (any such S is called a *Cohen ring* for R), but not functorially in R. For example, if $R = \mathbb{F}_p((\overline{T}))$, then the p-adic completion S of $\mathbb{Z}_p((T))$ admits an isomorphism $S/pS \cong R$ taking the class of T to $\overline{T}$, but there are numerous automorphisms of S lifting the identity map on R; in fact, the group of such automorphisms acts simply transitively on the inverse image of $\overline{T}$.

One way to lift imperfect rings is to consider pairs (R, B) in which R is a reduced ring of characteristic p and B is a finite subset of R such that $\{\prod_{b \in B} b^{e_b} : e_b \in \{0, \ldots, p-1\}\}$ is a basis for R as an R^p-module. (Such a set B is called a *p-basis* of R; the existence of such a set implies that R is a finite projective R^p-module, which is to say that R is *F-split*.) Then one can functorially lift (R, B) to a pair $(S, \tilde{B})$ in which S is a p-adically complete flat $\mathbb{Z}_p$-algebra and $\tilde{B}$ is a finite subset of S lifting B.

DEFINITION 2.2.12. For R a perfect ring of characteristic p, let $W(R)$ denote the strict p-ring with residue ring R; concretely, $W(R)$ consists of sequences $(\overline{x}_0, \overline{x}_1, \ldots)$ in R which are identified with the convergent sums $\sum_{n=0}^{\infty} p^n [\overline{x}_n]$. By functoriality (i.e., by Lemma 2.2.7), the absolute Frobenius φ on R lifts to a unique automorphism of R. For $I \subseteq R$ a perfect ideal, let $W(I)$ denote the ideal of $W(R)$ described in Remark 2.2.9.

REMARK 2.2.13. For conceptual purposes, it is sometimes useful to imagine the ring $W(R)$ as "the ring of power series in the variable p with coefficients in R." This point of view must of course be abandoned when one attempts to make any arguments involving calculations in $W(R)$; however, Remark 2.2.8 gives some control over the "carries" that occur in these calculations.

2.3. Tilting and untilting. In order to say more about perfectoid rings, we describe a fundamental construction that relates perfectoid rings to rings in characteristic p. This construction has its roots in the foundations of p-adic Hodge theory (see Remark 2.3.18), and the definition of perfectoid rings is in turn motivated by the construction.

DEFINITION 2.3.1. For (A, A^+) a perfectoid pair, define the *tilt* of A, denoted $A^\flat$, as the set $\varprojlim_{x \mapsto x^p} A$; it carries the structure of a monoid under multiplication.

We equip $A^\flat$ with the inverse limit topology, so that in particular the map $\sharp : A^\flat \to A$ which projects onto the final component is continuous; this gives $A^\flat$ the structure of a topological monoid. (We sometimes write $x^\sharp$ instead of $\sharp(x)$.) Let $A^{\flat+}$ be the submonoid $\varprojlim_{x \mapsto x^p} A^+$ of $A^\flat$, which is the preimage of $A^{\flat+}$ under $\sharp$.

We will see later (Theorem 2.3.9) that the formula

$$(2.3.1.1) \qquad (x_n)_n + (y_n)_n = (z_n)_n, \qquad z_n := \lim_{m \to \infty} (x_{m+n} + y_{m+n})^{p^m}$$

defines a ring structure on $A^\flat$ with respect to which it is a perfectoid ring of characteristic p, in such a way that $A^{\flat+}$ is a ring of integral elements. Another interpretation of the ring structure on $A^{\flat+}$ will come from the bijection

$$(2.3.1.2) \qquad A^{\flat+} \cong \varprojlim_{x \mapsto x^p} (A^+/I)$$

for any ideal of definition I as in Definition 2.1.1; note that the right-hand side of (2.3.1.2) is obviously a perfect ring of characteristic p.

In order to fill in the details of the previous construction, we describe an inverse construction using Witt vectors.

DEFINITION 2.3.2. Let (R, R^+) be a perfectoid pair in characteristic p. We will make frequent use of the ring $W(R^+)$, which is commonly denoted $\mathbf{A}_{\inf}(R, R^+)$ (although we will not use this notation until the next lecture).

Let $W^b(R)$ denote the subset of $W(R)$ consisting of series $\sum_{n=0}^\infty p^n[\overline{x}_n]$ for which the set $\{\overline{x}_n : n = 0, 1, \dots\}$ is bounded in R. By Remark 2.2.8, this forms a subring of $W(R)$ containing $W(R^+)$. We equip $W^b(R)$ with the topology of uniform convergence in the coordinates (see Remark 2.6.3); any continuous map $R \to S$ of perfectoid rings in characteristic p induces a homomorphism $W^b(R) \to W^b(S)$.

REMARK 2.3.3. If R is Tate, then so is $W^b(R)$: for any pseudouniformizer ϖ in R, $[\varpi]$ is a pseudouniformizer in $W^b(R)$. However, if R is analytic, it is not clear that $W^b(R)$ is analytic; compare Lemma 2.6.13.

The following construction provides something analogous to a Weierstrass-prepared power series over a nonarchimedean field.

DEFINITION 2.3.4. An element $z = \sum_{n=0}^\infty p^n[\overline{z}_n] \in W(R^+)$ is *primitive of degree 1* (or *primitive* for short) if $\overline{z}_0$ is topologically nilpotent and $\overline{z}_1$ is a unit in R^+; in other words, $z = [\overline{z}_0] + pz_1$ where z_1 is a unit in $W(R^+)$. Note that multiplying a primitive element by a unit gives another such element (e.g., using Remark 2.2.8); we say that an ideal of $W(R^+)$ is *primitive (of degree 1)* if it is principal with some (hence any) generator being a primitive element.

The primitive elements will play a role analogous to that played by the ideal $(T - p)$ in the isomorphism $\mathbb{Z}[\![T]\!]/(T-p) \cong \mathbb{Z}_p$. In particular, they admit a form of Euclidean division which is quite useful for getting control of elements of perfectoid rings; this will be studied extensively in §2.6.

REMARK 2.3.5. If $z_1, z_2 \in W(R^+)$ are primitive elements such that $z_1 = yz_2$ for some $y \in W(R^+)$, then by Remark 2.2.8, $\overline{z}_{1,1} - \overline{y}_0 \overline{z}_{2,1}$ is topologically nilpotent. It follows that y is a unit in $W(R^+)$.

EXERCISE 2.3.6. If R is Tate, then in some sources a primitive element of $W(R^+)$ is assumed to have the form $p + [\varpi]\alpha$ where $\varpi \in R$ is a topologically nilpotent unit and $\alpha \in W(R^+)$ is arbitrary. Show that if z is a primitive element in the sense of Definition 2.3.4, then it has some associate of the form $p + [\varpi]\alpha$ but need not have this form itself.

Before stating a general theorem, let us discuss a couple of key examples.

EXAMPLE 2.3.7. Put $R := \mathbb{F}_p((\overline{T}^{p^{-\infty}}))$ (i.e., the ring obtained by taking the $\overline{T}$-adic completion of $\mathbb{F}_p[\overline{T}^{p^{-\infty}}]$ and then inverting $\overline{T}$) and $R^+ := R^\circ$. The element $z := p - [\overline{T}]$ is primitive, and $W^b(R)/(z)$ is the completion of $\mathbb{Q}_p(p^{p^{-\infty}})$, which is a perfectoid field.

EXAMPLE 2.3.8. Put $R := \mathbb{F}_p((\overline{T}^{p^{-\infty}}))$, $R^+ := R^\circ$. The element $z := \sum_{i=0}^{p-1}[1+\overline{T}]^i$ is primitive (because it maps to p under $W(R^+) \to W(\mathbb{F}_p)$), and $W^b(R)/(z)$ is the completion of $\mathbb{Q}_p(\mu_{p^\infty})$, which is a perfectoid field.

THEOREM 2.3.9. *The formula*

$$(2.3.9.1) \qquad (R, R^+, I) \mapsto (A := W^b(R)/IW^b(R), A^+ := W(R^+)/I)$$

defines a equivalence of categories from triples (R, R^+, I), in which (R, R^+) is a perfectoid pair of characteristic p and I is a primitive ideal of $W(R^+)$, to perfectoid pairs (A, A^+). (A morphism $(R, R^+, I) \to (S, S^+, J)$ in this category is a morphism $(R, R^+) \to (S, S^+)$ of Huber pairs carrying I into J; in fact, by Remark 2.3.5 the image always equals J.) Furthermore, there is a quasi-inverse functor which takes (A, A^+) to (A^b, A^{b+}, I) with the ring structure on A^b given by (2.3.1.1).

PROOF. By Lemma 2.6.14, it will follow that the equation (2.3.9.1) gives a well-defined functor. By Lemma 2.7.9, we will obtain the quasi-inverse functor. □

COROLLARY 2.3.10. *Let A be a perfectoid ring, that is, A is a Huber ring such that (A, A^+) is a perfectoid pair for some ring of integral elements A^+. Then (A, A^+) is a perfectoid pair for every ring of integral elements A^+.*

PROOF. By Theorem 2.3.9, we can write $A = W^b(R)/I$ for some perfectoid ring R of characteristic p and some ideal I; more precisely, the ideal I admits a primitive generator in $W(R^+)$ for some ring of integral elements R^+ of R, and hence in $W(R^\circ)$. By Example 2.1.2, every ring of integral elements R^+ of R is perfect, and so (R, R^+) is a perfectoid pair which untilts to a perfectoid pair (A, A^+). It thus suffices to check that every A^+ arises in this fashion.

Since A is uniform, every ring of integral elements is contained in the ring A° of power-bounded elements and contains the set $A^{\circ\circ}$ of topologically nilpotent elements, which is an ideal of A°. In fact, the rings of integral elements are in bijection with integrally closed subrings of $A^\circ/A^{\circ\circ}$. Similarly, the rings of integral elements of R are in bijection with integrally closed subrings of $R^\circ/R^{\circ\circ}$. By Lemma 2.7.10, the rings $A^\circ/A^{\circ\circ}$ and $R^\circ/R^{\circ\circ}$ are isomorphic; this completes the proof. (See Remark 2.3.16 for a more refined version of the isomorphism $A^\circ/A^{\circ\circ} \cong R^\circ/R^{\circ\circ}$). □

COROLLARY 2.3.11. *A nonarchimedean field F is a perfectoid ring if and only if it is a perfectoid field.*

PROOF. One direction is Example 2.1.3. In the other direction, if F is a perfectoid ring, then by Corollary 2.3.10, $(F, \mathfrak{o}_F)$ is a perfectoid pair. If F is of characteristic p, then F is perfect (Example 2.1.2) and the valuation on F is nontrivial (hence not discrete by perfectness), so F is a perfectoid field. If F is of characteristic

0, then (p) is an ideal of definition of $\mathfrak{o}_F$, so Remark 2.3.12 applies to show that $\varphi : \mathfrak{o}_F/(p) \to \mathfrak{o}_F/(p)$ is surjective. Since the valuation on $F^\flat$ is not discrete, neither is the valuation on F, so F is a perfectoid field. $\square$

REMARK 2.3.12. If (A, A^+) is a perfectoid pair, then the existence of a surjective morphism $W(R^+) \to A^+$ as in Theorem 2.3.9 implies that $\varphi : A^+/(p) \to A^+/(p)$ is surjective: every $x \in A^+/(p)$ lifts to some element $y = \sum_{n=0}^{\infty} p^n [\overline{y}_n] \in W(R^+)$, and the image of $[\overline{y}_0^{1/p}]$ in $A^+/(p)$ maps to x via φ.

DEFINITION 2.3.13. With notation as in Theorem 2.3.9, the perfectoid pair (A, A^+) corresponding to the triple (R, R^+, I) is called the *untilt* of (R, R^+) corresponding to the primitive ideal I.

DEFINITION 2.3.14. Theorem 2.3.9 implies that for any perfectoid pair (A, A^+), there is a surjective map $W(A^{\flat+}) \to A^+$ whose kernel is primitive (and in particular principal), which extends to a map $W^b(A^\flat) \to A$. These maps are traditionally denoted by θ.

Note that for any $\overline{x} \in R$, the sequence $(\theta([\overline{x}^{p^{-n}}]))_n$ forms an element of $\varprojlim_{x \mapsto x^p} A = A^\flat$. In the course of proving Theorem 2.3.9, we will see that the identification of $A^\flat$ with R identifies this sequence with $\overline{x}$. In other words,
$$\sharp = \theta \circ [\bullet].$$

EXAMPLE 2.3.15. With notation as in Theorem 2.3.9, the pair
$$(A\langle T_1^{p^{-\infty}}, \ldots, T_n^{p^{-\infty}}\rangle, A^+\langle T_1^{p^{-\infty}}, \ldots, T_n^{p^{-\infty}}\rangle)$$
is the untilt of
$$(R\langle \overline{T}_1^{p^{-\infty}}, \ldots, \overline{T}_n^{p^{-\infty}}\rangle, R^+\langle \overline{T}_1^{p^{-\infty}}, \ldots, \overline{T}_n^{p^{-\infty}}\rangle)$$
corresponding to the primitive ideal generated by I, with $\sharp(\overline{T}_i) = T_i$. Similarly, the pair
$$(A\langle T_1^{\pm p^{-\infty}}, \ldots, T_n^{\pm p^{-\infty}}\rangle, A^+\langle T_1^{\pm p^{-\infty}}, \ldots, T_n^{\pm p^{-\infty}}\rangle)$$
is the untilt of
$$(R\langle \overline{T}_1^{\pm p^{-\infty}}, \ldots, \overline{T}_n^{\pm p^{-\infty}}\rangle, R^+\langle \overline{T}_1^{\pm p^{-\infty}}, \ldots, \overline{T}_n^{\pm p^{-\infty}}\rangle)$$
corresponding to the primitive ideal generated by I, again with $\sharp(\overline{T}_i) = T_i$.

REMARK 2.3.16. With notation as in Theorem 2.3.9, let z be a generator of I. For J an ideal (resp. ideal of definition) of A^+ containing p, the inverse image of J in $W(R^+)$ is an ideal containing p and z, and so is also the inverse image of an ideal (resp. ideal of definition) $J^\flat$ of $A^{\flat+}$ containing $\overline{z}$. Conversely, every ideal (resp. ideal of definition) of $A^{\flat+}$ containing $\overline{z}$ arises in this fashion. From the construction, we have a canonical isomorphism
$$A^+/J \cong R^+/J^\flat.$$

We mention some related facts here.

- A perfectoid ring A is Tate if and only if $A^\flat$ is Tate (Corollary 2.6.16). This implies that if A is Tate, then A^+ admits a principal ideal of definition J satisfying $p \in J^p$.

- For J an ideal of definition of A^+ with $p \in J^p$, let $J^{(p)}$ be the ideal of A^+ generated by p together with x^p for each $x \in J$; this ideal is contained in J^p, but may be strictly smaller unless J is principal. Then the surjective map $\varphi: A^+/J \to A^+/J^p$ induces a bijective map $A^+/J \to A^+/J^{(p)}$.

REMARK 2.3.17. For A, B two perfectoid rings, it is not true in general that any homomorphism $f^\flat : A^\flat \to B^\flat$ lifts to a homomorphism $f : A \to B$, because the image of $\ker(\theta : W^b(A^\flat) \to A)$ need not be contained in $\ker(\theta : W^b(B^\flat) \to B)$. However, this image is always a primitive ideal, so given $A, B^\flat, f^\flat$ there is a unique choice of an untilt B of $B^\flat$ for which $f^\flat$ does lift to a homomorphism $f : A \to B$. In other words, the categories of perfectoid A-algebras and perfectoid $A^\flat$-algebras are equivalent. For example, this comment applies to the setting of [156], in which the only perfectoid rings are considered are algebras over some fixed perfectoid field.

Again, we collect some historical notes.

REMARK 2.3.18. The operation $A \mapsto A^\flat$ was originally introduced by Fontaine–Wintenberger [69, 70, 178, 179] in the case where A is the completion of an algebraic extension of $\mathbb{Q}_p$ having a certain property (of being *strictly arithmetically profinite*) which implies that A is perfectoid; by a theorem of Sen [163], this includes the completion of any infinitely ramified Galois extension of $\mathbb{Q}_p$ whose Galois group is a p-adic Lie group. (See [120] for further discussion of this implication.) This construction is a key step in the Fontaine's construction of the de Rham and crystalline period rings [66, §2], [67].

The same construction for somewhat more general rings appeared in the work on Faltings on the crystalline comparison isomorphism [57, §2], [58, §2b], as a key step in constructive relative analogues of Fontaine's rings; this direction was further pursued by Andreatta [7]. The terminology of *tilting* and *untilting*, and the notations $\flat$ and $\sharp$, were introduced by Scholze in [156]; previously these constructions did not have commonly used names (the term *inverse perfection* for the construction $A^+ \mapsto \varprojlim A^+/I$ is used in [117]).

Like other concepts in the theory of perfectoid spaces, the notion of a *primitive element* can be found implicitly throughout the literature of p-adic Hodge theory, and somewhat more explicitly in [63] and [109]. However, it does not appear at all in [156] because no reference is made therein to Witt vectors: since only perfectoid algebras over a perfectoid field are considered, one implicitly untilts using the primitive ideal coming from the base field. In [117], the primitive elements considered (as per [117, Definition 3.3.4]) are those for which $\overline{z}_0$ is a unit in R; this level of generality corresponds to the restriction that perfectoid rings be $\mathbb{Q}_p$-algebras. The definition of primitive elements used here is the one introduced by Fontaine in [68], and adopted by Kedlaya–Liu in [118, Definition 3.2.3].

The theta map (Definition 2.3.14) first appeared in the case where A is a completed algebraic closure of $\mathbb{Q}_p$, in Fontaine's construction of the ring $\mathbf{B}_{\mathrm{dR}}$ of de Rham periods [66, §2] (see also Definition 4.6.5); it appears again in the work of Andreatta (see above).

Theorem 2.3.9 was established by Scholze [156, Theorem 5.2] for perfectoid rings over a perfectoid field, (but without reference to Witt vectors; see above) and independently by Kedlaya–Liu [117, Theorem 3.6.5] for perfectoid rings over $\mathbb{Q}_p$. It was extended to Tate rings by Kedlaya–Liu [118, Theorem 3.3.8]. The extension to analytic rings is original to these notes.

2.4. Algebraic aspects of tilting.
We next describe the extent to which tilting is compatible with certain algebraic properties of morphisms of perfectoid rings.

The following argument shows that the category of perfectoid spaces admits fiber products, which is not known for the full category of adic spaces (because a completed tensor product of sheafy Huber rings is not known to again be sheafy).

THEOREM 2.4.1. *Let $A \to B, A \to C$ be two morphisms of perfectoid rings. Then $B \widehat{\otimes}_A C$ is again a perfectoid ring, and its formation commutes with tilting.*

PROOF. See Lemma 2.8.7. □

THEOREM 2.4.2. *Let $f : A \to B$ be a morphism of perfectoid rings. Then f is a strict inclusion if and only if $f^\flat$ is a strict inclusion. (A similar statement holds for nonstrict inclusions, but lies much deeper; see Corollary 2.9.13.)*

PROOF. From the definition of $A^\flat$ as a topological space, it is obvious that if f is a strict inclusion, then so is $f^\flat$. Conversely, if $f^\flat$ is a strict inclusion, then so is $W^b(A^\flat) \to W^b(B^\flat)$. By Theorem 2.3.9, we have $A \cong W^b(A^\flat)/I$ and $B \cong W^b(B^\flat)/IW^b(B^\flat)$ for some primitive ideal I of $W^b(A^\flat)$. Using Lemma 2.8.1, we deduce that f is a strict inclusion. □

THEOREM 2.4.3. *Let $f : A \to B$ be a morphism of Huber rings in which A is perfectoid.*

(a) *If B is uniform and f has dense image, then B is perfectoid.*
(b) *If B is perfectoid, then f has dense image if and only if $f^\flat$ has dense image.*

PROOF. Part (a) will follow from Lemma 2.8.4. To check (b) in one direction, note that if $f^\flat$ has dense image, then the composition $W^b(A^\flat) \to W^b(B^\flat) \to B$ has dense image, so f has dense image. The other direction will again follow from Lemma 2.8.4. □

THEOREM 2.4.4. *Let $f : A \to B$ be a morphism of Huber rings in which A is perfectoid.*

(a) *If B is uniform and f is surjective, then B is perfectoid.*
(b) *If B is perfectoid, then f is surjective if and only if $f^\flat$ is surjective.*

PROOF. Part (a) is a consequence of Theorem 2.4.3(a). To check (b) in one direction, note that if $f^\flat$ is surjective, then the composition $W^b(A^\flat) \to W^b(B^\flat) \to B$ is surjective, so f is surjective. In the other direction, if f is surjective, then Lemma 2.8.6 implies that $f^\flat$ is surjective. □

COROLLARY 2.4.5. *For A a perfectoid ring, the map $I \mapsto I^\flat$ defines a bijection between closed ideals of A with A/I uniform and closed perfect ideals of $A^\flat$.*

PROOF. In light of Theorem 2.4.4, it suffices to check that if R is a perfectoid ring of characteristic p and I is a perfect ideal of R, then R/I is uniform (this being the case of the desired statement where $A = A^\flat = R$). To this end, promote R to a uniform Banach ring as per Remark 1.5.4. Then note that if $\tilde{x} \in R$ lifts $x \in R/I$, then $\tilde{x}^{1/p}$ lifts $x^{1/p} \in R/I$, so $|x^{1/p}| \leq |x|^{1/p}$. From Definition 1.5.11, it follows that R/I is uniform. □

COROLLARY 2.4.6. *For A a perfectoid ring and I a closed ideal of A, A/I is uniform (and hence perfectoid) if and only if there exists some subset S of $A^\flat$ such that I is the closure of the ideal generated by $\sharp(x^{p^{-n}})$ for all $x \in S$ and all nonnegative integers n.*

PROOF. Suppose that A/I is uniform; it is then perfectoid by Theorem 2.4.4(a). By Corollary 2.4.5, $I^\flat$ is a closed perfect ideal of $A^\flat$, so $A^\flat/I^\flat$ is perfectoid and $(A/I)^\flat \cong A^\flat/I^\flat$. We may then take S to be the set $\{\sharp(x) : x \in I^\flat\}$.

Conversely, suppose that there exists S of the specified form. Let J be the closure of the ideal of $A^\flat$ generated by $x^{p^{-n}}$ for all $x \in S$ and all nonnegative integers n; note that $x^{p^{-n}} \in I^\flat$, so $J \subseteq I^\flat$. Since the set of generators of J is stable under taking p-th roots, J is perfect, so $A^\flat/J$ is perfectoid. By Theorem 2.4.4(b), the morphism $A^\flat \to A^\flat/J$ untilts to a surjective morphism, whose kernel is the closed ideal of A generated by $\sharp(x^{p^{-n}})$ for all $x \in S$ and all nonnegative integers n. It follows that this kernel equals I, so A/I is perfectoid. □

EXERCISE 2.4.7. Using Corollary 2.4.6, adapt Example 1.5.7 to give an example of a perfectoid ring in characteristic p which is analytic but not Tate. Note that no such example can exist over $\mathbb{Q}_p$, but one can construct mixed-characteristic examples by untilting.

EXERCISE 2.4.8. In this exercise, we exhibit a perfectoid ring over $\mathbb{Q}_p$ which is not an algebra over a perfectoid field.
  (a) Prove that the completions of $\mathbb{Q}_p(p^{p^{-\infty}})$ and $\mathbb{Q}_p(\mu_{p^\infty})$ have no common subfield larger than $\mathbb{Q}_p$. One way to do this is to use the Ax–Sen–Tate theorem [10] to show that any complete subfield of either of these two fields is itself the completion of an algebraic extension of $\mathbb{Q}_p$.
  (b) Let F be the completed perfect closure of $\mathbb{F}_p((\overline{T}_1))$. Put $R := F\langle \overline{T}_2^{p^{-\infty}}\rangle$, $R^+ := R^\circ$. Let R_1 be the quotient of R by the closure of the ideal $(\overline{T}_2^{p^{-n}} : n = 0, 1, \ldots)$ and put $R_1^+ = R_1^\circ$. Let R_2 be the quotient of R by the closure of the ideal $(\overline{T}_2^{p^{-n}} - 1 : n = 0, 1, \ldots)$ and put $R_2^+ = R_2^\circ$. Prove that there exists a primitive element $z \in W(R^+)$ which maps to $p - [\overline{T}_1]$ in $W(R_1^+)$ and to $\sum_{i=0}^{p-1}[1 + \overline{T}_1]^i$ in $W(R_2^+)$.
  (c) Combine (a), (b), Example 2.3.7, and Example 2.3.8 to obtain the desired example.

In Corollary 2.4.5, it is easy to see that the condition that A/I be uniform is necessary, by the following trivial example.

EXAMPLE 2.4.9. Let K be a perfectoid field. The quotient of $K\langle T^{p^{-\infty}}\rangle$ by the closed ideal (T) is not uniform (or even reduced), and hence not perfectoid. By contrast, the quotient by the closure of the ideal $(T^{p^{-n}} : n = 0, 1, \ldots)$ is the field K again.

A far less trivial example is the following.

EXAMPLE 2.4.10. Let $\mathbb{C}_p$ be the completion of an algebraic closure of $\mathbb{Q}_p$ and choose an element $\epsilon = (\ldots, \zeta_p, 1) \in \mathbb{C}_p^\flat$ in which ζ_{p^n} is a primitive p^n-th root of unity. Take $A := \mathbb{C}_p\langle T^{\pm p^{-\infty}}\rangle$ and let I be the ideal $(T - 1)$ of A. The ideal I is closed, so the quotient A/I is a Banach ring. Let B be the uniform completion of

A/I (i.e., the completion with respect to the spectral seminorm); we may identify B with the ring $\mathrm{Cont}(\mathbb{Z}_p, \mathbb{C}_p)$ of continuous functions from $\mathbb{Z}_p$ to $\mathbb{C}_p$ in such a way that the natural map $f: A \to B$ takes $T^{p^{-n}}$ to the function $\gamma \to \zeta_{p^n}^\gamma$.

These rings were analyzed, in different language, by Fresnel and de Mathan [71]. In particular, they showed [71, Theorems 1–3] that f is surjective and the kernel of the induced map $A/I \to B$ consists of the closure of the nilradical. In particular, we have the highly nonobvious fact that I is not a radical ideal.

REMARK 2.4.11. Theorem 2.4.1 was proved by Scholze [156, Proposition 6.18] for perfectoid rings over a perfectoid field, and independently by Kedlaya–Liu [117, Theorem 3.6.11] for perfectoid rings over $\mathbb{Q}_p$. This was extended to Tate perfectoid rings by Kedlaya–Liu [118, Theorem 3.3.13]. The extension to analytic rings is original to these notes.

Theorem 2.4.3 and Theorem 2.4.4 were proved for perfectoid rings over $\mathbb{Q}_p$ in [117, Theorem 3.6.17], and for Tate rings in [118, Theorem 3.3.18]. The extensions to analytic rings are original to these notes.

2.5. Geometric aspects of tilting. We now describe the interaction of the perfectoid condition with rational and étale localization. This will allow us to define perfectoid spaces and the étale topology on them.

THEOREM 2.5.1. *For (A, A^+) a perfectoid pair, the formula $v \mapsto v \circ \sharp$ defines a bijection $\mathrm{Spa}(A, A^+) \cong \mathrm{Spa}(A^\flat, A^{\flat+})$ which identifies rational subspaces on both sides; in particular, this map is a homeomorphism.*

PROOF. By Theorem 2.3.9, (A, A^+) is the untilt of $(A^\flat, A^{\flat+})$ corresponding to some primitive ideal. We may thus apply Lemma 2.6.12 to show that the map is well-defined, and Corollary 2.6.15 to show that it is bijective. It is easy to see that the rational subspace of $\mathrm{Spa}(A^\flat, A^{\flat+})$ defined by the parameters $\overline{f}_1, \ldots, \overline{f}_n, \overline{g}$ corresponds to the rational subspace of $\mathrm{Spa}(A, A^+)$ defined by the parameters $\sharp(\overline{f}_1), \ldots, \sharp(\overline{f}_n), \sharp(\overline{g})$; in the other direction, Lemma 2.6.17 implies that every rational subspace of $\mathrm{Spa}(A, A^+)$ corresponds to a rational subspace of $\mathrm{Spa}(A^\flat, A^{\flat+})$. □

REMARK 2.5.2. One of the remarkable features of Theorem 2.5.1 is the fact that $\sharp: A^\flat \to A$ is not a ring homomorphism (it is multiplicative but not additive), and yet pullback by $\sharp$ defines a morphism of spectra. We like to think of $\sharp$ as defining a "homotopy equivalence" between $A^\flat$ and A instead of a true morphism.

THEOREM 2.5.3. *Let (A, A^+) be a perfectoid pair.*
(a) *For $(A, A^+) \to (B, B^+)$ a rational localization, (B, B^+) is again a perfectoid pair.*
(b) *The functor $(B, B^+) \to (B^\flat, B^{\flat+})$ defines an equivalence of categories between rational localizations of (A, A^+) and of $(A^\flat, A^{\flat+})$.*

PROOF. For (R, R^+) a perfectoid pair of characteristic p and $(R, R^+) \to (S, S^+)$ a rational localization, we may apply the universal property of a rational localization to the composition $(R, R^+) \xrightarrow{\varphi^{-1}} (R, R^+) \to (S, S^+)$ to refactor it as $(R, R^+) \to (S, S^+) \to (S, S^+)$. Composing the second morphism on either side with $\varphi: (S, S^+) \to (S, S^+)$ yields a morphism which fixes the image of R in S, and by extension the inverse of any element of this image which is a unit in S; by continuity, the composition is the identity on S, so (S, S^+) is a perfectoid pair.

By the previous paragraph plus Theorem 2.5.1, to prove both (a) and (b) it suffices to check that the untilt of any rational localization is again a rational localization. This requires some argument because the universal property of a rational localization is quantified over arbitrary Huber pairs, not just perfectoid pairs or uniform pairs; see Lemma 2.8.8. □

COROLLARY 2.5.4. *Any perfectoid ring is stably uniform, and hence sheafy by Theorem* 1.2.13.

COROLLARY 2.5.5. *Let A be a Huber ring admitting a continuous homomorphism $A \to B$ to a perfectoid ring which splits in the category of topological A-modules (a/k/a a sousperfectoid ring; see Remark* 1.2.19*). Then A is stably uniform.*

PROOF. Combine Corollary 2.5.4 with Lemma 1.2.18. □

DEFINITION 2.5.6. In light of Corollary 2.5.4, for any perfectoid pair (A, A^+) the space $\mathrm{Spa}(A, A^+)$ admits the structure of an adic space. An adic space locally of this form is called a *perfectoid space*.

COROLLARY 2.5.7. *For (A, A^+) a perfectoid pair, the residue field of every point of $\mathrm{Spa}(A, A^+)$ is a perfectoid field. (See also Corollary* 2.9.14 *for a related result.)*

PROOF. Let $x \in \mathrm{Spa}(A, A^+)$ be a point with residue field K. Then K contains as a dense subring the direct limit of B over all rational localizations $(A, A^+) \to (B, B^+)$ for which $x \in \mathrm{Spa}(B, B^+)$. (More precisely, x corresponds to a valuation on each B; these valuations are compatible with localizations, and K is the completion of the direct limit with respect to the topology defined by the induced valuation.) By Theorem 2.5.3, each pair (B, B^+) is again perfectoid.

From the previous paragraph, one may deduce that K is a perfectoid ring. One way to see this is to note that the completion in K of the direct limit of the rings B^+ is contained in K°, and the quotient is killed by any topologically nilpotent unit of K (that is, it is *almost zero*; see Definition 2.9.4); one may then check the perfectoid property by a direct calculation. Alternatively, form the product of B over all rational localizations $(A, A^+) \to (B, B^+)$ for which $x \in \mathrm{Spa}(B, B^+)$, take the A-subalgebra generated by the direct sum, and complete for the supremum norm; this yields a perfectoid ring which maps to K with dense image, and K is uniform because it is a nonarchimedean field, so K is a perfectoid ring by Theorem 2.4.3 (compare [**118**, Corollary 3.3.22]).

In any case, K is both a nonarchimedean field and a perfectoid ring, so by Corollary 2.3.11 it is a perfectoid field. □

EXERCISE 2.5.8. Here is a rare example of a ring which can be shown to be stably uniform despite not being (directly) susceptible to Corollary 2.5.5. Let K be an algebraically closed perfectoid field of characteristic $p > 2$. Put
$$A_0 := K\langle T^{p^{-\infty}}\rangle, \qquad A := A_0[T^{1/2}], \qquad A' := K\langle (T^{1/2})^{p^{-\infty}}\rangle.$$
Equip these rings with the Gauss norm.
  (a) Show that for $i \in \frac{1}{2}\mathbb{Z}[p^{-1}]$, if $i \geq \frac{1}{2}$ then $T^i \in A$. Deduce that the natural map $A \to A'$ does not split in the category of A-modules.
  (b) Show that the map $\mathrm{Spa}(A', A'^\circ) \to \mathrm{Spa}(A, A^\circ)$ is a homeomorphism and identifies rational subspaces on both sides.

(c) Show that $\mathrm{Spa}(A, A^\circ)$ contains a unique point v_0 with $v_0(T) = 0$, whose complement is a perfectoid space. Note that the residue field of v_0 is K, so A satisfies the conclusion of Corollary 2.5.7 but not the hypothesis.

(d) Using (b), show that a general rational subspace of $\mathrm{Spa}(A, A^\circ)$ containing v_0 has the form
$$\{v \in \mathrm{Spa}(A, A^\circ) : v(\lambda_0 T^{1/2}) \leq 1, v(\lambda_1 T^{1/2} - \mu_1) \geq 1, \ldots, v(\lambda_n T^{1/2} - \mu_n) \geq 1\}$$
for some nonnegative integer n and some $\lambda_0, \ldots, \lambda_n, \mu_1, \ldots, \mu_n \in K$ with $\lambda_i \geq 1$ for $i \geq 0$ and $\lambda_i \geq \mu_i \geq 1$ for $i > 0$.

(e) Let $(A, A^\circ) \to (B, B^+)$ be the rational localization corresponding to a rational subspace as in (d) with $\lambda_0 = 1$. (It turns out that $B^+ = B^\circ$, but this isn't crucial for what follows.) A general element of B can be written in the form
$$a_0 + \sum_{i=1}^{n} \sum_{j=1}^{\infty} a_{i,j}(\lambda_i T^{1/2} - \mu_i)^{-j}$$
for some $a_0, a_{i,j} \in A_0$. Prove that each $a_{i,j}$ can be replaced by an element with all exponents in $[0, 1]$ without increasing the quotient norm (i.e., the maximum of the norms of a_0 and all of the $a_{i,j}$).

(f) Put $B' = B \widehat{\otimes}_A A'$. Show that for
$$x = a_0 + \sum_{i=1}^{n} \sum_{j \in \mathbb{Z}[p^{-1}], j > 0} a_{i,j}(\lambda_i T^{1/2} - \mu_i)^{-j} \in B'$$
with $a_0 \in A_0, a_{i,j} \in K$, the spectral norm of x is equal to the maximum of the norms of a_0 and the $a_{i,j}$.

(g) Show that $B \to B \widehat{\otimes}_A A'$ is a strict inclusion. Deduce that A is stably uniform.

THEOREM 2.5.9. *Let A be a perfectoid ring.*

(a) *For $A \to B$ a finite étale morphism, B is again a perfectoid ring for its natural topology as an A-module. (Note that Lemma 1.10.1 implies that B is uniform.)*

(b) *The functor $B \mapsto B^\flat$ defines an equivalence of categories between finite étale algebras over A and over $A^\flat$.*

PROOF. See Lemma 2.8.11. □

COROLLARY 2.5.10. *For (A, A^+) a perfectoid pair, there is a functorial homeomorphism $\mathrm{Spa}(A, A^+)_{\mathrm{et}} \cong \mathrm{Spa}(A^\flat, A^{\flat+})_{\mathrm{et}}$.*

REMARK 2.5.11. It was conjectured in [158, Conjecture 2.16] that if (A, A^+) is a Huber pair over a perfectoid field and $\mathrm{Spa}(A, A^+)$ is a perfectoid space, then A is a perfectoid ring. This is refuted by the first example of Buzzard–Verberkmoes cited in Remark 1.2.24.

However, it is possible that a similar question with slightly different hypotheses does admit an affirmative answer. For an example of a partial result in this direction, Theorem 1.2.22 implies that if (A, A^+) is a Huber pair such that $\mathrm{Spa}(A, A^+)$ is a perfectoid space and $A = H^0(\mathrm{Spa}(A, A^+), \mathcal{O})$, then A is sheafy.

EXERCISE 2.5.12. Extend the proof of Corollary 2.5.7 to show that for any nonempty subset S of $\mathrm{Spa}(A, A^+)$, the completion (for the supremum norm over S) of the stalk of the structure sheaf of $\mathrm{Spa}(A, A^+)$ at S is a perfectoid ring.

REMARK 2.5.13. For A a perfectoid ring promoted to a Banach ring, the homeomorphism $\mathcal{M}(A) \cong \mathcal{M}(A^\flat)$ induced by Theorem 2.5.1 (by restricting to valuations of height 1) was first described in [109, Corollary 7.2] (and alluded to in [107]).

Theorem 2.5.3 was proved for perfectoid rings over a perfectoid field by Scholze [156, Theorem 6.3], and independently for perfectoid rings over $\mathbb{Q}_p$ by Kedlaya–Liu [117, Theorem 3.6.14]. It was generalized to Tate rings by Kedlaya–Liu [118, Theorem 3.3.18]. The extension to analytic rings is original to these notes, but uses similar methods.

For perfectoid fields, Theorem 2.5.9 generalizes the *field of norms correspondence* of Fontaine–Wintenberger [69, 70]. The result in this case originated from a private communication between this author and Brian Conrad after the 2009 Clay Mathematics Institute summer school on p-adic Hodge theory; this argument was subsequently reproduced in [110, Theorem 1.5.6] and [117, Theorem 3.5.6]. (The key special case of algebraically closed perfectoid fields amounts to an argument we learned from Robert Coleman in 1998, as documented in [101, §4].) The result was obtained independently by Scholze using a different approach based on almost ring theory; see [156, Theorem 3.7] for a side-by-side treatment of both approaches.

Theorem 2.5.3 is a generalization of (part of) the *almost purity theorem* of Faltings, which appears implicitly in [56] and somewhat more explicitly in [58]. It was proved for perfectoid rings over a perfectoid field by Scholze [156, Theorem 7.9], and independently for perfectoid rings over $\mathbb{Q}_p$ by Kedlaya–Liu [117, Theorem 3.6.21]. It was generalized to Tate rings by Kedlaya–Liu [118, Theorem 3.3.18]. The extension to analytic rings is original to these notes, but uses similar methods. (See [117, Remark 5.5.10] for some additional discussion.)

2.6. Euclidean division for primitive ideals. In order to prove most of our main results, we need to establish a version of Euclidean division for primitive elements. Our presentation of this construction follows [109].

HYPOTHESIS 2.6.1. Throughout §2.6, let (R, R^+) be a perfectoid pair of characteristic p, and let $z \in W(R^+)$ be a primitive element. Promote R to a uniform Banach ring as per Remark 1.5.4.

Note that by definition, these hypotheses imply that R is analytic. However, we will only need that hypothesis starting with Lemma 2.6.13; the results before that require only that R be perfect and uniform. This will be important in §2.7, where we must make some calculations with a putative perfectoid ring of characteristic p before establish its analyticity.

DEFINITION 2.6.2. Define the *Gauss norm* on $W^b(R)$ by the formula
$$\left| \sum_{n=0}^{\infty} p^n [\overline{x}_n] \right| = \sup\{|\overline{x}_n| : n = 0, 1, \ldots, \};$$
note that the supremum is in general not achieved. Using Remark 2.2.8, it can be shown that the Gauss norm is a power-multiplicative norm, or even a multiplicative norm in case the norm on R is multiplicative [109, Lemma 4.2]; moreover, $W(R^+)$ and $W^b(R)$ are both complete with respect to this norm.

For z a primitive element, for the Gauss norm we have $|pz_1| = 1 > |z - pz_1|$, so for all $x \in W^b(R)$, we have $|zx| = |x|$. Consequently, the ideals $zW(R^+)$ and $zW^b(R)$ are closed in their respective rings, and using the quotient norms we may equip $W(R^+)/(z)$ and $W^b(R)/(z)$ with the structure of Banach (and Huber) rings.

Note also that $zW^b(R) \cap W(R^+) = zW(R^+)$, so the map $W(R^+)/(z) \to W^b(R)/(z)$ is injective.

REMARK 2.6.3. The topology induced by the Gauss norm on $W^b(R)$ can be interpreted as the topology of uniform convergence in the Teichmüller coordinates. On $W(R^+)$, this can also be interpreted as the I-adic topology for $I = ([\overline{x}_1], \ldots, [\overline{x}_n])$ where $\overline{x}_1, \ldots, \overline{x}_n \in R^+$ generate the unit ideal in R, although it requires some care to show this when $n > 1$.

As in [109, 117, 118], one can also consider *weighted Gauss norms* on $W^b(R)$ given by formulas of the form

$$\left| \sum_{n=0}^{\infty} p^n [\overline{x}_n] \right|_\rho = \max\{\rho^{-n} |\overline{x}_n| : n = 0, 1, \ldots, \}$$

for some $\rho \in (0, 1)$. (Note that the supremum becomes a maximum as soon as $\rho < 1$.) On $W(R^+)$, the topology induced by a weighted Gauss norm can be interpreted as the (p, I)-adic topology for I as above.

For z a primitive element, the quotient norms on $W^b(R)/(z)$ induced by the Gauss norm and any weighted Gauss norm coincide. This will follow from the fact that every nonzero element of the quotient admits a prepared representative (Lemma 2.6.9).

For primitive elements, we have a useful analogue of Euclidean division. The following discussion is taken from [109, §5].

DEFINITION 2.6.4. Let $z = [\overline{z}_0] + pz_1 \in W(R^+)$ be primitive. For $x = \sum_{n=0}^{\infty} p^n [\overline{x}_n] \in W^b(R)$, define the *Euclidean quotient and remainder* of x modulo z as the pair (q, r) where

$$x_1 := p^{-1}(x - [\overline{x}_0]), \quad q := z_1^{-1} x_1, \quad r := x - qz = [\overline{x}_0] - [\overline{z}_0] \sum_{n=0}^{\infty} z_1^{-1} p^n [\overline{x}_{n+1}].$$

DEFINITION 2.6.5. An element $x = \sum_{n=0}^{\infty} p^n [\overline{x}_n] \in W^b(R)$ is *prepared* if $|\overline{x}_0| \geq |\overline{x}_n|$ for all $n > 0$. (This corresponds to the definition of *stable* in [109, §5], but we need to save that term for another meaning later; this terminology is meant to suggest the *Weierstrass preparation theorem.*)

LEMMA 2.6.6. *For $z \in W(R^+)$ primitive, no nonzero multiple of z in $W^b(R)$ is prepared.*

PROOF. Suppose by way of contradiction that $x = \sum_{n=0}^{\infty} p^n [\overline{x}_n] \in W^b(R)$ is nonzero and zx is prepared. On one hand, the reduction of zx equals $\overline{z}_0 \overline{x}_0$, so $\overline{x}_0 \neq 0$ and

$$|zx| = |\overline{z}_0 \overline{x}_0| < |\overline{x}_0| \leq |x|.$$

On the other hand, as described in Definition 2.6.2 we have $|zx| = |x|$, a contradiction. □

LEMMA 2.6.7. *Let $z \in W(R^+)$ be primitive. If $x \in W^b(R)$ is prepared, then the quotient norm of the class of x in $W^b(R)/(z)$ is equal to $|x|$.*

PROOF. Suppose to the contrary that there exists $y \in x + zW^b(R)$ with $|y| < |x|$. Then $x - y$ is a nonzero prepared multiple of z, so Lemma 2.6.6 yields a contradiction. □

COROLLARY 2.6.8. *Let $z \in W(R^+)$ be primitive. For $\overline{x} \in R$, the quotient norm of the class of $[\overline{x}]$ in $W^b(R)/(z)$ equals $|\overline{x}|$.*

LEMMA 2.6.9. *For z primitive and $x \in W^b(R)$ not divisible by z, form the sequence $x_0, x_1, \ldots$ in which $x_0 = x$ and x_{m+1} is the Euclidean remainder of x_m modulo z. Then for every sufficiently large m, x_m is prepared.*

PROOF. If $|x_{m+1}| > |\overline{z}_0| |x_m|$ for some m, then x_{m+1} is prepared; in addition, $|x_{m+2}| = |x_{m+1}|$, so $x_{m+2}, x_{m+3}, \ldots$ are also prepared. Otherwise, for q_m the Euclidean quotient of x_m modulo z, the sum $\sum_{m=0}^{\infty} q_m$ converges to a limit q satisfying $x = qz$, so x represents the zero class in the quotient ring, contradiction. □

COROLLARY 2.6.10. *For z primitive, the quotient norm on $W^b(R)/(z)$ is power-multiplicative. If in addition the norm on R is multiplicative, then the quotient norm on $W^b(R)/(z)$ is multiplicative.*

PROOF. Combine Lemma 2.6.7 with Lemma 2.6.9. (Note that we are implicitly using parts (a) and (b) of Exercise 2.6.11 below.) □

EXERCISE 2.6.11. Let $x_1, \ldots, x_n \in W^b(R)$ be prepared elements.

(a) Suppose that the norm on R is multiplicative. Show that $x_1 \cdots x_n$ is prepared.
(b) Suppose that $x_1 = \cdots = x_n$. Show that $x_1 \cdots x_n$ is prepared. (Hint: combine (a) with Lemma 1.5.21.)
(c) Give an example in which $x_1 \cdots x_n$ is not prepared. (Hint: take R to be a product of two perfectoid fields.)

LEMMA 2.6.12. *For z primitive, $A := W^b(R)/(z)$, $\pi : W^b(R) \to A$ the quotient map, $A^+ = W(R^+)/(z)$, and $v \in \mathrm{Spa}(A, A^+)$ arbitrary, we have $v \circ \pi \circ [\bullet] \in \mathrm{Spa}(R, R^+)$.*

PROOF. The nontrivial point is that $v \circ \pi \circ [\bullet]$ satisfies the strong triangle inequality; this follows from Remark 2.2.8. □

Up to now, none of the arguments have required the hypothesis that R be analytic. We add that hypothesis now.

LEMMA 2.6.13. *For R analytic and z primitive, $W^b(R)/(z)$ is analytic.*

PROOF. Put $A = W^b(R)/(z)$, $A^+ = W(R^+)/(z)$. Let $\pi : W^b(R) \to A$ denote the quotient map. For $v \in \mathrm{Spa}(A, A^+)$, by Lemma 2.6.12 the formula $x \mapsto v(\pi([x]))$ defines a valuation $w \in \mathrm{Spa}(R, R^+)$. By Lemma 1.1.3, there exists $x \in R$ such that $w(x) \neq 0$; hence $v(\pi([x])) \neq 0$. By Lemma 1.1.3 again, A is analytic. □

LEMMA 2.6.14. *The formula (2.3.9.1) defines a functor from triples (R, R^+, I), in which (R, R^+) is a perfectoid pair of characteristic p and I is a primitive ideal of $W(R^+)$, to perfectoid pairs (A, A^+).*

PROOF. By Corollary 2.6.10 and Lemma 2.6.13, A is uniform and analytic. Let z be a generator of I. Let J be an ideal of definition of R^+ such that $\overline{z}_0 \in J^p$ (this exists because R is perfect). Then the set of $x = \sum_{n=0}^{\infty} p^n [\overline{x}_n] \in W(R^+)$ with $\overline{x}_0 \in J$ maps to an ideal of definition $\tilde{J}$ of A^+ with $p \in \tilde{J}^p$ such that $\varphi : A^+/\tilde{J} \to A^+/\tilde{J}^p$ is surjective. Hence (A, A^+) is a perfectoid pair. □

COROLLARY 2.6.15. *For (R, R^+, I) corresponding to (A, A^+) as in Lemma 2.6.14, the construction of Lemma 2.6.12 defines a bijective map* $\mathrm{Spa}(A, A^+) \to \mathrm{Spa}(R, R^+)$.

PROOF. For $v \in \mathrm{Spa}(R, R^+)$ corresponding to the pair (K, K^+), the triple $(K, K^+, IW(K^+))$ corresponds via Lemma 2.6.14 to a pair (L, L^+). By Corollary 2.6.8, L is an analytic field; this pair then corresponds to the unique valuation in $\mathrm{Spa}(A, A^+)$ mapping to v. □

COROLLARY 2.6.16. *For (A, A^+) a uniform analytic Huber pair, the following conditions are equivalent.*
  (a) *There exists a pseudouniformizer $\varpi \in A$ such that ϖ^p divides p in A^+ and $\varphi: A^+/(\varpi) \to A^+/(\varpi^p)$ is surjective.*
  (b) *The ring A is Tate and perfectoid.*
  (c) *The ring A is perfectoid and the ring $A^\flat$ is Tate.*
  (d) *The ring A is perfectoid and there exists a uniformizer $\overline{\varpi} \in A^\flat$ such that $\sharp(\overline{\varpi})^p$ divides p in A^+ and $\varphi: A^+/(\sharp(\overline{\varpi})) \to A^+/(\sharp(\overline{\varpi}^p))$ is surjective.*

PROOF. Since all four conditions imply that (A, A^+) is perfectoid (using Corollary 2.3.10), we may assume this from the outset. It is clear that (a) implies (b), (c) implies (a), and (d) implies (c); it thus remains to check that (b) implies (d). Let $\varpi \in A$ be any pseudouniformizer. Lift ϖ to some $x_0 \in W(A^{\flat+})$, then form the sequence $x_0, x_1, \ldots$ as in Lemma 2.6.9. Write $x_m = \sum_{n=0}^\infty p^n [\overline{x}_{m,n}]$ with $\overline{x}_{m,n} \in A^{\flat+}$. For each $v \in \mathrm{Spa}(A, A^+)$, corresponding to $w \in \mathrm{Spa}(A^\flat, A^{\flat+})$ via Corollary 2.6.15, we may apply Lemma 2.6.7 and Lemma 2.6.9 to see that for m sufficiently large, $w(\overline{x}_{m,0}) = v(\varpi) \neq 0$; in particular, there exists a neighborhood U of w in $\mathrm{Spa}(A^\flat, A^{\flat+})$ on which $\overline{x}_{m,0}$ does not vanish. Since $\mathrm{Spa}(A^\flat, A^{\flat+})$ is quasicompact, we may make a uniform choice of m for which $\overline{x}_{m,0}$ vanishes nowhere on $\mathrm{Spa}(A^\flat, A^{\flat+})$. By Corollary 1.5.20, $\overline{x}_{m,0}$ is a pseudouniformizer in $A^\flat$, and we may take $\overline{\varpi} = \overline{x}_{m,0}^{p^{-k}}$ for k sufficiently large to achieve the desired result. □

LEMMA 2.6.17. *For (R, R^+, I) corresponding to (A, A^+) as in Lemma 2.6.14, under the bijection $\mathrm{Spa}(A, A^+) \to \mathrm{Spa}(R, R^+)$ of Corollary 2.6.15, every rational subspace of $\mathrm{Spa}(A, A^+)$ arises from some rational subspace of $\mathrm{Spa}(R, R^+)$.*

PROOF. Choose $f_1, \ldots, f_n, g \in A$ generating the unit ideal. By Exercise 1.2.2, for $\epsilon > 0$ sufficiently small, perturbing $f_1, \ldots, f_n, g$ by elements of norm at most ϵ does not change the resulting rational subspace. For $y = f_1, \ldots, f_n, g$ in turn, choose $x_0 \in W^\flat(R)$ lifting y and define the sequence $x_0, x_1, \ldots$ as in Lemma 2.6.9; then for sufficiently large m we have

$$\max\{(\alpha \circ \sharp)(y), \epsilon\} = \max\{\alpha(\overline{x}_m), \epsilon\} \qquad (\alpha \in \mathcal{M}(R)).$$

By replacing y with $\sharp(\overline{x}_m)$ in the list of parameters for our rational subspace, we achieve the desired result. □

REMARK 2.6.18. The idea of performing Euclidean division in rings of Witt vectors has a long history. For Witt vectors over a valuation ring, some early instances can be found in work of the author [**103**, Lemma 3.28], [**105**, Lemma 2.6.3]; the thread was then picked up by Fargues–Fontaine (see [**63**, Preface, §1.1] for further historical context). For Witt vectors over more general topological rings, an early source is [**109**, Lemma 5.5]; the point of view taken therein led to the treatment of perfectoid fields given in [**110**] and the treatment of perfectoid rings in [**117**, **118**].

By contrast, the original work of Scholze [**156**] did not rely on Euclidean division for Witt vectors, this being supplanted by systematic use of *almost commutative algebra* in the sense of Faltings.

2.7. Primitive ideals and tilting. We now show that the construction of Lemma 2.6.14 accounts for all perfectoid pairs, completing the proof of Theorem 2.3.9.

LEMMA 2.7.1. *For (A, A^+) a uniform Huber pair in which p is topologically nilpotent, topologize the set $A^\flat := \varprojlim_{x \mapsto x^p} A$ as in Definition 2.3.1.*

(a) *Let $(x_n)_n, (y_n)_n \in A^\flat$ be elements. The limit in the formula*

$$(2.7.1.1) \qquad (x_n)_n + (y_n)_n = \left(\lim_{m \to \infty} (x_{m+n} + y_{m+n})^{p^m} \right)_n$$

exists and defines an element of $A^\flat$.

(b) *Using (2.7.1.1) to define addition, $A^\flat$ is a perfect uniform Huber ring of characteristic p.*

(c) *The subset $A^{\flat+} := \varprojlim_{x \mapsto x^p} A^+$ is a subring of $A^\flat$. Moreover, for any ideal of definition I of A^+ for which $p \in I^p$, the map*

$$A^{\flat+} = \varprojlim_{x \mapsto x^p} A^+ \to \varprojlim_{x \mapsto x^p} (A^+/I)$$

of topological rings, for the inverse limit of discrete topologies on the target, is an isomorphism.

PROOF. In the ring $W(\mathbb{F}_p[x^{p^{-\infty}}, y^{p^{-\infty}}])$, we have

$$(2.7.1.2) \qquad [x + y] = \lim_{m \to \infty} ([x^{p^{-m}}] + [y^{p^{-m}}])^{p^m};$$

from this equality, we easily deduce (a).

In $A^\flat$, it is obvious that addition is commutative, multiplication distributes over addition, the p-power map is a bijection, and adding something to itself p times gives zero; using (2.7.1.2), we also see that addition is associative and continuous. It follows that $A^\flat$ is a perfect topological ring of characteristic p and that $x \mapsto |x^\sharp|$ is a norm defining the topology of $A^\flat$; in particular, $A^\flat$ is a uniform Huber ring. This proves (b), from which (c) follows easily using Lemma 2.2.4. □

HYPOTHESIS 2.7.2. For the remainder of §2.7, let (A, A^+) be a perfectoid pair. Keep in mind that we do not yet know that $A^\flat$ is analytic; this will be established in Lemma 2.7.8. Consequently, we need to be a bit wary about applying results from §2.6 to avoid creating a vicious circle (see Hypothesis 2.6.1).

DEFINITION 2.7.3. By Lemma 2.2.7, there exists a unique homomorphism $\theta : W(A^{\flat+}) \to A^+$ satisfying $\theta([x]) = \sharp(x)$ for all $x \in A^{\flat+}$. Since $A^+/((p) + I) \to A^+/((p) + I^p)$ is surjective for an ideal of definition I as in Definition 2.1.1 (see Remark 2.1.5), θ is surjective. The map θ extends to a homomorphism $W^\flat(A^\flat) \to A$ which we also call θ; however, we do not yet know that this map is surjective (this will follow from Lemma 2.7.9).

LEMMA 2.7.4. *For any ideal of definition I of A^+ with $p \in I^p$, there exist topologically nilpotent elements $\overline{x}_1, \ldots, \overline{x}_n$ of $A^{\flat+}$ such that $\sharp(\overline{x}_1), \ldots, \sharp(\overline{x}_n)$ generate I.*

PROOF. Choose generators $x_1, \ldots, x_n$ of I. The surjectivity of θ implies that $\overline{x}_1, \ldots, \overline{x}_n$ can be chosen so that $x_i - \sharp(\overline{x}_i) \in I^p$ for $i = 1, \ldots, n$; this yields the claim. □

LEMMA 2.7.5. *The ideal* $\ker(\theta) \subseteq W(A^{\flat+})$ *is primitive.*

PROOF. It suffices to exhibit a single primitive generator. Choose $\overline{x}_1, \ldots, \overline{x}_n$ as in Lemma 2.7.4. Since $p \in I^p$, we can write p in the form $\sum_{i=1}^{n} y_i \sharp(\overline{x}_i)$ for some $y_i \in A^+$. Lift each y_i to $\tilde{y}_i \in W(A^{\flat+})$ and put

$$z := p - \sum_{i=1}^{n} \tilde{y}_i[\overline{x}_i] \in W(A^{\flat+});$$

then z is evidently primitive.

It remains to show that z generates $\ker(\theta)$. If on the contrary $y \in \ker(\theta)$ is not divisible by z, then we may apply Lemma 2.6.9 (which does not require $A^\flat$ to be analytic) to produce a prepared element $y' \in W(A^{\flat+})$ congruent to y modulo z; in particular, y' also belongs to $\ker(\theta)$ and is not zero. We now obtain a contradiction by a variant of the proof of Lemma 2.6.6. Let $\overline{y}' \in A^{\flat+}$ be the reduction of y'; since y' is prepared, we have $|p^{-1}(y' - [\overline{y}'])| \leq |\overline{y}'|$. Applying θ, we obtain $|-p^{-1}\sharp(\overline{y}')| \leq |\sharp(\overline{y}')|$, which is only possible if $\sharp(\overline{y}') = 0$ and hence $\overline{y}' = 0$. This contradicts the assertion that y' is nonzero and prepared. □

LEMMA 2.7.6. *The map* $v \mapsto v \circ \sharp$ *defines an injective map* $\mathrm{Spa}(A, A^+) \to \mathrm{Spa}(A^\flat, A^{\flat+})$.

PROOF. The map is well-defined by Lemma 2.7.5 and Lemma 2.6.12 (or simply imitating the proof of the latter). Since the image of $\sharp : A^{\flat+} \to A^+$ generates a dense $\mathbb{Z}$-subalgebra of A^+, the map $\mathrm{Spa}(A^+, A^+) \to \mathrm{Spa}(A^{\flat+}, A^{\flat+})$ is injective. Combining this observation with Corollary 1.1.4 yields the injectivity of $\mathrm{Spa}(A, A^+) \to \mathrm{Spa}(A^\flat, A^{\flat+})$. □

LEMMA 2.7.7. *Suppose that A is Tate and admits a pseudouniformizer ϖ such that ϖ^p divides p in A^+ and $\varphi : A^+/(\varpi) \to A^+/(\varpi^p)$ is surjective. Then $A^\flat$ is Tate (hence perfectoid) and the map of Lemma 2.7.6 is bijective.*

PROOF. By the proof of Lemma 2.7.4 in the case $n = 1$, we can find $\overline{\varpi} \in A^{\flat+}$ such that $\varpi - \sharp(\overline{\varpi}) \in \varpi^p A^+$ and hence $\varpi A^+ = \sharp(\overline{\varpi})A^+$. It follows that $\overline{\varpi}$ is a pseudouniformizer of $A^\flat$, so $A^\flat$ is Tate. By Lemma 2.7.1, $(A^\flat, A^{\flat+})$ is a perfectoid pair; in fact, the triple $(A^\flat, A^{\flat+}, \ker(\theta))$ corresponds to (A, A^+) as in Lemma 2.6.14. By Corollary 2.6.15, the map of Lemma 2.7.6 is bijective. □

LEMMA 2.7.8. *With notation as in Lemma 2.7.4, the elements $\overline{x}_1, \ldots, \overline{x}_n$ generate the unit ideal in $A^\flat$. In particular, $A^\flat$ is analytic.*

PROOF. Apply Lemma 2.7.5 to construct a primitive generator z of $\ker(\theta)$. Promote A to a uniform Banach ring as per Remark 1.5.4; pulling back along $\sharp$ then provides a norm promoting $A^\flat$. Since $\sharp(\overline{x}_1), \ldots, \sharp(\overline{x}_n)$ generate the unit ideal in A, we may form the associated standard rational covering of $\mathrm{Spa}(A, A^+)$; namely, for $i = 1, \ldots, n$, let $(A, A^+) \to (B_i, B_i^+)$ be the rational localization defined by the parameters $\sharp(\overline{x}_1), \ldots, \sharp(\overline{x}_n), \sharp(\overline{x}_i)$. By Lemma 2.7.1, $B_i^\flat$ is a uniform Huber ring containing $\overline{x}_i$ as a pseudouniformizer (because $\sharp(\overline{x}_i)$ is invertible in B_i), and hence a Tate perfectoid ring of characteristic p. (This did not yet require

Lemma 2.7.7 because we already had a choice of $\overline{\varpi}$ in mind.) The surjective map $W^b(A^\flat)\langle T_1, \ldots, T_n\rangle \to A\langle T_1, \ldots, T_n\rangle \to B$ factors through $W^b(B_i^\flat) \to B_i$ via the map taking T_i to $[\overline{f}_i/\overline{g}]$, so $W^b(B_i^\flat) \to B_i$ is surjective. Let $(B_i', B_i'^+)$ be the untilt of $(B_i^\flat, B_i^{\flat+})$ corresponding to the ideal (z), which is perfectoid by Lemma 2.6.14; we now have a surjective map $B_i' \to B_i$. The map $\mathrm{Spa}(B_i, B_i^+) \to \mathrm{Spa}(B_i', B_i'^+)$ is thus a closed immersion, and hence a homeomorphism of $\mathrm{Spa}(B_i, B_i^+)$ onto a closed subset of $\mathrm{Spa}(B_i', B_i'^+)$. On the other hand, by Corollary 2.6.15, the image of $\mathrm{Spa}(B_i', B_i'^+) \to \mathrm{Spa}(A, A^+)$ consists entirely of points v for which $v(\sharp(\overline{x}_j)) \leq v(\sharp(\overline{x}_i))$ for $j = 1, \ldots, n$, and hence is contained in $\mathrm{Spa}(B_i, B_i^+)$. It follows that $B_i' \to B_i$ must in fact be an isomorphism, and so (B_i, B_i^+) is perfectoid.

For some suitably large m, $\varpi := \sharp(\overline{g}^{p^{-m}})$ is a pseudouniformizer of B_i such that ϖ^p divides p in B_i^+ and $\varphi : B_i^+/(\varpi) \to B_i^+/(\varpi^p)$ is bijective. We may thus apply Lemma 2.7.7 to deduce that $\mathrm{Spa}(B_i, B_i^+) \to \mathrm{Spa}(B_i^\flat, B_i^{\flat+})$ is bijective, and hence that $\mathcal{M}(B_i) \to \mathcal{M}(B_i^\flat)$ is bijective.

For $\alpha \in \mathcal{M}(A^\flat)$ with $\alpha(\overline{x}_1), \ldots, \alpha(\overline{x}_n)$ not all zero, we can find an index $i \in \{1, \ldots, n\}$ for which $\max\{\alpha(\overline{x}_1), \ldots, \alpha(\overline{x}_n)\} = \alpha(\overline{x}_i) \neq 0$, and then α belongs to the image of $\mathcal{M}(B_i^\flat)$. In particular, the joint zero locus Z of $\overline{x}_1, \ldots, \overline{x}_n$ in $\mathcal{M}(A^\flat)$, which is closed, has as complement the image of $\mathcal{M}(B_1^\flat \oplus \cdots \oplus B_n^\flat)$ in $\mathcal{M}(A^\flat)$, which is also closed (because the image of a continuous map from a quasicompact space to a Hausdorff space is closed). Consequently, Z is a closed-open subset of $\mathcal{M}(A^\flat)$; since $A^\flat$ is a Banach algebra over the trivially valued field $\mathbb{F}_p$, we may apply [**13**, Theorem 7.4.1] to realize Z as the zero locus of some idempotent element $\overline{e} \in A^\flat$. Put $e = \sharp(\overline{e})$; then e is an idempotent of A which vanishes nowhere, and so $e = 1$. It follows that $\overline{e} = 1$, $Z = \emptyset$, and $A^\flat$ is analytic. $\square$

LEMMA 2.7.9. *The formula*

$$(A, A^+) \mapsto (R := A^\flat, R^+ := A^{\flat+}, I := \ker(\theta : W(R^+) \to A^+))$$

defines a functor from perfectoid pairs (A, A^+) to triples (R, R^+, I), in which the pair (R, R^+) is a perfectoid pair of characteristic p and I is a primitive ideal of $W(R^+)$. This functor and the functor from Lemma 2.6.14 are quasi-inverses of each other, so they are both equivalences of categories.

PROOF. By Lemma 2.7.1 and Lemma 2.7.8, (R, R^+) is a perfectoid pair of characteristic p. By Lemma 2.7.5, I is a primitive ideal. It is evident that applying the functor from Lemma 2.6.14 followed by this functor yields a functor naturally isomorphic to the identity; to complete the proof, we must verify that the composition in the other direction is also naturally isomorphic to the identity.

Notate the composite functor as $(A, A^+) \mapsto (B, B^+)$; then the map $\theta : W^b(A^\flat) \to A$ factors through an injective morphism $B \to A$ which (by the discussion of Definition 2.7.3) restricts to an isomorphism $B^+ \to A^+$. Fix elements $\overline{x}_1, \ldots, \overline{x}_n \in A^{\flat+}$ as in Lemma 2.7.4 (for some ideal of definition); for any $y \in A$, there exists a nonnegative integer m such that $\sharp(\overline{x}_1)^m y, \ldots, \sharp(\overline{x}_n)^m y \in A^+ = B^+ \subseteq B$. Since $\sharp(\overline{x}_1), \ldots, \sharp(\overline{x}_n)$ generate the unit ideal in B, it follows that $y \in B$, and so $B \to A$ is surjective. This proves the claim. $\square$

We have now established Theorem 2.3.9, and thus are free to invoke it in subsequent proofs.

LEMMA 2.7.10. *The map $\theta : W^b(R) \to A$ induces a surjection $W(R^\circ) \to A^\circ$ and an isomorphism $R^\circ/R^{\circ\circ} \to A^\circ/A^{\circ\circ}$. (Recall that $A^{\circ\circ}$ denotes the set of topologically nilpotent elements of A.)*

PROOF. It is clear that $\theta(W(R^\circ)) \subseteq A^\circ$; we establish the reverse implication as follows. Start with an element $y \in A^\circ$, choose a lift $x \in W^b(R)$ of y, form the sequence $x_0, x_1, \ldots$ as in Lemma 2.6.9, and choose m for which x_m is prepared. The reduction $\overline{x}_m$ of x_m satisfies $|p^{-1}(x_m - [\overline{x}_m])| \leq |\overline{x}_m|$, and so $|y - \sharp(\overline{x}_m)| < |\sharp(\overline{x}_m)|$. This forces $|\overline{x}_m| = |y| \leq 1$; since x_m is prepared, this in turn implies $x_m \in W(R^\circ)$.

This produces the surjection $W(R^\circ) \to A^\circ$; since p is topologically nilpotent in A, the map $W(R^\circ) \to A^\circ \to A^\circ/A^{\circ\circ}$ factors through a surjection $R^\circ/R^{\circ\circ} \to A^\circ/A^{\circ\circ}$. Using Lemma 2.6.9 again, we see that this map is injective. $\square$

2.8. More proofs. We continue to establish the basic properties of perfectoid rings.

LEMMA 2.8.1. *Let $f : (A, A^+) \to (B, B^+)$ be a morphism of perfectoid Huber pairs such that $f^\flat$ is strict. (By Theorem 2.4.2 this is equivalent to f being strict, but this lemma is used in the proof of that statement.) Let $z \in W(A^{\flat+})$ be a generator of $\ker(\theta)$. Then within $W^b(B^\flat)$ we have the equalities*

$$zW(B^{\flat+}) \cap W(A^{\flat+}) = zW(A^{\flat+}), \qquad zW^b(B^\flat) \cap W^b(A^\flat) = zW(B^\flat)$$

PROOF. We check the first assertion, the second being similar. If $x \in W(A^{\flat+})$ can be written in $W(B^{\flat+})$ as yz for some y, then from the shape of z we see that y is congruent modulo $[\overline{z}]$ to an element of $W(A^{\flat+})$. Writing $y = w_0 + [\overline{z}]y_1$ with $w_0 \in W(A^{\flat+})$, we may repeat the argument to see that y_1 is congruent modulo $[\overline{z}]$ to an element of $W(A^{\flat+})$, and so on. Since $W(A^{\flat+})$ is complete for the $[\overline{z}]$-adic topology, we deduce the claim. $\square$

We have now established Theorem 2.4.2, and thus are free to use it in subsequent proofs.

LEMMA 2.8.2. *Let (A, A^+) be a perfectoid Huber pair. Then for any positive integer m, there exists an ideal of definition I_m of A^+ such that $p \in I_m^{p^m}$.*

PROOF. The case $m = 1$ is included in Definition 2.1.1. Given an ideal of definition I_m such that $p \in I^{p^m}$, choose generators $x_1, \ldots, x_n$ of I_m. By Remark 2.1.5, there exist elements $y_1, \ldots, y_n$ of A^+ such that $y_i^p \equiv x_i \pmod{I_m^p}$; it follows easily that $y_1^p, \ldots, y_m^p$ are topologically nilpotent and generate I_m. Hence the ideal I_{m+1} of A^+ generated by $y_1, \ldots, y_m$ is also an ideal of definition and satisfies $I_{m+1}^p = I_m$; hence $p \in I^{p^{m+1}}$ as desired. $\square$

The following metric criterion for the perfectoid property is adapted from [**117**, Proposition 3.6.2].

LEMMA 2.8.3. *Let (A, A^+) be a uniform Huber pair and let I be an ideal of definition of A^+ such that $p \in I^p$. Using the ideal I, promote A to a uniform Banach ring as per Remark 1.5.4. Then A is perfectoid if and only if there exists some $c \in (0, 1)$ such that for every $x \in A$, there exists $y \in A$ with $|x - y^p| \leq c|x|$.*

PROOF. If there exists some c as described, then for m sufficiently large, the ideal of definition I_m given by Lemma 2.8.2 has the property that $\varphi : A^\circ/I_m \to A^\circ/I_m^p$ is surjective, so A is perfectoid. Conversely, suppose that A is perfectoid.

By Theorem 2.3.9, the map $\theta : W^b(A^\flat) \to A$ is surjective with kernel generated by some primitive element $z \in W(A^{\flat+})$. Put $c := |\overline{z}_0|$. By Lemma 2.6.9, for each $x \in A$ there exists $\overline{y} \in A^\flat$ such that $|x - \sharp(\overline{y})| \leq c|x|$; we may then take $y = \sharp(\overline{y}^{1/p})$. □

LEMMA 2.8.4. *Let $f : A \to B$ be a morphism of uniform Banach rings with dense image. If A is perfectoid, then B is perfectoid and $f^\flat$ has dense image.*

PROOF. By arguing as in Lemma 2.7.1, we see that $B^\flat$ is a uniform Huber ring which is perfect of characteristic p; it is also analytic because it receives the continuous map $f^\flat$ from $A^\flat$. By Theorem 2.3.9, we have $A \cong W^b(A^\flat)/(z)$ for some primitive element $z \in W(A^{\flat+})$. Let $B_0^\flat$ be the closure of the image of $A^\flat$ in $B^\flat$, which is a perfectoid ring of characteristic p, and set $B_0 := W^b(B_0^\flat)/(z)$. Since the composition $W^b(A^\flat) \to A \to B$ has dense image, so does the induced map $B_0 \to B$. We may thus use B_0 to verify that B satisfies the condition of Lemma 2.8.3, so B is perfectoid. The map $B_0^\flat \to B^\flat$ is a strict inclusion, as then is $B_0 \to B$ by Theorem 2.4.2; since the latter map also has dense image, it is an isomorphism. Hence $f^\flat$ has dense image. □

LEMMA 2.8.5. *Let $f : A \to B$ be a surjective morphism of perfectoid Banach rings. Then the quotient norm induced by the spectral norm on A coincides with the spectral norm on B. (In other words, $B^+/f(A^+)$ is an almost zero B^+-module; see Definition 2.9.4 below.)*

PROOF. We adapt the argument from [117, Proposition 3.6.9(c)]. Since f is surjective, the induced map $\mathcal{M}(B) \to \mathcal{M}(A)$ is injective; by Lemma 1.5.21, it follows that f has operator norm at most 1 (i.e., it is *submetric*). By Theorem 1.1.9, the quotient norm on B is bounded by some constant $c > 1$ times the given norm. It will suffice to check that c can be replaced by $c^{1/p}$; in fact, it further suffices to check that for every $b \in B$, there exists $a \in A$ such that $|a| \leq c|b|$ and $|b - f(a)| \leq p^{-1/p}|b|$ (as we may then iterate the construction).

To begin, lift b^p to $a' \in A$ with $|a'| \leq c|b^p|$. Choose some lift x of a' to $W^b(A^\flat)$, then construct the sequence x_m as in Lemma 2.6.9 with respect to a primitive generator of $\ker(\theta : W^b(A^\flat) \to A)$ provided by Lemma 2.7.5. For m sufficiently large,

$$(2.8.5.1) \qquad \alpha(a' - \sharp(\overline{x}_{m,0})) \leq p^{-1} \max\{\alpha(a'), |b^p|\} \qquad (\alpha \in \mathcal{M}(A)).$$

We claim that $a := \sharp(\overline{x}_{m,0}^{1/p})$ has the desired property. To see this, we may use Lemma 1.5.21 to reformulate the desired inequality as

$$(2.8.5.2) \qquad \beta(b - f(a)) \leq p^{-1/p}\beta(b) \qquad (\beta \in \mathcal{M}(B)).$$

To check this, we first deduce from (2.8.5.1) that $\beta(b^p - f(a)^p) \leq p^{-1}|b^p|$. If $\beta(p) > 0$, we may deduce (2.8.5.2) from a simple analysis of the p-th power map in a mixed-characteristic nonarchimedean field [108, Lemma 10.2.2]. If instead $\beta(p) = 0$, we may instead deduce (2.8.5.2) more trivially. □

LEMMA 2.8.6. *Let $f : A \to B$ be a surjective morphism of perfectoid rings. Then $f^\flat$ is also surjective.*

PROOF. We adapt the argument from [117, Proposition 3.6.9(d)]. Promote A and B to uniform Banach rings (equipped with their spectral norms) as per Remark 1.5.4. It will suffice to check that each $\overline{b} \in B^\flat$ can be lifted to some $\overline{a} \in A^\flat$

with $|\bar{a}| \leq p^{1/2}|\bar{b}|$; in fact, it further suffices to check that there exists $\bar{a}$ with $|\bar{a}| \leq p^{1/2}|\bar{b}|$ and $|\bar{b} - f^\flat(\bar{a})| \leq p^{-1/2}|\bar{b}|$ (as we may then iterate the construction).

Apply Lemma 2.8.5, we may lift $\sharp(\bar{b}) \in B$ to $a' \in A$ with $|a'| \leq p^{1/2}|\bar{b}|$. Using Lemma 2.6.9 as in the proof of Lemma 2.8.5, we can lift a' to $x \in W^b(A^\flat)$ so that
$$|a' - \sharp(\bar{x}_0)| \leq p^{-1}\max\{|a'|, |\bar{b}|\}.$$
We claim that $\bar{a} := \bar{x}_0$ has the desired property. To see this, apply Remark 2.2.8 to deduce that
$$\left|\sharp(\bar{b}) - \sharp(f^\flat(\bar{a})) - \sharp(\bar{b} - f^\flat(\bar{a}))\right| \leq p^{-1/2}|\bar{b}|.$$
From this, it follows that
$$\left|\bar{b} - f^\flat(\bar{a})\right| = \left|\sharp(\bar{b} - f^\flat(\bar{a}))\right| \leq p^{-1/2}|\bar{b}|$$
as desired. $\square$

LEMMA 2.8.7. *Let (A, A^+) be a perfectoid pair. Let z be a generator of $\ker(\theta: W(A^{\flat+}) \to A^+)$. Let $(A, A^+) \to (B, B^+), (A, A^+) \to (C, C^+)$ be two morphisms of perfectoid pairs. Then the pair $(B\widehat{\otimes}_A C, B^+\widehat{\otimes}_{A^+} C^+)$ is the untilt of the pair $(B^\flat\widehat{\otimes}_{A^\flat} C^\flat, B^{\flat+}\widehat{\otimes}_{A^{\flat+}} C^{\flat+})$ corresponding to the ideal (z).*

PROOF. This is clear in the case where C is equal to the perfectoid Tate algebra $A\langle T_s^{p^{-\infty}} : s \in S\rangle$ for some possibly infinite index set S (i.e., the completion of the perfect polynomial ring $A[T_s^{p^{-\infty}} : s \in S]$ for the Gauss norm) and $C^+ = A^+\langle T_s^{p^{-\infty}} : s \in S\rangle$, as then the completed tensor product is obtained by substituting B for A, and likewise on the tilt side.

To handle the general case, put
$$\tilde{C} = A\langle T_s^{p^{-\infty}} : s \in C^{\flat+}\rangle, \qquad \tilde{C}^+ = A^+\langle T_s^{p^{-\infty}} : s \in C^{\flat+}\rangle;$$
then there is a morphism $(\tilde{C}, \tilde{C}^+) \to (C, C^+)$ sending $T_s^{p^{-n}}$ to $\sharp(s^{p^{-n}})$. The tilt of this morphism has image containing all of $C^{\flat+}$, and hence is surjective (compare the proof of Lemma 2.7.9); it is thus surjective by Theorem 2.4.4. By the previous paragraph, $(B\widehat{\otimes}_A \tilde{C}, B^+\widehat{\otimes}_{A^+} \tilde{C}^+)$ is the untilt of $(B^\flat\widehat{\otimes}_{A^\flat} \tilde{C}^\flat, B^{\flat+}\widehat{\otimes}_{A^{\flat+}} \tilde{C}^{\flat+})$ corresponding to the ideal (z). Since $\tilde{C}^\flat \to C^\flat$ is strict surjective by Theorem 1.1.9, so is $B^\flat\widehat{\otimes}_{A^\flat} \tilde{C}^\flat \to B^\flat\widehat{\otimes}_{A^\flat} C^\flat$; by Theorem 2.4.4(b), we can thus untilt the surjective morphism $(B^\flat\widehat{\otimes}_{A^\flat} \tilde{C}^\flat, B^{\flat+}\widehat{\otimes}_{A^{\flat+}} \tilde{C}^{\flat+}) \to (B^\flat\widehat{\otimes}_{A^\flat} C^\flat, B^{\flat+}\widehat{\otimes}_{A^{\flat+}} C^{\flat+})$ to get a surjective morphism $(B\widehat{\otimes}_A \tilde{C}, B^+\widehat{\otimes}_{A^+} \tilde{C}^+) \to (D, D^+)$. By the same token, $\tilde{C} \to C$ is strict surjective, as then is $B\widehat{\otimes}_A \tilde{C} \to \widehat{\otimes}_A C$; we thus obtain a surjective morphism $(B\widehat{\otimes}_A C, B^+\widehat{\otimes}_{A^+} C^+) \to (D, D^+)$.

To see that this map is an isomorphism, it suffices to do so at the level of the rings of integral elements. To do this, first observe that D^+ is isomorphic to the completion of $B^+\otimes_{A^+} C^+$ with respect to an ideal of definition of A^+ (this being obviously true with C replaced by $\tilde{C}$; the general case follows because $\tilde{C}^+ \to C^+$ is surjective). It follows that D^+ is isomorphic to the completion of $B^+\otimes_{A^+} C^+$ with respect to an ideal of definition of A^+. This in turn implies that we get the same answer when completing with with respect to the spectral norm, which yields the desired result. (Compare [**118**, Theorem 3.3.13], or [**156**, Proposition 6.18] for a different approach using almost commutative algebra.) $\square$

LEMMA 2.8.8. *Suppose that* $\overline{f}_1, \ldots, \overline{f}_n, \overline{g} \in A^\flat$ *are elements such that* $\sharp(\overline{f}_1), \ldots,$ $\sharp(\overline{f}_n), \sharp(\overline{g})$ *generate the unit ideal in* A. *Let* $(A, A^+) \to (B, B^+)$ *be the rational localization defined by these parameters. Then*

$$B \cong A\langle T_1^{p^{-\infty}}, \ldots, T_n^{p^{-\infty}}\rangle/(\sharp(\overline{g}^{p^{-j}})T_i^{p^{-j}} - \sharp(\overline{f}^{p^{-j}}) : i = 1, \ldots, n; j = 0, 1, \ldots)^\wedge.$$

In particular, (B, B^+) *is an untilt of the localization of* $(A^\flat, A^{\flat+})$ *defined by* $\overline{f}_1, \ldots, \overline{f}_n, \overline{g}$.

PROOF. We emulate [**117**, Remark 3.6.16]. Denote the quotient being compared to B as B'. Choose $h_1, \ldots, h_n, k \in A$ such that $h_1\sharp(\overline{f}_1) + \cdots + h_n\sharp(\overline{f}_n) + k\sharp(\overline{g}) = 1$. For $i_1, \ldots, i_n \in \mathbb{Z}[p^{-1}]_{\geq 0}$, $T_1^{i_1} \cdots T_n^{i_n}$ represents the same class in B' as

$$(k + h_1T_1 + \cdots + h_nT_n)^n \sharp(\overline{f}_1^{i_1 - \lfloor i_1 \rfloor})$$
$$\cdots \sharp(\overline{f}_n^{i_n - \lfloor i_n \rfloor})\sharp(\overline{g}^{n-(i_1-\lfloor i_1\rfloor]+\cdots+i_n-\lfloor i_n\rfloor)})T_1^{\lfloor i_1 \rfloor} \cdots T_n^{\lfloor i_n \rfloor}.$$

We thus construct an inverse of the map $B \to B'$. □

LEMMA 2.8.9. *Theorem 2.5.9 holds in the case where A is a perfectoid field. Moreover, the tilting operation preserves the degrees of field extensions.*

PROOF. See [**177**, Lecture 2] or the references in Remark 2.5.13. □

LEMMA 2.8.10. *Let $R \to S$ be a finite (resp. finite étale) morphism of perfectoid rings of characteristic p. Then any untilt of this morphism is finite (resp. finite étale).*

PROOF. Let $A \to B$ be an untilt of $R \to S$. Choose $\overline{x}_1, \ldots, \overline{x}_n$ which generate S as an R-module. By the open mapping theorem, the resulting map $R^n \to S$ is strict; it follows easily from this that $[\overline{x}_1], \ldots, [\overline{x}_n]$ generate $W^\flat(S)$ over $W^\flat(R)$, and hence that $\sharp(\overline{x}_1), \ldots, \sharp(\overline{x}_n)$ generate B over A.

Suppose now that $R \to S$ is finite étale. Since A is uniform, as in [**117**, Proposition 2.8.4] we may check that $A \to B$ is finite flat by checking that its rank is locally constant, which follows from Lemma 2.8.9. (Note that this argument uses essentially the fact that A is reduced; compare [**53**, Exercise 20.13].)

Now recall that $R \to S$ is finite étale if and only if both $R \to S$ and $S \otimes_R S \to S$ are finite flat. (The "only if" direction is obvious. The "if" direction holds because $S \otimes_R S \to S$ being flat implies that $R \to S$ is formally unramified [**166**, Tag 092M]. See also the discussion of *weakly étale morphisms* in [**166**, Tag 092A].) By the compatibility of untilting with tensor products (Lemma 2.8.7), we may repeat the previous argument to see that $B \otimes_A B \to B$ is finite flat, and then deduce that $A \to B$ is finite étale. □

LEMMA 2.8.11. *Let (A, A^+) be a perfectoid pair. Let z be a generator of* $\ker(\theta : W(A^{\flat+}) \to A^+)$. *Then the functor $B^\flat \mapsto W^\flat(B^\flat)/(z)$ defines an equivalence of categories between finite étale $B^\flat$-algebras and finite étale B-algebras.*

PROOF. The functor is well-defined by Lemma 2.8.10, and fully faithful by Theorem 2.3.9; it thus suffices to check essential surjectivity. In the case where A is a perfectoid field, essential surjectivity follows from Lemma 2.8.9. In the general case, given a finite étale morphism $A \to B$, we may combine the field case with the henselian property of local rings (Remark 1.2.5) to produce a rational covering $\{(A, A^+) \to (A_i, A_i^+)\}$ such that for each i, $B \otimes_A A_i$ is the untilt of some finite étale A_i-algebra. By full faithfulness, these modules collate to give a finite étale $\mathcal{O}$-module on $\mathrm{Spa}(A^\flat, A^{\flat+})$. Since $\mathrm{Spa}(A^\flat, A^{\flat+})$ is sheafy by Corollary 2.5.4, we may apply Theorem 1.4.2 (or Remark 1.4.3) to obtain a finite étale $A^\flat$-module $B^\flat$ which untilts to B. In particular, B is perfectoid. □

EXERCISE 2.8.12. Let A be a perfectoid ring. Note that in general, not every finite A-algebra is perfectoid, even if we restrict to characteristic p (trivially by adjoining nilpotents to destroy uniformity, or less trivially as in Exercise 2.5.8 where the result is still uniform). Nonetheless, show that $B \to B^\flat$ defines an equivalence of categories between *perfectoid* finite A-algebras and perfectoid finite $A^\flat$-algebras.

2.9. Additional results about perfectoid rings.
We mention some additional results whose proofs lie outside the scope of these notes.

THEOREM 2.9.1 (Kedlaya). *Any perfectoid ring whose underlying ring is a field is a perfectoid field. (That is, it is not necessary to assume in advance that the topology is given by a multiplicative norm.)*

PROOF. Any such ring is Tate (not just analytic), so [**113**, Theorem 4.2] applies. □

COROLLARY 2.9.2. *Let A be a perfectoid ring. Let I be a maximal ideal of A (which is automatically closed; see Remark 1.1.1). If A/I is uniform, then it is a perfectoid field.*

PROOF. If A/I is uniform, it is again a perfectoid ring by Theorem 2.4.4, and so Theorem 2.9.1 applies. □

The following corollary is analogous to a standard fact about perfect rings.

COROLLARY 2.9.3. *Any noetherian perfectoid ring is a finite direct sum of perfectoid fields.*

PROOF. Let A be a noetherian perfectoid ring. For $\overline{x} \in A^\flat$, the sequence of ideals $(\sharp(\overline{x}^{p^{-n}}))_n$ of A forms an ascending chain, and hence must stabilize. That is, there exists a positive integer n such that for $y = \sharp(\overline{x}^{p^{-n}})$, we have $y = wy^p$ for some $w \in A$. For such w, $y^{p-1}w$ is an idempotent in A, which defines a splitting $A \cong A_1 \oplus A_2$ of perfectoid rings by projecting onto A_1. Since $y(y^{p-1}w) = y$, y must project to a unit in A_1 and to zero in A_2; consequently, $\overline{x}$ must project to a unit in $A_1^\flat$ and to zero in $A_2^\flat$. In other words, every element of $A^\flat$ equals a unit times an idempotent. Since idempotent ideals in $A^\flat$ satisfy the ascending chain condition (as seen by applying $\sharp$), we deduce that $A^\flat$ is a finite direct sum of fields, each of which must be a perfectoid field by Theorem 2.9.1. □

DEFINITION 2.9.4. For A a perfectoid ring, an A^+-module M is *almost zero* if it is annihilated by every topologically nilpotent element of A^+. Such modules form a thick Serre subcategory of the category of A^+-modules, so one may form the quotient category.

THEOREM 2.9.5. *Let (A, A^+) be a perfectoid pair. Then for each $i > 0$, the A^+-modules $H^i(\mathrm{Spa}(A, A^+), \mathcal{O}^+)$ and $H^i(\mathrm{Spa}(A, A^+)_{\mathrm{et}}, \mathcal{O}^+)$ are almost zero.*

PROOF. See [**156**, Lemma 6.3(iv)] in the case where A is an algebra over a perfectoid field, [**117**, Lemma 9.2.8] in the case where A is an algebra over $\mathbb{Q}_p$, or [**118**, Corollary 3.3.20] in the case where A is Tate. The analytic case is similar. □

In the Tate case, the following statement is [**118**, Theorem 3.7.4].

EXERCISE 2.9.6. A *seminormal* ring is a ring R in which the map
$$R \to \{(y, z) \in R \times R : y^3 = z^2\}, \qquad x \mapsto (x^2, x^3)$$
is an isomorphism. This definition is due to Swan [**168**]. Using Theorem 2.9.5, show that any perfectoid ring is seminormal. (Hint: work locally around a point $v \in \mathrm{Spa}(A, A^+)$, distinguishing between the cases where $v(y), v(z)$ are both zero or both nonzero.)

THEOREM 2.9.7. *Let A be a uniform analytic Huber ring such that some faithfully finite étale (i.e., faithfully flat and finite étale) A-algebra is perfectoid. Then A is perfectoid.*

PROOF. In the case where A is Tate, this is [**118**, Theorem 3.3.25]; the analytic case is similar. □

PROBLEM 2.9.8. Does Theorem 2.9.7 remain true if "étale" is weakened to "flat"?

THEOREM 2.9.9. *Let (A, A^+) be a (not necessarily sheafy) Huber pair in which A is Tate and p is topologically nilpotent, and put $X := \mathrm{Spa}(A, A^+)$. Then there exists a directed system $(A, A^+) \to (A_i, A_i^+)$ of faithfully finite étale morphisms such that the completion of $\varinjlim_i A_i$ for the seminorm induced by the spectral seminorm on each A_i is a perfectoid ring. (Note that the transition morphisms are isometric for the spectral seminorms.)*

PROOF. For A a Huber ring over $\mathbb{Q}_p$, this follows from an argument of Colmez: for X affinoid, it suffices to repeatedly adjoin p-power roots of units. See [**157**, Proposition 4.8] (nominally in the locally noetherian case, but the argument does not depend on this) or [**117**, Lemma 3.6.26, Lemma 9.2.5]. In the Tate case, a modification of Colmez's argument by Scholze applies; see [**118**, Lemma 3.3.28]. □

REMARK 2.9.10. It is far from clear whether Theorem 2.9.9 remains true if we assume only that A is analytic, rather than Tate; there is no obvious mechanism to ensure in this case that A has "enough" finite étale extensions. Nonetheless, Theorem 2.9.9 implies that in any analytic adic space (or preadic space; see Definition 1.11.2) on which p is topologically nilpotent, in the pro-étale topology (to be introduced in Weinstein's third lecture [**177**, Lecture 3], but see also Definition 3.8.1) there exists a neighborhood basis consisting of perfectoid spaces. This fact underpins the use of perfectoid spaces in p-adic Hodge theory, as in the lectures of Bhatt [**17**] and Caraiani [**28**]; it also gives rise to the functor from analytic adic spaces in which p is topologically nilpotent to diamonds (Definition 4.3.1).

We next establish an observation of Bhatt about quotients of perfectoid rings, using a key lemma of André from the proof of the direct summand conjecture (see Remark 4.3.19).

LEMMA 2.9.11 (André). *Let A be a perfectoid ring, let $\varpi \in A$ be a pseudouniformizer, and let $g \in A$ be an element. Let B be the completion of*
$$\varinjlim_{n \to \infty} A\langle T^{p^{-\infty}}\rangle \left\langle \frac{T-g}{\varpi^n} \right\rangle$$

for the infimum of the spectral norms (which does not depend on ϖ). Then B is perfectoid and $A^\circ/(\varpi) \to B^\circ/(\varpi)$ is almost faithfully flat; that is, for any $A^\circ/(\varpi)$-module M, the groups $\operatorname{Tor}_i^{A^\circ/(\varpi)}(M, B^\circ/(\varpi))$ are almost zero A°-modules for all $i > 0$.

PROOF. See [**6**, §2.5] or [**16**, Theorem 2.3]. □

THEOREM 2.9.12 (Bhatt). *Let A be a perfectoid ring, let I be a closed ideal of A, and let B be the uniform completion of A/I. Then B is perfectoid and the natural map $f: A \to B$ is surjective.*

PROOF. By Theorem 2.4.3(a), B is perfectoid. Choose a pseudouniformizer $\overline{\varpi} \in A^\flat$ such that $\varpi := \sharp(\overline{\varpi})$ divides p in A°. By repeated application of Lemma 2.9.11, we may construct a morphism $A \to A'$ with A' perfectoid such that $A^\circ/(\varpi) \to A'^\circ/(\varpi)$ is almost faithfully flat and every $g \in I$ admits a coherent sequence of p-power roots $\{g^{p^{-n}}\}_n$ in A'. Let J be the closure of the ideal of A' generated by the $g^{p^{-n}}$ for all g and n; by Corollary 2.4.6, A'/J is perfectoid.

By Theorem 2.4.1, $B' := B \widehat{\otimes}_A A'$ is again perfectoid. Note that B' is the uniform completion of the quotient of A' by the closed ideal generated by I; since each $g^{p^{-n}}$ projects to a nilpotent element of A'/IA', B is also the uniform completion of A'/J. Since the latter is already perfectoid, it is in fact isomorphic to B'; that is, $f' : A' \to B'$ is surjective.

By Theorem 2.4.4(b), $f'^\flat : A'^\flat \to B'^\flat$ is also surjective. By the open mapping theorem (Theorem 1.1.9) and perfectness, the map $A'^{\flat\circ} \to B'^{\flat\circ}$ is almost surjective; the map $A'^{\flat\circ}/(\overline{\varpi}) \to B'^{\flat\circ}/(\overline{\varpi})$ is thus almost surjective. This is the same map as $A'^\circ/(\varpi) \to B'^\circ/(\varpi)$; since $A^\circ/(\varpi) \to A'^\circ/(\varpi)$ is almost faithfully flat and $B'^\circ/(\varpi)$ is almost isomorphic to $B^\circ/(\varpi) \otimes_{A^\circ/(\varpi)} A'^\circ/(\varpi)$, we deduce that $A^\circ/(\varpi) \to B^\circ/(\varpi)$ is almost surjective. This is the same map as $A^{\flat\circ}/(\overline{\varpi}) \to B^{\flat\circ}/(\overline{\varpi})$; applying this repeatedly, we deduce that $A^{\flat\circ} \to B^{\flat\circ}$ is almost surjective. Consequently, $f^\flat$ is surjective; by Theorem 2.4.4, f is also surjective. □

COROLLARY 2.9.13. *Let $f : A \to B$ be a morphism of perfectoid rings. Then f is injective if and only if $f^\flat : A^\flat \to B^\flat$ is injective.*

PROOF. It is obvious that if $f^\flat$ fails to be injective, then so does f: if $\overline{x} \in \ker(f^\flat)$, then $\sharp(\overline{x}) \in \ker(f)$. Conversely, suppose that f has nonzero kernel I. Since B is uniform, any homomorphism $A/I \to B$ of Banach rings factors uniquely through the uniform completion of A/I; however, by Theorem 2.9.12, the latter has the form A/J for some closed ideal J containing I, and is again perfectoid. By Theorem 2.4.4, $A^\flat \to (A/J)^\flat$ is surjective, and cannot be an isomorphism because $J \neq 0$; it thus has nonzero kernel by Corollary 2.4.5. Since $f^\flat$ factors as $A^\flat \to (A/J)^\flat \to B^\flat$, it too has nonzero kernel. □

We obtain the following refinement of Theorem 2.9.1, which was left as an open problem in [**113**].

COROLLARY 2.9.14. *Let A be a perfectoid ring. Then for every maximal ideal I of A, the quotient A/I is a perfectoid field. In particular, there is exactly one point of $\mathcal{M}(A)$ with residue field A/I.*

PROOF. Since I is maximal, it is closed (Remark 1.1.1), so A/I is a nonzero Banach ring. In particular, the uniform completion of A/I is also nonzero (Corollary 1.5.22); by Theorem 2.9.12, it has the form A/J for some closed ideal J of A containing I. By maximality, we must have $I = J$; hence A/I is a perfectoid ring. By Theorem 2.9.1, A/I is a perfectoid field. □

REMARK 2.9.15. Theorem 2.9.12 implies that in the language of [159], there is no difference between *closed immersions* and *strongly closed immersions* of perfectoid spaces (though the two concepts do differ for uniform adic spaces). That is, for any perfectoid pair (A, A^+), any closed subset of $\mathrm{Spa}(A, A^+)$ which is the zero locus of some ideal of A has the form $\mathrm{Spa}(B, B^+)$ where B is a perfectoid quotient of A. Nonetheless, there remain some open questions about the zero loci of perfectoid power series; for further discussion, see the student project of Weinstein [177].

REMARK 2.9.16. In Theorem 2.9.12, we do not know whether the ideal J/I of A/I is always equal to the closure of the nilradical. See Example 2.4.10 for a nontrivial example where this does in fact occur.

REMARK 2.9.17. We postpone one more result until we have discussed fundamental groups: an amazing recent theorem of Achinger that asserts that adic affinoid spaces on which p is topologically nilpotent have no higher étale homotopy groups. See Theorem 4.1.26 and Corollary 4.1.27.

REMARK 2.9.18. For discussion of various foundational problems concerning perfectoid rings and spaces in the spirit of the Scottish Book[11], see [115]. Another apt analogue in point-set topology is the book [167].

3. Sheaves on Fargues–Fontaine curves

We next pick up on a topic introduced in Weinstein's lectures [177]: the construction of Fargues–Fontaine which gives rise to a "moduli space of untilts" of a given perfectoid space. In this lecture, we study vector bundles and coherent sheaves on Fargues–Fontaine curves (associated to a perfectoid field) and relative Fargues–Fontaine curves (associated to a perfectoid ring or space), and a profound relationship between these sheaves and étale local systems. We will see in the final lecture how these results can be formally recast in a more suggestive manner that suggests how to put the analogy between mixed and equal characteristic on a firm footing.

Whereas in the first two lectures these notes constitute a fairly self-contained treatment of the material, some of the material in the last two lectures is far beyond the scope of what can be treated here. We thus revert to a more conventional order of presentation, in which we either prove statements on the spot or defer to external references.

---

[11]The Scottish Book is an artifact of the Lwów School of Mathematics, which was active in the 1930s and included Stefan Banach, Stanisław Mazur, Hugo Steinhaus, Stanisław Ulam, and other Polish mathematicians. The book records many problems that arose in the development of modern functional analysis. Mazur famously offered a live goose as the reward for the solution of a particular problem; when this problem was finally resolved by Per Enflo in the 1970s, Mazur dutifully presented Enflo with the promised prize in a ceremony broadcast on Polish television. The book was created at and named for the Scottish Café in Lwów; the city is now Lviv, Ukraine and the site of the Café now houses the Szkocka restaurant, where one can view a replica of the original book (as confirmed by the author in 2018).

3.1. Absolute and relative Fargues–Fontaine curves.
We begin by recalling the construction of Fargues–Fontaine [**60–63**] associated to a perfectoid field, and its generalization to perfectoid rings and spaces by Kedlaya–Liu [**117**, §8.7–8.8]. This generalization appears in a somewhat different guise also in [**162**, §11.2].

HYPOTHESIS 3.1.1. *Throughout §3.1, let (R, R^+) be a perfectoid pair of characteristic p and put $S = \mathrm{Spa}(R, R^+)$. Note that only the case where R is Tate is treated in* [**117**].

DEFINITION 3.1.2. Define the ring $\mathbf{A}_{\inf} := W(R^+)$. It is complete for the adic topology defined by the inverse image of some ideal of definition of R^+.

LEMMA 3.1.3. *Choose topologically nilpotent elements $\overline{x}_1, \ldots, \overline{x}_n \in R^+$ which generate the unit ideal in R.*

(a) *For the p-adic topology on $\mathbf{A}_{\inf}$, the ring $\mathbf{A}_{\inf}[p^{-1}]\langle \frac{[\overline{x}_1]}{p}, \ldots, \frac{[\overline{x}_n]}{p} \rangle$ is stably uniform.*

(b) *For $i = 1, \ldots, n$, for the $[\overline{x}_i]$-adic topology on $\mathbf{A}_{\inf}$, the ring*

$$\mathbf{A}_{\inf}[[\overline{x}_i]^{-1}]\left\langle \frac{p}{[\overline{x}_i]}, \frac{[\overline{x}_1]}{[\overline{x}_i]}, \ldots, \frac{[\overline{x}_n]}{[\overline{x}_i]} \right\rangle$$

is stably uniform.

PROOF. We will verify both claims using Corollary 2.5.5. Let S be the p-adic completion of $\mathbb{Z}_p[p^{p^{-\infty}}]$; the natural morphism $\mathbb{Z}_p \to S$ is split in the category of $\mathbb{Z}_p$-modules by the morphism $S \to \mathbb{Z}_p$ taking all fractional powers of p to zero. By the same token, every element of the p-adic completion of $\mathbf{A}_{\inf} \otimes_{\mathbb{Z}_p} S$ has a unique representation as a sum

$$x = \sum_{t \in \mathbb{Z}[p^{-1}]_{\geq 0}} x_t p^t \qquad (x_t \in \mathbf{A}_{\inf})$$

where for every positive integer m, there are only finitely many indices t for which $x_t \not\equiv 0 \pmod{p^m}$. From this, it follows that the p-adic topology is induced by a power-multiplicative norm, namely the one taking x to the maximum of p^{-m-t} over all $t \in \mathbb{Z}[p^{-1}]_{\geq 0}$ and all nonnegative integers m for which x_t is not divisible by p^m.

Note that the Frobenius map on $(\mathbf{A}_{\inf} \otimes_{\mathbb{Z}_p} S)/(p)$ is surjective, On onc hand, this implies that the ring $(\mathbf{A}_{\inf} \widehat{\otimes}_{\mathbb{Z}_p} S)[p^{-1}]$, taking the completed tensor product for the p-adic topology, is uniform (by the previous paragraph) and hence perfectoid. We may thus tensor the morphism $S \to \mathbb{Z}_p$ over $\mathbb{Z}_p$ with $\mathbf{A}_{\inf}[p^{-1}]$ to get a splitting of $\mathbf{A}_{\inf}[p^{-1}] \to (\mathbf{A}_{\inf} \widehat{\otimes}_{\mathbb{Z}_p} S)[p^{-1}]$. By Corollary 2.5.5, $\mathbf{A}_{\inf}[p^{-1}]$ is stably uniform, yielding (a).

On the other hand, the ring $(\mathbf{A}_{\inf}\langle\frac{p}{[\overline{x}_i]}\rangle \widehat{\otimes}_{\mathbb{Z}_p} S)[\overline{x}_i^{-1}]$, taking the completed tensor product for the $[\overline{x}_i]$-adic topology, is also perfectoid (note that this is an example of a perfectoid ring which is Tate but not a $\mathbb{Q}_p$-algebra); we may thus tensor the morphism $S \to \mathbb{Z}_p$ over $\mathbb{Z}_p$ with $\mathbf{A}_{\inf}\langle\frac{p}{[\overline{x}_i]}\rangle[\overline{x}_i^{-1}]$ to get a splitting of $\mathbf{A}_{\inf}\langle\frac{p}{[\overline{x}_i]}\rangle[\overline{x}_i^{-1}] \to (\mathbf{A}_{\inf}\langle\frac{p}{[\overline{x}_i]}\rangle \widehat{\otimes}_{\mathbb{Z}_p} S)[\overline{x}_i^{-1}]$. By Corollary 2.5.5, $\mathbf{A}_{\inf}\langle\frac{p}{[\overline{x}_i]}\rangle[\overline{x}_i^{-1}]$ is stably uniform, yielding (b). (Compare [**162**, Proposition 11.2.1] for a similar argument.) □

REMARK 3.1.4. The ring appearing in Lemma 3.1.3(b) can be viewed as a subring of $W(R_i)$ where $(R, R^+) \to (R_i, R_i^+)$ is the rational localization with parameters $\overline{x}_1, \ldots, \overline{x}_n, \overline{x}_i$. In particular, every element has a unique expansion $\sum_{n=0}^{\infty} p^n [\overline{y}_n]$

with $\overline{y}_n \in R_i$. By contrast, elements of the ring appearing in Lemma 3.1.3(a) do not necessarily admit expansions of the form $\sum_{n \in \mathbb{Z}} p^n [\overline{y}_n]$.

DEFINITION 3.1.5. For the topology described in Definition 3.1.2, the space $\mathrm{Spa}(\mathbf{A}_{\mathrm{inf}}, \mathbf{A}_{\mathrm{inf}})$ is not analytic; the analytic locus $\mathrm{Spa}(\mathbf{A}_{\mathrm{inf}}, \mathbf{A}_{\mathrm{inf}})^{\mathrm{an}}$ consists of those v for which either $v(p) \neq 0$ *or* there exists a topologically nilpotent element $\overline{x}$ of R^+ for which $v([\overline{x}]) \neq 0$. For $\overline{x}_1, \ldots, \overline{x}_n$ topologically nilpotent elements of R^+ which generate the unit ideal in R, $\mathrm{Spa}(\mathbf{A}_{\mathrm{inf}}, \mathbf{A}_{\mathrm{inf}})^{\mathrm{an}}$ can also be described as the set of v for which $v(p), v([\overline{x}_1]), \ldots, v([\overline{x}_n])$ are not all zero. By Lemma 3.1.3, $\mathrm{Spa}(\mathbf{A}_{\mathrm{inf}}, \mathbf{A}_{\mathrm{inf}})^{\mathrm{an}}$ is a stably uniform adic space.

Let Y_S be the subspace of $\mathrm{Spa}(\mathbf{A}_{\mathrm{inf}}, \mathbf{A}_{\mathrm{inf}})^{\mathrm{an}}$ consisting of those v for which $v(p) \neq 0$ *and* there exists a topologically nilpotent element $\overline{x}$ of R^+ for which $v([\overline{x}]) \neq 0$. Again, the latter condition need only be tested for $\overline{x}$ running over a finite set of elements which generate the unit ideal in R.

The action of φ on Y_S is properly discontinuous. The quotient space $X_S := Y_S/\varphi^{\mathbb{Z}}$ in the category of locally ringed spaces is the *adic (relative) Fargues–Fontaine curve* over S, which we also denote by FF_S (especially in cases where we want to use X to mean another space).

EXERCISE 3.1.6. Prove that for any perfectoid space X over $\mathbb{Q}_p$, $X \times_{\mathbb{Q}_p} \mathrm{FF}_S$ is a perfectoid space.

REMARK 3.1.7. With some effort, it can be shown that $\mathbf{A}_{\mathrm{inf}}$ is sheafy and hence $\mathrm{Spa}(\mathbf{A}_{\mathrm{inf}}, \mathbf{A}_{\mathrm{inf}})$ is itself a (nonanalytic) adic space (see Problem A.6.2). For the case where R is a nonarchimedean field, see Remark 3.1.10.

REMARK 3.1.8. When one develops the theory of relative Fargues–Fontaine curves, it is generally necessary to also consider the quotient of Y_S by $\varphi^{n\mathbb{Z}}$ for n a positive integer; this gives a finite étale covering of X_S with Galois group $\mathbb{Z}/n\mathbb{Z}$. To simplify the exposition, we (mostly) omit further mention of this construction.

REMARK 3.1.9. Suppose that R is Tate, and let $\varpi \in R$ be a pseudouniformizer. We can then make the description of X_S somewhat more explicit. To begin with, Y_S is the subspace of $v \in \mathrm{Spa}(\mathbf{A}_{\mathrm{inf}}, \mathbf{A}_{\mathrm{inf}})$ for which $v(p[\varpi]) \neq 0$. This space can be covered by the subspaces
$$U_n := \{v \in Y_S : v(p)^{cp^n} \leq v(\varpi) \leq v(p)^{p^n}\},$$
$$V_n := \{v \in Y_S : v(p)^{p^{n+1}} \leq v(\varpi) \leq v(p)^{cp^n}\} \qquad (n \in \mathbb{Z}),$$
where $c \in (1, p) \cap \mathbb{Q}$ is arbitrary. The action of φ permutes the U_n (among themselves) and the V_n (among themselves), and hence is properly discontinuous. The spaces U_0 and V_0 map isomorphically to their images in X_S and cover the latter. In particular, X_S can be covered by two affinoid subspaces, so for every pseudocoherent sheaf $\mathcal{F}$ on X_S we have $H^i(X_S, \mathcal{F}) = 0$ for all $i > 1$.

REMARK 3.1.10. Suppose that $R = F$ is a nonarchimedean field and $R^+ = \mathfrak{o}_F$. Then for any pseudouniformizer ϖ of F, the ring $\mathbf{A}_{\mathrm{inf}}[[\varpi]^{-1}]\langle \frac{p}{[\varpi]} \rangle$ admits euclidean division, and hence is a principal ideal domain; see [111, Corollary 2.10]. (This result appeared previously in [105, Lemma 2.6.3], but the proof contains several errors; see the online errata.) In addition, the ring $\mathbf{A}_{\mathrm{inf}}[[\varpi]^{-1}]\langle \frac{p}{[\varpi]} \rangle$ is strongly noetherian [111, Theorem 3.2]; consequently, in this case X_S is a noetherian adic space.

There is a useful illustration of the space $\mathrm{Spa}(\mathbf{A}_{\mathrm{inf}}, \mathbf{A}_{\mathrm{inf}})$ in the lectures of Bhatt [17]. To summarize, as in Example 1.6.11 there is a unique point $v \in \mathrm{Spa}(\mathbf{A}_{\mathrm{inf}}, \mathbf{A}_{\mathrm{inf}})$

which is not analytic, namely the one with $v(p) = v([\varpi]) = 0$, and the only rational subspace containing v is the whole of $\operatorname{Spa}(\mathbf{A}_{\inf}, \mathbf{A}_{\inf})$. We may thus deduce that $\operatorname{Spa}(\mathbf{A}_{\inf}, \mathbf{A}_{\inf})$ is an adic space.

By contrast, if R is not a finite direct sum of perfectoid fields, then R itself cannot be noetherian (Corollary 2.9.3), and it is easily shown that X_S is not noetherian either. Also, one should not expect a particularly explicit description of X_S in this case, by analogy with the structure of Berkovich analytic spaces: these are reasonable to describe combinatorially in dimension 1 (or dimension 2 over a trivially valued base field) and unreasonable in higher dimensions.

REMARK 3.1.11. The space Y_S is not affinoid, because it is not quasicompact. However, it is a *quasi-Stein space* in the category of adic spaces: it is a direct limit of affinoid subspaces where the transition maps induce dense inclusions of coordinate rings. For example, if R contains a pseudouniformizer ϖ, then the subspaces $\{v \in Y_S : v(p) \leq v(\varpi)^n, v(\varpi) \leq v(p)^n\}$ for $n = 0, 1, \ldots$ form an ascending sequence of the desired form. Quasi-Stein spaces behave somewhat like affinoid spaces in that certain sheaves on them can be interpreted in terms of modules over coordinate rings. See [118, §2.6] for a detailed discussion.

The category of vector bundles on X_S can be interpreted as the category of φ-equivariant vector bundles on Y_S. In light of the previous paragraph, the latter can be interpreted as finite projective $\mathcal{O}(Y_S)$-modules equipped with φ-action.

REMARK 3.1.12. As described in [162, Example 11.2.2], each point of X_S corresponds to a Huber pair (K, K^+) in which K is a perfectoid field; the tilting operation thus defines a map $X_S \to S$ which turns out to be a projection of topological spaces. (See [162, Proposition 11.2.1] for a description of the corresponding map $Y_S \to S$, keeping in mind that φ acts trivially on $|S|$.) This construction commutes with base change; in particular, for F a nonarchimedean field and $\operatorname{Spa}(F, \mathfrak{o}_F) \to S$ a morphism, the fiber of X_S over $\operatorname{Spa}(F, \mathfrak{o}_F)$ coincides with the adic Fargues–Fontaine curve over F. In this sense, X_S is a "family of curves" over S. However, this projection map is not a morphism of adic spaces. (It will acquire an interpretation in the language of diamonds; see Definition 4.3.7.)

Given an untilt (A, A^+) of (R, R^+), Theorem 2.3.9 produces a primitive element z of $W(R^{\flat})$ such that $W(R^+)/(z) \cong A^+$ via the theta map. If A is a $\mathbb{Q}_p$-algebra, then the element z gives rise to a closed immersion of $\operatorname{Spa}(A^+, A^+)$ into $\operatorname{Spa}(\mathbf{A}_{\inf}, \mathbf{A}_{\inf})$, which restricts (using Lemma 1.6.5) to a closed immersion of $\operatorname{Spa}(A, A^+)$ into X_S. If we identify $\operatorname{Spa}(A, A^+)$ with $\operatorname{Spa}(R, R^+)$ via Theorem 2.5.1, then this closed immersion becomes a section of the projection $|X_S| \to |S|$ in the category of topological spaces.

REMARK 3.1.13. Promote R to a uniform Banach ring with norm $|\bullet|$ as per Remark 1.5.4. The coordinate ring $\mathcal{O}(Y_S)$ can then be interpreted as the Fréchet completion of $\mathbf{A}_{\inf}[p^{-1}]$ for the family of Gauss norms (as in Definition 2.6.2) corresponding to the norms $|\bullet|^r$ for all $r > 0$. In [117], this ring appears under the notation $\tilde{\mathcal{R}}_R^\infty$ and is an example of an *extended Robba ring*, named by analogy with Remark 3.5.4.

Using the norm on R, one can construct a deformation retract of the maximal Hausdorff quotient of the space X_S (see Remark 1.5.15) onto a suitable section of $|X_S| \to |S|$; this implies that $|X_S| \to |S|$ has contractible fibers. See [109, Theorem 7.8].

DEFINITION 3.1.14. Let $\mathcal{O}(1)$ be the line bundle on X_S corresponding to the trivial line bundle on Y_S on a generator $\mathbf{v}$, with the isomorphism $\varphi^*\mathcal{O}(1) \cong \mathcal{O}(1)$ given by $1 \otimes \mathbf{v} \mapsto p^{-1}\mathbf{v}$.

Via the following theorem, the sheaf $\mathcal{O}(1)$ may be viewed as an ample line bundle on X_S. This makes it possible to compare X_S to a schematic construction.

THEOREM 3.1.15. *Suppose either that R is Tate and $\mathcal{F}$ is a vector bundle on X_S, or that $(R, R^+) = (F, \mathfrak{o}_F)$ for some nonarchimedean field F and $\mathcal{F}$ is a coherent sheaf on X_S. For $n \in \mathbb{Z}$, define the twisted sheaf $\mathcal{F}(n) := \mathcal{F} \otimes \mathcal{O}(1)^{\otimes n}$. Then for all sufficiently large n, the following statements hold.*
   (a) *We have $H^1(X_S, \mathcal{F}(n)) = 0$.*
   (b) *The sheaf $\mathcal{F}(n)$ is generated by finitely many global sections.*

PROOF. See [**117**, Lemma 8.8.4, Proposition 8.8.6]. □

DEFINITION 3.1.16. Define the graded ring
$$P_S := \bigoplus_{n=0}^{\infty} P_{S,n}, \qquad P_{S,n} := H^0(X_S, \mathcal{O}(n)).$$
The scheme $\mathrm{Proj}(P_S)$ is called the *schematic Fargues–Fontaine curve* over S. By construction, there is a morphism $X_S \to \mathrm{Proj}(P_S)$ of locally ringed spaces.

By analogy with Serre's GAGA theorem for complex algebraic varieties [**164**], we have the following result.

THEOREM 3.1.17. *The morphism $X_S \to \mathrm{Proj}(P_S)$ has the following properties.*
   (a) *Suppose that R is Tate. Then pullback of vector bundles from $\mathrm{Proj}(P_S)$ to X_S defines an equivalence of categories.*
   (b) *Suppose that $(R, R^+) = (F, \mathfrak{o}_F)$ for some nonarchimedean field F. Then pullback of coherent sheaves from $\mathrm{Proj}(P_S)$ to X_S is an equivalence of categories.*
   (c) *In both (a) and (b), the pullback functor preserves sheaf cohomology.*

PROOF. As in the usual GAGA theorem, the strategy is to first prove preservation of H^1, then preservation of H^0, then full faithfulness of the pullback functor, then essential surjectivity of the pullback functor. At each stage, one uses Theorem 3.1.15 to reduce to considering the sheaves $\mathcal{O}(n)$ for $n \in \mathbb{Z}$, which one studies by comparing $\mathcal{O}(n)$ with $\mathcal{O}(n+1)$. For more details, see [**117**, Theorem 6.3.12] for (a) and [**118**, Theorem 4.7.4] for (b). □

LEMMA 3.1.18. *Let $s \in H^0(X_S, \mathcal{O}(1))$ be a section which does not vanish on any fiber of X_S (that is, its pullback to $X_{\mathrm{Spa}(F, \mathfrak{o}_F)}$ does not vanish for any nonarchimedean field F). Let $\mathcal{I} \subset \mathcal{O}$ be the image of $s \otimes \mathcal{O}(-1)$ in $\mathcal{O}$. Then the zero locus Z of $\mathcal{I}$ is an untilt of S, and the projection $|X_S| \to |S|$ restricts to a homeomorphism $|Z(\mathcal{I})| \cong |S|$.*

PROOF. We may work locally on S. For starters, we may assume S admits a pseudouniformizer ϖ; as per Remark 3.1.9, we may further assume that Z is contained in the affinoid subspace $U = \{v \in Y_S : v(p)^c \leq v(\varpi) \leq v(p)\}$ for some $c \in (1, p) \cap \mathbb{Q}$. We may also assume that Z is cut out by a single element $f \in H^0(U, \mathcal{O})$.

In the case where S is a point, we can perform a Weierstrass factorization using the Newton polygon of f, as in [**105**, Lemma 2.6.7], to write f as a multiple of a primitive element; using this primitive element, we may realize Z as an untilt of S. In the general case, we may make the same argument in a suitably small neighborhood of any particular $x \in S$, and thus deduce the claim. (See also [**162**, Proposition 11.3.1].) □

REMARK 3.1.19. It is unclear whether Theorem 3.1.15 and Theorem 3.1.17 extend to the case where R is analytic, not just Tate. One serious difficulty in that case is that by Corollary 2.6.16, if R is not Tate then it admits no untilts over $\mathbb{Q}_p$, so by Lemma 3.1.18, $H^0(X_S, \mathcal{O}(1))$ cannot contain an element which does not vanish identically on any fiber. As a result, even the nonvanishing of $H^0(X_S, \mathcal{O}(n))$ for n large is unclear.

REMARK 3.1.20. Suppose that $R = F$ is a nonarchimedean field and $R^+ = \mathfrak{o}_F$. Let K be an untilt of F. As in Remark 3.1.12, the map $\theta : W(\mathfrak{o}_F) \to \mathfrak{o}_K$ is surjective with kernel generated by some primitive element z. If K is of characteristic p, then $K = F$ and we may take $z = p$.

Suppose hereafter that K is of characteristic 0. The zero locus of z in Y_S is a single point; projecting this point from Y_S to X_S amounts to forgetting the difference between F and its images under powers of Frobenius. From an algebraic point of view, this makes sense to do because Frobenius commutes with all automorphisms, and so this forgetting does not mess up any functoriality.

However, not all points of X_S arise in this fashion, even if F is algebraically closed. Consider by way of analogy the points on the adic projective line over F. There, the points of height 1 are conventionally divided into four types (following [**13**, Example 1.4.3]):

1. rigid-analytic points over a completed algebraic closure of F;
2. generic points of (virtual) closed discs of rational radius;
3. generic points of (virtual) closed discs of irrational radius;
4. points which witness the failure of F to be spherically complete (i.e., equivalence classes of descending chains of closed discs with empty rigid-analytic intersection).

The points of higher height are considered to be a fifth type; the type 5 points are specializations of type 2 points (see [**156**, Example 2.20] for an illustration).

The structure of X_S is quite analogous to this. For example, see [**109**, Theorem 8.17] for a classification of the height 1 points which reproduces many features of the Berkovich classification.

REMARK 3.1.21. The previous construction globalizes to give an adic (relative) Fargues–Fontaine curve over any perfectoid space of characteristic p. Better yet, one may take the base space to be a suitably nice stack on the category of perfectoid spaces, such as a diamond; see Definition 4.3.7.

REMARK 3.1.22. Although [**63**] shows a 2018 publication date, the original construction of Fargues–Fontaine dates back[12] to 2006. As this origin precedes

---

[12]The history given in the preface to [**63**] starts in 2009; however, it is my understanding that the initial discussions between Fargues and Fontaine began at a conference in Venice in 2006, at which I was also present (but not privy to their conversations).

the general promulgation of the theory of perfectoid fields, the original construction involved only algebraically closed nonarchimedean fields, and yielded only the schematic curves (Definition 3.1.16); the published version of [**63**] includes both the schematic and adic viewpoints.

The relative Fargues–Fontaine curves, in both the schematic and adic versions over a Tate base ring, were introduced by Kedlaya–Liu in [**117**].

3.2. An analogy: vector bundles on Riemann surfaces.

We continue with an analogy from classical algebraic geometry that will inform our work.

DEFINITION 3.2.1. Let X be a smooth proper curve over a field k. Recall that the *degree* of a line bundle on X is defined as the degree of the divisor associated to any nonzero rational section (noting that any two such divisors differ by a principal divisor, whose degree is 0). Define the *degree* of a vector bundle V of rank n as the degree of $\wedge^n V$, denoted $\deg(V)$. Define the *slope* of a nonzero vector bundle V as the ratio
$$\mu(V) := \frac{\deg(V)}{\operatorname{rank}(V)};$$
we say V is *semistable* (resp. *stable*) if V contains no proper nonzero subbundle V' with $\mu(V') > \mu(V)$ (resp. $\mu(V') \geq \mu(V)$). Every semistable bundle is a successive extension of stable bundles of the same slope.

One can think of stable vector bundles as the building blocks out of which arbitrary vector bundles are built. One result in this direction is a theorem of Harder–Narasimhan [**88**], to the effect that every bundle admits a certain canonical filtration with semistable quotients. We will see a more general version of this result later (Theorem 3.4.11).

REMARK 3.2.2. When $X \cong \mathbf{P}_k^1$, then $\operatorname{Pic}(X) \cong \mathbb{Z}$ with the inverse map being $n \mapsto \mathcal{O}(n)$, and a theorem of Grothendieck [**81, 92**] states that every vector bundle splits (nonuniquely) as a direct sum of line bundles. A nonzero vector bundle V is semistable if and only if it splits (noncanonically) as a direct sum $\mathcal{O}(n)^{\oplus m}$ for some m, n; in particular, every semistable bundle has integral slope. (These statements were extended by Harder [**87**] to G-bundles on $\mathbf{P}_k^1$, for G a split reductive algebraic group.)

This example is somewhat misleading in its simplicity. In general, not every semistable bundle has integral slope; in particular, not every bundle splits as a direct sum of line bundles. For example, if X is a curve of genus 1 and k is algebraically closed of characteristic 0, a theorem of Atiyah [**9**] implies that for any line bundle L on X of odd degree, there is a unique stable vector bundle V of rank 2 such that $\wedge^2 V \cong L$.

REMARK 3.2.3. Because the definitions of stable and semistable vector bundles are nonexistence criteria, rather than existence criteria, they can be problematic to work with. For example, it is not apparent from the definition that the pullback of a (semi)stable bundle along a morphism of curves is again (semi)stable: the pullback bundle may have a subbundle that witnesses the failure of (semi)stability but is not itself the pullback of a subbundle on the original curve. For another example, for any two nonzero bundles V, V' on X we have
$$\mu(V \otimes V') = \mu(V) + \mu(V'),$$

but it is not apparent that the tensor product of two semistable bundles is again semistable.

The subtlety of this point is illustrated by the fact that the situation depends crucially on the characteristic of k. In characteristic 0, both pullback and tensor product preserve semistability; this can be proved either algebraically (see [99, Chapter 3]) or by using the Lefschetz principle to reduce to the case $k = \mathbb{C}$, then appealing to a "positive" but transcendental characterization of semistability (see Theorem 3.2.4). By contrast, in characteristic p, semistability is preserved by pullback along separable morphisms [99, Lemma 3.2.2] but not along Frobenius (see [148] for further discussion of this phenomenon); semistability is also not preserved by tensor products, as first shown by Gieseker [77].

In characteristic 0, the issues in the previous remark are resolved by the following theorem of Narasimhan–Seshadri [143], later reproved by Donaldson [46].

THEOREM 3.2.4. *Assume that $k = \mathbb{C}$, and choose a closed point $x_0 \in X$. Then a vector bundle V of rank n on X is stable of slope 0 if and only if it admits a connection $\nabla : V \to V \otimes_{\mathcal{O}_X} \Omega_{X/k}$ whose holonomy representation $\rho : \pi_1(X^{\mathrm{an}}, x_0) \to \mathrm{GL}_n(\mathbb{C})$ (i.e., the one from the Riemann–Hilbert correspondence, obtained by analytic continuation of local sections in the kernel of ∇) is irreducible and unitary. In the latter case, ∇ and ρ are uniquely determined by V.*

REMARK 3.2.5. The use of the terms *stable* and *semistable* in this manner stems from the original context in which these notions were studied, via geometric invariant theory. Assume (for simplicity) that k is of characteristic 0. For $G \subseteq \mathrm{GL}(n)_k$ a reductive k-algebraic group, a point $x \in \mathbf{A}_k^n$ is *stable* for the action of G if its stabilizer in G is finite and its G-orbit is closed in $\mathbf{A}_k^n$ (the finite-stabilizer condition is comparable to the definition of a *stable curve* as one with a finite automorphism group) Given a vector bundle V of rank n of a particular rank and degree on X, one can tensor with a suitable ample line bundle to obtain a bundle generated by global sections; this gives a point x in a certain affine space carrying a linear action of $G = \mathrm{GL}(n)_k$, which is stable for the action if and only if V is stable as a bundle. This makes it possible to construct and study moduli spaces of stable bundles by quotienting a certain orbit space for the action of G.

3.3. The formalism of slopes. We next describe a more general framework in which slopes and (semi)stability can be considered. The presentation is based on [150]; see André [4] for an alternate point of view based on tannakian categories.

DEFINITION 3.3.1. A *slope category* consists of the following data.
- An exact faithful functor $F : \mathcal{C} \to \mathcal{D}$ for some exact category $\mathcal{C}$ and some abelian category $\mathcal{D}$ such that for every $V \in \mathcal{C}$, the category of admissible monomorphisms into V (i.e., monomorphisms which occur as kernels of epimorphisms) is equivalent via F to the category of monomorphisms into $F(V)$.
- An assignment rank $: \mathcal{D} \to \mathbb{Z}_{\geq 0}$ which is constant on isomorphism classes, additive on short exact sequences, and takes only the zero object to 0.
- An assignment deg $: \mathcal{C} \to \Gamma$ (to some totally ordered abelian group Γ) which is constant on isomorphism classes, additive on short exact sequences, and with the property that for every morphism $f : V_1 \to V_2$ in $\mathcal{C}$ for which $F(f)$ is an isomorphism, we have $\deg(V_1) \leq \deg(V_2)$ with equality if and only if f is an isomorphism.

In order to parse this definition, we first translate the motivating example of vector bundles on curves into this framework.

EXAMPLE 3.3.2. Let X be a smooth proper algebraic curve over a field k with generic point η. Let $\mathcal{C}$ be the exact category of vector bundles on X; a monomorphism $V' \to V$ is admissible if and only if V' is isomorphic to a *saturated* subbundle of V, i.e., one for which the quotient V/V' is torsion-free. Let $\mathcal{D}$ be the category of finite-dimensional $\kappa(\eta)$-vector spaces; there is an obvious exact faithful functor $F : \mathcal{C} \to \mathcal{D}$ taking a bundle V to its stalk V_η. For $V \in \mathcal{C}$ and $K \to F(V)$ a monomorphism, the subsheaf of V given by $U \mapsto \ker(V(U) \to F(V)/K)$ is an admissible subobject of V because $\mathcal{O}(U)$ is a Dedekind domain. Let rank : $\mathcal{D} \to \mathbb{Z}_{\geq 0}$ be the dimension function.

Take $\Gamma = \mathbb{Z}$ and let $\deg : \mathcal{C} \to \Gamma$ be the usual degree function: for $\mathcal{F} \in \mathcal{C}$ of rank $n > 0$, $\deg(\mathcal{F})$ is the degree of the divisor defined by any nonzero rational section of the line bundle $\wedge^n \mathcal{F}$. (The unambiguity of this definition relies on the fact that any principal divisor on X has degree 0.) The fact that this is additive in short exact sequences comes down to the fact that if $0 \to V' \to V \to V'' \to 0$ is exact, then there is a natural isomorphism

$$(3.3.2.1) \qquad \wedge^{\text{rank}(V)} V \cong \wedge^{\text{rank}(V')} V' \otimes \wedge^{\text{rank}(V'')} V''.$$

If $f : V \to V'$ is a morphism in $\mathcal{C}$ for which $F(f)$ is an isomorphism, then $\text{rank}(V)$ and $\text{rank}(V')$ are equal to a common value n, $\wedge^n f : \wedge^n V \to \wedge^n V'$ is injective with cokernel supported on some finite set S of closed points, and $\deg(V) - \deg(V')$ is a nonnegative linear combination of the degrees of the points in S. (The difference $\deg(V) - \deg(V')$ can also be interpreted as $\dim_k H^0(X, \text{coker}(\wedge^n f))$, but this interpretation will not persist for abstract curves as in Definition 3.3.4.)

DEFINITION 3.3.3. For $k = \mathbb{C}$, let X^{an} denote the *analytification* of X, which is a compact Riemann surface. There is a canonical morphism $X^{\text{an}} \to X$ in the category of locally ringed spaces; by (a very special case of) Serre's GAGA theorem [164], pullback along this morphism equates the categories of vector bundles (and coherent sheaves) on X and X^{an} and preserves sheaf cohomology. We can thus formally restate Example 3.3.2 in terms of vector bundles on X^{an}, or even in terms of Γ-equivariant vector bundles on Y where $Y \to X^{\text{an}}$ is a Galois covering space map with deck transformation group Γ. For example, if X is of genus at least 2, then the universal covering space is an open unit disc with deck transformations by the fundamental group of X^{an}.

Our subsequent discussion will involve a generalization of Example 3.3.2.

DEFINITION 3.3.4. An *abstract curve* is a connected, separated, noetherian scheme X which is regular of dimension 1; any such scheme has a unique generic point η which is also the unique nonclosed point. An *abstract complete curve* is an abstract curve X equipped with a nonzero map $\deg : \text{Div}(X) \to \mathbb{Z}$ which is nonnegative on effective divisors and zero on principal divisors. For X an abstract complete curve, we may emulate Example 3.3.2 to obtain a slope category with $\mathcal{C}$ being the category of vector bundles on X.

This generalization is sufficient to discuss Fargues–Fontaine curves. However, we will also introduce some additional examples of the slope formalism in §3.5.

3.4. Harder–Narasimhan filtrations.
Fix now a formalism of slopes. We now define and construct the Harder–Narasimhan filtrations associated to objects of $\mathcal{C}$.

DEFINITION 3.4.1. Define the *slope* of a nonzero object $V \in \mathcal{C}$ as the ratio
$$\mu(V) := \frac{\deg(V)}{\operatorname{rank}(F(V))} \in \Gamma \otimes_{\mathbb{Z}} \mathbb{Q}.$$
If $0 \to V' \to V \to V'' \to 0$ is an exact sequence in $\mathcal{C}$ with $V', V'' \neq 0$, then
$$\min\{\mu(V'), \mu(V'')\} \leq \mu(V) \leq \max\{\mu(V'), \mu(V'')\}$$
with equality if and only if $\mu(V') = \mu(V'')$.

A nonzero object $V \in \mathcal{C}$ is *semistable* (resp. *stable*) if V contains no proper nonzero subobject V' with $\mu(V') > \mu(V)$ (resp. $\mu(V') \geq \mu(V)$); this implies that V admits no proper quotient V'' with $\mu(V'') < \mu(V)$ (resp. $\mu(V'') \leq \mu(V)$). Note that our hypotheses ensure that any rank 1 object is stable and that any twist of a (semi)stable object by a rank-1 object is again (semi)stable. (It would be reasonable to treat the zero object as being semistable of *every* slope, but we won't do this.)

DEFINITION 3.4.2. For $f: V \to V'$ a monomorphism in $\mathcal{C}$, $F(f)$ lifts to an admissible monomorphism $\tilde{f}: \tilde{V} \to V'$ in $\mathcal{C}$ through which f factors. We call $\tilde{f}$ the *saturation* of f, and call $\tilde{V}$ the *saturation* of V in V'.

Note that V and $\tilde{V}$ have the same rank and $\wedge^{\operatorname{rank}(V)} V$ is a subobject of the rank-1 object $\wedge^{\operatorname{rank}(V)} \tilde{V}$. Since rank-1 objects are stable, we have $\deg(V) \leq \deg(\tilde{V})$ and (if $\operatorname{rank}(V) > 0$) $\mu(V) \leq \mu(\tilde{V})$, with equality if and only if $V = \tilde{V}$.

REMARK 3.4.3. For $f: V \to V'$ an arbitrary morphism in $\mathcal{C}$, the kernel of $F(f)$ corresponds to an admissible subobject of V which is a kernel of f. By the same token, the cokernel of this admissible monomorphism is an image of f.

The poset of subobjects of a given object in $\mathcal{C}$ is a lattice. For $V_1 \to V', V_2 \to V'$ two monomorphisms in $\mathcal{C}$, we write $V_1 \cap V_2$ and $V_1 + V_2$ for the meet and join, respectively (mimicking the notation for vector bundles); these fit into an exact sequence
$$0 \to V_1 \cap V_2 \to V_1 \oplus V_2 \to V_1 + V_2 \to 0.$$
Beware that the join of two admissible subobjects need not be admissible.

REMARK 3.4.4. A consequence of the previous discussion is that for any admissible subobject V' of V, if we form the associated exact sequence
$$0 \to V' \to V \to V'' \to 0$$
and take W to be another (not necessarily admissible) subobject of V, we have another short exact sequence
$$0 \to W' \to W \to W'' \to 0$$
where $W' = V' \cap W$ is a subobject of V' (and an admissible subobject of W) and W'' is a subobject of V''.

LEMMA 3.4.5. *If $V, V' \in \mathcal{C}$ are semistable and $\mu(V) > \mu(V')$, then $\operatorname{Hom}_{\mathcal{C}}(V, V') = 0$.*

PROOF. Suppose by way of contradiction that $f: V \to V'$ is a nonzero morphism. Let W be the image of V in V' (in the sense of Remark 3.4.3); then $\mu(V) \leq \mu(W) \leq \mu(V')$, a contradiction. $\square$

LEMMA 3.4.6. *Let*
$$0 \to V' \to V \to V'' \to 0$$
be a short exact sequence of nonzero objects in $\mathcal{C}$. If V', V'' are semistable of the same slope μ, then so is V.

PROOF. For any nonzero subobject W of V, with notation as in Remark 3.4.4 we have
$$\deg(W) = \deg(W') + \deg(W'') \leq \mu \operatorname{rank}(W') + \mu \operatorname{rank}(W'') \leq \mu \operatorname{rank}(W),$$
so $\mu(W) \leq \mu$ (even in the corner cases where $W' = 0$ or $W'' = 0$). $\square$

COROLLARY 3.4.7. *For any $\mu \in \Gamma \otimes_{\mathbb{Z}} \mathbb{Q}$, the objects of $\mathcal{C}$ which are semistable of slope μ (plus the zero object) form an exact abelian subcategory of $\mathcal{C}$ which is closed under extensions.*

PROOF. Augment the previous lemma with the observation that if $V \in \mathcal{C}$ is semistable of slope μ, any subobject W of V which is semistable of slope μ is admissible: otherwise, the saturation of W would witness the failure of semistability of V. $\square$

DEFINITION 3.4.8. For $V \in \mathcal{C}$, a *Harder–Narasimhan filtration* (or *HN filtration*) of V is a filtration
$$(3.4.8.1) \qquad 0 = V_0 \subset \cdots \subset V_l = V$$
such that each inclusion $V_{i-1} \to V_i$ is admissible with cokernel being semistable of some slope μ_i, and $\mu_1 > \cdots > \mu_l$. By convention, the trivial filtration of the zero object is an HN filtration. If $V \neq 0$, then the sequence
$$0 = V_1/V_1 \subset \cdots \subset V_l/V_1 = V/V_1$$
constitutes an HN filtration of V/V_1; this provides the basis for various inductive arguments.

In order to better digest this definition, we give an alternate characterization of the first step of the HN filtration.

LEMMA 3.4.9. *Suppose that $V \in \mathcal{C}$ is nonzero and admits an HN filtration labeled as in (3.4.8.1). Then μ_1 is the maximum slope of any nonzero subobject of V, and V_1 is the maximal subobject of V of slope μ_1.*

PROOF. We proceed by induction on $\operatorname{rank}(V)$. There is nothing to check if V is semistable. Otherwise, for any nonzero subobject W of V, set notation as in Remark 3.4.4 with $V' = V_1$. Using the semistability of V_1 and applying the induction hypothesis to V/V_1, we see that
$$\deg(W) \leq \deg(W') + \deg(W'') \leq \mu_1 \operatorname{rank}(W') + \mu(V_2/V_1) \operatorname{rank}(W'') \leq \mu_1 \operatorname{rank}(W),$$
with strict inequality whenever $W'' \neq 0$. This proves the claim. $\square$

We now turn around and construct the object with the properties of V_1 identified in Lemma 3.4.9. It is relatively easy to see that the possible slopes of subobjects of V are bounded above, but this would only imply that the maximum is achieved if Γ is discrete (because the slopes of subobjects of an object of rank n belong to $\frac{1}{1}\Gamma \cup \cdots \cup \frac{1}{n}\Gamma$, which would then itself be a discrete set). However, it is easy to give an alternate argument that works even when Γ is not discrete, and which even in the discrete case gives additional crucial information.

LEMMA 3.4.10. *Suppose that $V \in \mathcal{C}$ is nonzero. Then V admits a nonzero subobject V_1 of some slope $\mu_1 \geq \mu(V)$ such that μ_1 is the maximum slope of any nonzero subobject of V, and V_1 is the maximal subobject of V of slope μ_1.*

PROOF. We proceed by induction on $\text{rank}(V)$, with the case of V semistable serving as a trivial base case. If V is not semistable, then the set of nonzero proper subobjects W with $\mu(W) > \mu(V)$ is nonempty; by saturating, we may find an admissible subobject W of this form of maximal rank. (Note that we do not attempt to maximize the *slope* of W, just its rank; hence this is a priori a maximization over a finite set.) By the induction hypothesis, W admits a subobject V_1 of the claimed form; we will show that this subobject also has the desired effect for V. This amounts to showing that any subobject X of V satisfying $\mu(X) \geq \mu_1$ must be contained in W; to see this, write the exact sequence

$$0 \to W \cap X \to W \oplus X \to W + X \to 0,$$

note that $\mu(W \cap X) \leq \mu_1$ if $W \cap X \neq 0$, and then compute that

$$\begin{aligned}\deg(W + X) &= \deg(W) + \deg(X) - \deg(W \cap X) \\ &= \text{rank}(W)\mu(W) + \text{rank}(X)\mu(X) - \text{rank}(W \cap X)\mu(W \cap X) \\ &\geq \text{rank}(W)\mu(W) + (\text{rank}(X) - \text{rank}(W \cap X))\mu_1 \\ &= \text{rank}(W + X)\mu_1.\end{aligned}$$

Since $\mu_1 \geq \mu(W) > \mu(V)$, $W + X$ is a subobject of V of slope strictly greater than $\mu(V)$; its saturation is an admissible subobject with the same property. By the maximality of $\text{rank}(W)$, this is only possible if $\text{rank}(W + X) = \text{rank}(W)$, and hence if $W + X = W$ because W is admissible. $\square$

Putting the two preceding lemmas together gives us HN filtrations in general.

THEOREM 3.4.11 (after Harder–Narasimhan). *Every object $V \in \mathcal{C}$ admits a unique HN filtration.*

PROOF. We check both existence and uniqueness by induction on $\text{rank}(V)$, the case $V = 0$ serving as a trivial base case. To establish uniqueness, note that Lemma 3.4.9 implies that the choice of V_1 is uniquely determined, and then applying the induction hypothesis to V/V_1 forces the rest of the filtration. To establish existence, take V_1 as in Lemma 3.4.10; the maximal slope condition ensures that V_1 is semistable. For any subobject W'' of V/V_1, the inverse image W of W'' in V is strictly larger than V_1, so $\mu(W) < \mu(V_1)$ and so $\mu(W'') < \mu(V_1)$. Consequently, we obtain an HN filtration of V by starting with V_1, then lifting the terms of an HN filtration of V/V_1 produced by the induction hypothesis. $\square$

DEFINITION 3.4.12. Suppose now that $\Gamma \subseteq \mathbb{R}$. For $V \in \mathcal{C}$, with notation as in (3.4.8.1), the *slope multiset* of V is the multisubset of $\Gamma \otimes_{\mathbb{Z}} \mathbb{Q}$ of cardinality $\text{rank}(V)$ consisting of $\mu(V_i/V_{i-1})$ with multiplicity $\text{rank}(V_i/V_{i-1})$ for $i = 1, \dots, l$. As is typical when studying nonarchimedean fields, it is convenient and customary to repackage these values as the slopes of a piecewise affine function. We define the *HN polygon* of V, denoted $\text{HN}(V)$ to be the graph of the continuous, concave-down function from $[0, \text{rank}(V)]$ to $\mathbb{R}$ given by the formula

$$x \mapsto \deg(V_{i-1}) + (x - \text{rank}(V_{i-1}))\mu(V_i) \qquad (i = 1, \dots, l; \text{rank}(V_{i-1}) \leq x \leq \text{rank}(V_i)).$$

That is, start at $(0,0)$ and draw n segments of width 1 whose slopes are the elements of the HN multiset in *decreasing* order, counting multiplicities. See Figure 1 for an illustration.

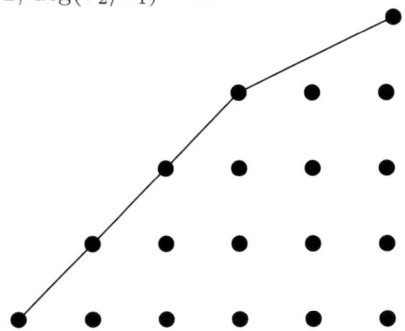

FIGURE 1. The HN polygon associated to an object V admitting a filtration $0 = V_0 \subset V_1 \subset V_2 = V$ with $\mathrm{rank}(V_1) = 3$, $\deg(V_1) = 3$, $\mathrm{rank}(V_2/V_1) = 2$, $\deg(V_2/V_1) = 1$.

LEMMA 3.4.13. *For $V, V' \in \mathcal{C}$, the slope multiset of $V \oplus V'$ is the multiset union of the slope multisets of V and V'. We may characterize this by writing $\mathrm{HN}(V \oplus V') = \mathrm{HN}(V) \oplus \mathrm{HN}(V')$.*

PROOF. We proceed by induction on $\mathrm{rank}(V) + \mathrm{rank}(V')$, with all cases where either summand is zero serving as base cases. If $V, V' \neq 0$, let V_1, V_1' be the first step in the respective HN filtrations, and let μ_1, μ_1' be the respective slopes. Without loss of generality suppose that $\mu_1 \geq \mu_1'$. By Lemma 3.4.9, the largest element of the slope multiset of $V_1 \oplus V_2$ is μ_1, with the corresponding subobject being V_1 if $\mu_1 > \mu_1'$ or $V_1 \oplus V_1'$ if $\mu_1 = \mu_1'$. In either case, we may then conclude using the induction hypothesis. □

REMARK 3.4.14. At this point, it is necessary to comment on two different sign conventions that we have implicitly adopted at this point. The first is the direction of the inequality in the definition of semistability (or equivalently, the choice of sign in the definition of the degree function): we are using the sign convention compatible with the literature on geometric invariant theory (Remark 3.2.5), which is incompatible with the literature on Dieudonné modules. The second is the choice of concavity (up or down) in the definition of the HN polygon (which can be interpreted as the choice to label filtrations in ascending order); we are using the sign convention compatible with the literature on algebraic groups, which is incompatible with the usual definition of Newton polygons. When comparing results between sources, it is important to keep track of both possible sign discrepancies.

We may characterize the HN polygon directly (without overt reference to the HN filtration) as follows.

LEMMA 3.4.15. *For $V \in \mathcal{C}$, the HN polygon is the boundary of the upper convex hull of the set of points $(\mathrm{rank}(W), \deg(W)) \in \mathbb{R}^2$ as W runs over all subobjects of V.*

PROOF. On one hand, the steps of the HN filtration show that the boundary of the upper convex hull lies on or above the HN polygon. We establish the reverse inequality by induction on rank(V). Given a subobject W of V, set notation as in Remark 3.4.4 with $V' = V_1$. By the definition of semistability, the point (rank(W'), deg(W'')) lies under the line $y = \mu(V_1)x$; by the induction hypothesis, the point (rank(W'), deg(W'')) lies on or below the HN polygon of V/V_1. This yields the claim. □

In terms of slope multisets, we may formally promote Lemma 3.4.5 as follows.

COROLLARY 3.4.16. *For $V, V' \in \mathcal{C}$, if the least element of the slope multiset of V is greater than the greatest element of the slope multiset of V', then $\mathrm{Hom}_\mathcal{C}(V, V') = 0$.*

PROOF. If V, V' are both semistable, then this is exactly the assertion of Lemma 3.4.5. If V is general and V' is semistable, then the first step of the slope filtration of V cannot map nontrivially to V'; we may thus deduce this case by induction on rank(V). If V, V' are both general, then V maps trivially to the final quotient of the slope filtration of V'; we may thus deduce this case by induction on rank(V'). □

Corollary 3.4.16 has various consequences about the possibilities for the set of slope multisets for the three terms in a short exact sequence. A full analysis of this in the case of the Fargues–Fontaine curves is part of the student project (see §A.1); we limit ourselves here to one simple observation.

LEMMA 3.4.17. *If $0 \to V' \to V \to V'' \to 0$ is a short exact sequence, then $\mathrm{HN}(V) \leq \mathrm{HN}(V' \oplus V'')$ with the same endpoint.*

PROOF. For every subobject W of V, Remark 3.4.4 gives rise to a subobject $W' \oplus W''$ of $V' \oplus V''$ of the same degree and rank; using the criterion from Lemma 3.4.15, we deduce the claim. □

COROLLARY 3.4.18. *Suppose that $0 = V_0 \subset \cdots \subset V_m = V$ is a filtration of V by admissible subobjects such that each quotient V_i/V_{i-1} is semistable of some slope μ_i. Then $\mathrm{HN}(V) \leq \mathrm{HN}(V_1/V_0 \oplus \cdots \oplus V_m/V_{m-1})$ with the same endpoint.*

REMARK 3.4.19. So far, we have said nothing about tensor products; in fact, we did not even include a symmetric monoidal structure on the category $\mathcal{C}$ in the definition of a slope formalism. In practice, all of the examples we will consider in this lecture admit such a structure which satisfies the property

(3.4.19.1) $$\mathrm{rank}(V \otimes V') = \mathrm{rank}(V)\,\mathrm{rank}(V'),$$
$$\deg(V \otimes V') = \deg(V)\,\mathrm{rank}(V') + \deg(V')\,\mathrm{rank}(V).$$

For V, V' nonzero, we again have

$$\mu(V \otimes V') = \mu(V) + \mu(V'),$$

but it is again not apparent that the tensor product of two semistable bundles is again semistable. When this occurs, the HN filtrations are *determinantal* in the sense that slopes of objects behave like the determinants of linear transformations (or more precisely, their images under some valuation). This is the same arrangement that one encounters initially in the study of the Weil conjectures, as in [39]: before one can define the *weights* of a coefficient object, one must work with the ultimately equivalent concept of *determinantal weights*, whose definition is based on the fact that there is no ambiguity about weights for objects of rank 1.

REMARK 3.4.20. A thoroughly modern twist on slope formalisms comes from the work of Bridgeland [**26**], which gives rise to slope formalisms for *triangulated* categories; the motivating example is the bounded derived category of coherent sheaves on an algebraic variety. In this setting, one assigns to each object a complex number in the upper half-plane (called a *central charge* for presently irrelevant physical reasons), with the argument of this value playing the role of the slope; under fairly mild hypothesis, every object has a Harder–Narasimhan filtration.

3.5. Additional examples of the slope formalism. To provide some indication of the power of this formalism, we describe some other classes of examples. In each case, the key question is whether or not the tensor product of semistable objects is semistable; the example of vector bundles on an algebraic curve in positive characteristic shows that this is not guaranteed by the slope formalism (see Remark 3.2.3).

EXAMPLE 3.5.1. Let R be an integral domain in which every finitely generated ideal is principal (a *Bézout domain*), or more generally an integral domain in which every finitely generated ideal is projective (a *Prüfer domain*); note that for such a ring, every finitely presented torsion-free R-module is projective. Let Φ be a monoid acting on R via ring homomorphisms. Let $\mathcal{C}$ be the category of finite projective R-modules with Φ-actions: that is, one must specify an underlying module M together with isomorphisms $\varphi^* M \cong M$ for each $\varphi \in \Phi$ compatible with composition (with the identity element of Φ acting via the identity map). Let $\mathcal{D}$ be the category of finite-dimensional $\mathrm{Frac}(R)$-vector spaces with Φ-actions. Let $\mathrm{rank} : \mathcal{D} \to \mathbb{Z}_{\geq 0}$ be the dimension function.

Let $v : H^1(\Phi, R^\times) \to \Gamma$ be a homomorphism with the following property: if $x \in R$ is nonzero and satisfies $\varphi(x)/x \in R^\times$ for all $\varphi \in \Phi$, then the cocycle c taking φ to $\varphi(x)/x$ satisfies $v(c) \geq 0$ with equality if and only if $x \in R^\times$. Define $\deg : \mathcal{C} \to \Gamma$ as follows: for $V \in \mathcal{C}$ of rank n, choose a generator $\mathbf{v}$ of $\wedge^n V$; let $c : \Phi \to R^\times$ be the cocycle taking φ to the element r for which the specified isomorphism $\varphi^* \wedge^n V = (\wedge^n V) \otimes_{R,\varphi} R \to \wedge^n V$ takes $\mathbf{v} \otimes 1$ to $r\mathbf{v}$; and put $\deg(V) = v(c)$.

It is obvious that deg is constant on isomorphism classes and (from (3.3.2.1)) additive in short exact sequences. If $f : V_1 \to V_2$ is a morphism in $\mathcal{C}$ for which $F(f)$ is an isomorphism, then $\mathrm{rank}(V_1)$ and $\mathrm{rank}(V_2)$ are equal to a common value n, $\wedge^n f : \wedge^n V_1 \to \wedge^n V_2$ is injective with cokernel isomorphic to R/xR for some $x \in R$ with $\varphi(x)/x \in R^\times$, and $\deg(V_2) - \deg(V_1) = v(x) \geq 0$ with equality only if $V_1 = V_2$.

REMARK 3.5.2. Example 3.5.1 is formulated so as to bring to mind the following case. Take R to be the holomorphic functions on the open unit disc in $\mathbb{C}$; this is a Bézout domain which is not noetherian (consider an infinite sequence accumulating at the boundary, and form the ideal of functions which vanish at all but finitely many of these points). Take Φ to be the deck transformation group for a Teichmüller uniformization of a Riemann surface X of genus at least 2; then the objects of $\mathcal{C}$ are precisely the vector bundles on X, and it is straightforward to reverse-engineer the map v so as to recover the usual degree function.

REMARK 3.5.3. Another possibly familiar context for Example 3.5.1 is where R is the fraction field of the ring of Witt vectors over a perfect field k of characteristic

p, and Φ is the monoid generated by the Witt vector Frobenius map; in this case, $\mathcal{C}$ is the category of *isocrystals* over k. While this example is important in the theory of Dieudonné modules, from the point of view of the slope formalism it is misleadingly simple: if $f : V_1 \to V_2$ is a morphism in $\mathcal{C}$ for which $F(f)$ is an isomorphism, then f is itself an isomorphism. In any case, one can use the standard Dieudonné–Manin classification theorem in the case where k is algebraically closed (see Theorem 3.6.19) to show that for arbitrary k, the tensor product of semistable objects is semistable.

Somewhere between the two previous examples, we find the following example of great interest in p-adic Hodge theory.

REMARK 3.5.4. Let F be a complete discretely valued field with valuation $v_0 : F^\times \to \mathbb{Z}$. Let R be the ring consisting of all formal Laurent series $\sum_{n \in \mathbb{Z}} c_n t^n$ over F which converge in some region of the form $* < |t| < 1$ (where $*$ depends on the series). By analogy with the complex-analytic case, results of [132] (on the theory of divisors in rigid-analytic discs) show that R is a Bézout domain; note that the units in R consist precisely of the nonzero series with bounded coefficients, so v_0 induces a valuation map $v : R^\times \to \mathbb{Z}$. This construction occurs commonly in the theory of p-adic differential equations, where it is commonly known as the *Robba ring* over F (in the variable t).

Let Φ be the monoid generated by a single endomorphism $\varphi : R \to R$ given as a substitution $\sum_{n \in \mathbb{Z}} c_n t^n \mapsto \sum_{n \in \mathbb{Z}} c_n \varphi(t)^n$, where $\varphi(t) = t^m u$ for some integer $m > 1$ and some $u \in R^\times$ with $v(u) = 0$. Let $v : H^1(\Phi, R^\times) \to \mathbb{Z}$ be the homomorphism taking the cocycle c to $v(c(\varphi))$; again using the results of [132], one sees that this satisfies the condition of Example 3.5.1.

When F is of residue characteristic p and $m = p$, it is known that the tensor product of semistable objects is semistable, by a classification theorem similar to the one we will give for vector bundles on the Fargues–Fontaine curve. See [106].

A closely related example from p-adic Hodge theory is the following.

EXAMPLE 3.5.5. Let k be a perfect field of characteristic p. Let K be the fraction field of the ring of Witt vectors over k. Let L be a finite totally ramified extension of K. Let $\mathcal{C}$ be the category of *filtered φ-modules* over L; such an object is a finite-dimensional K-vector space V equipped with a φ-action, for φ the Witt vector Frobenius, plus an exhaustive separated $\mathbb{Z}$-indexed filtration on $V \otimes_K L$ by L-subspaces. Morphisms in $\mathcal{C}$ are maps which are φ-equivariant and respect the filtration. Note that there are monomorphisms which are not admissible, because the inverse image of the target filtration is the "wrong" filtration on the source.

We define deg by using exterior powers to reduce to the case of one-dimensional spaces. For V of dimension 1, the filtration jumps at a unique integer i, the action of Frobenius on a generator is multiplication by some $r \in K^\times$, and we set $\deg(V) = i - v_p(r)$. This yields a slope formalism in which preservation of semistability by tensor product was shown by Faltings [55] using the relationship to Galois representations which are *crystalline* in Fontaine's sense, and more directly by Totaro [172]; another approach can be obtained by going through Remark 3.5.4 using a construction of Berger [12]. (Here the semistable objects are commonly called *weakly admissible* objects.)

A related construction is that of *filtered (φ, N)-modules*, in which one adds to the data a (necessarily nilpotent) K-linear endomorphism N of V satisfying

$N\varphi = p\varphi N$. Again, Berger's method can be used to show that the tensor product of semistable objects is semistable, ultimately using the relationship to Galois representations which are *log-crystalline*[13] in Fontaine's sense.

EXAMPLE 3.5.6 (suggested by Sean Howe). Let L/K be an arbitrary extension of fields. Let $\mathcal{C}$ be the category whose objects are pairs (V, W) in which V is a finite-dimensional K-vector space and W is a finite-dimensional L-vector subspace of $V \otimes_K L$; a morphism $(V, W) \to (V', W')$ is a morphism $f : V \to V'$ of K-vector spaces for which $f(W) \subseteq W'$. Let F be the functor $(V, W) \mapsto V$. We take the rank to be $\dim_K V$ and the degree to be $\dim_L W$.

An important special case of this example is the case where $K = \mathbb{Q}_p, L = \mathbb{C}_p$. In this case, a result of Scholze–Weinstein [**161**, Theorem 5.2.1] shows that the category $\mathcal{C}$ is equivalent to the isogeny category of p-divisible groups over $\mathcal{O}_{\mathbb{C}_p}$; the rank and degree then correspond to the height and dimension of the associated p-divisible group. Note also that $\mathcal{C}$ can be viewed as a subcategory of the category described in Example 3.5.5 with $k = \mathbb{F}_p$, taking the trivial φ-action on V and the filtration taking the value 0 at indices less than 0, W at index 0, and $V \otimes_K L$ at indices greater than 0.

REMARK 3.5.7. Another closely related example is that of *Banach-Colmez spaces*. Roughly speaking, for F a complete algebraically closed nonarchimedean field of mixed characteristics, a Banach-Colmez space is a Banach space over $\mathbb{Q}_p$ which is obtained by forming an extension of a finite-dimensional F-vector space by a finite-dimensional $\mathbb{Q}_p$-vector space, then quotienting by another finite-dimensional $\mathbb{Q}_p$-vector space. In this setup, the rank is given by the dimension of the F-vector space and the degree by the difference between the dimensions of the two $\mathbb{Q}_p$-vector spaces.

In order to make this definition precise, we need to formulate the construction in a way that fixes the F-dimension without fixing the F-linear structure (which we do not want morphisms to respect). This is most naturally done using the pro-étale topology on the category of perfectoid spaces of characteristic p; see Definition 3.7.4.

There is a close relationship between this example, Example 3.5.5, and vector bundles on Fargues–Fontaine curves: they all arise from different t-structures on the same derived category. See [**133**] for a detailed discussion.

In all of the preceding examples, the group Γ is discrete. Let us end with a few examples where Γ is not discrete.

EXAMPLE 3.5.8. Let k be a perfect field of characteristic p. Let X be the scheme obtained by glueing together the rings $\operatorname{Spec} k[T^{1/p^\infty}]$ and $\operatorname{Spec} k[T^{-1/p^\infty}]$ together along their common open subscheme $\operatorname{Spec} k[T^{\pm 1/p^\infty}]$. By emulating the argument for the usual projective line, one can exhibit a homomorphism $\mathbb{Z}[p^{-1}] \to \operatorname{Pic}(X)$ taking $n \in \mathbb{Z}[p^{-1}]$ to a line bundle $\mathcal{O}(n)$ whose global sections are homogeneous polynomials of degree n (when $n \geq 0$), and show that this is an isomorphism (see for example [**33**] or [**47**]). We may emulate Example 3.3.2 to obtain a slope formalism on vector bundles whose degree function takes values in $\mathbb{Z}[p^{-1}]$.

If we view X as the inverse limit of $\mathbf{P}_k^1$ along Frobenius, then every vector bundle on X is the pullback of a vector bundle on some copy of $\mathbf{P}_k^1$, and so by

---

[13] The term *semistable* is more common here, and its etymology in this usage is entirely defensible, but the ensuing terminological conflict renders the term *log-crystalline* a preferred alternative.

Grothendieck's theorem splits as a direct sum of line bundles. Beware however that one cannot derive this splitting by directly imitating the proof for $\mathbf{P}^1_k$: in that argument, it is crucial that every exact sequence of the form
$$0 \to \mathcal{O} \to V \to \mathcal{O}(1) \to 0$$
splits, but that fails here. The corresponding Ext group is spanned by homogeneous monomials in x, y of total degree -1 in which each variable occurs with degree strictly less than 0, and hence is nonzero.

In any case, we see that the tensor product of semistable bundles is semistable.

EXAMPLE 3.5.9. Let K be a nonarchimedean field of residue characteristic p. Let X be the adic space obtained by glueing $\mathrm{Spa}(K\langle T^{p^{-\infty}}\rangle, K\langle T^{p^{-\infty}}\rangle^\circ)$ and $\mathrm{Spa}(K\langle T^{-p^{-\infty}}\rangle, K\langle T^{-p^{-\infty}}\rangle^\circ)$ together along $\mathrm{Spa}(K\langle T^{\pm p^{-\infty}}\rangle, K\langle T^{\pm p^{-\infty}}\rangle^\circ)$. Using Exercise 1.5.27, we see that every line bundle on either $\mathrm{Spa}(K\langle T^{p^{-\infty}}\rangle, K\langle T^{p^{-\infty}}\rangle^\circ)$ or $\mathrm{Spa}(K\langle T^{-p^{-\infty}}\rangle, K\langle T^{-p^{-\infty}}\rangle^\circ)$ is trivial; using this, one can imitate the argument in [33] to see that $\mathrm{Pic}(X) \cong \mathbb{Z}[p^{-1}]$.

By contrast with the previous example, it is not the case that every vector bundle is a direct sum of line bundles! We illustrate this by showing that for $p > 2$, there is a vector bundle V of rank 2 with $\wedge^2 V \cong \mathcal{O}(1)$ which cannot be written as the direct sum of two line bundles. (To cover $p = 2$, one should be able to construct a vector bundle V of rank 3 with $\wedge^3 V \cong \mathcal{O}(1)$ which does not split as a direct sum of a line bundle and another bundle.)

To construct V, identify $\mathrm{Ext}^1_\mathcal{C}(\mathcal{O}(1), \mathcal{O})$ with the completion for the supremum norm of the K-vector space on the monomials $x^{-i}y^{-1+i}$ for $i \in \mathbb{Z}[p^{-1}] \cap (0, 1)$, and take V to be an extension corresponding to an element of this space of the form $s = \sum_{n=0}^\infty c_n x^{-i_n} y^{-1+i_n}$ where c_n is a null sequence in K and i_n is an increasing sequence in $\mathbb{Z}[p^{-1}] \cap (0, \frac{1}{2})$ with limit $\frac{1}{2}$. If we had an isomorphism $V \cong \mathcal{O}(j) \oplus \mathcal{O}(1-j)$ for some $j \in \mathbb{Z}[p^{-1}]$, we would have to have $j \in (0, 1)$ (otherwise s would be forced to split), and without loss of generality $j \in (0, \frac{1}{2})$ (since $p > 2$ we have $\frac{1}{2} \notin \mathbb{Z}[p^{-1}]$). Moreover, we would have $\mathrm{Hom}_\mathcal{C}(\mathcal{O}, V(j-1)) = K$. However, from the exact sequence
$$0 \to \mathcal{O}(j-1) \to V(j-1) \to \mathcal{O}(j) \to 0$$
we obtain an exact sequence
$$0 \to \mathrm{Hom}_\mathcal{C}(\mathcal{O}, V(j-1)) \to \mathrm{Hom}_\mathcal{C}(\mathcal{O}, \mathcal{O}(j)) \to \mathrm{Ext}_\mathcal{C}(\mathcal{O}, \mathcal{O}(j-1))$$
where the last arrow is a connecting homomorphism. If we represent the source and target of this map as homogeneous sums of degree j and $j-1$, then the map between them is given by multiplication by s. More precisely, the source is (topologically) spanned by monomials $x^i y^{j-i}$ for $i \in \mathbb{Z}[p^{-1}] \cap [0, j]$, while the target is obtained by quotienting out by monomials in which either x or y occurs with nonnegative degree. For $t = \sum_{0 < i < j} d_i x^i y^{j-i}$ in the source, the corresponding element of the target is $\sum_{i, n : i < i_n} c_n d_i x^{i-i_n} y^{j-1-i+i_n}$ (the exponent of y is always negative because $i_n < 1/2 < 1 - j$); if t represents an element of the kernel of the connecting homomorphism, in $K\langle T^{p^{-\infty}} \rangle$ we must have
$$\left(\sum_{0 < i < j} d_i T^i\right)\left(\sum_{n=0}^\infty c_n T^{1-i_n}\right) \equiv 0 \pmod{T}.$$

However, by considering Newton polygons, we see that this congruence cannot even hold modulo $T^{1/2+j+\epsilon}$ for any $\epsilon > 0$ unless $t = 0$. This yields the desired contradiction.

Unfortunately, this example does not give rise to a slope formalism for general K, because the underlying rings are not Bézout domains unless K is discretely valued. In that case, we may emulate Example 3.3.2 to obtain a slope category whose degree function takes values in $\mathbb{Z}[p^{-1}]$. For $p > 2$, the above example is a semistable object of rank 2 and degree $\frac{1}{2}$.

We expect that if V is semistable of rank r and degree d, then $V(-n)$ is spanned by horizontal sections whenever $d > rn$. If so, this would imply (using the fact that $\mathbb{Z}[p^{-1}]$ is not discrete in $\mathbb{R}$) that the tensor product of semistable objects is semistable.

REMARK 3.5.10. When K is perfectoid of characteristic p, the two preceding examples are related via an analytification morphism from the adic space to the scheme. However, the discrepancies between the two cases make it clear that there is no version of the GAGA theorem applicable to this morphism.

In connection with the analogy between archimedean and p-adic Hodge theory (see §A.4), Sean Howe has suggested the following example.

EXERCISE 3.5.11. Consider the category $\mathcal{C}$ of $\mathbf{G}_m$-equivariant vector bundles on $\mathbf{P}^1$, equipped with the fiber functor

$$V \mapsto \{\mathbf{G}_m\text{-invariant sections of } V \text{ over } \mathbf{P}^1 \setminus \{0, \infty\}\}.$$

Equip $\mathbb{Z}^2$ with the lexicographic ordering. Consider the degree function $\mathcal{C} \to \mathbb{Z}^2$ induced by the function $L \mapsto (n, p)$ on $\mathbf{G}_m$-equivariant line bundles, in which n is the usual degree of L and p is the order of vanishing at 0 of an invariant section over $\mathbf{P}^1 \setminus \{0, \infty\}$.

(a) Show that this gives a slope formalism.
(b) What are the stable and semistable bundles?
(c) Classify the $\mathbf{G}_m$-equivariant vector bundles on $\mathbf{P}^1$.
(d) Give an equivalence between $\mathbf{G}_m$-vector bundles on $\mathbf{P}^1$ and a linear-algebraic category, and describe the slope formalism in these terms. Then give a linear-algebraic description of the subcategory of objects of slope $(n, *)$ for a fixed n.

We conclude with an exotic example coming from Arakelov theory (compare [4, §3.2.1]).

EXAMPLE 3.5.12. Let $\mathcal{C}$ be the category of Euclidean lattices, i.e., finite free $\mathbb{Z}$-modules equipped with positive-definite inner products, in which morphisms are homomorphisms of lattices which have operator norm at most 1 with respect to the inner products. Let deg be the function assigning to a lattice L the quantity $-\log \det L$. This gives an example of the slope formalism; the Harder–Narasimhan filtrations in this context were previously known as *Grayson–Stuhler filtrations* before the analogy between them was observed. The preservation of semistability by tensor products has been conjectured by Bost and known in some cases.

One may similarly replace $\mathbb{Z}$ with $\mathfrak{o}_K$ for K a number field, considering finite projective $\mathfrak{o}_K$-modules equipped with Hermitian norms with respect to all real and complex embeddings. This again gives a slope filtration in which the preservation of semistability by tensor products is conjectured by Bost and known in some cases; see [4, §3.2.1] for further discussion.

3.6. Slopes over a point. We now consider the slope formalism associated to vector bundles on Fargues–Fontaine curves, as treated in [63]. This provides an improved perspective on some of my previous work on φ-modules over the Robba ring [103, 105].

HYPOTHESIS 3.6.1. *Throughout §3.6, let F be a perfectoid field of characteristic p and take $S = \mathrm{Spa}(F, \mathfrak{o}_F)$.*

THEOREM 3.6.2 (Fargues–Fontaine). *The scheme $\mathrm{Proj}(P_S)$ is an abstract curve. Every closed point x has residue field which is an untilt of a finite extension of F; setting $\deg(x)$ to be the degree of that finite extension gives $\mathrm{Proj}(P_S)$ the structure of an abstract complete curve. (In particular, if F is algebraically closed, then every closed point has residue field which is an untilt of F itself.)*

PROOF. See [63, §6] in the case where F is algebraically closed and [63, §7] in the general case. □

REMARK 3.6.3. For any pseudouniformizer $\overline{\varpi} \in F$, the series
$$\sum_{n \in \mathbb{Z}} p^{-n} [\overline{\varpi}^{p^n}]$$
converges to an element $\mathbf{v}$ of $H^0(Y_S, \mathcal{O})$ satisfying $\varphi(\mathbf{v}) = p\mathbf{v}$, and hence to an element of $H^0(X_S, \mathcal{O}(1))$. This section vanishes at a single closed point whose residue field is an untilt of F itself. It follows that $\deg(\mathcal{O}(1)) = 1$.

In light of Theorem 3.6.2, we may apply Definition 3.3.4 to obtain a slope formalism on the vector bundles on $\mathrm{Proj}(P_S)$, or equivalently (by Theorem 3.1.17) on the vector bundles on FF_S. These obey an analogue of Grothendieck's theorem, although with slightly more basic objects than just the powers of $\mathcal{O}(1)$.

DEFINITION 3.6.4. For $d = \frac{r}{s}$ a rational number in lowest terms (which is to say $r, s \in \mathbb{Z}$, $s > 0$, and $\gcd(r,s) = 1$), let $\mathcal{O}(d)$ be the vector bundle on X_S corresponding to the trivial vector bundle on Y_S of rank s on the basis $\mathbf{v}_1, \ldots, \mathbf{v}_s$ with the isomorphism $\varphi^* \mathcal{O}(d) \to \mathcal{O}(d)$ sending $1 \otimes \mathbf{v}_1, \ldots, 1 \otimes \mathbf{v}_s$ to $\mathbf{v}_2, \ldots, \mathbf{v}_s, p^{-r}\mathbf{v}_1$. This bundle is the pushforward of the line bundle $\mathcal{O}(r)$ on the s-fold cover of FF_S described in Remark 3.1.8.

The following is an easy variant of Remark 3.6.3.

EXERCISE 3.6.5. For $d > 0$, $H^0(\mathrm{FF}_S, \mathcal{O}(d))$ is an infinite-dimensional $\mathbb{Q}_p$-vector space. (By contrast, $H^0(\mathrm{FF}_S, \mathcal{O})) = \mathbb{Q}_p$.)

EXERCISE 3.6.6. Suppose that F is algebraically closed. For $d, d' \in \mathbb{Q}$, $\mathcal{O}(d) \otimes \mathcal{O}(d')$ is isomorphic to a direct sum of copies of $\mathcal{O}(d + d')$. If you don't see how to check this "by hand", use the Dieudonné–Manin classification theorem (see Theorem 3.6.19).

COROLLARY 3.6.7. *For $d = \frac{r}{s}$ in lowest terms, $\mathcal{O}(d)$ is stable of slope d and degree r.*

PROOF. All of the claims reduce to the case where F is algebraically closed. We start with the degree statement. For $d \in \mathbb{Z}$, the claim follows from Remark 3.6.3.

For $d \notin \mathbb{Z}$, using (3.4.19.1) we reduce to checking that $\deg(\mathcal{O}(d)^{\otimes s}) = ds^{s+1}$; this follows from the previous case using Exercise 3.6.6.

We next check semistability. Suppose that $\mathcal{F}$ is a nonzero subobject of $\mathcal{O}(d)$ of slope greater than d. Then $\mathcal{F}^{\otimes s}$ is a subobject of $\mathcal{O}(d)^{\otimes s}$ of slope greater than $ds = r$; the first step $\mathcal{G}$ in the HN filtration of $\mathcal{F}^{\otimes s}$ is a semistable subobject of slope greater than r. By Exercise 3.6.6, $\mathcal{O}(d)^{\otimes s} \cong \mathcal{O}(r)^{\oplus s^s}$, so the existence of a nonzero map $\mathcal{G} \to \mathcal{O}(d)^{\otimes s}$ implies the existence of a nonzero map $\mathcal{G}(-r) \to \mathcal{O}$. Since $\mathcal{G}(-r)$ is semistable of degree $\mu(\mathcal{G}) - r > 0$, this is a contradiction against Lemma 3.4.5.

We finally note that since $\gcd(r, s) = 1$, $\mathcal{O}(d)$ cannot admit any nonzero proper submodule of slope exactly d. It follows that $\mathcal{O}(d)$ is stable. □

REMARK 3.6.8. The stability of $\mathcal{O}(d)$ is not preserved by base extension from $\mathbb{Q}_p$ to a larger field. See Remark 4.3.11.

COROLLARY 3.6.9. *Suppose that F is algebraically closed. For $d, d' \in \mathbb{Q}$, the following statements hold.*
  (a) *If $d \leq d'$, then $\mathrm{Hom}(\mathcal{O}(d), \mathcal{O}(d')) \neq 0$.*
  (b) *If $d > d'$, then $\mathrm{Hom}(\mathcal{O}(d), \mathcal{O}(d')) = 0$.*

PROOF. Using Exercise 3.6.6 and the identification
$$\mathrm{Hom}(\mathcal{F}, \mathcal{F}') \cong H^0(\mathrm{FF}_S, \mathcal{F}^\vee \otimes \mathcal{F}'),$$
this reduces to checking that $H^0(\mathrm{FF}_S, \mathcal{O}(d)) \neq 0$ whenever $d \geq 0$, which is Exercise 3.6.5; and that $H^0(\mathrm{FF}_S, \mathcal{O}(d)) = 0$ whenever $d < 0$, which follows from Corollary 3.6.7 (again, there are no nonzero maps from $\mathcal{O}$ to a semistable bundle of negative degree). □

EXERCISE 3.6.10. For $\mathcal{F}, \mathcal{F}'$ vector bundles on FF_S, produce a canonical isomorphism
$$\mathrm{Ext}^1(\mathcal{F}, \mathcal{F}') \cong H^1(\mathrm{FF}_S, \mathcal{F}^\vee \otimes \mathcal{F}').$$
Deduce that if F is algebraically closed, then for $d, d' \in \mathbb{Q}$ with $d \geq d'$, we have $\mathrm{Ext}^1(\mathcal{O}(d), \mathcal{O}(d')) = 0$. (For general F, this remains true when $d > d'$.)

EXERCISE 3.6.11. For $d \in \mathbb{Q}$, show that as a (noncommutative) $\mathbb{Q}$-algebra, $\mathrm{End}(\mathcal{O}(d))$ is isomorphic to the division algebra over $\mathbb{Q}_p$ of invariant d. Remember (from local class field theory) that this algebra is split by *every* degree-d extension of $\mathbb{Q}_p$.

EXERCISE 3.6.12. Suppose that F is algebraically closed. Prove that $\mathrm{Pic}(\mathrm{FF}_S) \cong \mathbb{Z}$, i.e., every line bundle on FF_S is isomorphic to $\mathcal{O}(d)$ for some $d \in \mathbb{Z}$ (namely its degree). This is not true for general F; see Corollary 3.6.17 for the reason why.

THEOREM 3.6.13 (Kedlaya, Fargues–Fontaine). *Suppose that F is algebraically closed. Then every vector bundle on FF_S splits as a direct sum of vector bundles of the form $\mathcal{O}(d_i)$ for some $d_i \in \mathbb{Q}$.*

PROOF. Using the alternate formulation in terms of finite projective modules over an extended Robba ring equipped with a semilinear φ-action (as in Remark 3.1.13), this result first appears in [**105**, Theorem 4.5.7]; the case where F is the completed algebraic closure of a power series field was previously treated in [**103**, Theorem 4.16] using similar methods. Using the formulation in terms of vector bundles on $\mathrm{Proj}(P_S)$, a different proof of this result was obtained by Fargues–Fontaine [**63**, §8].

The general strategy of both arguments can be characterized as follows. (This is further axiomatized in [**63**, §5] into a theory of *generalized Riemann spheres*, but we give only a summary here.) We may proceed by induction on rank($\mathcal{F}$), the rank 1 case being Exercise 3.6.12. Given a bundle $\mathcal{F}$, one knows from Theorem 3.1.15 that it admits a filtration in which each successive quotient splits as a direct sum of copies of $\mathcal{O}(d)$ for a single value of d. By Corollary 3.4.18, the HN polygon of the associated graded module is an upper bound on HN($\mathcal{F}$). Since the degree function is discretely valued in this setting, we may choose a filtration whose associated graded module has minimal HN polygon (the minimal polygon need not *a priori* be unique but this doesn't matter); the crux of the argument is to show that any such filtration must split. By the induction hypothesis, this reduces to the case of a two-step filtration

$$0 \to \mathcal{O}(d) \to \mathcal{F} \to \mathcal{O}(d') \to 0.$$

By Exercise 3.6.10, there is nothing to check unless $d < d'$; in this case, one must show that either the sequence splits or the filtration is not minimal. The former condition is equivalent to the HN polygon of $\mathcal{F}$ having slopes d and d'; see Remark 3.6.14.

Before discussing the proof of this further, we make a motivating observation. Write $d = \frac{r}{s}, d' = \frac{r'}{s'}$ in lowest terms, so that $\mu(\mathcal{F}) = \frac{r+r'}{s+s'}$. Now consider the subset of $\mathbb{Z}^2$ obtained by taking all of the points under HN($\mathcal{O}(d) \oplus \mathcal{O}(d')$) *except* for the interior vertex (s', r'); the upper convex envelope is an upper bound for HN($\mathcal{F}$) as long as the latter is not equal to HN($\mathcal{O}(d) \oplus \mathcal{O}(d')$). Let d'' be the least slope of this envelope. (See Figure 2 for two illustrated examples of this definition.)

FIGURE 2. The effect of removing the interior vertex of the HN polygons of $\mathcal{O}(d) \oplus \mathcal{O}(d')$ with $(d, d') = \left(\frac{1}{3}, \frac{3}{2}\right)$, in which case $d'' = \frac{1}{2}$; and $(d, d') = \left(-\frac{2}{3}, \frac{3}{2}\right)$, in which case $d'' = -\frac{1}{2}$.

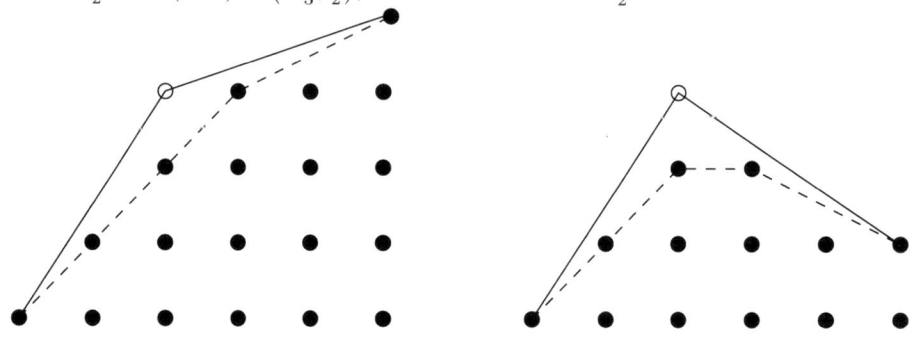

In particular, if the filtration is not minimal, then $\text{Hom}(\mathcal{O}(d''), \mathcal{F}) \neq 0$; in fact, this also holds if the filtration splits because $\text{Hom}(\mathcal{O}(d''), \mathcal{O}(d')) \neq 0$ by Corollary 3.6.9. This suggests that our next step should be to prove that $\text{Hom}(\mathcal{O}(d''), \mathcal{F}) \neq 0$ in all cases; using the induction hypothesis, it is not hard to show that this is in fact sufficient to complete the proof (by showing that either $\mathcal{F}$ splits or the original filtration was not minimal).

Note that d, d'' are the slopes of two sides of a triangle with vertices at lattice points and containing no lattice points in its interior; if we write $d'' = \frac{r''}{s''}$ in lowest terms, this implies that $rs'' - r''s = 1$. By considering $\mathcal{F} \otimes \mathcal{O}(-d'')$ and invoking

Exercise 3.6.6, we reduce to checking the following special case: for any short exact sequence

(3.6.13.1) $$0 \to \mathcal{O}\left(-\frac{1}{n}\right) \to \mathcal{F} \to \mathcal{O}(1) \to 0$$

one has $H^0(\mathrm{FF}_S, \mathcal{F}) \neq 0$, or equivalently the connecting homomorphism

$$H^0(\mathrm{FF}_S, \mathcal{O}(1)) \to H^1(\mathrm{FF}_S, \mathcal{O}\left(-\frac{1}{n}\right))$$

is not injective. This can be done in several ways.

- One option is to check this directly using an *ad hoc* calculation, as in [**103**, Proposition 4.15] or [**91**, Proposition 9.5].
- The explicit calculation can be simplified by allowing passage from F to a suitable extension field F'. For instance, one may take F' to be a (spherically complete) field of Hahn–Mal'cev–Neumann generalized power series, leading to a somewhat easier computation [**106**, Proposition 2.1.6]. Descent of the conclusion from F' to F is implicit in the proof of Theorem 3.7.5.
- A cleaner option is to use the dimension theory for Banach–Colmez spaces as in [**63**, §8.4]. This ultimately depends on some calculations of Colmez [**31**, §7]. (Conversely, Theorem 3.6.13 can be used to establish the dimension theory for Banach–Colmez spaces; we will not work this out here.)
- A somewhat more conceptual approach, described in [**63**, §8.3], is to identify the moduli space of nonsplit extension as in (3.6.13.1) with a certain moduli space of p-divisible groups considered in [**161**] via the (rational) Dieudonné module functor. This identification builds upon results of Hartl [**90**] and Faltings [**59**].

Using any of these approaches, the proof is completed. □

REMARK 3.6.14. In Theorem 3.6.13, the multiset consisting of each d_i with multiplicity rank($\mathcal{O}(d_i)$) equals the slope multiset of the bundle, and hence is independent of the choice of the decomposition. Moreover, for each μ, the sum of all summands $\mathcal{O}(d_i)$ with $d_i \geq \mu$ is a step of the HN filtration, and hence independent of all choices. We will exploit these observations in the following corollaries.

COROLLARY 3.6.15. *For any inclusion $F \to F'$ of perfectoid fields, base extension from $X_{\mathrm{Spa}(F, \mathfrak{o}_F)}$ to $X_{\mathrm{Spa}(F', \mathfrak{o}_{F'})}$ preserves semistability of vector bundles.*

COROLLARY 3.6.16. *For F arbitrary, the tensor product of semistable vector bundles is semistable.*

PROOF. Immediate from Exercise 3.6.6 and Theorem 3.6.13. □

By Theorem 3.6.13, if F is algebraically closed, then a vector bundle is semistable of degree 0 if and only if it is trivial. This formally promotes to a statement directly analogous to the Narasimhan–Seshadri theorem.

COROLLARY 3.6.17. *For F arbitrary, the category of vector bundles on FF_S which are semistable of degree 0 is equivalent to the category of continuous representations of G_F on finite-dimensional $\mathbb{Q}_p$-vector spaces via the functor $\mathcal{F} \mapsto H^0(X_{\mathrm{Spa}(\mathbb{C}_F, \mathfrak{o}_{\mathbb{C}_F})}, \mathcal{F})$ for $\mathbb{C}_F$ a completed algebraic closure of F.*

REMARK 3.6.18. Corollary 3.6.17 is closely related to the theory of (φ, Γ)-*modules*, an important tool in p-adic Hodge theory which gives a useful description of the category of continuous representations of G_F, for F a finite extension of $\mathbb{Q}_p$, on finite-dimensional $\mathbb{Q}_p$-vector spaces. See [120] for a detailed discussion of how this older theory fits into the framework of perfectoid fields and spaces.

It is worth comparing Theorem 3.6.13 with the Dieudonné–Manin classification theorem.

THEOREM 3.6.19. *In the notation of Remark 3.5.3, suppose that k is algebraically closed. Then every object of $\mathcal{C}$ splits as a direct sum of objects of the form $\mathcal{O}(d_i)$ for various $d_i \in \mathbb{Q}$, where for $d_i = \frac{r}{s}$ the object $\mathcal{O}(d_i)$ is a vector space on the generators $\mathbf{v}_1, \ldots, \mathbf{v}_s$ equipped with the φ-action sending $\mathbf{v}_1, \ldots, \mathbf{v}_s$ to $\mathbf{v}_2, \ldots, \mathbf{v}_s, p^{-r}\mathbf{v}_1$.*

PROOF. The original references are [45, 136]. Alternatively, see [41, §4.4] or [108, Theorem 14.6.3]. □

REMARK 3.6.20. The main distinction between Theorem 3.6.13 and Theorem 3.6.19 is that when $d < d'$, the group $\mathrm{Hom}(\mathcal{O}(d), \mathcal{O}(d'))$ vanishes in the category of isocrystals but not in the category of vector bundles on FF_S. This implies that in the category of isocrystals over some perfect field k, the HN filtration splits uniquely; whereas in the category of vector bundles over FF_S, the HN filtration splits (by Exercise 3.6.10) but not uniquely.

REMARK 3.6.21. In [63], the construction of FF_S is generalized from what we consider here by allowing the role of the field $\mathbb{Q}_p$ to be played by an arbitrary local field E of residue characteristic p. When E is of mixed characteristic, this does not give any essentially new results compared to the case we have considered. When E is of positive characteristic, one gets a distinct but closely analogous situation originally considered by Hartl–Pink [91], who proved the analogue of Theorem 3.6.13. See Remark 3.7.7 for more discussion.

3.7. Slopes in families. We next indicate how the slope formalism behaves in families, i.e., for vector bundles on relative Fargues–Fontaine curves.

DEFINITION 3.7.1. Let **Pfd** be the category of perfectoid spaces of characteristic p. For $S \in \mathbf{Pfd}$ and $\mathcal{F}$ a vector bundle on FF_S, for any morphism $\mathrm{Spa}(F, F^+) \to S$ with F a perfectoid field, we may pull back $\mathcal{F}$ to $X_{\mathrm{Spa}(F,F^+)}$ and compute its HN polygon. By Corollary 3.6.15, the result depends only on the image of $\mathrm{Spa}(F, F^+)$ in S; moreover, the construction does not depend on the choice of F^+ at all. We thus get a well-defined function $\mathrm{HN}(\mathcal{F}, \bullet)$ on the maximal Hausdorff quotient of S; by restriction, we view this as a function on S which is constant under specialization.

THEOREM 3.7.2 (Kedlaya–Liu). *For $S \in \mathbf{Pfd}$, let $\mathcal{F}$ be a vector bundle on FF_S.*

(a) *The function $\mathrm{HN}(\mathcal{F}, \bullet)$ is upper semicontinuous. That is, for any given polygon P, the set of $x \in S$ for which $\mathrm{HN}(\mathcal{F}, x) \leq P$ is open; moreover, this set is partially proper (i.e., stable under generization).*

(b) *If $\mathrm{HN}(\mathcal{F}, \bullet)$ is constant on S, then $\mathcal{F}$ admits a filtration which pulls back to the HN filtration at any point.*

PROOF. For (a), see [**117**, Theorem 7.4.5]; note that what is asserted in [**117**] is *lower* semicontinuity because of the sign convention therein that HN polygons are concave up, not concave down (Remark 3.4.14). For (b), see [**117**, Corollary 7.4.10]. □

COROLLARY 3.7.3. *For $S \in $ **Pfd** and $\mathcal{F} \in $ **Vec**$_{FF_S}$, the set of points in S at which $\mathrm{HN}(\mathcal{F}, \bullet)$ has all slopes equal to zero is an open (and partially proper) subset of S, called the étale locus of $\mathcal{F}$.*

DEFINITION 3.7.4. A morphism $f: Y \to X$ in **Pfd** is *pro-étale* if locally on Y it is of the form $\mathrm{Spa}(A_\infty, A_\infty^+) \to \mathrm{Spa}(A, A^+)$ where (A_∞, A_∞^+) is the completion (for the spectral norm) of $\varinjlim_i (A_i, A_i^+)$ for some filtered system of finite étale morphisms $(A, A^+) \to (A_i, A_i^+)$. For $S \in $ **Pfd**, let S_{proet} denote the *pro-étale site* as defined in [**177**, Lecture 3]; that is, a pro-étale morphism $f: Y \to X$ is a covering if it is surjective and every quasicompact open subset of X is contained in the image of a quasicompact open subset of Y (compare Definition 3.8.5).

For $S^\sharp$ an untilt of S, Corollary 2.5.10 induces a functorial homeomorphism $\sharp^*: S^\sharp_{\mathrm{proet}} \cong S_{\mathrm{proet}}$; if $S^\sharp$ is a space over $\mathbb{Q}_p$, this map factors through a functorial map $\sharp^*: X_{S,\mathrm{proet}} \to S_{\mathrm{proet}}$. We emphasize that this is *not* the pullback along a genuine map of spaces $X_S \to S$, although there is such a map in the category of diamonds (see Definition 4.3.7).

For $S \in $ **Pfd**, by an *étale $\mathbb{Q}_p$-local system* on S, we will mean a sheaf of $\mathbb{Q}_p$-modules on S_{proet} which is locally finite free. For $S = \mathrm{Spa}(F, \mathfrak{o}_F)$, this is equivalent to a continuous representation of G_F on a finite-dimensional $\mathbb{Q}_p$-vector space; for S connected, there is a similar interpretation in terms of continuous representations of the étale fundamental group of S (see Remark 4.1.6).

THEOREM 3.7.5 (Kedlaya–Liu). *For $S \in $ **Pfd**, the functor $V \mapsto \sharp^{-1}(V) \otimes_{\mathbb{Q}_p} \mathcal{O}_{FF_S}$ defines an equivalence of categories between étale $\mathbb{Q}_p$-local systems on S and vector bundles on FF_S which at every point of S are semistable of degree 0; more precisely, there is a quasi-inverse functor taking $\mathcal{F}$ to the sheaf $\sharp_* \mathcal{F}$ given by $U \mapsto H^0(FF_U, \mathcal{F})$. Moreover, this equivalence of categories equates sheaf cohomology groups on both sides.*

PROOF. In light of Definition 3.8.1 below, this follows from [**117**, Theorem 9.3.13] (for the equivalence of categories) and [**117**, Theorem 8.7.13, Theorem 9.4.5] (for the comparison of cohomology). □

REMARK 3.7.6. For S a point, every étale $\mathbb{Q}_p$-local system can be realized as the base extension of an étale $\mathbb{Z}_p$-local system, by using the compactness of G_F to obtain a stable lattice in the associated Galois representation. By contrast, for general (or even affinoid) S, an étale $\mathbb{Q}_p$-local system only *locally* admits a stable lattice; this is unsurprising if one thinks of examples of étale coverings of rigid analytic spaces with noncompact groups of deck transformations, such as the Tate uniformization of an elliptic curve or the Lubin-Tate period mapping.

REMARK 3.7.7. Suppose one were to construct a "moduli space" of vector bundles on Fargues–Fontaine curves with a certain property (which really just means a particular vector bundle on the curve over a particular base space). Then Theorem 3.7.2 would give rise to a locally closed stratification of the moduli space by

HN polygons, and Theorem 3.7.5 would give rise to an étale $\mathbb{Q}_p$-local system over the (possibly empty) open stratum corresponding to the zero polygon.

In the next lecture, we will be interested precisely in moduli spaces of this type, parametrizing vector bundles with certain additional structures (reductions of the structure group, modifications along certain sections of the structure morphism). The category of diamonds provides a substantive (i.e., not meaninglessly formal) context in which such moduli spaces can be constructed, providing an approach to emulating certain constructions in positive characteristic which provide a geometric approach to the Langlands correspondence.

These developments are largely motivated by developments in the analogous setting in equal positive characteristic (see Remark 3.6.21), particularly the work of Hartl [**89**, **90**] and Genestier–Lafforgue [**76**].

Theorem 3.7.5 has various applications beyond the scope of these notes. For example, it is an ingredient (together with the methods of [**157**], the properties of pseudocoherent sheaves such as Theorem 1.4.18, and a number of additional ideas) into a finiteness theorem for the cohomology of étale $\mathbb{Q}_p$-local systems on smooth proper rigid analytic spaces; see [**119**].

3.8. More on exotic topologies. We have just seen and used one example of a topology on adic spaces finer than the étale topology, the *pro-étale topology* on the category of perfectoid spaces of characteristic p. In fact, there are quite a few exotic topologies at work in the theory of perfectoid spaces; we focus on a couple that will occur in the last lecture.

DEFINITION 3.8.1. Recall the definition of the *pro-étale topology* from Definition 3.7.4. This definition was formulated for perfectoid spaces of characteristic p, but may also be used for any adic space (or even any preadic space; see §1.11).

This construction is the natural "pro" analogue[14] of the étale topology according to the general discussion of pro-categories in SGA 4 [**8**, Exposé I]. However, this is not the *pro-étale topology* as originally introduced in [**157**, §3] for locally noetherian spaces, then generalized to arbitrary adic spaces in [**117**, §9.1]. In that definition, one only considers inverse systems in which eventually all of the morphisms are finite and surjective.[15] This definition has the advantage of retaining certain features of the étale topology, such as the fact that a pro-étale morphism in this sense induces an open map of underlying topological spaces [**157**, Lemma 3.10(iv)], [**117**, Lemma 9.1.6(b)]. More seriously, under some conditions, the ring morphism associated to a pro-étale morphism of adic affinoid spaces is either flat, or at least preserves the category of complete pseudocoherent modules; for example, this holds when the base ring is perfectoid [**118**, Theorem 3.4.6], or when the morphism of spaces is a perfectoid subdomain of a seminormal (Exercise 2.9.6) affinoid space over a mixed-characteristic nonarchimedean field (see [**157**, Lemma 8.7(ii)] for the case where the base affinoid is smooth, and [**118**, Lemma 8.3.3] in the general case).

For this last reason, we propose to retronymically refer to the older version of the pro-étale topology as the *flattening pro-étale topology*.

---

[14]This is an argument for spelling *pro-étale* with a hyphen: even if you share my preference for suppressing hyphens on prefixes, the *pro-étale topology* is really the *pro-(étale topology)*.

[15]It should be possible to replace this condition with a Mittag-Leffler condition without changing the resulting topos.

THEOREM 3.8.2. *For (A, A^+) a perfectoid pair, the structure presheaf on $\mathrm{Spa}(A, A^+)_{\mathrm{proet}}$ is an acyclic sheaf. The same is also true if we replace the pro-étale topology with the v-topology (Definition 3.8.5).*

PROOF. For A Tate, this (in both cases) is a consequence of [118, Theorem 3.5.5]. The general case follows from this statement plus Theorem 1.3.4. □

REMARK 3.8.3. By Theorem 3.8.2, the pro-étale topology on the category of perfectoid spaces is subcanonical (i.e., representable functors are sheaves; see Definition 1.11.1). By contrast, the pro-étale topology on other types of adic spaces is often not subcanonical. For instance, for K a nonarchimedean field of mixed characteristics, the largest subcategory of the category of rigid analytic spaces over K on which the pro-étale topology is subcanonical is the category of *seminormal* spaces (Exercise 2.9.6); see [118, Theorem 8.2.3].

REMARK 3.8.4. In the category of schemes, it is useful to refine the étale topology to the *fpqc*[16] topology, in which any faithfully flat quasicompact morphism is treated as a covering. It would be useful to do something similar for adic spaces, but flatness is a tricky concept to deal with in the presence of topological completions (compare Remark 1.3.5).

However, there is an even finer topology for schemes which does admit a suitable adic analogue: the *h-topology* introduced by Voevodsky for use in $\mathbf{A}^1$-homotopy theory [174], in which the coverings are the universal topological epimorphisms (e.g., blowups). This topology is so fine that it is not even subcanonical on the full category of schemes; analogously to Remark 3.8.3, for excellent schemes over $\mathbb{Q}$, the maximal subcategory on which the h-topology is subcanonical is the category of seminormal schemes [174, Proposition 3.2.10], [98, Proposition 4.5], [118, Proposition 1.4.22]. For nonnoetherian schemes, a more useful variant of this topology has been introduced by Rydh [154].

DEFINITION 3.8.5. By analogy with the h-topology, in [162] one finds consideration of the *v-topology*[17] on **Pfd**, in which a morphism $f: Y \to X$ is considered as a covering if f is surjective and every quasicompact open subset of X is contained in the image of a quasicompact open subset of Y. The v-topology on **Pfd** is subcanonical by Theorem 3.8.2.

DEFINITION 3.8.6. By a *vector bundle* on an adic space with respect to the pro-étale topology or the v-topology, we mean a sheaf of $\mathcal{O}$-modules which is locally finite free. It is not reasonable to try to work with pseudocoherent sheaves at this level of generality, due to the use of blatantly nonflat covers; however, in [118] one does find such a notion for the flattening pro-étale topology.

THEOREM 3.8.7. *For (A, A^+) a perfectoid pair, the pullback functor $\mathbf{FPMod}_A \to \mathbf{Vec}_{\mathrm{Spa}(A, A^+)_{\mathrm{proet}}}$ is an equivalence of categories. The same is also true if we replace the pro-étale topology with the v-topology (Definition 3.8.5).*

PROOF. For A Tate, this (in both cases) is a consequence of [118, Theorem 3.5.8]. The general case follows from this statement plus Theorem 1.4.2. □

---

[16] Acronym for *fidèlement plat quasi-compact*.
[17] In the original version of [162], this was called the *faithful topology*.

REMARK 3.8.8. Theorem 3.8.2 and Theorem 3.8.7, in the case of the v-topology, are analogous to certain results of Gabber about the h-topology on perfect schemes. See [**21**, §3], [**20**, Theorem 1.2].

REMARK 3.8.9. For $S \in \mathbf{Pfd}$, the category of étale $\mathbb{Q}_p$-local systems on S is equivalent (via pullback) to the category of sheaves of $\mathbb{Q}_p$-modules for the v-topology which are locally finite free. See [**117**, Remark 1.4.2], [**118**, Remark 4.5.2].

4. Shtukas

The Langlands correspondence describes a relationship between Galois representations and automorphic forms extending class field theory, appropriately formulated as a statement about the algebraic group $\mathbf{G}_m$, to more general algebraic groups. In the setting where the Galois group in question is that of a function field over a finite field, there is a geometric approach pioneered by Drinfeld [**48**] (for the group GL_2) and subsequently extended by L. Lafforgue [**128**] (for the group GL_n) and V. Lafforgue [**130**] (for more general groups).

In this final lecture, we give some hints as to how the preceding discussion can be reformulated, using the language of *diamonds* introduced in [**177**, Lecture 3], in a manner that is consistent with geometric Langlands. This amounts to a segue into Scholze's Berkeley lecture notes [**162**].

4.1. Fundamental groups.
We first review basic facts about the profinite fundamental groups of schemes. A standard introduction to this topic is the book of Murre [**142**]; our presentation draws heavily on a course of de Jong [**37**].

DEFINITION 4.1.1. For X a scheme, let $\mathbf{FEt}(X)$ denote the category of finite étale coverings of X; for A a ring, we write $\mathbf{FEt}(A)$ as shorthand for $\mathbf{FEt}(\mathrm{Spec}(A))$ and confuse an object of this category with its coordinate ring. The following observations about $\mathbf{FEt}$ will be useful.

(a) If $A = \varinjlim_i A_i$ in the category of rings, then the base extension functor from the 2-direct limit $\varinjlim_i \mathbf{FEt}(A_i)$ to the category $\mathbf{FEt}(A)$ is an equivalence of categories. (By [**166**, Tag 01ZC], the functor is fully faithful. By [**166**, Tag 00U2], any $B \in \mathbf{FEt}(A)$ is the base extension of some étale A_i-algebra B_i for some i. By [**166**, Tags 01ZO, 07RR], we may increase i to ensure that B_i is also finite and faithfully flat over A_i; hence the functor is essentially surjective.)

(b) If $f : Y \to X$ is a proper surjective morphism of schemes, then the functor from $\mathbf{FEt}(X)$ to descent data with respect to f (i.e., objects of $\mathbf{FEt}(Y)$ equipped with isomorphisms of their two pullbacks to $Y \times_X Y$ satisfying the cocycle condition on $Y \times_X Y \times_X Y$) is an equivalence of categories. For a similar statement with a much weaker hypothesis on f, see [**154**, Theorem 5.17].

The concept of a *Galois category* was originally introduced in SGA 1 [**83**, Exposé V]. We instead take the approach of [**166**, Tag 0BMQ], starting with the definition from [**166**, Tag 0BMY].

DEFINITION 4.1.2. Let $\mathcal{C}$ be a category and let $F : \mathcal{C} \to \mathbf{Set}$ be a covariant functor. We say that $(\mathcal{C}, F)$ is a *Galois category* if the following conditions hold.

(a) The category $\mathcal{C}$ admits finite limits and finite colimits.

(b) Every object of $\mathcal{C}$ is a (possibly empty) finite coproduct of connected objects. (Here $X \in \mathcal{C}$ is *connected* if it is not initial and every monomorphism $Y \to X$ is either a monomorphism or a morphism out of an initial object.)
(c) For every $X \in \mathcal{C}$, $F(X)$ is finite.
(d) The functor F is exact and reflects isomorphisms.

We often refer to F in this context as a *fiber*[18] *functor* by analogy with the primary example (Definition 4.1.3).

A key property of this definition is its relationship with profinite groups. Let G be the automorphism group of the functor F; then G is a profinite group and the action of G on F induces an equivalence of categories between $\mathcal{C}$ and the category of finite G-sets [**166**, Tag 0BN4]. This abstracts the usual construction of the absolute Galois group of a field; see Remark 4.1.4.

DEFINITION 4.1.3. For X a connected scheme, the category $\mathbf{FEt}(X)$ is a Galois category [**166**, Tag 0BNB]. For $\overline{x}$ a geometric point of X (i.e., a scheme over X of the form $\mathrm{Spec}(k)$ for k some algebraically closed field), the *profinite fundamental group* $\pi_1^{\mathrm{prof}}(X, \overline{x})$ is the automorphism group of the functor $\mathbf{FEt}(X) \to \mathbf{Set}$ taking Y to $|Y \times_X \overline{x}|$ (noting that $Y \times_X \overline{x}$ is a disjoint union of copies of $\overline{x}$); the point $\overline{x}$ is called the *basepoint* in this definition.

From the construction, we obtain a natural functor from $\mathbf{FEt}(X)$ to the category of finite sets equipped with $\pi_1^{\mathrm{prof}}(X, \overline{x})$-actions. Using properties of Galois categories, we see that $\pi_1^{\mathrm{prof}}(X, \overline{x})$ is profinite with a neighborhood basis of open subgroups given by the point stabilizers in $|Y \times_X \overline{x}|$ for each $Y \in \mathbf{FEt}(X)$. Moreover, the previous functor defines an equivalence between $\mathbf{FEt}(X)$ and the category of *finite $\pi_1^{\mathrm{prof}}(X, \overline{x})$-sets* for the profinite topology on $\pi_1^{\mathrm{prof}}(X, \overline{x})$ (i.e., finite sets with the discrete topology carrying continuous group actions).

REMARK 4.1.4. For $X = \mathrm{Spec}(K)$ with K a field, a geometric point $\overline{x}$ of X amounts to a field embedding $K \to L$ with L algebraically closed, and $\pi_1^{\mathrm{prof}}(X, \overline{x})$ is the absolute Galois group of K acting on the separable closure of K in L. Similarly, for general X, if $\overline{x} = \mathrm{Spec}(L)$ is a geometric point lying over $x = \mathrm{Spec}(K) \in X$, then $\pi_1^{\mathrm{prof}}(X, \overline{x})$ remains (naturally) unchanged if we replace $\overline{x}$ by the spectrum of the separable or algebraic closure of K in L.

REMARK 4.1.5. In Definition 4.1.3, the definition of $\pi_1^{\mathrm{prof}}(X, \overline{x})$ is independent of the choice of $\overline{x}$, but only in a weak sense: any two choices of basepoints gives a pair of groups and an isomorphism between them, but the latter is only specified up to composition with an inner automorphism. This includes the familiar fact that "the" absolute Galois group of a field F is only functorial up to inner automorphism, as its definition depends on the choice of an algebraic closure of F. It also corresponds to an analogous ambiguity for topological fundamental groups, arising from the fact that changing the basepoint of a loop requires choosing a particular isotopy class of paths from one point to the other. For this reason, the choice of an isomorphism $\pi_1^{\mathrm{prof}}(X, \overline{x}_1) \cong \pi_1^{\mathrm{prof}}(X, \overline{x}_2)$ is sometimes referred to as a *path* (in French, *chemin*) between the two basepoints $\overline{x}_1$ and $\overline{x}_2$.

---

[18]Also frequently spelled *fibre*, but this is due more to the influence of Francophones on the early development of this topic than to the standard discrepancies between US-style and UK-style spelling.

REMARK 4.1.6. The profinite fundamental group of a scheme is often called the *étale fundamental group* and denoted $\pi_1^{\mathrm{et}}(X, \overline{x})$. We avoid this terminology here for the following reasons.

For a topological space X (which is connected, locally path-connected, and locally simply connected) and a point $x \in X$, the fundamental group $\pi_1(X, x)$ (or retronymically, the *topological fundamental group*) can be interpreted in terms of deck transformations of covering space maps which need not be finite. If one uses only the finite covering space maps as in Definition 4.1.3, one instead obtains the profinite completion of $\pi_1(X, x)$, which we call the *profinite fundamental group* of X (with basepoint x) and denote by $\pi_1^{\mathrm{prof}}(X, x)$.

For a rigid analytic space X (or Berkovich space or adic space) and a geometric point $\overline{x}$ of X, one can again define a profinite fundamental group $\pi_1^{\mathrm{prof}}(X, x)$ using finite étale coverings as in Definition 4.1.3. However, there are interesting étale coverings which are not finite, such as the Tate uniformizations of elliptic curves of bad reduction. To account for this, de Jong [35] defines the étale fundamental group $\pi_1^{\mathrm{et}}(X, \overline{x})$ in terms of coverings which locally-on-the-target[19] split as disjoint unions of finite étale coverings. Again, the profinite completion of this group yields the profinite fundamental group. Despite this, though, the profinite fundamental group fails to detect many interesting examples; for instance, the Hodge–Tate period map discussed in [161] (which reinterprets the Gross-Hopkins period map [80]; see also [28]) gives rise to a connected étale covering of $\mathbf{P}^{1,\mathrm{an}}_{\mathbb{C}_p}$ with deck transformations by $\mathrm{PSL}_2(\mathbb{Q}_p)$, a group with no nontrivial finite quotients (consistent with the triviality of the profinite fundamental group of $\mathbf{P}^{1,\mathrm{an}}_{\mathbb{C}_p}$).

Let us now return to the case of schemes. Motivated by the previous examples, let us define the étale fundamental group in terms of deck transformations of coverings which are locally-on-the-target the disjoint unions of finite étale coverings. For X a normal connected scheme, X is irreducible and we may thus choose the base point $\overline{x}$ to lie over the generic point η of X; to compute fundamental groups, there is no harm in replacing X with its reduced closed subscheme, which has the same finite étale covers. We may then argue (see [166, Tag 0BQM]) that $\pi_1^{\mathrm{et}}(X, \overline{x})$ is a quotient of the absolute Galois group of η (i.e., the automorphism group of the integral closure of $\kappa(\eta)$ in $\kappa(\overline{x})$), hence is profinite, hence coincides with $\pi_1^{\mathrm{prof}}(X, \overline{x})$.

By contrast, if X is a scheme which is not normal, then its étale fundamental group need not be profinite. For example, let X be a nodal cubic curve in $\mathbf{P}^2_{\mathbb{C}}$. Let Y be the normalization of X, and let y_1, y_2 be the two distinct points in Y mapping to the node in X. Then for any basepoint $\overline{x}$, $\pi_1^{\mathrm{et}}(X, \overline{x})$ is isomorphic to $\mathbb{Z}$, with the corresponding cover being the "helical" covering of X obtained from the disjoint union $\bigsqcup_{n \in \mathbb{Z}} Y_n$ of $\mathbb{Z}$-many copies of Y by identifying $y_2 \in Y_n$ with $y_1 \in Y_{n+1}$ for each $n \in \mathbb{Z}$. (Similar considerations apply when X is the scheme obtained by glueing two copies of $\mathbf{P}^1_{\mathbb{C}}$ along two distinct closed points.)

REMARK 4.1.7. In order to construct the non-profinite fundamental groups described in Remark 4.1.6 using the formalism of Galois categories, one must modify the definition of a Galois category by relaxing some of the finiteness hypotheses. One candidate for a replacement definition is the concept of an *infinite Galois theory* given in [19, Definition 7.2.1]; this generalizes a construction of Noohi [144].

---

[19]This is not the same as defining this condition locally on the source. However, in the context of topological covering spaces the two would be equivalent.

REMARK 4.1.8. Another possible name for the profinite algebraic group is the *algebraic fundamental group*, but this terminology has at least two defects of its own. One is that in the context of complex manifolds, it may be interpreted as referring to the pro-algebraic completion with respect to the images of finite-dimensional linear representations; see for example [54]. The other is that it may be confused with Nori's *fundamental group scheme* of a variety over a field [145, 146].

REMARK 4.1.9. Remark 4.1.6 is consistent with the behavior of étale $\mathbb{Q}_p$-local systems, which for analytic spaces correspond to representations of the étale fundamental group rather than the profinite fundamental group. This is also true for schemes for any natural definition of étale $\mathbb{Q}_p$-local systems, e.g., as locally finite free modules over the locally constant sheaf $\underline{\mathbb{Q}_p}$ on the pro-étale topology of X in the sense of Bhatt–Scholze [19].

REMARK 4.1.10. Let $Y \to X$ be a morphism of connected schemes. Suppose that for every connected $Z \in \mathbf{FEt}(X)$, the scheme $Y \times_X Z$ is connected. Then for any geometric point $\overline{y}$ of Y, the map $\pi_1^{\mathrm{prof}}(Y, \overline{y}) \to \pi_1^{\mathrm{prof}}(X, \overline{y})$ is surjective.

LEMMA 4.1.11. *Let $k \to k'$ be an extension of algebraically closed fields. Let X be a connected scheme over k. Then $X_{k'}$ is also connected.*

PROOF. See [82, EGA IV.2, Théorème 4.4.4]. □

DEFINITION 4.1.12. We would like to think of the profinite fundamental group of a scheme as a "topological invariant", but this goal is hampered by a fundamental defect: it is not stable under base change. More precisely, if $k \to k'$ is an extension of algebraically closed fields and X is a connected scheme over k, then $X_{k'}$ is again connected by Lemma 4.1.11; for any geometric point $\overline{x}$ of $X_{k'}$, the morphism $\pi_1^{\mathrm{prof}}(X_{k'}, \overline{x}) \to \pi_1^{\mathrm{prof}}(X, \overline{x})$ is surjective. However, it is easy to exhibit examples where this map fails to be injective; see Example 4.1.13. If (X is connected and) $\pi_1^{\mathrm{prof}}(X_{k'}, \overline{x}) \to \pi_1^{\mathrm{prof}}(X, \overline{x})$ is an isomorphism for any $k', \overline{x}$, we say that the morphism $X \to k$ is π_1-*proper*; this (highly nonstandard!) terminology is motivated by the fact that proper morphisms with connected total space have this property (Corollary 4.1.19).

EXAMPLE 4.1.13. Let $k \to k'$ be an extension of algebraically closed fields of characteristic $p > 0$ and put $X := \mathrm{Spec}(k[T])$. For any geometric point $\overline{x}$ of X, the Artin–Schreier construction provides an identification
$$\mathrm{Hom}_{\mathbf{TopGp}}(\pi_1^{\mathrm{prof}}(X, \overline{x}), \mathbb{Z}/p\mathbb{Z}) \cong \bigoplus_{n>0, n \neq 0 \pmod{p}} kT^i$$
(for $\mathbf{TopGp}$ the category of topological groups). This group is not invariant under enlarging k.

EXAMPLE 4.1.14. Let $k \to k'$ be an extension of algebraically closed fields of characteristic $p > 0$. Let X be a smooth, projective, connected curve of genus g over k. Then for any geometric point $\overline{x}$ of $X_{k'}$, $\mathrm{Hom}(\pi_1^{\mathrm{prof}}(X, \overline{x}), \mathbb{Z}/p\mathbb{Z})$ is a finite free $\mathbb{Z}/p\mathbb{Z}$-module of rank equal to the p-rank of X. This rank can be computed in terms of the geometric points of the p-torsion subscheme of the Jacobian, and thus is invariant under base change from k to k'. Thus the argument of Example 4.1.13 does not apply in this case, and indeed Corollary 4.1.19 below will imply that $\pi_1^{\mathrm{prof}}(X_{k'}, \overline{x}) \to \pi_1^{\mathrm{prof}}(X, \overline{x})$ is an isomorphism.

It turns out that the essential feature of Example 4.1.14 which separates it from Example 4.1.13 is properness. We show this through a series of arguments.

LEMMA 4.1.15. *Let $f : Y \to X$ be a morphism of schemes which are qcqs (quasi-compact and quasiseparated). Suppose that the base change functor $\mathbf{FEt}(X) \to \mathbf{FEt}(Y)$ is an equivalence of categories.*
  (a) *The map $\pi_0(X) \to \pi_0(Y)$ is a homeomorphism.*
  (b) *Suppose that one of X or Y is connected. Then so is the other, and for any geometric point $\overline{y}$ of Y the map $\pi_1^{\mathrm{prof}}(Y, \overline{y}) \to \pi_1^{\mathrm{prof}}(X, \overline{y})$ is a homeomorphism.*

PROOF. See [**166**, Tag 0BQA]. □

LEMMA 4.1.16. *Let $k \to k'$ be an extension of algebraically closed fields of characteristic 0. Let X be a k-scheme.*
  (a) *The base change functor $\mathbf{FEt}(X) \to \mathbf{FEt}(X_{k'})$ is an equivalence of categories.*
  (b) *If X is connected (as then is $X_{k'}$ by Lemma 4.1.11), then for any geometric point $\overline{x}$ of $X_{k'}$, the map $\pi_1^{\mathrm{prof}}(X_{k'}, \overline{x}) \to \pi_1^{\mathrm{prof}}(X, \overline{x})$ is a homeomorphism. That is, the morphism $X \to k$ is π_1-proper.*

PROOF. We start with some initial reductions. We need only prove (a), as then (b) follows from Lemma 4.1.15. We may assume that X is affine. By writing the coordinate ring A of X as a direct limit of finitely generated k-subalgebras A_i and applying Definition 4.1.1(a) to both A and to $A \otimes_k k' = \varinjlim_i (A_i \otimes_k k')$, we may further reduce to the case where X is of finite type over k. By forming a hypercovering of X by smooth varieties using resolution of singularities and applying Definition 4.1.1(b), we may also assume that X is smooth. Using the Lefschetz principle, we may also assume that k and k' are contained in $\mathbb{C}$; we may then assume without loss of generality that $k' = \mathbb{C}$.

If X is connected, then so is $X_{\mathbb{C}}$ by Lemma 4.1.11, as then is $X_{\mathbb{C}}^{\mathrm{an}}$ by [**83**, SGA 1, Exposé X, Proposition 2.4]; from this, it follows that $\mathbf{FEt}(X) \to \mathbf{FEt}(X_{\mathbb{C}})$ is fully faithful. To prove essential surjectivity, apply resolution of singularities to construct a compactification $\overline{X}$ of X whose boundary is a divisor Z of simple normal crossings. Given a finite étale cover of $X_{\mathbb{C}}$, we obtain a corresponding $\mathbb{Z}$-local system on $X_{\mathbb{C}}^{\mathrm{an}}$ with finite global monodromy; by the Riemann–Hilbert correspondence plus GAGA, this gives rise to a vector bundle on $\overline{X}_{\mathbb{C}}$ equipped with an integrable connection having regular logarithmic singularities along $Z_{\mathbb{C}}$. The moduli stack of such objects is the base extension from k to $\mathbb{C}$ of a corresponding stack of finite type over k; since the base extension must consist of discrete points, these points coincide with the connected components of the stack, which remain invariant under base extension (Lemma 4.1.11 again). We thus obtain a vector bundle with integrable meromorphic connection on $\overline{X}$ itself; the sheaf of sections of this bundle is the underlying $\mathcal{O}_X$-module of a finite étale $\mathcal{O}_X$-algebra descending the original cover of $X_{\mathbb{C}}$. □

REMARK 4.1.17. From the proof of Lemma 4.1.16, we see that if X is a smooth scheme over an algebraically closed field k of characteristic 0, $\pi_1^{\mathrm{prof}}(X, \overline{x})$ can be computed as the profinite completion of $\pi_1(X_{\mathbb{C}}^{\mathrm{an}}, \overline{x})$ for any embedding $k \to \mathbb{C}$ (and any closed point $\overline{x}$ of $X_{\mathbb{C}}$). However, even if X is projective, the group $\pi_1(X_{\mathbb{C}}^{\mathrm{an}}, \overline{x})$ is not in general independent of the choice of the embedding $k \to \mathbb{C}$, as first observed by Serre [**165**].

LEMMA 4.1.18. *Let A be a henselian local ring with residue field κ. Let $f : X \to S := \mathrm{Spec}(A)$ be a proper morphism of schemes. Then the base change functor $\mathbf{FEt}(X) \to \mathbf{FEt}(X \times_S \mathrm{Spec}(\kappa))$ is an equivalence of categories.*

PROOF. This is a relatively easy argument in terms of relatively difficult theorems (on algebraization and approximation). See [**166**, Tag 0A48]. □

COROLLARY 4.1.19. *Let $k \to k'$ be an extension of algebraically closed fields (of any characteristic). Let X be a proper k-scheme.*
  (a) *The base change functor $\mathbf{FEt}(X) \to \mathbf{FEt}(X_{k'})$ is an equivalence of categories.*
  (b) *If X is connected (as then is $X_{k'}$ by Lemma 4.1.11), then for any geometric point $\overline{x}$ of $X_{k'}$, the map $\pi_1^{\mathrm{prof}}(X_{k'}, \overline{x}) \to \pi_1^{\mathrm{prof}}(X, \overline{x})$ is a homeomorphism. That is, the morphism $X \to k$ is π_1-proper.*

PROOF. Part (a) is obtained from Lemma 4.1.18 by writing k' as a direct limit of finitely generated k-algebras; see [**166**, Tag 0A49]. Given (a), (b) follows from Lemma 4.1.15. □

REMARK 4.1.20. If $k = \mathbb{C}$ and X is proper over k, then the GAGA theorem, as extended to the proper case in SGA 1 [**83**, Exposé XII], implies that any finite covering space map of the analytification X^{an} of X is in fact the analytification of a finite étale cover of X. Hence if $\overline{x}$ is a geometric point lying over a closed point x of X, then $\pi_1^{\mathrm{prof}}(X, \overline{x})$ can be interpreted as the profinite completion of $\pi_1(X^{\mathrm{an}}, x)$.

We now turn to analogues of the homotopy exact sequence of a fiber bundle of topological spaces. The following result, similar in spirit to Stein factorization, is a refinement of [**83**, SGA 1, Exposé X, Corollaire 1.3] adapted from a similar result for diamonds [**162**, Proposition 16.3.3].

LEMMA 4.1.21. *Let $X \to S$ be a qcqs morphism of schemes with connected, π_1-proper geometric fibers. Assume in addition that for every geometric point $\overline{s}$ of S, every connected finite étale covering of $X \times_S \overline{s}$ extends to a finite étale covering of $X \times_S U$ with connected geometric fibers over some étale neighborhood U of $\overline{s}$ in S. Then for any finite étale morphism $X' \to X$, there exists a commutative diagram*

$$\begin{array}{ccc} X' & \longrightarrow & X \\ \downarrow & & \downarrow \\ S' & \longrightarrow & S \end{array}$$

such that $S' \to S$ is finite étale and $X' \to S'$ has geometrically connected fibers. Additionally, this diagram is initial among diagrams

$$\begin{array}{ccc} X' & \longrightarrow & X \\ \downarrow & & \downarrow \\ T & \longrightarrow & S \end{array}$$

where $T \to S$ is finite étale; in particular, it is unique up to unique isomorphism.

PROOF. In light of the uniqueness statement, the claim is fpqc-local on S; by Lemma 4.1.11, the hypothesis is also fpqc-local on S. We may thus assume first that S is affine and reduced (since replacing S by its reduced closed subscheme does not change its étale site), and second that S is *strictly w-local* in the sense of [**19**]; in particular, every finite étale covering of a closed-open subspace of S splits. In this case, the uniqueness property is vacuously true, and we need only check existence; this amounts to showing that X' splits as a finite disjoint union of closed-open subspaces, each of which maps to some closed-open subspace of S with geometrically connected fibers.

It suffices to work étale-locally around some geometric point $\overline{s} \in S$. By the qcqs hypothesis, the functor

$$(4.1.21.1) \qquad 2\text{-}\varinjlim_{U \ni \overline{s}} \mathbf{FEt}(X \times_S U) \to \mathbf{FEt}(X \times_S \overline{s}),$$

where U runs over étale neighborhoods of $\overline{s}$ in S, is an equivalence of categories. We may thus reduce to the case where $X \times_S \overline{s}$ is connected, in which case we must produce U so that $X' \times_S U$ has connected geometric fibers over U. By shrinking U, we may first ensure that $X' \times_S \overline{s}$ lifts to some finite étale cover of $X \times_S U$ with connected geometric fibers over U (by hypothesis), and second that this cover is isomorphic to X' (again because (4.1.21.1) is an equivalence). $\square$

This then yields a variant of [**83**, SGA1, Exposé X, Corollaire 1.4] adapted from [**162**, Proposition 16.3.6].

COROLLARY 4.1.22. *With notation and hypotheses as in Lemma* 4.1.21, *suppose in addition that S is connected. Then X is connected, and for any geometric point $\overline{x}$ of X mapping to the geometric point $\overline{s}$ of S, the sequence*

$$\pi_1^{\mathrm{prof}}(X \times_S \overline{s}, \overline{x}) \to \pi_1^{\mathrm{prof}}(X, \overline{x}) \to \pi_1^{\mathrm{prof}}(S, \overline{s}) \to 1$$

is exact.

PROOF. We first check that X is connected. It is apparent that $X \neq \emptyset$. Suppose by way of contradiction that X disconnects as $X_1 \sqcup X_2$. For any geometric point $\overline{s} \in S$, $X \times_S \overline{s}$ is connected by hypothesis, so one of $X_1 \times_S \overline{s}, X_2 \times_S \overline{s}$ must be empty. Suppose that $X_1 \times_S \overline{s}$ is empty; since X_1 is qcqs, this space can be rewritten as the inverse limit $\varprojlim_U X_1 \times_S U$ for U running over étale neighborhoods of $\overline{s}$ in S. At the level of topological spaces, we have an inverse limit of spectral spaces and spectral morphisms, which can only be empty if it is empty at some term. (For the constructible topologies, this is an inverse limit of compact Hausdorff spaces, which by Tikhonov's theorem cannot be empty if none of the terms is empty.) It follows that $\{s \in S : X_{1,s} = \emptyset\}$ is open, as then is $\{s \in S : X_{2,s} = \emptyset\}$. Since these sets cannot overlap, they form a disconnect of S, a contradiction.

By the previous paragraph, if $S' \to S$ is finite étale and S' is connected, then so is $X \times_S S'$. By Remark 4.1.10, $\pi_1^{\mathrm{prof}}(X, \overline{x}) \to \pi_1^{\mathrm{prof}}(S, \overline{s})$ is surjective.

Let G be a finite quotient of $\pi_1^{\mathrm{prof}}(X, \overline{x})$ corresponding to $X' \in \mathbf{FEt}(X)$. Let $G \to H$ be the quotient corresponding to a Galois cover $S' \to S$ as produced by Lemma 4.1.21 (the uniqueness property of that result implies the Galois property of the cover). Since $X' \to S'$ has geometrically connected fibers, the map $\pi_1^{\mathrm{prof}}(X \times_S \overline{s}, \overline{x}) \to \ker(G \to H)$ must be surjective. This completes the proof of exactness. $\square$

This in turn yields a variant of [**83**, SGA 1, Exposé X, Corollaire 1.7], giving a Künneth formula for fundamental groups of products.

COROLLARY 4.1.23. *Let k be an algebraically closed field and put $S := \mathrm{Spec}(k)$. Let $X \to S, Y \to S$ be morphisms such that Y is connected and $X \to S$ is* *qcqs and π_1-proper. (The π_1-proper condition holds if k is of characteristic 0, by Lemma 4.1.16, or if $X \to S$ is proper, by Corollary 4.1.19.) Then $Z := X \times_S Y$ is connected, and for any geometric point $\bar{z}$ of Z the map*

$$\pi_1^{\mathrm{prof}}(Z, \bar{z}) \to \pi_1^{\mathrm{prof}}(X, \bar{z}) \times \pi_1^{\mathrm{prof}}(Y, \bar{z})$$

is an isomorphism of topological groups.

PROOF. Apply Corollary 4.1.22 to the morphism $Z \to Y$; both hypotheses of Lemma 4.1.21 are satisfied because $X \to S$ is π_1-proper. We then have a commutative diagram of groups

$$\pi_1^{\mathrm{prof}}(Z_{\bar{s}}, \bar{z}) \longrightarrow \pi_1^{\mathrm{prof}}(Z, \bar{z}) \longrightarrow \pi_1^{\mathrm{prof}}(Y, \bar{z}) \longrightarrow 1$$
$$\searrow \quad \downarrow$$
$$\pi_1^{\mathrm{prof}}(X, \bar{z})$$

in which the top row is exact. This proves the claim. □

Although we do not use it here, we wish to point out the following recent result of Achinger [3, Theorem 1.1.1].

DEFINITION 4.1.24. For X a connected scheme, we say that X is a $K(\pi, 1)$ *scheme* if for some (hence any) geometric point $\bar{x}$ of X, for every locally constant sheaf of finite abelian groups $\mathcal{F}$ on X_{et}, the natural maps

(4.1.24.1) $$H^*(\pi_1^{\mathrm{prof}}(X, \bar{x}), \mathcal{F}_{\bar{x}}) \to H^*(X_{\mathrm{et}}, \mathcal{F})$$

are isomorphisms. This is analogous to the corresponding definition in topology, which can be formulated as the assertion that the higher homotopy groups of X all vanish. We may similarly define the concept of a $K(\pi, 1)$ *adic space*.

REMARK 4.1.25. The usual definition of a $K(\pi, 1)$ scheme imposes the condition on (4.1.24.1) only for torsion sheaves whose order is invertible on X (see for example [147, Definition 5.3], [1, Definition 9.20]). We need the stronger restriction here in order to pass the condition through the tilting equivalence.

THEOREM 4.1.26 (Achinger). *Let X be a connected affine scheme over $\mathbb{F}_p$. Then X is a $K(\pi, 1)$ scheme.*

As in [3, Theorem 6.4.2], this yields the following corollary.

COROLLARY 4.1.27. *Let $X := \mathrm{Spa}(A, A^+)$ be a connected Tate adic affinoid space on which p is topologically nilpotent. Then X is a $K(\pi, 1)$ adic space.*

PROOF. In case X is affinoid perfectoid, the statements follow by applying Corollary 2.5.10 to reduce to the case of an affinoid perfectoid space in characteristic p, then reducing to Theorem 4.1.26 via an algebraization argument (see [3, Proposition 6.4.1]). This implies the general case using Theorem 2.9.9. □

REMARK 4.1.28. In Theorem 4.1.26, the isomorphism in (4.1.24.1) is easy to verify for p-torsion coefficients using the Artin–Schreier construction. The subtle part is to extend this argument to all coefficients; this makes use of certain very strong results on the presentation of schemes of finite type over a positive-characteristic field as finite étale covers of affine spaces, in the spirit of [102, 104]. (The one-dimensional cases of such results may be viewed as positive-characteristic analogues of Belyi's theorem on covers of $\mathbf{P}^1$ ramified over three points, as in [78, §4].)

REMARK 4.1.29. In Corollary 4.1.27, the condition that p be topologically nilpotent is essential: there exist affinoid spaces over $\mathbb{C}((t))$ which are not $K(\pi, 1)$ spaces. An explicit example is the closed subspace of the unit 3-ball in x, y, z cut out by the equation $xy = z^2 - t$; see [2, §7] for a closely related example.

4.2. Drinfeld's lemma. We next introduce a fundamental result of Drinfeld[20] which gives a replacement for the Künneth formula for fundamental groups (Corollary 4.1.23) for products of schemes in characteristic p. More precisely, the original result of Drinfeld [50, Theorem 2.1], [51, Proposition 6.1] gives a key special case (see Remark 4.2.13); the general case is due to E. Lau [131, Theorem 8.1.4], except for a superfluous restriction to schemes of finite type. See also [162, Theorem 17.2.4].

DEFINITION 4.2.1. For any scheme X over $\mathbb{F}_p$, let $\varphi_X : X \to X$ be the *absolute Frobenius* morphism, induced by the p-th power map on rings. For $f : Y \to X$ a morphism of schemes, define the *relative Frobenius* $\varphi_{Y/X} : Y \to \varphi_X^* Y$ to be the unique morphism making the diagram

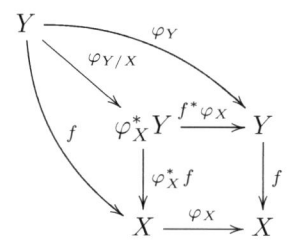

$$\begin{array}{ccc} Y & & \\ & \searrow \varphi_Y & \\ \varphi_{Y/X} \downarrow & & \\ f \searrow \quad \varphi_X^* Y & \xrightarrow{f^* \varphi_X} & Y \\ & \varphi_X^* f \downarrow & \downarrow f \\ & X \xrightarrow{\varphi_X} & X \end{array}$$

commute.

The following argument is similar in style to the proof of Serre's GAGA theorem [164].

LEMMA 4.2.2. *Let X be a projective scheme over $\mathbb{F}_p$. Let k be a separably closed field of characteristic p. Then pullback along $X_k \to X$ defines an equivalence of categories between coherent sheaves on X and coherent sheaves on X_k equipped with isomorphisms with their φ_k-pullbacks. Moreover, for $\mathcal{F}$ a coherent sheaf on X, the induced maps*

$$H^i(X, \mathcal{F}) \otimes_{\mathbb{F}_p} k \to H^i(X_k, \mathcal{F})$$

are φ-equivariant isomorphisms.

PROOF. The assertion about comparison of cohomology is a consequence of flat base change (this step is trivial compared to the analogous step in GAGA), and immediately implies that the pullback functor is fully faithful (by forming internal Homs and comparing H^0 groups).

It thus remains to prove essential surjectivity. In the case $R = \mathbb{F}_p$, this is a result of Lang, as reported by Katz in SGA 7 [40, Exposé XXII, Proposition 1.1]. We summarize the argument in the style of [117, Lemma 3.2.6]: if V is a vector space with basis $\mathbf{e}_1, \ldots, \mathbf{e}_n$ over k equipped with the action of φ_k taking $\mathbf{e}_j$ to

---

[20] A more accurate transliteration of Дринфельд would be *Drinfel'd*, but this would lead to the typographical monstrosity of *Drinfel'd's lemma*.

$\sum_i A_{ij}\mathbf{e}_i$, then the closed subscheme X of $\operatorname{Spec} k[U_{ij} : i,j = 1, \ldots, n]$ cut out by the matrix equation $\varphi(U) = A^{-1}U$ is finite (evidently) and étale (by the Jacobian criterion) over $\operatorname{Spec}(k)$, and so splits as a disjoint union of k-rational points (because k is separably closed). Projecting to a component of this disjoint union, we obtain elements $\mathbf{v}_1, \ldots, \mathbf{v}_n$ of V defined by $\mathbf{v}_j = \sum_i U_{ij}\mathbf{e}_i$ which are fixed by φ; for a suitable choice of component, these elements form a basis of V.

To treat the general case, fix an ample line bundle $\mathcal{O}(1)$ on X; we can then identify X with the Proj of the graded ring $\bigoplus_{n=0}^\infty \Gamma(X, \mathcal{O}(n))$, X_k with the Proj of the graded ring $\bigoplus_{n=0}^\infty \Gamma(X_k, \mathcal{O}(n))$, and $\mathcal{F}$ with the sheaf associated to the graded module $\bigoplus_{n=0}^\infty \Gamma(X_k, \mathcal{F}(n))$. Each graded piece of this module is a finite-dimensional k-vector space, so we may apply the previous paragraph to write it as $S_n \otimes_{\mathbb{F}_p} k$ for $S_n = \Gamma(X_k, \mathcal{F}(n))^{\varphi_k}$. The sheaf $\mathcal{F}$ then arises as the pullback of the sheaf on X associated to the graded module $\bigoplus_{n=0}^\infty S_n$. (Compare [51, Proposition 1.1], [127, I.3, Lemme 3], [131, Lemma 8.1.1], [162, Lemma 17.2.6].) $\square$

REMARK 4.2.3. As with the GAGA theorem (see [83, SGA 1, Expose XII]), using Chow's lemma one can immediately promote Lemma 4.2.2 to the case where X is proper over $\mathbb{F}_p$. However, it does not hold if we only require X to be of finite type over $\mathbb{F}_p$. For example, take $X = \operatorname{Spec}(k[T^\pm])$ and $\mathcal{F} = \tilde{M}$ for M the free module on the single generator $\mathbf{v}$ equipped with the φ_k-action taking $\mathbf{v}$ to $T\mathbf{v}$; then M cannot have a φ_k-invariant element.

Using the previous argument, we may show that "quotienting by relative Frobenius" can be used to mitigate failures of π_1-properness.

DEFINITION 4.2.4. For X a scheme and Γ a group of automorphisms of X, let $\mathbf{FEt}(X/\Gamma)$ denote the category of finite étale coverings Y equipped with an action of Γ. That is, we must specify isomorphisms $Y \to \gamma^*Y$ for each $\gamma \in \Gamma$, subject to the condition that for $\gamma_1, \gamma_2 \in Y$, composing the γ_1-pullback of $Y \to \gamma_2^*Y$ with $Y \to \gamma_1^*Y$ yields the chosen map $Y \to (\gamma_1\gamma_2)^*Y$.

We say that X is Γ-connected if X is nonempty and its only Γ-stable closed-open subsets are itself and the empty set. If X is Γ-connected, then for any geometric point $\overline{x}$ of X, the category $\mathbf{FEt}(X/\Gamma)$ equipped with the fiber functor $Y \mapsto |Y \times_X \overline{x}|$ is a Galois category in the sense of Definition 4.1.2; the argument is the same as in [166, Tag 0BNB] except for condition (b), in which the Γ-connected hypothesis is used. We then write $\pi_1^{\mathrm{prof}}(X/\Gamma, \overline{x})$ for the automorphism group of this fiber functor.

In these notations, when Γ is generated a single element γ, we will typically write X/γ in place of X/Γ.

We need the following variant of Definition 4.1.1(a).

LEMMA 4.2.5. Let $X = \operatorname{Spec}(A)$ be an affine scheme over $\mathbb{F}_p$. Let k be a field of characteristic p. Write A as a filtered direct limit of finitely generated $\mathbb{F}_p$-subalgebras A_i. Then the base extension functor

$$2\text{-}\varprojlim \mathbf{FEt}((A_i \otimes_{\mathbb{F}_p} k)/\varphi_k) \to \mathbf{FEt}((A \otimes_{\mathbb{F}_p} k)/\varphi_k)$$

is an equivalence of categories.

PROOF. By the same argument as in Definition 4.1.1(a), the functor

$$2\text{-}\varprojlim \mathbf{FEt}(A_i \otimes_{\mathbb{F}_p} k) \to \mathbf{FEt}(A \otimes_{\mathbb{F}_p} k)$$

is an equivalence of categories. This implies immediately that the given functor is fully faithful. To establish essential surjectivity, note that for $B \in \mathbf{FEt}((A \otimes_{\mathbb{F}_p} k)/\varphi_k)$, we know that for some index i, B descends to $B_i \in \mathbf{FEt}(A_i \otimes_{\mathbb{F}_p} k)$ while $\varphi_k^* B$ descends to $\varphi_k^* B_i'$ for some $B_i' \in \mathbf{FEt}(A_i \otimes_{\mathbb{F}_p} k)$. In addition, the isomorphisms

$$B_i \otimes_{A_i} A \cong B_i' \otimes_{A_i} A, \qquad \varphi_k^*(B_i' \otimes_{A_i} A) \cong B_i \otimes_{A_i} A$$

both descend to $\mathbf{FEt}(A_j \otimes_{\mathbb{F}_p} k)$ for some j. This proves the claim. □

LEMMA 4.2.6. *Let X be a scheme over $\mathbb{F}_p$. Let k be an algebraically closed field of characteristic p. Then the base extension functor*

$$\mathbf{FEt}(X) \to \mathbf{FEt}(X_k/\varphi_k)$$

is an equivalence of categories, with the quasi-inverse functor being given by taking φ_k-invariants.

PROOF. We first reduce to the case where X is affine. Using Lemma 4.2.5, we further reduce to the case where X is of finite type over $\mathbb{F}_p$. Applying Definition 4.1.1(b) to a suitable covering, we further reduce[21] to the case where X is normal and connected. Choose an open immersion $X \to X'$ with X' normal and projective over $\mathbb{F}_p$. Now note that the following categories are equivalent (using Lemma 4.2.2 between (b) and (c)):

(a) finite étale morphisms $Y \to X_k$ with isomorphisms $\varphi_k^* Y \cong Y$;
(b) finite morphisms $Y \to X_k'$ with Y normal and étale over X_k with isomorphisms $\varphi_k^* Y \cong Y$;
(c) finite morphisms $Y \to X'$ with Y normal and étale over X;
(d) finite étale morphisms $Y \to X$.

This proves the claim. (Compare [**127**, IV.2, Théorème 4], [**131**, Lemma 8.1.2], [**162**, Lemma 17.2.6].) □

EXAMPLE 4.2.7. Let k be an algebraic closure of $\mathbb{F}_p$ and put $X = \operatorname{Spec}(k)$. Then X_k is highly disconnected: there is a natural homeomorphism $\pi_0(X_k) \cong \operatorname{Gal}(k/\mathbb{F}_p) \cong \widehat{\mathbb{Z}}$. However, the action of φ_k on $\pi_0(X)$ is via translations by the dense subgroup $\mathbb{Z}$ of $\widehat{\mathbb{Z}}$; consequently, there is no φ_k-stable disconnection of X, as predicted by Lemma 4.2.6.

COROLLARY 4.2.8. *Let X be a connected scheme over $\mathbb{F}_p$. Let k be an algebraically closed field of characteristic p.*

(a) *The scheme X_k is φ_k-connected.*
(b) *For any geometric point $\overline{x}$ of X, the map*

$$\pi_1^{\mathrm{prof}}(X, \overline{x}) \to \pi_1^{\mathrm{prof}}(X_k/\varphi_k, \overline{x})$$

is a homeomorphism of profinite groups.

PROOF. Let k_0 be the integral closure of $\mathbb{F}_p$ in k; by Lemma 4.1.11, we have $\pi_0(X_{k_0}) = \pi_0(X_k)$. We may thus argue as in Example 4.2.7, i.e., by identifying $\pi_0(X_{k_0})$ with a quotient of $\widehat{\mathbb{Z}}$ on which φ_k acts via translation by the dense subgroup $\mathbb{Z}$. This proves (a). Given (a), (b) follows immediately from Lemma 4.2.6. □

---

[21] Had it been helpful to do so, we could have added de Jong's alterations theorem [**36**] into this argument to further reduce to the case where X is smooth and admits a compactification with good boundary.

This then leads to a corresponding mitigation for products of varieties.

REMARK 4.2.9. For X a connected scheme over $\mathbb{F}_p$, if we view $\mathbf{FEt}(X/\varphi)$ as the category of pairs (Y, σ) where $Y \in \mathbf{FEt}(X)$ and $\sigma : Y \to \varphi_X^* Y$ is a single isomorphism, then the forgetful functor $\mathbf{FEt}(X/\varphi) \to \mathbf{FEt}(X)$ admits a distinguished section taking Y to $(Y, \varphi_{Y/X})$. However, this section is not an equivalence: whereas every *connected* finite étale cover Y of X admits only the action by $\varphi_{Y/X}$ (which commutes with all automorphisms of Y over X), for a disconnected cover this action may be twisted by an automorphism of Y over X that permutes connected components. From this, one deduces that for $\overline{x}$ a geometric point of X, there is a canonical isomorphism
$$\pi_1^{\mathrm{prof}}(X/\varphi, \overline{x}) \cong \pi_1^{\mathrm{prof}}(X, \overline{x}) \times \widehat{\mathbb{Z}} \cong \pi_1^{\mathrm{prof}}(X, \overline{x}) \times G_{\mathbb{F}_p}.$$

DEFINITION 4.2.10. Let $X_1, \ldots, X_n$ be schemes over $\mathbb{F}_p$ and put $X := X_1 \times_{\mathbb{F}_p} \cdots \times_{\mathbb{F}_p} X_n$. Write φ_i as shorthand for φ_{X_i}. Define the category
$$\mathbf{FEt}(X/\Phi) := \mathbf{FEt}(X/\langle \varphi_1, \ldots, \varphi_n \rangle) \times_{\mathbf{FEt}(X/\varphi_X)} \mathbf{FEt}(X)$$
via the functor $\mathbf{FEt}(X) \to \mathbf{FEt}(X/\varphi_X)$ described in Remark 4.2.9. In other words, an object of $\mathbf{FEt}(X/\varphi)$ is a finite étale covering $Y \to X$ equipped with commuting isomorphisms $\beta_i : Y \cong \varphi_i^* Y$ whose composition is $\varphi_{Y/X}$. (Here "composition" and "commuting" must be interpreted suitably: by the "composition" $\beta_i \circ \beta_j$, we really mean $(\beta_j^* \beta_i) \circ \beta_j$.) Note that for any $i \in \{1, \ldots, n\}$, there is a canonical equivalence of categories
$$\mathbf{FEt}(X/\Phi) \cong \mathbf{FEt}(X/\langle \varphi_1, \ldots, \widehat{\varphi_i}, \ldots, \varphi_n \rangle).$$
In case $X_1, \ldots, X_n$ are connected, by Lemma 4.2.11 we may obtain a Galois category in the sense of Definition 4.1.2 by considering the usual fiber functor defined by any geometric point $\overline{x}$ of X; we denote the corresponding group by $\pi_1^{\mathrm{prof}}(X/\Phi, \overline{x})$.

LEMMA 4.2.11. *With notation as in Definition 4.2.10, if $X_1, \ldots, X_n$ are connected, then X is $\langle \varphi_1, \ldots, \widehat{\varphi_i}, \ldots, \varphi_n \rangle$-connected for any $i \in \{1, \ldots, n\}$. We say for short that X is Φ-connected.*

PROOF. We adapt the proof of Corollary 4.1.22, proving the claim by induction on n with trivial base case $n = 1$. For the induction step, it is apparent that $X \neq \emptyset$. Suppose by way of contradiction that X admits a Φ-invariant disconnection $U_1 \sqcup U_2$. Put $X' := X_1 \times_{\mathbb{F}_p} \cdots \times_{\mathbb{F}_p} X_{n-1}$. For each geometric point $\overline{s} \in X_n$, Corollary 4.2.8 implies that the φ_n-invariant closed-open subsets of $X \times_{X_n} \overline{s}$ are just the pullbacks of the closed-open subsets of X'; consequently, the Φ-invariant closed-open subsets of $X \times_{X_n} \overline{s}$ are just the pullbacks of the $\langle \varphi_2, \ldots, \varphi_{n-1} \rangle$-invariant closed-open subsets of X'. By the induction hypothesis, this implies that one of $U_1 \times_{X_n} \overline{s}, U_2 \times_{X_n} \overline{s}$ must be empty. Consequently, the sets $\{s \in X_n : U_1 \times_{X_n} s = \emptyset\}$ and $\{s \in X_n : U_2 \times_{X_n} s = \emptyset\}$ form a set-theoretic partition of X_n; using Tikhonov's theorem, we see that each of these sets is open. This yields a disconnection of X_n, and thus a contradiction. □

THEOREM 4.2.12 ("Drinfeld's lemma"). *Let $X_1, \ldots, X_n$ be connected qcqs schemes over $\mathbb{F}_p$ and put $X := X_1 \times_{\mathbb{F}_p} \cdots \times_{\mathbb{F}_p} X_n$. Then for any geometric point $\overline{x}$ of X, the map*
$$\pi_1^{\mathrm{prof}}(X/\Phi, \overline{x}) \to \prod_{i=1}^n \pi_1^{\mathrm{prof}}(X_i, \overline{x})$$
is an isomorphism of topological groups.

PROOF. In light of Definition 4.2.10, we may rewrite the group on the left as
$$\pi_1^{\mathrm{prof}}(X_1 \times_{\mathbb{F}_p} (X_2/\varphi) \times_{\mathbb{F}_p} \cdots \times_{\mathbb{F}_p} (X_n/\varphi), \overline{x}).$$
We may then proceed by induction on n, with the base case $n = 1$ being trivial. The induction step follows from Lemma 4.2.6 as in the proof of Corollary 4.1.23. $\square$

REMARK 4.2.13. The original result of Drinfeld [51, Proposition 6.1] is somewhat more restrictive than Theorem 4.2.12; it treats the case where $n = 2$ and $X_1 = X_2 = \mathrm{Spec}(F)$ where F is the function field of a curve over a finite field. See [130, Lemme 8.2] for further discussion of this case, including additional references.

In the spirit of the theory of diamonds, one may reinterpret Drinfeld's lemma as follows.

REMARK 4.2.14. Let **Perf** denote the category of perfect schemes over $\mathbb{F}_p$. Identify each $X \in$ **Perf** with the representable functor $h_X :$ **Perf** $\to$ **Set** taking Y to $\mathrm{Hom}_{\mathbf{Perf}}(Y, X)$, which is a sheaf for the Zariski, étale, and fpqc topologies. Let $X/\varphi :$ **Pfd** $\to$ **Set** be the functor taking $Y \in$ **Perf** to the set of pairs (f, g) where $f : Y \to X$ is a morphism and $g : Y \to \varphi_X^* Y$ is an isomorphism (using f to define $\varphi_X^* Y$). Beware that X/φ is no longer a sheaf for any of the topologies in question; it is only a *stack* over **Perf** in the sense of [166, Tag 026F]. (See Problem A.6.3.)

For a suitable definition of the étale topology on stacks (as in Definition 1.11.2), Lemma 4.2.6 asserts an equivalence
$$\mathbf{FEt}(X) \to \mathbf{FEt}(X \times_{\mathrm{Spec}(\mathbb{F}_p)} (\mathrm{Spec}(k)/\varphi)) \qquad (X \in \mathbf{Perf}),$$
which we may think of as formally defining an isomorphism
$$\pi_1^{\mathrm{prof}}(X, \overline{x}) \to \pi_1^{\mathrm{prof}}(X \times_{\mathrm{Spec}(\mathbb{F}_p)} (\mathrm{Spec}(k)/\varphi), \overline{x}).$$
Similarly, let $X_1, \ldots, X_n$ be connected schemes over $\mathbb{F}_p$ and put $X := X_1 \times_{\mathbb{F}_p} \cdots \times_{\mathbb{F}_p} X_n$. Let $X/\Phi :$ **Perf** $\to$ **Set** be the functor taking $Y \in$ **Perf** to the set of tuples $(f, \beta_1, \ldots, \beta_n)$ where $f : Y \to X$ is a morphism and $\beta_i : Y \to \varphi_{X_i}^* Y$ are commuting isomorphisms which compose to $\varphi_{Y/X}$. We may then think of Theorem 4.2.12 as formally defining an isomorphism
$$\pi_1^{\mathrm{prof}}\left(\left(\prod_{i=1}^n X_i\right)/\Phi, \overline{x}\right) \to \prod_{i=1}^n \pi_1^{\mathrm{prof}}(X_i, \overline{x}).$$

REMARK 4.2.15. Following up on Remark 4.2.14, using Remark 4.2.9 we obtain an isomorphism of
$$\pi_1^{\mathrm{prof}}((X_1/\varphi) \times_{\mathrm{Spec}(\mathbb{F}_p)/\varphi} \cdots \times_{\mathrm{Spec}(\mathbb{F}_p)/\varphi} (X_n/\varphi), \overline{x})$$
with the limit (i.e., fiber product) of the diagram

$$\begin{array}{ccccc} \pi_1^{\mathrm{prof}}(X_1/\varphi, \overline{x}) & & \cdots & & \pi_1^{\mathrm{prof}}(X_n/\varphi, \overline{x}) \\ & \searrow & \downarrow & \swarrow & \\ & & \pi_1^{\mathrm{prof}}(\mathrm{Spec}(\mathbb{F}_p)/\varphi, \overline{x}). & & \end{array}$$

This statement admits a highly suggestive topological analogue. Namely, let $X_1 \to S, \ldots, X_n \to S$ be Serre fibrations of topological spaces, and let x be a basepoint of $X := X_1 \times_S \cdots \times_S X_n$ mapping to $x_1, \ldots, x_n, s$ in $X_1, \ldots, X_n, S$. Suppose further that S is a $K(\pi, 1)$ (this being analogous to the algebro-geometric situation, e.g., in light of Theorem 4.1.26). Since $\pi_2(S) = 0$, we may combine the long exact sequence of homotopy groups associated to a fibration with the formula for the fundamental group of an ordinary product to deduce that $\pi_1(X, x)$ is the limit of the diagram

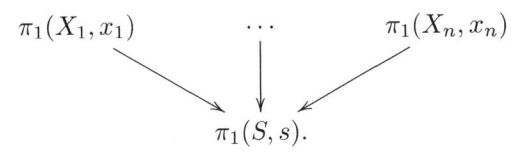

$$\begin{array}{ccccc} \pi_1(X_1, x_1) & \cdots & \pi_1(X_n, x_n) \\ & \searrow \downarrow \swarrow & \\ & \pi_1(S, s). & \end{array}$$

4.3. Drinfeld's lemma for diamonds. We now establish an analogue of Drinfeld's lemma for diamonds (and somewhat more general sheaves). This involves a reinterpretation of relative Fargues–Fontaine curves in the language of diamonds (already discussed in [**177**, Lecture 4]), which can be taken as a retroactive justification for their construction.

The foundations of the theory of diamonds require a fair bit of care, and substantial parts of [**160**] and [**162**] are devoted to these foundations. Our discussion here should be taken as no more than a brief summary of these treatments.

DEFINITION 4.3.1. Let **Pfd** again denote the category of perfectoid spaces of characteristic p. Identify each $S \in$ **Pfd** with the representable functor $h_X :$ **Pfd** $\to$ **Set**; the latter is a pro-étale sheaf (see Remark 3.8.3). As per [**162**, Definition 8.3.1] and [**160**, Definition 11.1], a *diamond* is a pro-étale sheaf of sets on **Pfd** which is a quotient of an object of **Pfd** by a pro-étale equivalence relation. These form a category via natural transformations of functors.

For X a perfectoid space (not necessarily of characteristic p), let $X^\diamond$ be the representable functor $h_{X^\flat}$. Using Remark 2.9.10, we may extend this construction to a functor $X \mapsto X^\diamond$ from analytic adic spaces on which p is topologically nilpotent to diamonds: explicitly, for $Y \in$ **Pfd**, $X^\diamond(Y)$ consists of isomorphism classes of pairs $(Y^\sharp, f)$ in which $Y^\sharp$ is an untilt of Y (i.e., a perfectoid space equipped with an isomorphism $(Y^\sharp)^\flat \cong Y$) and $f : Y^\sharp \to X$ is a morphism of adic spaces. Beware that this functor is not fully faithful (see Remark 3.8.3).

For (A, A^+) a Huber pair in which p is topologically nilpotent (with A analytic as usual), we write $\mathrm{Spd}(A, A^+)$ (the "diamond spectrum") as shorthand for $\mathrm{Spa}(A, A^+)^\diamond$. Furthermore, if $A = F$ is a nonarchimedean field and $A^+ = \mathfrak{o}_F$, we usually just write $\mathrm{Spd}(F)$.

We will also need a more permissive construction.

DEFINITION 4.3.2. Recall that the pro-étale topology is refined by the *v-topology* (see Definition 3.8.5), which is still subcanonical on **Pfd**. We may formally promote the v-topology to diamonds.

A *small v-sheaf* is a sheaf on **Pfd** which admits a surjective morphism from some perfectoid space; any diamond is a small v-sheaf. Using small v-sheaves, we may extend the functor $(A, A^+) \to \mathrm{Spd}(A, A^+)$ to some non-analytic Huber pairs. For example, $\mathrm{Spd}(\mathbb{F}_p)$ is a terminal object in the category of small v-sheaves. For another example, by analogy with the interpretation of $\mathrm{Spd}(\mathbb{Q}_p)$ as the functor taking $S \in$ **Pfd** to the set of isomorphism classes of untilts of S over $\mathbb{Q}_p$ (see [**177**, Lecture 3]), one can interpret $\mathrm{Spd}(\mathbb{Z}_p)$ as the functor taking $S \in$ **Pfd** to the set of isomorphism classes of untilts of S, or more precisely of pairs $(S^\sharp, \iota)$ in which $S^\sharp$ is a perfectoid space and $\iota : (S^\sharp)^\flat \cong S$ is an isomorphism. (Note that $\mathrm{Spd}(\mathbb{Z}_p((T))) \to \mathrm{Spd}(\mathbb{Z}_p)$ is an admissible covering for the v-topology.)

REMARK 4.3.3. Note that the definition of a small v-sheaf does not include any properties on an equivalence relation (or any relative representability condition). Somewhat surprisingly, such conditions are superfluous! Namely, if $Y \to X$ is a surjective morphism from a diamond (e.g., a perfectoid space) to a small v-sheaf, then $Y \times_X Y$ is also a diamond. See [160] for further discussion.

REMARK 4.3.4. For X a perfectoid space not necessarily of characteristic p, the functor $X \mapsto X^\diamond$ depends only on $X^\flat$, and thus loses information. However, X also determines a morphism $X^\diamond \to \mathrm{Spd}(\mathbb{Z}_p)$ of small v-sheaves, and the resulting functor from X to small v-sheaves over $\mathrm{Spd}(\mathbb{Z}_p)$ is fully faithful.

As in Remark 4.2.14, we consider quotients by Frobenius.

DEFINITION 4.3.5. For X a small v-sheaf, let $X/\varphi : \mathbf{Pfd} \to \mathbf{Set}$ denote the functor taking $Y \in \mathbf{Pfd}$ to the set of pairs (f,g) where $f : Y \to X$ is a morphism of diamonds and $g : Y \to \varphi_X^* Y$ is an isomorphism (using f to define $\varphi_X^* Y$). In general this is not a sheaf for either the pro-étale or v-topologies, but it is a stack over $\mathbf{Pfd}$ for these topologies. However, if X is "sufficiently nontrivial" then X/φ is a sheaf for the v-topology (and hence a small v-sheaf, since $X \to X/\varphi$ is surjective); for instance, this happens if X arises from an analytic adic space, or more generally if X is *locally spatial* in the sense of [162, Definition 17.3.1] (meaning roughly that the "underlying topological space" of X is well-behaved).

We now reinterpret the construction of Fargues–Fontaine curves in the language of diamonds and small v-sheaves, starting with a calculation adapted from [162, Proposition 11.2.2].

LEMMA 4.3.6. *For $S = \mathrm{Spa}(R,R^+) \in \mathbf{Pfd}$, put $\mathbf{A}_{\inf} := \mathbf{A}_{\inf}(R,R^+)$, let $\overline{x}_1,\ldots,\overline{x}_n \in R^+$ be topologically nilpotent elements which generate the unit ideal in R, and put*

$$U_S := \{v \in \mathrm{Spa}(\mathbf{A}_{\inf},\mathbf{A}_{\inf}) : v([\overline{x}_i]) \neq 0 \text{ for some } i \in \{1,\ldots,n\}\}.$$

Then there is a natural (in S) isomorphism of small v-sheaves

$$S^\diamond \times \mathrm{Spd}(\mathbb{Z}_p) \cong U_S^\diamond.$$

PROOF. For $Y \in \mathbf{Pfd}$, $(S^\diamond \times \mathrm{Spd}(\mathbb{Z}_p))(Y)$ consists of pairs $(f, Y^\sharp)$ in which $f : Y \to S$ is a morphism in $\mathbf{Pfd}$ and $Y^\sharp$ is an isomorphism class of untilts of Y. For $Y = \mathrm{Spa}(R',R'^+)$, such data correspond to a primitive ideal I of $W(R'^+)$ for which $Y^\sharp = \mathrm{Spa}(W^b(R')/I, W(R'^+)/I)$ and a morphism $(R,R^+) \to (R',R'^+)$ of Huber rings. The latter induces a map $W(R^+) \to W(R'^+)$ and hence a map $W(R^+) \to W^b(R')/I$ under which the images of $[\overline{x}_1],\ldots,[\overline{x}_n]$ generate the unit ideal. We thus obtain a map $Y^\sharp \to U_S$ and hence a morphism $Y^\diamond \to U_S^\diamond$.

In the other direction, $U_S^\diamond(Y)$ consists of pairs $(Y^\sharp, f)$ in which $Y^\sharp$ is an isomorphism class of untilts of Y and $f : Y^\sharp \to U_S$ is a morphism of adic spaces. The latter gives rise to a map $W(R^+) \to W^b(R')/I$ under which the images of $[\overline{x}_1],\ldots,[\overline{x}_n]$ generate the unit ideal; we may thus tilt to obtain a map $R^+ \to R'$ which extends to R. We thus obtain a morphism $Y^\diamond \to S^\diamond \times \mathrm{Spd}(\mathbb{Z}_p)$. □

DEFINITION 4.3.7. Recall that for $S \in \mathbf{Pfd}$, the relative Fargues–Fontaine curve over S is defined as the quotient

(4.3.7.1) $$\mathrm{FF}_S := Y_S/\varphi_S$$

where φ_S is the map induced by the Witt vector Frobenius. Using Lemma 4.3.6, we have natural isomorphisms of diamonds

$$Y_S^\diamond \cong S^\diamond \times \mathrm{Spd}(\mathbb{Q}_p), \qquad \mathrm{FF}_S^\diamond \cong Y_S^\diamond \cong (S^\diamond/\varphi) \times \mathrm{Spd}(\mathbb{Q}_p).$$

In particular, there is now a natural projection map $\mathrm{FF}_S^\diamond \to S^\diamond/\varphi$. Since φ acts trivially on the underlying topological space $|S|$ and on the étale site S_{et}, this projection induces the map $|\mathrm{FF}_S| \to |S|$ seen in Remark 3.1.12 and the map $\mathrm{FF}_{S,\mathrm{et}} \to S_{\mathrm{et}}$ of étale sites seen in Definition 3.7.4.

In light of the previous constructions, it is natural to define the *relative Fargues–Fontaine curve* over any small v-sheaf X as the stack

$$\mathrm{FF}_X := (X/\varphi) \times \mathrm{Spd}(\mathbb{Q}_p);$$

in light of (4.3.7.1), FF_X is a small v-sheaf, and even a diamond if X is a diamond. Taking $X = \mathrm{Spd}(\mathbb{F}_p)$ yields an object which one might call the *absolute Fargues–Fontaine curve*.

Recalling the setup of Drinfeld's lemma, we make the following observation and definition.

DEFINITION 4.3.8. Let $X_1, \ldots, X_n$ be small v-sheaves and put $X := X_1 \times \cdots \times X_n$. Write φ_i as shorthand for φ_{X_i}. Define the category

$$\mathbf{FEt}(X/\Phi) := \mathbf{FEt}(X/\langle\varphi_1, \ldots, \varphi_n\rangle) \times_{\mathbf{FEt}(X/\varphi)} \mathbf{FEt}(X)$$

where $\mathbf{FEt}(X) \to \mathbf{FEt}(X/\varphi)$ is the canonical section of the forgetful functor $\mathbf{FEt}(X/\varphi) \to \mathbf{FEt}(X)$ (see Remark 4.2.9). For any $i \in \{1, \ldots, n\}$, there is a canonical equivalence of categories

$$\mathbf{FEt}(X/\Phi) \cong \mathbf{FEt}(X/\langle\varphi_1, \ldots, \widehat{\varphi_i}, \ldots, \varphi_n\rangle).$$

DEFINITION 4.3.9. For X a small v-sheaf, from Definition 4.3.8 we have

$$\mathbf{FEt}((X \times \mathrm{Spd}(\mathbb{Q}_p))/\Phi) \cong \mathbf{FEt}(\mathrm{FF}_X) \cong \mathbf{FEt}(X \times (\mathrm{Spd}(\mathbb{Q}_p)/\varphi)).$$

The small v-sheaf $X \times (\mathrm{Spd}(\mathbb{Q}_p)/\varphi)$ (which is a diamond if X is, because $\mathrm{Spd}(\mathbb{Q}_p)/\varphi$ is a diamond; see Definition 4.3.5) is an object we have not previously seen; following Fargues [65, Formulation of Fargues' conjecture], we call it the *mirror curve*[22] over X. Note that this object does not project naturally to $\mathrm{Spd}(\mathbb{Q}_p)$ unless X is equipped with such a projection.

We now obtain the following analogue of Lemma 4.2.6.

LEMMA 4.3.10. *Let X be a small v-sheaf. Let F be an algebraically closed nonarchimedean field of characteristic p. Then the base extension functor*

$$\mathbf{FEt}(X) \to \mathbf{FEt}(X \times (\mathrm{Spd}(F)/\varphi)) \cong \mathbf{FEt}((X/\varphi) \times \mathrm{Spd}(F))$$

is an equivalence of categories. (The final equivalence comes from Definition 4.3.8.)

PROOF. We reduce immediately to the case where $X = \mathrm{Spd}(A, A^+)$ for some perfectoid pair (A, A^+) of characteristic p. Choose an untilt K of F of characteristic 0 (which is itself algebraically closed by Lemma 2.8.9); using the isomorphism

$$\mathbf{FEt}((X/\varphi) \times \mathrm{Spd}(F)) = \mathbf{FEt}(\mathrm{FF}_X^\diamond \times_{\mathrm{Spd}(\mathbb{Q}_p)} \mathrm{Spd}(F))$$

---

[22] As far as I know, this terminology is not meant to refer specifically to *mirror symmetry* in mathematical physics.

from Definition 4.3.7, we reduce to showing that the functor

(4.3.10.1) $$\mathbf{FEt}(X) \to \mathbf{FEt}(\mathrm{FF}_X \times_{\mathbb{Q}_p} K), \qquad X' \mapsto \mathrm{FF}_{X'} \times_{\mathbb{Q}_p} K$$

is an equivalence of categories. (Recall that $\mathrm{Spd}(K)$ is just $\mathrm{Spd}(F)$ equipped with a particular morphism to $\mathrm{Spd}(\mathbb{Q}_p)$ that identifies the choice of the untilt.) This claim reduces to the case where A is an algebraically closed field: one first applies this reduction to full faithfulness, then using full faithfulness one applies the reduction again to essential surjectivity.

Suppose first that $K = \mathbb{C}_p$. In this case, the argument is due independently to Fargues–Fontaine (see [**61**, §5.2] for the original announcement and [**63**, Théorème 8.6.1] for the proof) and Weinstein [**176**, Theorem 3.4.3]. Using Remark 1.2.5, we see that

$$2\text{-}\varinjlim_E \mathbf{FEt}(\mathrm{FF}_X \times_{\mathbb{Q}_p} E) \to \mathbf{FEt}(\mathrm{FF}_X \times_{\mathbb{Q}_p} K)$$

is an equivalence for E running over finite extensions of $\mathbb{Q}_p$ within K. That is, any connected finite étale covering $f: Y \to \mathrm{FF}_X \times_{\mathbb{Q}_p} K$ can be realized as the base extension of some connected finite étale covering $f_0: Y_0 \to \mathrm{FF}_X \times_{\mathbb{Q}_p} E$ for some finite extension E of $\mathbb{Q}_p$. The vector bundle $f_{0*}\mathcal{O}_{Y_0}$ carries an $\mathcal{O}_{\mathrm{FF}_X \times_{\mathbb{Q}_p} E}$-algebra structure. Apply Theorem 3.6.13 to the vector bundle $f_{0*}\mathcal{O}_{Y_0}$, and let μ be the largest slope that occurs in the decomposition. If $\mu > 0$, then any element of a copy of $\mathcal{O}(\mu)$ occurring in the decomposition corresponds to a square-zero element of $H^0(Y, \mathcal{O}_Y)$, which does not exist because Y is connected. It follows that $\mu = 0$; similarly, the smallest slope that occurs in the decomposition cannot be negative. Hence $f_{0*}\mathcal{O}_{Y_0}$ is a trivial bundle of rank equal to the degree of f, as then is $f_*\mathcal{O}_Y$. Now $H^0(Y, \mathcal{O}_Y) = H^0(\mathrm{FF}_X \times_{\mathbb{Q}_p} K, f_*\mathcal{O}_Y)$ is a *connected* finite étale K-algebra, and hence isomorphic to K itself because K is algebraically closed; this proves the claim.

Unfortunately, we do not know how to handle general K using the previous approach, as the proof of Theorem 3.6.13 relies crucially on the degree function taking discrete values. The general case requires some arguments outside of the scope of these notes, so we will only give a sketch here and defer to [**116**] for more details. By an inductive argument, we may reduce to the situation where K is a completed algebraic closure of $K_0(t)$ for some perfectoid field K_0 for which the desired conclusion is already known. Let $\rho: \pi_1^{\mathrm{prof}}(\mathrm{FF}_X \times_{\mathbb{Q}_p} K, \overline{x}) \to V$ be a discrete representation on a finite-dimensional $\mathbb{C}_p$-vector space trivialized by some connected finite étale covering $f: Y \to \mathrm{FF}_X \times_{\mathbb{Q}_p} K$. Using Remark 1.2.5, we can realize f as the base extension of a covering $f_0: Y_0 \to \mathrm{FF}_X \times_{\mathbb{Q}_p} \mathrm{Spa}(B, B^+)$ for some one-dimensional affinoid algebra B over F_0 and some point $x \in \mathrm{Spa}(B, B^+)$ whose residue field K_1 admits K as a completed algebraic closure. We may then regard $f_{0*}\mathcal{O}_{Y_0}$ as a vector bundle on $\mathrm{FF}_X \times_{\mathbb{Q}_p} \mathrm{Spa}(B, B^+)$ equipped with an $\mathcal{O}_{\mathrm{FF}_X}$-linear connection. By regarding $\mathrm{FF}_X \times_{\mathbb{Q}_p} \mathrm{Spa}(B, B^+)$ as a family of one-dimensional affinoids over the points of $\mathrm{FF}_X \times_{\mathbb{Q}_p} K_0$ and studying the ramification of f_0 along these affinoids using the theory of p-adic differential equations, one establishes the existence of a ramification filtration on V with properties analogous to those of the ramification filtration of a discrete representation of G_{K_1}. In particular, the unramified component of V induces a representation of $\pi_1^{\mathrm{prof}}(\mathrm{Spec}(F) \times_{\mathrm{Spec}(\mathbb{F}_p)} \mathrm{Spec}(\kappa_K)/\Phi)$, which by the algebraic Drinfeld's lemma (Theorem 4.2.12) must be trivial. In particular, if ρ is nontrivial, then any irreducible component V_0 of V has the property that $V_0^\vee \otimes V_0$ is trivial; this is only possible if V_0 is one-dimensional. However, by directly analyzing the structure of line bundles on $\mathrm{FF}_X \times_{\mathbb{Q}_p} K$, one may show that V_0 corresponds to a subbundle of rank 1 of $f_* \mathcal{O}_Y$, and one can show directly that any line bundle over $\mathrm{FF}_X \times_{\mathbb{Q}_p} K$ is trivial; thus the action of ρ on V_0 is induced by a necessarily nontrivial finite abelian representation of $\pi_1^{\mathrm{prof}}(\operatorname{Spec} K, \overline{x})$, a contradiction. □

REMARK 4.3.11. For $S \in \mathbf{Pfd}$, the space FF_S has a family of cyclic finite étale covers corresponding to replacing the quotient by φ with the quotient by a power of φ (Remark 3.1.8). If $S = \operatorname{Spa}(F, \mathfrak{o}_F)$ for F a perfectoid field, these covers are all connected.

However, suppose that K is an algebraically closed perfectoid field over $\mathbb{Q}_p$. Then one consequence of Lemma 4.3.10 is that the corresponding covers of $K \times_{\mathbb{Q}_p} \mathrm{FF}_S$ are all split! This can be seen more explicitly using the fact that for $d \in \mathbb{Q}$, the bundle $\mathcal{O}(d)$ on FF_S is indecomposable (Corollary 3.6.7) but its pullback to $K \times_{\mathbb{Q}_p} \mathrm{FF}_S$ splits as a direct sum of line bundles.

REMARK 4.3.12. Lemma 4.3.10 asserts that for any diamond X, the *geometric* profinite fundamental group of FF_X coincides with the profinite fundamental group of X. For $X = \operatorname{Spd}(\mathbb{Q}_p)$, this recovers the interpretation of $G_{\mathbb{Q}_p}$ as the profinite fundamental group of a diamond, as originally described in [176]. A variation on this theme is the interpretation of the absolute Galois groups of certain fields as topological fundamental groups by Kucharczyk–Scholze [126].

REMARK 4.3.13. The proof of Lemma 4.3.10 described above relies crucially on Lemma 4.2.6. In [116], it is shown that not only can this dependence be removed, but one can then turn around and recover Drinfeld's lemma for schemes from Lemma 4.3.10 (or more precisely from Theorem 4.3.14).

THEOREM 4.3.14. *Let $X_1, \ldots, X_n$ be connected spatial (and in particular qcqs) diamonds. Then $X := X_1 \times \cdots \times X_n$ is Φ-connected and, for any geometric point $\overline{x}$ of X, the map*

$$\pi_1^{\mathrm{prof}}(X/\Phi, \overline{x}) \to \prod_{i=1}^n \pi_1^{\mathrm{prof}}(X_i, \overline{x})$$

is an isomorphism of profinite groups.

PROOF. As in Theorem 4.2.12, we rewrite the group on the left as

$$\pi_1^{\mathrm{prof}}(X_1 \times (X_2/\varphi) \times \cdots \times (X_n/\varphi), \overline{x})$$

and then induct on n, the base case $n = 1$ being trivial. Again, to prove the induction step, we use Lemma 4.3.10 to imitate the proof of Corollary 4.1.23; see [116] for more details. □

REMARK 4.3.15. As in Remark 4.2.15, we may reformulate Theorem 4.3.14 to say that

$$\pi_1^{\mathrm{prof}}((X_1/\varphi) \times_{\operatorname{Spd}(\mathbb{F}_p)/\varphi} \cdots \times_{\operatorname{Spd}(\mathbb{F}_p)/\varphi} (X_n/\varphi), \overline{x})$$

may be naturally identified with the limit of the diagram

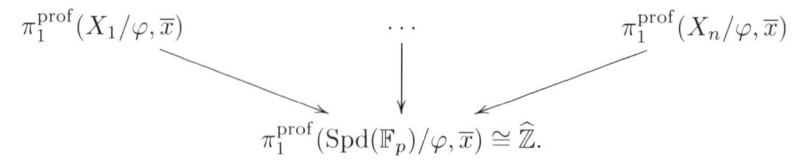

$$\pi_1^{\mathrm{prof}}(X_1/\varphi, \overline{x}) \quad \cdots \quad \pi_1^{\mathrm{prof}}(X_n/\varphi, \overline{x})$$
$$\searrow \quad \downarrow \quad \swarrow$$
$$\pi_1^{\mathrm{prof}}(\mathrm{Spd}(\mathbb{F}_p)/\varphi, \overline{x}) \cong \widehat{\mathbb{Z}}.$$

As a concrete illustration of Drinfeld's lemma, we highlight a corollary relevant to the study of multidimensional (φ, Γ)-modules, as in work of Zábrádi [180, 181] and Pal–Zábrádi [149]. A more detailed exposition of the argument appears in [29].

COROLLARY 4.3.16. *Let $F_1, \ldots, F_n$ be perfectoid fields of characteristic p, each equipped with a multiplicative norm. Let R^+ be the completion of $\mathfrak{o}_{F_1} \otimes_{\mathbb{F}_p} \cdots \otimes_{\mathbb{F}_p} \mathfrak{o}_{F_n}$ for the $(\varpi_1, \ldots, \varpi_n)$-adic topology, where $\varpi_i \in \mathfrak{o}_{F_i}$ is a pseudouniformizer, and put*

$$R := R^+[\varpi_1^{-1}, \ldots, \varpi_n^{-1}].$$

(Note that the ultimate definitions of R^+ and R do not depend on the choices of the ϖ_i.) Then the category of continuous representations of $G_{F_1} \times \cdots \times G_{F_n}$ on finite-dimensional $\mathbb{F}_p$-vector spaces is equivalent to the category of finite projective R-modules equipped with commuting semilinear actions of $\varphi_{F_1}, \ldots, \varphi_{F_n}$.

PROOF. Fix algebraic closures $\overline{F}_i$ of F_i, identify G_{F_i} with $\mathrm{Gal}(\overline{F}_i/F_i)$, let $\overline{R}^+$ be the completion of $\mathfrak{o}_{\overline{F}_1} \otimes_{\mathbb{F}_p} \cdots \otimes_{\mathbb{F}_p} \mathfrak{o}_{\overline{F}_n}$ for the $(\varpi_1, \ldots, \varpi_n)$-adic topology, and put

$$\overline{R} := \overline{R}^+[\varpi_1^{-1}, \ldots, \varpi_n^{-1}].$$

Equip $\overline{R}$ with the obvious action of $G_{F_1} \times \cdots \times G_{F_n}$. The functor in question then takes a representation V to

$$D(V) := (V \otimes_{\mathbb{F}_p} \overline{R})^{G_{F_1} \times \cdots \times G_{F_n}}$$

for the diagonal action on the tensor product, with $D(V)$ inheriting an action of φ_{F_i} from the canonical action on $\overline{R}$ and the trivial action on V. For any given V, we can also write this as

$$D(V) := (V \otimes_{\mathbb{F}_p} S)^{\mathrm{Gal}(E_1/F_1) \times \cdots \times \mathrm{Gal}(E_n/F_n)}$$

for some finite Galois extensions E_i of F_i within $\overline{F}_i$ and $S := E_1 \widehat{\otimes}_{\mathbb{F}_p} \cdots \widehat{\otimes}_{\mathbb{F}_p} E_n$. (More precisely, by analogy with the definition of R, we may write

$$S = S^+[\varpi_1^{-1}, \ldots, \varpi_n^{-1}]$$

where S^+ is the completion of $\mathfrak{o}_{E_1} \widehat{\otimes}_{\mathbb{F}_p} \cdots \widehat{\otimes}_{\mathbb{F}_p} \mathfrak{o}_{E_n}$ for the $(\varpi_1, \ldots, \varpi_n)$-adic topology.) Note that $\mathrm{Spec}(S \otimes_R S)$ splits into the graphs of the various maps $\mathrm{Spec}(S) \to \mathrm{Spec}(S)$ induced by $\mathrm{Gal}(E_1/F_1) \times \cdots \times \mathrm{Gal}(E_n/F_n)$; consequently, the action of this product on $V \otimes_{\mathbb{F}_p} S$ gives rise to a descent datum with respect to the faithfully flat homomorphism $R \to S$. By faithfully flat descent for modules [166, Tag 023N], we deduce that $D(V)$ is a finite projective R-module and the natural map

(4.3.16.1) $$D(V) \otimes_R S \to V \otimes_{\mathbb{F}_p} S$$

is an isomorphism.

To check that this functor is fully faithful, using internal Homs we reduce to checking that

$$V^{G_{F_1} \times \cdots \times G_{F_n}} = D(V)^{\varphi_{F_1}, \ldots, \varphi_{F_n}};$$

this follows by taking simultaneous Galois and Frobenius invariants on both sides of (4.3.16.1) and using the equality
$$\overline{R}^{\varphi_{F_1},\ldots,\varphi_{F_n}} = \mathbb{F}_p.$$
To check essential surjectivity, set
$$X_i := \mathrm{Spd}(F_i), \qquad \overline{x}_i := \mathrm{Spd}(\overline{F}_i)$$
and let $\overline{x}$ be a geometric point of $X := X_1 \times \cdots \times X_n$ lying over each $\overline{x}_i$. By Theorem 4.3.14, the map

(4.3.16.2) $$\pi_1^{\mathrm{prof}}(X/\Phi, \overline{x}) \to \prod_{i=1}^n \pi_1^{\mathrm{prof}}(X_i, \overline{x}) = G_{F_1} \times \cdots \times G_{F_n}.$$

is an isomorphism of profinite groups. Now let D be a finite projective R-module equipped with commuting semilinear actions of $\varphi_{F_1}, \ldots, \varphi_{F_n}$. By composing these actions, we get an action of the absolute Frobenius map φ_R; as in the proof of Lemma 4.2.2, we may invoke [117, Lemma 3.2.6] to see that the sheaf of φ_R-invariants of D on the finite étale site of $\mathrm{Spec}(R)$ is represented by $\mathrm{Spec}(S)$ for some faithfully finite étale R-algebra S. Since D carries actions of $\varphi_{F_1}, \ldots, \varphi_{F_n}$ composing to absolute Frobenius, S does likewise.

Now note that there is a natural morphism
$$\mathrm{Spd}(F_1) \times \cdots \times \mathrm{Spd}(F_n) \to \mathrm{Spd}(R, R^+)$$
which identifies the source with the diamond associated to the adic space
$$Y := \{v \in \mathrm{Spa}(R, R^+) : v(\varpi_1), \ldots, v(\varpi_n) < 1\};$$
this identification yields additional identifications

(4.3.16.3) $$R^+ = H^0(Y, \mathcal{O}^+), \qquad R = \bigcup_{m_1,\ldots,m_n=0}^{\infty} \varpi_1^{-m_1} \cdots \varpi_n^{-m_n} R^+$$

(see Remark 4.3.18 for an explicit example). Let S^+ be the integral closure of R^+ in S. By pulling back $\mathrm{Spa}(S, S^+)$ from $\mathrm{Spa}(R, R^+)$ to Y and invoking (4.3.16.2), we obtain a representation of $G_{F_1} \times \cdots \times G_{F_n}$ which we claim gives rise to D. By replacing each F_i with a suitable finite extension, we reduce to checking this in the case where this representation is trivial. That is, we may assume that $\mathrm{Spa}(S, S^+) \to \mathrm{Spa}(R, R^+)$ splits completely after pullback to Y and we must check that $R \to S$ itself splits completely.

We may view Y as a Stein space by writing it as the union of the ascending sequence $\{U_m\}$ of affinoid subspaces of R given by
$$U_m = \{v \in \mathrm{Spa}(R, R^+) : v(\varpi_i)^m \leq v(\varpi_j) \quad (i, j = 1, \ldots, n)\}.$$
Put $R_m := H^0(U_m, \mathcal{O})$, $R_m^+ := H^0(U_m, \mathcal{O}^+)$; the ring R^+ may then be (almost) identified with the inverse limit of the R_m^+. Similarly, the ring R maps to the inverse limit of the R_m, but this map is not an isomorphism. (The difference between R and $\varprojlim_m R_m$ is analogous to the difference between the ring $\mathbb{Z}_p[\![T]\!][p^{-1}]$ and the ring of rigid analytic functions on the open unit disc over $\mathbb{Q}_p$; the latter contains such elements as $\log(1 + T)$ which do not occur in the former.)

Let $\tilde{S}$ be the sheaf on $\mathrm{Spa}(R, R^+)$ associated to S (viewed as a finite R-algebra). For each m, put $S_m := H^0(U_m, \tilde{S})$; this ring coincides with $S \otimes_R R_m$, and hence is a finite étale R_m-algebra. The fact that $\mathrm{Spa}(S, S^+) \to \mathrm{Spa}(R, R^+)$ splits completely after pullback to Y means that for each m, we have a family of orthogonal idempotents in S_m which split it into copies of R_m, and that the formation of this family is compatible with base change among the R_m. In particular, these idempotents belong to the inverse limit of the S_m, and we must show that they actually belong to S.

Choose a presentation of S as a direct summand of a finite free R-module. This gives a distinguished choice of "coordinates" in R associated to each element of S, and similarly a choice of coordinates in $\varprojlim_m R_m$ associated to each element of $\varprojlim S_m$; we wish to show that the coordinates associated to one of our chosen idempotent elements belong to R.

Choose a connected rational subspace U of $\mathrm{Spa}(R, R^+)$ within Y containing a fundamental domain for the action of $\Phi = \prod_{i=1}^n \varphi_{F_i}^{\mathbb{Z}}$ on Y; a concrete example would be
$$U = \{v \in Y : v(\varpi_1)^p \leq v(\varpi_i) \leq v(\varpi_1) \qquad (i = 2, \ldots, n)\}.$$
(Note that $\varphi_R = \varphi_{F_1} \cdots \varphi_{F_n} \in \Phi$ acts trivially on Y.) We can then find some nonnegative integer m_1 such that restricting each of our idempotents to $H^0(U, \tilde{S})$ gives an element with coordinates in $\varpi_1^{-m_1} H^0(U, \mathcal{O}^+)$. However, this remains true, with the same value of m_1, upon replacing U with its translate by any $\gamma \in \prod_{i=2}^n \varphi_{F_i}^{\mathbb{Z}}$; note that these translates already cover Y because $\varphi_{F_1} \cdots \varphi_{F_n}$ acts trivially on Y. It follows that the coordinates in question belong to $\varpi_1^{-m_1} R^+$ and hence to R. □

REMARK 4.3.17. With notation as in Corollary 4.3.16, one may similarly show that the category of continuous representations of $G_{F_1} \times \cdots \times G_{F_n}$ on finite free $\mathbb{Z}_p$-modules is equivalent to the category of finite projective $W(R)$-modules equipped with commuting semilinear actions of $\varphi_{F_1}, \ldots, \varphi_{F_n}$.

REMARK 4.3.18. We describe the morphism $\mathrm{Spd}(F_1) \times \mathrm{Spd}(F_2) \to \mathrm{Spd}(R, R^+)$ in Corollary 4.3.16 more explicitly in the case where F_1, F_2 are the completed perfected closures of $\mathbb{F}_p((T_1)), \mathbb{F}_p((T_2))$, respectively. In this case, $\mathrm{Spd}(F_1) \times \mathrm{Spd}(F_2)$ is the diamond associated to the adic space
$$Y = \{v \in \mathrm{Spa}(F_1\langle T_2^{p^{-\infty}}\rangle, \mathfrak{o}_{F_1}\langle T_2^{p^{-\infty}}\rangle) : 0 < v(T_2) < 1\};$$
the ring
$$R^+ = H^0(Y, \mathcal{O}^+) = \mathbb{F}_p[\![T_1^{p^{-\infty}}, T_2^{p^{-\infty}}]\!]$$
is the (T_1, T_2)-adic completion of $\mathbb{F}_p[\![T_1, T_2]\!][T_1^{p^{-\infty}}, T_2^{p^{-\infty}}]$; and $H^0(Y, \mathcal{O})$ is the ring of formal sums $\sum_{m_1, m_2 \in \mathbb{Z}[p^{-1}]} c_{m_1 m_2} T_1^{m_1} T_2^{m_2}$ with $c_{m_1 m_2} \in \mathbb{F}_p$ whose support
$$S = \{(m_1, m_2) \in \mathbb{R}^2 : m_1, m_2 \in \mathbb{Z}[p^{-1}], c_{m_1 m_2} \neq 0\}$$
satisfies the following conditions.

- For any $x_0, y_0 \in \mathbb{R}$, the intersection
  $$S \cap \{(x, y) \in \mathbb{R}^2 : x \leq x_0, y \leq y_0\}$$
  is finite.
- The *lower convex hull* of S, i.e., the convex hull of the set
  $$\bigcup_{(m_1, m_2) \in S} \{(x, y) \in \mathbb{R}^2 : x \geq m_1, y \geq m_2\},$$
  admits a supporting line of slope $-s$ for each $s > 0$.

REMARK 4.3.19. In Remark 4.3.18, the ring R can be interpreted as the subring of $H^0(Y, \mathcal{O})$ consisting of functions which are bounded for $v(T_2)$ close to 1 and of *polynomial growth* for $v(T_2)$ close to 0. This suggests a close relationship (made more explicit in [29]) between the identification (4.3.16.3) and the perfectoid analogue of the *Riemann extension theorem (Hebbarkeitsatz)* introduced by Scholze [159, §2.3] to study the boundaries of perfectoid Shimura varieties. This result has been (refined and) used by André [5, 6] and Bhatt [16] to resolve a long-standing open problem in commutative algebra, the *direct summand conjecture* of Hochster: if $R \to S$ is a finite morphism of noetherian rings and R is regular, then $R \to S$ splits in the category of R-modules. (See [94] for several equivalent formulations and consequences.

REMARK 4.3.20. One probably cannot hope to have an analogue of Theorem 4.3.14 for étale fundamental groups in the sense of de Jong (Remark 4.1.6). In particular, if F is an algebraically closed perfectoid field of characteristic p and K is an algebraically closed perfectoid field of characteristic 0, then $\mathrm{Spd}(K) \times (\mathrm{Spd}(F)/\varphi) \cong K \times_{\mathbb{Q}_p} \mathrm{FF}^{\diamond}_{\mathrm{Spd}(F)}$ admits no finite étale coverings, but it may admit some nonfinite étale coverings; see Problem 4.3.21 for one possible construction.

PROBLEM 4.3.21. Choose two sections of $\mathcal{O}(1)$ on $\mathrm{FF}_{\mathrm{Spa}(F,\mathfrak{o}_F)}$ with distinct zeroes, use these to define a morphism $\mathrm{FF}_{\mathrm{Spa}(F,\mathfrak{o}_F)} \to \mathbf{P}^1_{\mathbb{Q}_p}$, and pull back the Hodge-Tate period mapping. Does the resulting covering split completely?

4.4. Shtukas in positive characteristic. We now arrive at the fundamental concept introduced by Drinfeld as a replacement for elliptic curves in positive characteristic; that is to say, the moduli spaces of such objects constitute a replacement for modular curves and Shimura varieties as a tool for studying Galois representations of a global function field in positive characteristic (which we are now prepared to think about as representations of profinite fundamental groups).

HYPOTHESIS 4.4.1. Throughout §4.4, let C be a smooth, projective, geometrically irreducible curve over a finite field $\mathbb{F}_q$ of characteristic p.

DEFINITION 4.4.2. Let S be a scheme over $\mathbb{F}_q$. A *shtuka* over S consists of the following data.
- A finite index set I and a morphism $(x_i)_{i \in I} : S \to C^I$.
- A vector bundle $\mathcal{F}$ over $C \times S$.
- An isomorphism of bundles

$$\Phi : (\varphi_S^* \mathcal{F})|_{(C \times S) \setminus \bigcup_{i \in I} \Gamma_{x_i}} \cong \mathcal{F}|_{(C \times S) \setminus \bigcup_{i \in I} \Gamma_{x_i}},$$

where $\Gamma_{x_i} \subset C \times S$ denotes the graph of x_i. The morphisms $x_i : S \to C$ are called the *legs* (in French, *pattes*[23]) of the shtuka.

REMARK 4.4.3. For Z a finite set of closed points of C, one may also consider shtukas with *level structure* at C; this amounts to insisting that the legs map S into $C \setminus Z$ and specifying a trivialization of $(\mathcal{F}, \Phi)$ over $Z \times S$.

REMARK 4.4.4. For K the function field of C and G a connected reductive algebraic group over K, Varshavsky [173] has introduced the notion of a *G-shtuka*,

---

[23]This word translates into English variously as *legs* or *paws*. However, the term *paw* is typically used only for mammalian feet, whereas the intended animal metaphor seems to be a caterpillar or millipede. We thus prefer the translation *legs*, following V. Lafforgue in [129].

the previous definition being the case $G = \mathrm{GL}_n$ for $n = \mathrm{rank}(\mathcal{F})$. In the case where G is *split* (i.e., G contains a split maximal torus), then the results of SGA 3 [42–44] imply that G extends canonically to a group scheme G_C over C, and we then insist that $\mathcal{F}$ be a G_C-torsor and Φ be an isomorphism of G_C-torsors.

REMARK 4.4.5. The word *shtuka* (in French, *chtouca*) is a transliteration of the Russian word штука, meaning a generic thing whose exact identity is unknown or irrelevant; it is probably derived from the German word *Stück* (meaning *piece*), although the Russian usage may[24] be influenced by the word что (meaning *what*). Some analogous terms in English are *widget, gadget, gizmo, doodad, whatchamacallit*; see Wikipedia on *placeholder names* for more examples.

REMARK 4.4.6. As pointed out in [79], one source of inspiration for the definition of shtukas is some work of Krichever on integrable systems arising from the Korteweg–de Vries (KdV) equation. The relationship between these apparently disparate topics has been exposed by Mumford [141].

4.5. Shtukas in mixed characteristic. It would be far outside the scope of these lectures to explain in any meaningful detail why shtukas are so important in the study of the Langlands correspondence over global function fields. Instead, we jump straight to the mixed-characteristic analogue, to illustrate a startling[25] convergence between shtukas and Fargues–Fontaine curves.

We take the approach to sheaves on stacks used in [166, Tag 06TF].

DEFINITION 4.5.1. Let $\mathcal{O} : \mathbf{Pfd} \to \mathbf{Ring}$ be the functor taking X to $\mathcal{O}(X)$. By Theorem 3.8.2 this functor is a sheaf of rings for the v-topology. For any small v-sheaf X, we may restrict $\mathcal{O}$ to the arrow category $\mathbf{Pfd}_X$ (i.e., the category of morphisms $S^\diamond \to X$ with $S \in \mathbf{Pfd}$) to obtain the *structure sheaf* on X.

A *vector bundle* on X is a locally finite free $\mathcal{O}_X$-module; let $\mathbf{Vec}_X$ denote the category of such objects. We avoid trying to define a *pseudocoherent sheaf* on a diamond or small v-sheaf due to the issues raised in §3.8.

This category of vector bundles lives entirely in characteristic p; we actually need something slightly different.

DEFINITION 4.5.2. Let $\mathcal{O}^\sharp : \mathbf{Pfd}_{\mathrm{Spd}(\mathbb{Z}_p)} \to \mathbf{Ring}$ be the functor taking X to $\mathcal{O}(X^\sharp)$, where $X^\sharp$ is the untilt of X corresponding to the structure morphism $X \to \mathrm{Spd}(\mathbb{Z}_p)$. By Theorem 3.8.2 again, this functor is a sheaf of rings for the v-topology. For any small v-sheaf X over $\mathrm{Spd}(\mathbb{Z}_p)$, we may restrict $\mathcal{O}^\sharp$ to $\mathbf{Pfd}_X$ to obtain the *untilted structure sheaf* on X. An *untilted vector bundle* on X is a locally finite free $\mathcal{O}_X^\sharp$-module; let $\mathbf{Vec}_X^\sharp$ denote the category of such objects.

As an immediate consequence of Theorem 3.8.7, we have the following.

THEOREM 4.5.3. *Let (A, A^+) be a perfectoid pair of characteristic p.*

(a) *The pullback functor $\mathbf{FPMod}_A \to \mathbf{Vec}_{\mathrm{Spd}(A,A^+)}$ is an equivalence of categories.*

---

[24] Beware that I have no evidence to support this speculative claim!

[25] To be fair, this convergence was anticipated well before the technology became available to make it overt. For example, as reported by Kisin in the introduction of [122], the analogy between Breuil–Kisin modules and shtukas was already manifest in the analogy between Kisin's work with that of Genestier–Lafforgue in positive characteristic [76].

(b) *Fix a morphism* $\mathrm{Spd}(A, A^+) \to \mathrm{Spd}(\mathbb{Z}_p)$ *corresponding to an untilt* $(A^\sharp, A^{\sharp+})$ *of* (A, A^+). *Then the pullback functor* $\mathbf{FPMod}_{A^\sharp} \to \mathbf{Vec}^\sharp_{\mathrm{Spd}(A,A^+)}$ *is an equivalence of categories.*

REMARK 4.5.4. Let X be an analytic adic space on which p is topologically nilpotent. In many cases of interest, the pushforward of $\mathcal{O}_{X_{\mathrm{proet}}}$ to X coincides with $\mathcal{O}_X$; in such cases, the base extension functor $\mathbf{Vec}_X \to \mathbf{Vec}^\sharp_{X^\diamond}$ is fully faithful. However, even in such cases, this functor is generally not essentially surjective (unless X is perfectoid, in which case Theorem 3.8.7 applies). For example, if $X = \mathrm{Spd}(\mathbb{Q}_p)$, then the source category consists of finite-dimensional $\mathbb{Q}_p$-vector spaces while the target consists of finite-dimensional $\mathbb{C}_p$-vector spaces equipped with continuous semilinear $G_{\mathbb{Q}_p}$-actions.

A related point is that if $X = \mathrm{Spa}(A, A^+)$ is not perfectoid, then objects of $\mathbf{Vec}^\sharp_{X^\diamond}$ need not be acyclic on X.

REMARK 4.5.5. The definition of a shtuka over a diamond (Definition 4.5.6) will refer to $\mathrm{Spd}(\mathbb{Z}_p) \times S$, but in order to formulate the definition correctly we must unpack this concept a bit in the case where $S \in \mathbf{Pfd}$. In this case, $\mathrm{Spd}(\mathbb{Z}_p) \times S^\diamond$ descends to an adic space in a canonical way: for $S = \mathrm{Spd}(R, R^+)$, writing $\mathbf{A}_{\mathrm{inf}}$ for $\mathbf{A}_{\mathrm{inf}}(R, R^+)$ we have

$$\mathrm{Spd}(\mathbb{Z}_p) \times S^\diamond \cong W_S^\diamond, \qquad W_S := \mathrm{Spa}(\mathbf{A}_{\mathrm{inf}}, \mathbf{A}_{\mathrm{inf}}) \setminus V([\overline{x}_1], \ldots, [\overline{x}_n]),$$

where $\overline{x}_1, \ldots, \overline{x}_n \in R$ are topologically nilpotent elements which generate the unit ideal. The space W_S has the property that the pushforward of $\mathcal{O}_{W_S,\mathrm{proet}}$ to W_S coincides with $\mathcal{O}_{W_S}$ (this can be seen from the explicit description given in the proof of Lemma 3.1.3), so we may view the vector bundles on W_S as a full subcategory of the untilted vector bundles on $\mathrm{Spd}(\mathbb{Z}_p) \times S^\diamond$ (with respect to the first projection).

DEFINITION 4.5.6. Let S be a diamond. A *shtuka* over S consists of the following data.
- A finite index set I and a morphism $(x_i)_{i \in I} : S \to \mathrm{Spd}(\mathbb{Z}_p)^I$.
- An untilted vector bundle $\mathcal{F}$ over $\mathrm{Spd}(\mathbb{Z}_p) \times S$ with respect to the first projection which locally-on-S arises from a vector bundle on the underlying adic space W_S (Remark 4.5.5).
- An isomorphism of bundles

$$\Phi : (\varphi_S^* \mathcal{F})|_{(\mathrm{Spd}(\mathbb{Z}_p) \times S) \setminus \bigcup_{i \in I} \Gamma_{x_i}} \cong \mathcal{F}|_{(\mathrm{Spd}(\mathbb{Z}_p) \times S) \setminus \bigcup_{i \in I} \Gamma_{x_i}},$$

where $\Gamma_{x_i} \subset \mathrm{Spd}(\mathbb{Z}_p) \times S$ denotes the graph of x_i. We also insist that Φ be meromorphic along $\bigcup_{i \in I} \Gamma_{x_i}$, this having been implicit in the schematic case.

Again, the morphisms $x_i : S \to \mathrm{Spd}(\mathbb{Z}_p)$ are called the *legs* of the shtuka.

REMARK 4.5.7. For $S \in \mathbf{Pfd}$, we could have defined a shtuka over S directly in terms of a vector bundle over W_S, without reference to untilted vector bundles. The point of the formulation used here is to encode the fact that shtukas satisfy descent for the v-topology, which does not immediately follow from Theorem 3.8.7 because W_S is not a perfectoid space.

To unpack this definition further, let us first consider the case of a shtuka with no legs.

REMARK 4.5.8. Suppose that $I = \emptyset$. A shtuka over S with no legs is simply an untilted vector bundle $\mathcal{F}$ over $\mathrm{Spd}(\mathbb{Z}_p) \times S$ (which locally-on-S descends to the underlying adic space W_S) equipped with an isomorphism with its φ-pullback.

In the case where $S = \mathrm{Spd}(R, R^+) \in \mathbf{Pfd}$, by restricting from W_S to Y_S and then quotienting by the action of φ, we obtain a vector bundle on the relative Fargues–Fontaine curve FF_S. However, not all vector bundles can arise in this fashion, for the following reasons.

- The resulting bundle is fiberwise semistable of slope 0.
- The associated étale $\mathbb{Q}_p$-local system (see Theorem 3.7.5) descends to an étale $\mathbb{Z}_p$-local system determined by the shtuka. (For a general étale $\mathbb{Q}_p$-local system, such a descent only exists locally on S; see [**117**, Corollary 8.4.7].)

In fact, the functor from shtukas over S with no legs to étale $\mathbb{Z}_p$-local systems on S is an equivalence of categories; this follows from a certain analogue of Theorem 3.7.5 (see [**117**, Theorem 8.5.3]).

REMARK 4.5.9. In the case where S is a geometric point, Remark 4.5.8 asserts that shtukas over S with no legs correspond simply to finite free $\mathbb{Z}_p$-modules. In particular, they extend canonically from W_S over all of $\mathrm{Spa}(\mathbf{A}_{\mathrm{inf}}, \mathbf{A}_{\mathrm{inf}})$.

A partial extension of this result is the following.

THEOREM 4.5.10 (Kedlaya). *Let (R, R^+) be a perfectoid Huber pair of characteristic p in which R is Tate, and write $\mathbf{A}_{\mathrm{inf}}$ for $\mathbf{A}_{\mathrm{inf}}(R, R^+)$.*

(a) *Let $\overline{x} \in R^+$ be a topologically nilpotent unit of R. Then the pullback functor from vector bundles on the scheme*

$$\mathrm{Spec}(\mathbf{A}_{\mathrm{inf}}) \setminus V(p, [\overline{x}])$$

to vector bundles on the analytic locus of $\mathrm{Spa}(\mathbf{A}_{\mathrm{inf}}, \mathbf{A}_{\mathrm{inf}})$ is an equivalence of categories.

(b) *Suppose that $R = F$ is a perfectoid field. Then both categories in (a) are equivalent to the category of finite free $\mathbf{A}_{\mathrm{inf}}$-modules and to the category of vector bundles on $\mathrm{Spec}(\mathbf{A}_{\mathrm{inf}})$.*

PROOF. See [**114**]. □

REMARK 4.5.11. It should be possible to extend Theorem 4.5.10 to the case where R is analytic. In this case, the statement would assert that for any topologically nilpotent elements $\overline{x}_1, \ldots, \overline{x}_n \in R^+$ generating the unit ideal in R, the pullback functor from vector bundles on the scheme

$$\mathrm{Spec}(\mathbf{A}_{\mathrm{inf}}) \setminus V(p, [\overline{x}_1], \ldots, [\overline{x}_n])$$

to vector bundles on the analytic locus of $\mathrm{Spa}(\mathbf{A}_{\mathrm{inf}}, \mathbf{A}_{\mathrm{inf}})$ is an equivalence of categories.

REMARK 4.5.12. Theorem 4.5.10(b) is analogous to the assertion that if R is a two-dimensional local ring, then the pullback functor from vector bundles on $\mathrm{Spec}(R)$ to vector bundles on the complement of the closed point is an equivalence of categories (because reflexive and projective R-modules coincide). By contrast, one does not have a similar assertion comparing, say, vector bundles on $\mathrm{Spec}(k[\![x, y, z]\!])$ (for k a field) with vector bundles on the complement of the locus where x and

y both vanish; similarly, Theorem 4.5.10(b) cannot be extended beyond the case where R is a perfectoid field.

Even if R is a perfectoid field, if R is not algebraically closed, then shtukas over S with no legs need not extend as bundles from W_S to the whole analytic locus of $\mathrm{Spa}(\mathbf{A}_{\mathrm{inf}}, \mathbf{A}_{\mathrm{inf}})$. Namely, the only ones that do so are the ones coming from étale local systems on S that extend to $\mathrm{Spa}(R^+, R^+)$, i.e., the ones corresponding to *unramified* Galois representations.

REMARK 4.5.13. As per Remark 4.5.12, for $S \in \mathbf{Pfd}$, the restriction functor on φ-equivariant vector bundles from the analytic locus of $\mathrm{Spa}(\mathbf{A}_{\mathrm{inf}}, \mathbf{A}_{\mathrm{inf}})$ to W_S is not essentially surjective. However, one does expect it to be fully faithful; see Lemma 4.5.17 for a special case of a related statement.

We now increase complexity slightly by considering shtukas with one leg.

LEMMA 4.5.14. *Suppose that $I = \{1\}$ is a singleton set and that the morphism x_1 factors through $\mathrm{Spd}(\mathbb{Q}_p)$. Then the following categories are canonically equivalent:*

(a) *shtukas over S with leg x_1;*
(b) *data $\mathcal{F}_1 \dashrightarrow \mathcal{F}_2$, where $\mathcal{F}_1$ is a φ-equivariant bundle over $\mathrm{Spd}(\mathbb{Z}_p) \times S$ (which locally-on-S descend to W_S), $\mathcal{F}_2$ is a φ-equivariant bundle over $\mathrm{Spd}(\mathbb{Q}_p) \times S$ (which locally-on-S descend to Y_S), and the arrow denotes a meromorphic φ-equivariant map over Y_S which is an isomorphism away from $\bigcup_{n \in \mathbb{Z}} \varphi^n(\Gamma_{x_1})$.*

PROOF. We start with the general idea: if one thinks of Φ as defining an isomorphism from $\varphi_S^* \mathcal{F}$ to $\mathcal{F}$ "up to a discrepancy," then $\mathcal{F}_1, \mathcal{F}_2$ are obtained by resolving the discrepancy respectively in favor of $\varphi_S^* \mathcal{F}, \mathcal{F}$.

We now make this explicit. Since we are constructing a canonical equivalence, we may assume that $S = \mathrm{Spd}(R, R^+)$ where (R, R^+) is a Tate perfectoid pair of characteristic p. Choose a pseudouniformizer $\varpi \in R$. Given a shtuka oven S with leg x_1, we obtain the bundle $\mathcal{F}_1$ by restricting $\varphi_S^* \mathcal{F}$ to $\{v \in W_S : v(p) \leq v([\varpi]^{p^{-n}})\}$ for sufficiently small n (ensuring that Γ_{x_i} does not meet this set; here we use the fact that x_1 factors through $\mathrm{Spd}(\mathbb{Q}_p)$), then using the isomorphism $\varphi_S^* \mathcal{F} \cong \mathcal{F}$ to enlarge n. Similarly, we obtain $\mathcal{F}_2$ by restricting $\mathcal{F}$ to $\{v \in W_S : v(p) \geq v([\varpi]^{p^n})\}$ for sufficiently small n, then using the isomorphism $\varphi_S^* \mathcal{F} \cong \mathcal{F}$ to enlarge n. (Note that in this second case, the union of these spaces is only Y_S, not W_S.) The meromorphic map $\varphi_S^* \mathcal{F} \dashrightarrow \mathcal{F}$ gives rise to the meromorphic map $\mathcal{F}_1 \dashrightarrow \mathcal{F}_2$. One may check that this construction is reversible and does not depend on ϖ. □

REMARK 4.5.15. Suppose that $S = \mathrm{Spd}(R, R^+) \in \mathbf{Pfd}$, $I = \{1\}$ is a singleton set, and that the morphism x_1 factors through $\mathrm{Spd}(\mathbb{Q}_p)$. From Lemma 4.5.14, we obtain a pair of vector bundles $\mathcal{G}_1, \mathcal{G}_2$ on FF_S and a meromorphic map $\mathcal{G}_1 \dashrightarrow \mathcal{G}_2$ which is an isomorphism away from the untilt corresponding to x_1. Of these, $\mathcal{G}_1$ arises from a shtuka with no legs, so it is fiberwise semistable of slope 0 and its associated étale $\mathbb{Q}_p$-local system descends to an étale $\mathbb{Z}_p$-local system determined by the shtuka (more precisely, by the restriction of $\mathcal{F}_1$ to the point $v(p) = 0$).

Over a point, we may relate this discussion back to previously studied concepts in p-adic Hodge theory.

DEFINITION 4.5.16. Let F be a perfectoid field of characteristic p and write $\mathbf{A}_{\inf}$ for $\mathbf{A}_{\inf}(F, \mathfrak{o}_F)$. Also fix a primitive element z of $\mathbf{A}_{\inf}$ corresponding to an untilt $F^\sharp$ of F of characteristic 0 (that is, z is not divisible by p). A *Breuil–Kisin module*[26] over $\mathbf{A}_{\inf}$ (with respect to z) is a finite free $\mathbf{A}_{\inf}$-module D equipped with an isomorphism $\Phi : (\varphi^* D)[z^{-1}] \cong D[z^{-1}]$. Let $x_1 : \operatorname{Spd}(F, \mathfrak{o}_F) \to \operatorname{Spd}(\mathbb{Z}_p)$ be the morphism corresponding to the untilt $F^\sharp$ of F.

The following result is due to Fargues [60], though the proof we obtain here is slightly different; it first appears in [162].

LEMMA 4.5.17. *With notation as in Definition 4.5.16, suppose that F is algebraically closed. Then restriction of φ-equivariant vector bundles along the inclusion*

$$Y_S \subset \{v \in \operatorname{Spa}(\mathbf{A}_{\inf}, \mathbf{A}_{\inf}) : v(p) \neq 0\}$$

is an equivalence of categories.

PROOF. Full faithfulness follows from a calculation using Newton polygons, which does not depend on F being algebraically closed or even a field (compare [162, Proposition 13.3.2]). Essential surjectivity is a consequence of Theorem 3.6.13. □

THEOREM 4.5.18 (Fargues). *With notation as in Definition 4.5.16, suppose that F is algebraically closed. Then the category of Breuil–Kisin modules over $\mathbf{A}_{\inf}$ is equivalent to the category of shtukas over $\operatorname{Spd}(F, \mathfrak{o}_F)$ with the single leg x_1.*

PROOF. By Lemma 4.5.14, a shtuka with one leg corresponds to a datum $\mathcal{F}_1 \dashrightarrow \mathcal{F}_2$ over Y_S. By Lemma 4.5.17, $\mathcal{F}_2$ extends uniquely over the point $v([\varpi]) = 0$ (for $\varpi \in F$ a pseudouniformizer). Meanwhile, by construction, $\mathcal{F}_1$ is already defined over the point $v(p) = 0$. By glueing, we obtain a vector bundle over the analytic locus of $\operatorname{Spa}(\mathbf{A}_{\inf}, \mathbf{A}_{\inf})$, which by Theorem 4.5.10 arises from a finite free $\mathbf{A}_{\inf}$-module. This proves the claim. □

REMARK 4.5.19. With notation as in Definition 4.5.16, Breuil–Kisin modules appear naturally in the study of crystalline representations. In fact, the crystalline comparison isomorphism in p-adic Hodge theory can be exhibited by giving a direct cohomological construction of suitable Breuil–Kisin modules from which the étale, de Rham, and crystalline cohomologies can be functorially recovered (the étale cohomology arising as in Remark 4.5.15). This is the approach taken in the work of Bhatt–Morrow–Scholze [18] (see also [139] and [17, Lecture 4]).

As noted above, the idea to formulate Definition 4.5.16 and relate it to shtukas with one leg as in Theorem 4.5.18 is due to Fargues [60]. This development was one of the primary triggers for both the line of inquiry discussed in this lecture and for [18].

REMARK 4.5.20. In light of the second part of Remark 4.5.12, Theorem 4.5.18 does not extend to the case where F is a general perfectoid field; the extra structure imposed by the existence of the Breuil–Kisin module restricts the étale $\mathbb{Z}_p$-local system arising from $\mathcal{F}_1$ in a nontrivial way. (When $F^\sharp$ is algebraic over $\mathbb{Q}_p$, this is

---

[26]The term *Breuil–Kisin module* originally referred to similar objects defined not over $\mathbf{A}_{\inf}$, but over a certain power series ring; see [122]. The relationship between this construction and the one we are now discussing is analogous to the relationship between *imperfect* and *perfect* (φ, Γ)-modules described in [120].

related to Fontaine's notion of a *crystalline* representation.) One can also try to consider *relative Breuil–Kisin modules* over more general base spaces, but then the first part of Remark 4.5.12 also comes into play.

4.6. Affine Grassmannians. The concept of an *affine Grassmannian* plays a central role in geometric Langlands, in enabling the construction of moduli spaces of shtukas. We describe three different flavors of the construction here; while these constructions operate with respect to more general algebraic groups, we restrict to the case of GL_n for the sake of exposition, deferring to [162, §19–21] for discussion of more general groups.

We start with the original affine Grassmannian of Beauville–Laszlo [11]. See [182, Lecture 1] for a detailed treatment.

DEFINITION 4.6.1. Fix a field k and a positive integer n. For R a k-algebra, a *lattice* in $R((t))^n$ is a finite projective $R[\![t]\!]$-submodule Λ such that the induced map
$$\Lambda \otimes_{R[\![t]\!]} R((t)) \to R((t))^n$$
is an isomorphism. The functor[27] Gr taking R to the set of lattices in $R((t))^n$ is a sheaf for the Zariski topology, so it extends to a sheaf on the category of k-schemes.

THEOREM 4.6.2 (Beauville–Laszlo). *The functor* Gr *is represented by an ind-projective k-scheme. More precisely, for each N, the subfunctor of lattices lying between $t^{-N} R[\![t]\!]^n$ and $t^N R[\![t]\!]^n$ is represented by a projective k-scheme* $\mathrm{Gr}^{(N)}$, *and the transition maps* $\mathrm{Gr}^{(N)} \to \mathrm{Gr}^{(N+1)}$ *are closed immersions.*

PROOF. See [182, Theorem 1.1.3]. □

The following analogue of the previous construction was originally considered at the pointwise level by Haboush [85], and in the following form by Kreidl [123].

DEFINITION 4.6.3. Let k be a perfect field of characteristic p. For R a perfect k-algebra, a *lattice* in $W(R)[p^{-1}]^n$ is a finite projective $W(R)$-submodule Λ such that the induced map
$$\Lambda \otimes_{W(R)} W(R)[p^{-1}] \to W(R)[p^{-1}]^n$$
is an isomorphism. Again, the functor Gr^{W} taking R to the set of lattices in $W(R)[p^{-1}]^n$ is a sheaf for the Zariski topology, so it extends to a sheaf on the category of perfect k-schemes.

The following statement is due to Bhatt–Scholze [20, Theorem 1.1], improving an earlier result of Zhu [183] which asserts $\mathrm{Gr}^{\mathrm{W},(N)}$ is represented by a proper algebraic space over k.

THEOREM 4.6.4 (Zhu, Bhatt–Scholze). *For each N, the functor* $\mathrm{Gr}^{\mathrm{W},(N)}$ *of lattices in $W(R)[p^{-1}]^n$ lying between $p^{-N} W(R)[p^{-1}]^n$ and $p^N W(R)[p^{-1}]^n$ is represented by the perfection of a projective k-scheme. The transition maps* $\mathrm{Gr}^{\mathrm{W},(N)} \to \mathrm{Gr}^{\mathrm{W},(N+1)}$ *are closed immersions.*

In the context of perfectoid spaces, it is natural to introduce the following variant of the previous construction.

---

[27]This use of the notation Gr conflicts with the notation for graded rings used in Definition 1.5.3, but we will not be using the latter in this lecture.

DEFINITION 4.6.5. Define the presheaves $\mathbf{B}_{\mathrm{dR}}^+, \mathbf{B}_{\mathrm{dR}}$ on the category of perfectoid Huber pairs whose values on (A, A^+) equal, respectively, the completion of $W^b(A^\flat)$ with respect to the principal ideal $\ker(\theta: W^b(A^\flat) \to A)$, and the localization of this ring with respect to a generator z of $\ker(\theta)$. These extend to sheaves on the category of perfectoid spaces with respect to the analytic topology, the étale topology, the pro-étale topology, and the v-topology.

For A a completed algebraic closure of $\mathbb{Q}_p$ and $A^+ = A^\circ$, $\mathbf{B}_{\mathrm{dR}}(A, A^+)$ is Fontaine's *ring of de Rham periods* [**66**, §2].

DEFINITION 4.6.6. For (A, A^+) a perfectoid pair, a *lattice* in $\mathbf{B}_{\mathrm{dR}}(A, A^+)^n$ is a finite projective $\mathbf{B}_{\mathrm{dR}}^+(A, A^+)$-submodule Λ such that the induced map

$$\Lambda \otimes_{\mathbf{B}_{\mathrm{dR}}^+(A,A^+)} \mathbf{B}_{\mathrm{dR}}(A, A^+) \to \mathbf{B}_{\mathrm{dR}}^+(A, A^+)^n$$

is an isomorphism. The functor $\mathrm{Gr}^{\mathrm{dR}}$ taking (A, A^+) to the set of lattices in $\mathbf{B}_{\mathrm{dR}}(A, A^+)^n$ is a sheaf for the analytic topology, so it extends to a sheaf on the category of perfectoid spaces. It is also a sheaf for the pro-étale topology (and even the v-topology), so it further extends to a sheaf on the category of small v-sheaves over $\mathrm{Spd}(\mathbb{Z}_p)$.

THEOREM 4.6.7 (Scholze). *For each N, the functor $\mathrm{Gr}^{\mathrm{dR},(N)}$ of lattices in $\mathbf{B}_{\mathrm{dR}}(A, A^+)^n$ lying between $z^{-N}\mathbf{B}_{\mathrm{dR}}^+(A, A^+)^n$ and $z^N\mathbf{B}_{\mathrm{dR}}^+(A, A^+)$ is a diamond. The transition maps $\mathrm{Gr}^{\mathrm{dR},(N)} \to \mathrm{Gr}^{\mathrm{dR},(N+1)}$ are closed immersions.*

PROOF. See [**162**, Theorem 19.2.4]. □

REMARK 4.6.8. The Beauville–Laszlo glueing theorem (Remark 1.9.19) was introduced to link affine Grassmannians with shtukas with one leg. To wit, let C be a curve over k, let V be a vector bundle of rank n on C, let $z \in C$ be a k-rational point, fix an identification of $\widehat{\mathcal{O}}_{C,z}$ with $k[\![t]\!]$, and fix a basis of V over $k[\![t]\!]$. Then the functor Gr may be identified with the functor taking a k-algebra R to the set of meromorphic morphisms $V \dashrightarrow V'$ from V to another vector bundle on C which are isomorphisms away from z.

One can do something similar in the context of Lemma 4.5.14. Suppose that $S = \mathrm{Spa}(R, R^+)$ is the tilt of (A, A^+); as per Remark 3.1.12, $\mathrm{Spa}(A, A^+)$ may be identified with a certain closed subspace of FF_S. Let $\mathcal{F}_1$ be a trivial bundle of rank n over $\mathrm{Spd}(\mathbb{Z}_p) \times S$ with a prescribed basis. Then a datum of the form $\mathcal{F}_1 \dashrightarrow \mathcal{F}_2$ gives rise to a lattice in $\mathbf{B}_{\mathrm{dR}}(R, R^+)^n$ by comparing the completions of $\mathcal{F}_1$ and $\mathcal{F}_2$ along the completion of $\mathrm{Spa}(A, A^+)$ (or rather, its preimage in $\mathrm{Spa}(\mathbf{A}_{\inf}, \mathbf{A}_{\inf})$). Using the Beauville–Laszlo theorem, we may reverse the construction to go from lattices to certain shtukas over S with leg x_1, namely those for which the étale $\mathbb{Z}_p$-local system associated to $\mathcal{F}_1$ (see Remark 4.5.15) is trivial. Since this always happens pro-étale locally, in this we may view $\mathrm{Gr}^{\mathrm{dR}}$ as the "moduli space of shtukas with leg x_1."

REMARK 4.6.9. In Definition 4.6.1, the ordinary Grassmannians of subspaces in an n-dimensional ambient space appears as truncations of the functor Gr in which we consider lattices between $R[\![t]\!]^n$ and $tR[\![t]\!]^n$. If we truncate Gr^W similarly, we lose sight of the distinction between mixed and equal characteristic, and so we get precisely the same thing as for Gr.

Somewhat analogously, if one truncates $\mathrm{Gr}^{\mathrm{dR}}$ by considering only lattices between $\mathbf{B}_{\mathrm{dR}}^+(A, A^+)^n$ and $z\mathbf{B}_{\mathrm{dR}}^+(A, A^+)$, then the "stacky" nature of the space goes away, and one is yet again considering an ordinary Grassmannian. From this point of view, the Grassmannian is closely related to a certain *Rapoport-Zink space*; such spaces were introduced as local analogues of Shimura varieties (see [153] for the original context and [152] for a more modern treatment).

One a similar note, the construction of Remark 4.6.8 provides a geometric interpretation for certain constructions in p-adic Hodge theory involving "modifications", as in [12].

REMARK 4.6.10. In the framework where GL_n is replaced by some more general group, truncations correspond to Schubert cells defined by cocharacters of the group, and the sort of truncation considered in Remark 4.6.9 corresponds to a *minuscule* cocharacter. Working with cocharacters which are not minuscule runs into subtleties of the sort described in Definition 4.6.11.

For truncations not covered by Remark 4.6.9, one can get some partial information by passing from lattices to their associated filtrations.

DEFINITION 4.6.11. With notation as in Definition 4.6.1, the lattice Λ in $R((t))^n$ gives rise to a descending filtration on R^n by taking $\mathrm{Fil}^m R^n$ to be the set of $\mathbf{v} \in R^n$ which occur as the reduction of some element of $t^m \Lambda \cap R[\![t]\!]^n$. This filtration uniquely determines Λ when Λ is sandwiched between $R[\![t]\!]^n$ and $tR[\![t]\!]^n$, but not in general.

The passage from lattices to filtrations corresponds to a morphism from Gr to a certain infinite flag variety. Taking truncations on both sides, we obtain a morphism from a Schubert cell to a flag variety. In the analogous construction for $\mathrm{Gr}^{\mathrm{dR}}$ (which we leave to the reader to imagine), the flag variety is again closely related to a Rapoport-Zink space, and is in some sense the best possible approximation to $\mathrm{Gr}^{\mathrm{dR}}$ that can be made within classical nonarchimedean analytic geometry; this demonstrates the necessity of passing to a world of more exotic objects (namely diamonds) in order to exhibit moduli spaces of shtukas in general.

REMARK 4.6.12. A similar subtlety to that described in Definition 4.6.11 occurs in classical Hodge theory: one can define period domains for Hodge structures as certain flag varieties, but in general one cannot construct universal variations of Hodge structures on these spaces. Classically, the obstruction to this is characterized in terms of the phenomenon of *Griffiths transversality*; the point of view taken above gives an alternate illustration of the obstruction that does not depend on differentials, and so makes sense in the perfectoid framework.

REMARK 4.6.13. So far we have only spoken about shtukas with one leg; however, in the context of the geometric Langlands correspondence one needs to consider shtukas with an arbitrary finite number of legs. For multiple disjoint legs, it is relatively straightforward to adapt the preceding discussion to obtain moduli spaces of shtukas. However, for crossing legs a more complicated construction is needed; in the classical framework, this construction is the *Beilinson–Drinfeld affine Grassmannian* [182, Lecture III]. For the analogue of Theorem 4.6.7 for crossings legs, see [162, Proposition 20.2.3].

REMARK 4.6.14. The preceding constructions can be thought of as vaguely analogous to the construction of classical moduli spaces in algebraic geometry using geometric invariant theory. As our earlier invocations of this analogy may suggest, the general strategy is to consider a suitable moduli space of vector bundles on relative Fargues–Fontaine curves, apply Theorem 3.7.2 to identify an open subspace of semistable bundles, apply Theorem 3.7.5 to upgrade these bundles to shtukas, then take a suitable quotient to remove unwanted rigidity. This quotient operation behaves poorly on the full moduli space of vector bundles, but somewhat better on the semistable locus.

4.7. Invitation to local Langlands in mixed characteristic. To conclude, we include a rough transcription of the closing lecture given by Scholze at the 2017 AWS, describing the Fargues–Scholze program for adapting methods from the geometric Langlands program to the local Langlands correspondence in mixed characteristic; we eschew the meticulous nature of the preceding notes in favor of the informal tone of Scholze's lecture. The definitive reference for this topic is to be the forthcoming (as of this writing) article [**64**]; in the interim, see [**177**, Lecture 4] and [**65**].

We first recall the approach of Drinfeld to the Langlands correspondence for global function fields in positive characteristic. Let C be a smooth, projective, geometrically irreducible curve over a finite field $\mathbb{F}_q$. Let G be a reductive group over $\mathbb{F}_q$ (e.g., take $G = \mathrm{GL}_2$ to be in the setting of Drinfeld). The moduli space of G-shtukas with one leg (see Remark 4.4.4) comes with a structure morphism to C given by the leg; we will denote this space for short by Sht and use it as an analogue of a Shimura variety over $\mathbb{Z}$.

We now consider a relative étale cohomology group $R^i f_* \overline{\mathbb{Q}}_\ell$ (for some prime ℓ not dividing q) for the morphism $f : \mathrm{Sht} \to C$; this is a local system on C, which is to say a representation of the fundamental group $\pi_1(C)$ (a/k/a the unramified Galois group of the function field F of C). There is also an action of the adelic group $G(\mathbf{A}_F)$ given by something like Hecke operators (i.e., correspondences defined by varying level structure).

The group $R^i f_* \overline{\mathbb{Q}}_\ell$ decomposes as a direct sum of certain automorphic representations π of $G(\mathbf{A}_F)$, each tensored with a certain representation $o(\pi)$ of G_F. The mapping $\pi \mapsto o(\pi)$ defines the global Langlands correspondence in some cases, but not all; Drinfeld realized that one could find the missing representations by considering shtukas with two legs rather than one.

Changing notation, now let Sht denote the moduli space of shtukas with two legs; the two legs now define a projection $f : \mathrm{Sht} \to C \times_{\mathbb{F}_q} C$. We again consider $R^i f_* \overline{\mathbb{Q}}_\ell$ (or more precisely $R^i f_! \overline{\mathbb{Q}}_\ell$, but we ignore the difference here); this now carries an action of $\pi_1(C \times_{\mathbb{F}_q} C)$. Moreover, the partial Frobenius action on shtukas allows us to factor this action through $\pi_1(C \times_{\mathbb{F}_q} C / \varphi_2^{\mathbb{Z}})$ where φ_2 is the partial Frobenius acting only on the second term; by Drinfeld's lemma (Theorem 4.2.12), the latter group is isomorphic to $\pi_1(C) \times \pi_1(C)$. One can then show that for good choices of data, $R^i f_* \overline{\mathbb{Q}}_\ell$ splits as a sum over certain automorphic representations π (including all cuspidal ones), each tensored with $o(\pi) \boxtimes o(\pi)^\vee$.

It is a grand dream to do something like this for number fields. In particular, one would like to consider the analogues of shtukas with two[28] legs; however, this would require introducing a suitable analogue of $C \times_{\mathbb{F}_q} C$. Such an analogue would be $\mathrm{Spec}(\mathbb{Z}) \times_? \mathrm{Spec}(\mathbb{Z})$ where the base of the fiber product is something mysterious (often called the "field of one element").

The magic of diamonds is that they provide meaningful interpretations of $\mathrm{Spec}(\mathbb{Q}_p) \times_? \mathrm{Spec}(\mathbb{Q}_p)$ and $\mathrm{Spec}(\mathbb{Z}_p) \times_? \mathrm{Spec}(\mathbb{Z}_p)$. To wit, the absolute product

---

[28]For simplicity, we only talk about shtukas with two legs, which suffice for dealing with GL_n and many other groups. However, in some cases more legs are needed, as in [**130**].

$\mathrm{Spd}(\mathbb{Q}_p) \times \mathrm{Spd}(\mathbb{Q}_p)$ exists in the category of diamonds and is "two-dimensional": writing one copy of $\mathrm{Spd}(\mathbb{Q}_p)$ as $\mathrm{Spd}(\mathbb{Q}_p(\mu_{p^\infty}))/\mathbb{Z}_p^\times$ and then tilting, we obtain

$$\mathrm{Spd}(\mathbb{Q}_p) \times \mathrm{Spd}(\mathbb{Q}_p) \cong \tilde{D}_{\mathbb{Q}_p}^\diamond/\mathbb{Z}_p^\times$$

where $\tilde{D}$ denotes the perfectoid punctured unit disc (see [162, Proposition 8.4.1] for a detailed exposition of this point). The appropriate analogue of Drinfeld's lemma in this case is Theorem 4.3.14; it implies that

$$\pi_1(\tilde{D}_{\mathbb{C}_p}^\diamond/\mathbb{Z}_p^\times) \cong G_{\mathbb{Q}_p}.$$

That is, as observed in [176], one has a description of the arithmetic Galois group $G_{\mathbb{Q}_p}$ as the *geometric* fundamental group of a certain space. (See [126] for some variations on this theme.)

We now pass to the mixed-characteristic setting. Now write Sht for the moduli space of mixed-characteristic local shtukas with one leg (in the sense of Definition 4.5.6); the leg defines a morphism $\mathrm{Sht} \to \mathrm{Spd}(\mathbb{Q}_p)$. As discussed in Remark 4.6.9, this space Sht generalizes the notion of a Rapoport-Zink space, and functions as a mixed-characteristic local analogue of a Shimura variety.

To put this in context, we recall how Rapoport-Zink spaces arise. Let H be a one-dimensional formal group of height n over $\overline{\mathbb{F}}_p$. Let $\mathfrak{X}_H$ denote the deformation space of H; it has the form $\mathrm{Spf}(W(\overline{\mathbb{F}}_p)[\![u_1, \ldots, u_{n-1}]\!])$. Let $\mathcal{M}_H$ be the generic fiber of this space; it is an $(n-1)$-dimensional open unit disc. By analogy with taking a tower of modular curves by increasing the level structure, one has a tower

$$\cdots \to \mathcal{M}_{H,1} \to \mathcal{M}_{H,0} \cong \mathcal{M}_H$$

where $\mathcal{M}_{H,n}$ classifies isomorphisms $\mathcal{H}[p^n] \cong (\mathbb{Z}/p^n\mathbb{Z})^h$ where $\mathcal{H}$ is the universal deformation of H. These spaces are somewhat mysterious; however, their inverse limit $\mathcal{M}_{H,\infty} \cong \varprojlim_n \mathcal{M}_{H,n}$ makes sense as a perfectoid space and admits a rather precise description (given in [161]): for C a complete algebraically closed extension of $\mathbb{Q}_p$ and $\infty \in \mathrm{FF}_{C^\flat}$ the corresponding point, $\mathcal{M}_{H,\infty}(C)$ corresponds naturally to the set of injective morphisms $\mathcal{O}^n \xrightarrow{f} \mathcal{O}(\frac{1}{n})$ with cokernel supported at ∞. This statement can now be translated into the language of shtukas with one leg (we leave this to the reader).

This is the limit of what can say without recourse to diamonds. However, using diamonds we may also form the moduli space Sht of shtukas with two legs, and we get a morphism $f: \mathrm{Sht} \to \mathrm{Spd}(\mathbb{Q}_p) \times \mathrm{Spd}(\mathbb{Q}_p)$. Now taking $R^i f_* \overline{\mathbb{Q}}_\ell$ gives an object with actions of $G_{\mathbb{Q}_p}^{\times 2}$ (by Drinfeld's lemma for diamonds) and $G(\mathbb{Q}_p)$; one hopes to decompose this object in order to obtain a realization of the local Langlands correspondence for reductive groups over p-adic fields.

Appendix A. Project descriptions

Of the projects described below, the following were resolved during the Winter School: Problem A.1.1 (see [22]), Problem A.3.1 (in preparation), Problem A.5.3 (in preparation).

A.1. Extensions of vector bundles and slopes (proposed by David Hansen). The primary project revolves around the following problem.

PROBLEM A.1.1. Let F be a perfectoid field. (Optionally, assume also that F is algebraically closed.) Determine the set of values taken by the triple (HN(V), HN(V'), HN(V'')) as $0 \to V' \to V \to V'' \to 0$ varies over all short exact sequences of vector bundles on the FF-curve X_F over F.

One part of this problem is of a combinatorial nature.

PROBLEM A.1.2. Determine the combinatorial constraints on (HN(V), HN(V'), HN(V'')) imposed by the slope formalism (e.g., the statement of Lemma 3.4.17).

In the other direction, we will consider some intermediate steps, such as the following.

PROBLEM A.1.3. Let V', V'' be two semistable vector bundles on X_F with $\mu(V') < \mu(V'')$. Show that a bundle V occurs in a short exact sequence $0 \to V' \to V \to V'' \to 0$ if and only if HN(V) lies between HN($V' \oplus V''$) and the straight line segment with the same endpoints as HN($V' \oplus V''$).

Addressing the problems discussed above requires some basic familiarity with Banach–Colmez spaces, which are described in [**177**, Lecture 4].

A.2. G-bundles. We next formulate a more general form of Problem A.1.1 (Problem A.2.5) in terms of algebraic groups. This requires giving a general description of G-objects in an exact tensor category.

DEFINITION A.2.1. For G an algebraic group over $\mathbb{Q}_p$, let $\mathbf{Rep}_{\mathbb{Q}_p}(G)$ denote the category of (algebraic) representations of G on finite-dimensional F-vector spaces.

Let $\mathcal{C}$ be an $\mathbb{Q}_p$-linear tensor category. (That is, $\mathcal{C}$ is an exact category where the morphism spaces are not just abelian groups but $\mathbb{Q}_p$-vector spaces, composition is not just additive but $\mathbb{Q}_p$-linear, $\mathcal{C}$ carries a symmetric monoidal structure which is yet again $\mathbb{Q}_p$-linear, and $\mathcal{C}$ carries a rank function which adds in short exact sequences and multiplies in tensor products.) By a *G-object in $\mathcal{C}$*, we will mean a covariant, $\mathbb{Q}_p$-linear, rank-preserving tensor functor $\mathbf{Rep}_{\mathbb{Q}_p}(G) \to \mathcal{C}$.

EXAMPLE A.2.2. Let $\mathcal{C}$ be the category of finite-dimensional $\mathbb{Q}_p$-vector spaces.
- For $G = \mathrm{GL}_n$, a G-object in $\mathcal{C}$ is the same as a vector space of dimension n. (This includes the case $G = \mathbf{G}_m$ by taking $n = 1$.)
- For $G = \mathrm{SL}_n$, a G-object in $\mathcal{C}$ is the same as a vector space V of dimension n plus a choice of generator of the one-dimensional space $\wedge^n V$.
- For $G = \mathrm{O}_n$ (resp. $G = \mathrm{Sp}_n$), a G-object in $\mathcal{C}$ is the same as a vector space V of dimension n plus the choice of a nondegenerate orthogonal (resp. symplectic) form on n.

EXAMPLE A.2.3. Let $\mathcal{C}$ be the category of vector bundles on an abstract curve C over $\mathbb{Q}_p$.
- For $G = \mathrm{GL}_n$, a G-object in $\mathcal{C}$ is the same as a vector bundle of rank n.
- For $G = \mathrm{SL}_n$, a G-object in $\mathcal{C}$ is the same as a vector bundle V of rank n plus a trivialization of $\wedge^n V$.
- For $G = \mathrm{O}_n$ (resp. $G = \mathrm{Sp}_n$), a G-object in $\mathcal{C}$ is the same as a vector bundle V of rank n plus a nondegenerate orthogonal (resp. symplectic) pairing $V \times V \to \mathcal{O}_C$.

REMARK A.2.4. The idea behind the definition of a G-object is that vector bundles on a scheme X (or for that matter, on a manifold X) correspond to elements of the pointed set $H^1(X, \mathrm{GL}_n)$. By contrast, if one replaces GL_n with a smaller group, the resulting vector bundle is not entirely generic: its construction respects certain extra structure, and the exact nature of that extra structure is encoded in the structure of the category $\mathbf{Rep}_{\mathbb{Q}_p}(G)$. This is closely related to the Tannaka–Krein duality theorem, which asserts that the group G can be reconstructed from the data of the category $\mathbf{Rep}_{\mathbb{Q}_p}(G)$ plus the fiber functor taking representations to their underlying $\mathbb{Q}_p$-vector spaces, by taking the automorphism group of the functor (just as in the definition of profinite fundamental groups).

We can now formulate a group-theoretic variant of Problem A.1.1; the statement of Problem A.1.1 constitutes the case of the following problem in which $G = \mathrm{GL}_n$ and H is a certain parabolic subgroup. For this problem, some relevant background is the classification of G-isocrystals by Kottwitz [124] (see also [125, 151]).

PROBLEM A.2.5. Suppose that F is algebraically closed. Let $H \to G$ be an inclusion of connected reductive algebraic groups over $\mathbb{Q}_p$. For a given H-bundle V on X_F, determine which (isomorphism classes of) G-bundles admit a reduction of structure to H.

REMARK A.2.6. There is a (perhaps fanciful) resemblance between Problem A.2.5 and some classic questions about numerical invariants (e.g., eigenvalues, singular values) of triples A, B, C of square matrices satisfying $A + B = C$. See [108, Chapter 4] as a starting point.

A.3. The open mapping theorem for analytic rings.

PROBLEM A.3.1. Write out a detailed proof of Theorem 1.1.9 for analytic rings, by modifying the argument in [93] for Tate rings.

The key substep is the following extension of [93, Proposition 1.9].

DEFINITION A.3.2. Let X be a topological space. A subset Z of X is *nowhere dense* if for every nonempty open subset U of X, there is a nonempty open subset V of U which is disjoint from Z. A subset Z of X is *meager* if it can be written as a countable union of nowhere dense subsets.

PROBLEM A.3.3. Let A be an analytic Huber ring. Let M, N be topological R-modules. Let $u : M \to N$ be an R-linear morphism whose image is not meager. Then for every neighborhood V of 0 in M, the closure of $u(V)$ is a neighborhood of 0 in N.

REMARK A.3.4. The key steps of [93, Proposition 1.9] are that if $x \in A$ is a topologically nilpotent unit, then for every neighborhood W of 0 in M

$$\bigcup_{n=1}^{\infty} x^{-n} W = M,$$

and each set $x^{-n} \overline{u(W)}$ is closed in N. For topologically nilpotent elements $x_1, \ldots, x_k \in A$ which generate the unit ideal, the correct analogue of the first statement is that

$$\bigcup_{n=1}^{\infty} W_n = M, \qquad W_n = \{m \in M : x_1^n m, \ldots, x_k^n m \in W\}.$$

The correct analogue of the second statement is that for each n,
$$\{m \in N : x_1^n m, \ldots, x_k^n m \in \overline{u(W)}\}$$
is closed in N. (If $\{m_i\}$ is a sequence in this set with limit m, then $x_j^n m_i \in \overline{u(W)}$ converges to $x_j^n m$ and so the latter is in $\overline{u(W)}$.)

PROBLEM A.3.5. Adapt the previous arguments to show that all of the results of [93], which are proved for topological rings containing a null sequence consisting of units, remain true for first-countable topological rings (not necessarily Huber rings) in which every open ideal is trivial. (Note that Remark 1.1.1 no longer applies, but this turns out not to be relevant to this argument.) What happens if one drops the first-countable hypothesis?

A.4. The archimedean Fargues–Fontaine curve (proposed by Sean Howe).

DEFINITION A.4.1. Let $\tilde{\mathbf{P}}$ be the projective curve in $\mathbf{P}_{\mathbb{R}}^2$ defined by the equation $x^2 + y^2 + z^2$. This is the unique nontrivial Brauer–Severi curve over $\mathbb{R}$. This object plays a fundamental role in archimedean Hodge theory (e.g., in the study of *mixed twistor $\mathcal{D}$-modules*).

We explore the analogy between $\tilde{\mathbf{P}}$ and the Fargues–Fontaine curve over an algebraically closed perfectoid field.

PROBLEM A.4.2. For an algebraic variety X over $\mathbb{R}$, write $\mathrm{FF}_X \times \mathbb{C}$ for the topological space $(X(\mathbb{C}) \times \mathbf{P}^1(\mathbb{C}))/c$, where c acts on $X(\mathbb{C})$ by the usual conjugation (fixing $X(\mathbb{R})$) and on $\mathbf{P}^1(\mathbb{C})$ by the antipode map $z \mapsto -\bar{z}^{-1}$. Can you formulate a precise archimedean analogue of Lemma 4.3.10? (Note that $\tilde{\mathbf{P}}$ is an algebraic analogue of $\mathbf{P}^1(\mathbb{C})/c$.)

DEFINITION A.4.3. Let $\tilde{W}$ be the Weil group of $\mathbb{R}$ modulo its center $\mathbb{R}^\times$: concretely, this group is a semidirect product $S^1 \rtimes \mathbb{Z}/2\mathbb{Z}$ where $\mathbb{Z}/2\mathbb{Z}$ acts by inversion on S^1. We view $\tilde{\mathbf{P}}$ as the projectized cone over the (scheme of) trace-zero, norm-zero elements in the quaternions $\mathbb{H}$, and identify $\tilde{W}$ with $\mathbb{C}^\times \sqcup j\mathbb{C}^\times \subset \mathbb{H}$, so that $\tilde{\mathbf{P}}$ has a natural action of $\tilde{W}$ with a unique fixed point p with residue field $\mathbb{C}$ on which $\tilde{W}$ acts through conjugation by $\mathbb{Z}/2\mathbb{Z}$.

For X an algebraic variety over $\mathbb{R}$, let $H^i(X(\mathbb{C}), \mathbb{R})$ denote the real singular cohomology of the topological space $X(\mathbb{C})$, equipped with its Hodge decomposition as $\bigoplus_{p+q=i} h^{p,q}$. We equip this $\mathbb{R}$-vector space with a representation of $\tilde{W}$ where S^1 acts as z^{-p+q} on $h^{p,q}$ and c acts by the automorphism induced by conjugation on $X(\mathbb{C})$; using this action, we equip the trivial vector bundle $\mathcal{O} \otimes H^i(X(\mathbb{C}), \mathbb{R})$ on $\tilde{\mathbf{P}}^1$ with a $\tilde{W}$-equivariant structure. We equip the algebraic de Rham cohomology $H^i_{\mathrm{dR}}(X)$ with the trivial $\tilde{W}$-action.

PROBLEM A.4.4. Retain notation as in Definition A.4.3.

(a) Prove the following de Rham comparison theorem: there is a natural identification
$$(\mathcal{O} \otimes H^i(X(\mathbb{C}), \mathbb{R}))_{\mathrm{Spec}(\mathrm{Frac}(\widehat{\mathcal{O}_p}))} \cong H^i_{\mathrm{dR}}(X) \otimes \mathrm{Frac}(\widehat{\mathcal{O}_p}),$$
as $\tilde{W}$-equivariant bundles over $\mathrm{Spec}(\mathrm{Frac}(\widehat{\mathcal{O}_p}))$, and in particular
$$(H^i(X(\mathbb{C}), \mathbb{R}) \otimes \mathrm{Frac}(\widehat{\mathcal{O}_p}))^{\tilde{W}} = H^i_{\mathrm{dR}}(X)$$

with the Hodge filtration corresponding to the filtration by order of poles (up to a change in the numbering). Compare with the p-adic de Rham comparison theorem.

(b) What are the corresponding modifications?

(c) Comparing with [170], you will find that we are not using the standard representation of the Weil group attached to a Hodge structure. For even weight, we've simply taken a Tate twist to land in weight 0, but for odd weight we've taken something genuinely different: our construction factors through the split version of the Weil group, while the the original representation does not. Can we fix this and/or should we want to? On a related note, is there a way to modify this construction so that we obtain the correct numbering on the Hodge filtration and the "natural" slopes for the modifications? In general, what can we do to make a stronger analogy with the p-adic case, and if we can't, how should we understand the difference?

A.5. Finitely presented morphisms.

DEFINITION A.5.1. Define a Huber ring A to be *strongly sheafy* if $A\langle T_1, \ldots, T_n \rangle$ is sheafy for every nonnegative integer n. For example, if A is strongly noetherian, then A is strongly sheafy by Theorem 1.2.11. For another example, if A is perfectoid, then we may see that A is strongly sheafy by applying Corollary 2.5.5 to the map $A\langle T_1, \ldots, T_n \rangle \to A\langle T_1^{p^{-\infty}}, \ldots, T_n^{p^{-\infty}} \rangle$.

DEFINITION A.5.2. Suppose A is strongly sheafy. A homomorphism $A \to B$ is *affinoid* if:

- it factors through a surjection $A\langle T_1, \ldots, T_n \rangle \to B$; and
- for some such factorization, $B \in \mathbf{PCoh}_{A\langle T_1, \ldots, T_n \rangle}$. Equivalently by Theorem 1.4.20, B is again sheafy.

For example, any rational localization is an affinoid morphism. In addition, any finite flat morphism, and in particular any finite étale morphism, is affinoid *provided* that the source is strongly sheafy and the target is sheafy (e.g., a finite étale morphism between perfectoid rings).

PROBLEM A.5.3. We previously gave an *ad hoc* definition of an étale morphism of adic spaces (Hypothesis 1.10.3). Use the concept of an affinoid morphism to give a definition in the strongly sheafy case closer to the one given by Huber in the strongly noetherian case [97, Definition 1.6.5].

PROBLEM A.5.4. Similarly, use the concept of an affinoid morphism to define *unramified* and *smooth* morphisms in the strongly sheafy case.

A.6. Additional suggestions.

PROBLEM A.6.1. Prove that for any (analytic) Huber pair (A, A^+), $\mathrm{Spa}(A\langle T \rangle, A^+\langle T \rangle) \to \mathrm{Spa}(A, A^+)$ is an open map.

PROBLEM A.6.2. Verify that for any perfectoid Huber pair (R, R^+) of characteristic p, the ring $\mathbf{A}_{\mathrm{inf}} := \mathbf{A}_{\mathrm{inf}}(R, R^+)$ is sheafy. See Remark 3.1.10 for the case where R is a nonarchimedean field. One possible approach is to show that $\mathbf{A}_{\mathrm{inf}}$ admits a split (in the category of topological $\mathbf{A}_{\mathrm{inf}}$-modules) embedding into a perfectoid (and hence sheafy) ring.

PROBLEM A.6.3. Find a "reasonable" (i.e., as small as possible) category of algebraic stacks in which Remark 4.2.14 can be interpreted.

References

[1] Ahmed Abbes and Michel Gros, *Covanishing topos and generalizations*, The p-adic Simpson correspondence, Ann. of Math. Stud., vol. 193, Princeton Univ. Press, Princeton, NJ, 2016, pp. 485–576. MR3444783

[2] Piotr Achinger, $K(\pi,1)$-*neighborhoods and comparison theorems*, Compos. Math. 151 (2015), no. 10, 1945–1964, DOI 10.1112/S0010437X15007319. MR3414390

[3] Piotr Achinger, *Wild ramification and* $K(\pi,1)$ *spaces*, Invent. Math. 210 (2017), no. 2, 453–499, DOI 10.1007/s00222-017-0733-5. MR3714509

[4] Yves André, *Slope filtrations*, Confluentes Math. 1 (2009), no. 1, 1–85, DOI 10.1142/S179374420900002X. MR2571693

[5] Yves André, *Le lemme d'Abhyankar perfectoide* (French, with French summary), Publ. Math. Inst. Hautes Études Sci. 127 (2018), 1–70, DOI 10.1007/s10240-017-0096-x. MR3814650

[6] Yves André, *La conjecture du facteur direct* (French, with French summary), Publ. Math. Inst. Hautes Études Sci. 127 (2018), 71–93, DOI 10.1007/s10240-017-0097-9. MR3814651

[7] Fabrizio Andreatta, *Generalized ring of norms and generalized* (ϕ, Γ)-*modules* (English, with English and French summaries), Ann. Sci. École Norm. Sup. (4) 39 (2006), no. 4, 599–647, DOI 10.1016/j.ansens.2006.07.003. MR2290139

[8] *Théorie des topos et cohomologie étale des schémas. Tome 1: Théorie des topos* (French), Lecture Notes in Mathematics, Vol. 269, Springer-Verlag, Berlin-New York, 1972. Séminaire de Géométrie Algébrique du Bois-Marie 1963–1964 (SGA 4); Dirigé par M. Artin, A. Grothendieck, et J. L. Verdier. Avec la collaboration de N. Bourbaki, P. Deligne et B. Saint-Donat. MR0354652

[9] M. F. Atiyah, *Vector bundles over an elliptic curve*, Proc. London Math. Soc. (3) 7 (1957), 414–452, DOI 10.1112/plms/s3-7.1.414. MR0131423

[10] James Ax, *Zeros of polynomials over local fields—The Galois action*, J. Algebra 15 (1970), 417–428, DOI 10.1016/0021-8693(70)90069-4. MR0263786

[11] Arnaud Beauville and Yves Laszlo, *Un lemme de descente* (French, with English and French summaries), C. R. Acad. Sci. Paris Sér. I Math. 320 (1995), no. 3, 335–340. MR1320381

[12] Laurent Berger, *Équations différentielles p-adiques et* (ϕ, N)-*modules filtrés* (French, with English and French summaries), Astérisque 319 (2008), 13–38. Représentations p-adiques de groupes p-adiques. I. Représentations galoisiennes et (ϕ, Γ)-modules. MR2493215

[13] Vladimir G. Berkovich, *Spectral theory and analytic geometry over non-Archimedean fields*, Mathematical Surveys and Monographs, vol. 33, American Mathematical Society, Providence, RI, 1990. MR1070709

[14] *Théorie des intersections et théorème de Riemann-Roch* (French), Lecture Notes in Mathematics, Vol. 225, Springer-Verlag, Berlin-New York, 1971. Séminaire de Géométrie Algébrique du Bois-Marie 1966–1967 (SGA 6); Dirigé par P. Berthelot, A. Grothendieck et L. Illusie. Avec la collaboration de D. Ferrand, J. P. Jouanolou, O. Jussila, S. Kleiman, M. Raynaud et J. P. Serre. MR0354655

[15] Bhargav Bhatt, *Algebraization and Tannaka duality*, Camb. J. Math. 4 (2016), no. 4, 403–461, DOI 10.4310/CJM.2016.v4.n4.a1. MR3572635

[16] Bhargav Bhatt, *On the direct summand conjecture and its derived variant*, Invent. Math. 212 (2018), no. 2, 297–317, DOI 10.1007/s00222-017-0768-7. MR3787829

[17] B. Bhatt, *The Hodge-Tate decomposition via perfectoid spaces*, Perfectoid Spaces: Lectures from the 2017 Arizona Winter School, Mathematical Surveys and Monographs, vol. 242, 193–244, American Mathematical Society, 2019.

[18] Bhargav Bhatt, Matthew Morrow, and Peter Scholze, *Integral p-adic Hodge theory*, Publ. Math. Inst. Hautes Études Sci. 128 (2018), 219–397, DOI 10.1007/s10240-019-00102-z. MR3905467

[19] Bhargav Bhatt and Peter Scholze, *The pro-étale topology for schemes* (English, with English and French summaries), Astérisque 369 (2015), 99–201. MR3379634

[20] Bhargav Bhatt and Peter Scholze, *Projectivity of the Witt vector affine Grassmannian*, Invent. Math. 209 (2017), no. 2, 329–423, DOI 10.1007/s00222-016-0710-4. MR3674218

[21] Bhargav Bhatt, Karl Schwede, and Shunsuke Takagi, *The weak ordinarity conjecture and F-singularities*, Higher dimensional algebraic geometry—in honour of Professor Yujiro Kawamata's sixtieth birthday, Adv. Stud. Pure Math., vol. 74, Math. Soc. Japan, Tokyo, 2017, pp. 11–39. MR3791207

[22] C. Birkbeck, T. Feng, D. Hansen, S. Hong, Q. Li, A. Wang, and L. Ye, Extensions of vector bundles on the Fargues–Fontaine curve, arXiv:1705.00710v3 (2018); to appear in *J. Inst. Math. Jussieu*.

[23] Siegfried Bosch, *Lectures on formal and rigid geometry*, Lecture Notes in Mathematics, vol. 2105, Springer, Cham, 2014. MR3309387

[24] S. Bosch, U. Güntzer, and R. Remmert, *Non-Archimedean analysis*, Grundlehren der Mathematischen Wissenschaften [Fundamental Principles of Mathematical Sciences], vol. 261, Springer-Verlag, Berlin, 1984. A systematic approach to rigid analytic geometry. MR746961

[25] Nicolas Bourbaki, *Espaces vectoriels topologiques. Chapitres 1 à 5* (French), New edition, Masson, Paris, 1981. Éléments de mathématique. [Elements of mathematics]. MR633754

[26] Tom Bridgeland, *Stability conditions on triangulated categories*, Ann. of Math. (2) **166** (2007), no. 2, 317–345, DOI 10.4007/annals.2007.166.317. MR2373143

[27] Kevin Buzzard and Alain Verberkmoes, *Stably uniform affinoids are sheafy*, J. Reine Angew. Math. **740** (2018), 25–39, DOI 10.1515/crelle-2015-0089. MR3824781

[28] A. Caraini, *Perfectoid Shimura varieties*, Perfectoid Spaces: Lectures from the 2017 Arizona Winter School, Mathematical Surveys and Monographs, vol. 242, 245–297, American Mathematical Society, 2019.

[29] A. Carter, K.S. Kedlaya, and G. Zábrádi, Drinfeld's lemma for perfectoid spaces and overconvergence of multivariate (φ, Γ)-modules, arXiv:1808.03964v1 (2018).

[30] Robert F. Coleman, *Classical and overconvergent modular forms*, Invent. Math. **124** (1996), no. 1-3, 215–241, DOI 10.1007/s002220050051. MR1369416

[31] Pierre Colmez, *Espaces de Banach de dimension finie* (French, with English and French summaries), J. Inst. Math. Jussieu **1** (2002), no. 3, 331–439, DOI 10.1017/S1474748002000099. MR1956055

[32] B. Conrad, Lecture notes on adic spaces available at http://math.stanford.edu/~conrad/Perfseminar/ (retrieved February 2017).

[33] Shaunak Das, *Vector Bundles on Perfectoid Spaces*, ProQuest LLC, Ann Arbor, MI, 2016. Thesis (Ph.D.)–University of California, San Diego. MR3553572

[34] Christopher Davis and Kiran S. Kedlaya, *On the Witt vector Frobenius*, Proc. Amer. Math. Soc. **142** (2014), no. 7, 2211–2226, DOI 10.1090/S0002-9939-2014-11953-8. MR3195748

[35] A. J. de Jong, *Étale fundamental groups of non-Archimedean analytic spaces*, Compositio Math. **97** (1995), no. 1-2, 89–118. Special issue in honour of Frans Oort. MR1355119

[36] A. J. de Jong, *Smoothness, semi-stability and alterations*, Inst. Hautes Études Sci. Publ. Math. **83** (1996), 51–93. MR1423020

[37] A.J. de Jong, Étale fundamental groups, Columbia lecture notes, fall 2015; notes by Pak-Hin Lee available at http://math.columbia.edu/~phlee/CourseNotes/EtaleFundamental.pdf (retrieved January 2017).

[38] Johan de Jong and Marius van der Put, *Étale cohomology of rigid analytic spaces*, Doc. Math. **1** (1996), No. 01, 1–56. MR1386046

[39] Pierre Deligne, *La conjecture de Weil. II* (French), Inst. Hautes Études Sci. Publ. Math. **52** (1980), 137–252. MR601520

[40] *Groupes de monodromie en géométrie algébrique. II* (French), Lecture Notes in Mathematics, Vol. 340, Springer-Verlag, Berlin-New York, 1973. Séminaire de Géométrie Algébrique du Bois-Marie 1967–1969 (SGA 7 II); Dirigé par P. Deligne et N. Katz. MR0354657

[41] Michel Demazure, *Lectures on p-divisible groups*, Lecture Notes in Mathematics, Vol. 302, Springer-Verlag, Berlin-New York, 1972. MR0344261

[42] *Schémas en groupes. I: Propriétés générales des schémas en groupes* (French), Séminaire de Géométrie Algébrique du Bois Marie 1962/64 (SGA 3). Dirigé par M. Demazure et A. Grothendieck. Lecture Notes in Mathematics, Vol. 151, Springer-Verlag, Berlin-New York, 1970. MR0274458

[43] *Schémas en groupes. II: Groupes de type multiplicatif, et structure des schémas en groupes généraux* (French), Séminaire de Géométrie Algébrique du Bois Marie 1962/64 (SGA 3). Dirigé par M. Demazure et A. Grothendieck. Lecture Notes in Mathematics, Vol. 152, Springer-Verlag, Berlin-New York, 1970. MR0274459

[44] *Schémas en groupes. III: Structure des schémas en groupes réductifs* (French), Séminaire de Géométrie Algébrique du Bois Marie 1962/64 (SGA 3). Dirigé par M. Demazure et A. Grothendieck. Lecture Notes in Mathematics, Vol. 153, Springer-Verlag, Berlin-New York, 1970. MR0274460

[45] Jean Dieudonné, *Groupes de Lie et hyperalgèbres de Lie sur un corps de caractéristique $p > 0$. VII* (French), Math. Ann. **134** (1957), 114–133, DOI 10.1007/BF01342790. MR0098146

[46] S. K. Donaldson, *A new proof of a theorem of Narasimhan and Seshadri*, J. Differential Geom. **18** (1983), no. 2, 269–277. MR710055

[47] G. Dorfsman-Hopkins, Projective geometry for perfectoid spaces, PhD thesis, University of Washington, 2019.

[48] V. G. Drinfel'd, *Elliptic modules* (Russian), Mat. Sb. (N.S.) **94(136)** (1974), 594–627, 656. MR0384707

[49] V. G. Drinfel'd, *Elliptic modules. II* (Russian), Mat. Sb. (N.S.) **102(144)** (1977), no. 2, 182–194, 325. MR0439758

[50] V. G. Drinfel'd, *Langlands' conjecture for* GL(2) *over functional fields*, Proceedings of the International Congress of Mathematicians (Helsinki, 1978), Acad. Sci. Fennica, Helsinki, 1980, pp. 565–574. MR562656

[51] V. G. Drinfel'd, *Cohomology of compactified moduli varieties of F-sheaves of rank 2* (Russian), Zap. Nauchn. Sem. Leningrad. Otdel. Mat. Inst. Steklov. (LOMI) **162** (1987), no. Avtomorfn. Funkts. i Teor. Chisel. III, 107–158, 189, DOI 10.1007/BF01099348; English transl., J. Soviet Math. **46** (1989), no. 2, 1789–1821. MR918745

[52] D. W. Dubois and A. Bukowski, *Real commutative algebra. II. Plane curves*, Rev. Mat. Hisp.-Amer. (4) **39** (1979), no. 4-5, 149–161. MR629099

[53] David Eisenbud, *Commutative algebra*, Graduate Texts in Mathematics, vol. 150, Springer-Verlag, New York, 1995. With a view toward algebraic geometry. MR1322960

[54] Hélène Esnault and Amit Hogadi, *On the algebraic fundamental group of smooth varieties in characteristic $p > 0$*, Trans. Amer. Math. Soc. **364** (2012), no. 5, 2429–2442, DOI 10.1090/S0002-9947-2012-05470-5. MR2888213

[55] Gerd Faltings, *Mumford-Stabilität in der algebraischen Geometrie* (German), Proceedings of the International Congress of Mathematicians, Vol. 1, 2 (Zürich, 1994), Birkhäuser, Basel, 1995, pp. 648–655. MR1403965

[56] Gerd Faltings, *p-adic Hodge theory*, J. Amer. Math. Soc. **1** (1988), no. 1, 255–299, DOI 10.2307/1990970. MR924705

[57] Gerd Faltings, *Crystalline cohomology and p-adic Galois-representations*, Algebraic analysis, geometry, and number theory (Baltimore, MD, 1988), Johns Hopkins Univ. Press, Baltimore, MD, 1989, pp. 25–80. MR1463696

[58] Gerd Faltings, *Almost étale extensions*, Astérisque **279** (2002), 185–270. Cohomologies p-adiques et applications arithmétiques, II. MR1922831

[59] Gerd Faltings, *Coverings of p-adic period domains*, J. Reine Angew. Math. **643** (2010), 111–139, DOI 10.1515/CRELLE.2010.046. MR2658191

[60] Laurent Fargues, *Quelques résultats et conjectures concernant la courbe* (French, with English and French summaries), Astérisque **369** (2015), 325–374. MR3379639

[61] Laurent Fargues and Jean-Marc Fontaine, *Vector bundles and p-adic Galois representations*, Fifth International Congress of Chinese Mathematicians. Part 1, 2, AMS/IP Stud. Adv. Math., 51, pt. 1, vol. 2, Amer. Math. Soc., Providence, RI, 2012, pp. 77–113. MR2908062

[62] Laurent Fargues and Jean-Marc Fontaine, *Vector bundles on curves and p-adic Hodge theory*, Automorphic forms and Galois representations. Vol. 2, London Math. Soc. Lecture Note Ser., vol. 415, Cambridge Univ. Press, Cambridge, 2014, pp. 17–104. MR3444231

[63] L. Fargues and J.-M. Fontaine, Courbes et fibrés vectoriels en théorie de Hodge p-adique, *Astérisque* **406** (2018).

[64] L. Fargues and P. Scholze, Geometrization of the local Langlands correspondence, in preparation.

[65] T. Feng, Notes from the Arbeitsgemeinschaft 2016: Geometric Langlands, http://web.stanford.edu/~tonyfeng/Arbeitsgemeinschaft2016.html.

[66] Jean-Marc Fontaine, *Sur certains types de représentations p-adiques du groupe de Galois d'un corps local; construction d'un anneau de Barsotti-Tate* (French), Ann. of Math. (2) **115** (1982), no. 3, 529–577, DOI 10.2307/2007012. MR657238

[67] Jean-Marc Fontaine, *Cohomologie de de Rham, cohomologie cristalline et représentations p-adiques* (French), Algebraic geometry (Tokyo/Kyoto, 1982), Lecture Notes in Math., vol. 1016, Springer, Berlin, 1983, pp. 86–108, DOI 10.1007/BFb0099959. MR726422

[68] Jean-Marc Fontaine, *Perfectoïdes, presque pureté et monodromie-poids (d'après Peter Scholze)* (French, with French summary), Astérisque **352** (2013), Exp. No. 1057, x, 509–534. Séminaire Bourbaki. Vol. 2011/2012. Exposés 1043–1058. MR3087355

[69] Jean-Marc Fontaine and Jean-Pierre Wintenberger, *Le "corps des normes" de certaines extensions algébriques de corps locaux* (French, with English summary), C. R. Acad. Sci. Paris Sér. A-B **288** (1979), no. 6, A367–A370. MR526137

[70] Jean-Marc Fontaine and Jean-Pierre Wintenberger, *Extensions algébriques et corps des normes des extensions APF des corps locaux* (French, with English summary), C. R. Acad. Sci. Paris Sér. A-B **288** (1979), no. 8, A441–A444. MR527692

[71] Jean Fresnel and Bernard de Mathan, *Algèbres L^1 p-adiques* (French, with English summary), Bull. Soc. Math. France **106** (1978), no. 3, 225–260. MR515402

[72] Jean Fresnel and Marius van der Put, *Rigid analytic geometry and its applications*, Progress in Mathematics, vol. 218, Birkhäuser Boston, Inc., Boston, MA, 2004. MR2014891

[73] Kazuhiro Fujiwara, Ofer Gabber, and Fumiharu Kato, *On Hausdorff completions of commutative rings in rigid geometry*, J. Algebra **332** (2011), 293–321, DOI 10.1016/j.jalgebra.2011.02.001. MR2774689

[74] Kazuhiro Fujiwara and Fumiharu Kato, *Foundations of rigid geometry. I*, EMS Monographs in Mathematics, European Mathematical Society (EMS), Zürich, 2018. MR3752648

[75] O. Gabber and L. Ramero, Foundations for almost ring theory – Release 7.5, arXiv:math/0409584v13 (2018).

[76] Alain Genestier and Vincent Lafforgue, *Théorie de Fontaine en égales caractéristiques* (French, with English and French summaries), Ann. Sci. Éc. Norm. Supér. (4) **44** (2011), no. 2, 263–360, DOI 10.24033/asens.2144. MR2830388

[77] David Gieseker, *Stable vector bundles and the Frobenius morphism*, Ann. Sci. École Norm. Sup. (4) **6** (1973), 95–101. MR0325616

[78] Wushi Goldring, *Unifying themes suggested by Belyi's theorem*, Number theory, analysis and geometry, Springer, New York, 2012, pp. 181–214, DOI 10.1007/978-1-4614-1260-1_10. MR2867918

[79] D. Goss, *What is... a shtuka?* Notices Amer. Math. Soc. **50** (2003), 36–37.

[80] M. J. Hopkins and B. H. Gross, *Equivariant vector bundles on the Lubin-Tate moduli space*, Topology and representation theory (Evanston, IL, 1992), Contemp. Math., vol. 158, Amer. Math. Soc., Providence, RI, 1994, pp. 23–88, DOI 10.1090/conm/158/01453. MR1263712

[81] A. Grothendieck, *Sur la classification des fibrés holomorphes sur la sphère de Riemann* (French), Amer. J. Math. **79** (1957), 121–138, DOI 10.2307/2372388. MR0087176

[82] A. Grothendieck, *Éléments de géométrie algébrique. IV. Étude locale des schémas et des morphismes de schémas. II* (French), Inst. Hautes Études Sci. Publ. Math. **24** (1965), 231. MR0199181

[83] A. Grothendieck, *Revêtements Étales et Groupe Fondamental (SGA 1)*, Lecture Notes in Math. 224, Springer-Verlag, Berlin, 1971. MR2017446

[84] Bernard Guennebaud, *Semi-normes multiplicatives sur une algèbre de Banach ultramétrique* (French), Séminaire de Théorie des Nombres, 1972–1973 (Univ. Bordeaux I, Talence), Exp. No. 13, Lab. Théorie des Nombres, Centre Nat. Recherche Sci., Talence, 1973, pp. 3. MR0405110

[85] William J. Haboush, *Infinite dimensional algebraic geometry: algebraic structures on p-adic groups and their homogeneous spaces*, Tohoku Math. J. (2) **57** (2005), no. 1, 65–117. MR2113991

[86] D. Hansen and K.S. Kedlaya, Sheafiness criteria for Huber rings, in preparation.

[87] Günter Harder, *Halbeinfache Gruppenschemata über vollständigen Kurven* (German), Invent. Math. **6** (1968), 107–149, DOI 10.1007/BF01425451. MR0263826

[88] G. Harder and M. S. Narasimhan, *On the cohomology groups of moduli spaces of vector bundles on curves*, Math. Ann. **212** (1974/75), 215–248, DOI 10.1007/BF01357141. MR0364254

[89] Urs Hartl, *Period spaces for Hodge structures in equal characteristic*, Ann. of Math. (2) **173** (2011), no. 3, 1241–1358, DOI 10.4007/annals.2011.173.3.2. MR2800715

[90] Urs Hartl, *On a conjecture of Rapoport and Zink*, Invent. Math. **193** (2013), no. 3, 627–696, DOI 10.1007/s00222-012-0437-9. MR3091977

[91] Urs Hartl and Richard Pink, *Vector bundles with a Frobenius structure on the punctured unit disc*, Compos. Math. **140** (2004), no. 3, 689–716, DOI 10.1112/S0010437X03000216. MR2041777

[92] Michiel Hazewinkel and Clyde F. Martin, *A short elementary proof of Grothendieck's theorem on algebraic vector bundles over the projective line*, J. Pure Appl. Algebra **25** (1982), no. 2, 207–211, DOI 10.1016/0022-4049(82)90037-8. MR662762

[93] T. Henkel, An Open Mapping Theorem for rings with a zero sequence of units, arXiv:1407.5647v2 (2014).

[94] Melvin Hochster, *Homological conjectures, old and new*, Illinois J. Math. **51** (2007), no. 1, 151–169. MR2346192

[95] R. Huber, *Continuous valuations*, Math. Z. **212** (1993), no. 3, 455–477, DOI 10.1007/BF02571668. MR1207303

[96] R. Huber, *A generalization of formal schemes and rigid analytic varieties*, Math. Z. **217** (1994), no. 4, 513–551, DOI 10.1007/BF02571959. MR1306024

[97] Roland Huber, *Étale cohomology of rigid analytic varieties and adic spaces*, Aspects of Mathematics, E30, Friedr. Vieweg & Sohn, Braunschweig, 1996. MR1734903

[98] Annette Huber and Clemens Jörder, *Differential forms in the h-topology*, Algebr. Geom. **1** (2014), no. 4, 449–478, DOI 10.14231/AG-2014-020. MR3272910

[99] Daniel Huybrechts and Manfred Lehn, *The geometry of moduli spaces of sheaves*, 2nd ed., Cambridge Mathematical Library, Cambridge University Press, Cambridge, 2010. MR2665168

[100] Christian Kappen, *Uniformly rigid spaces*, Algebra Number Theory **6** (2012), no. 2, 341–388, DOI 10.2140/ant.2012.6.341. MR2950157

[101] Kiran S. Kedlaya, *Power series and p-adic algebraic closures*, J. Number Theory **89** (2001), no. 2, 324–339, DOI 10.1006/jnth.2000.2630. MR1845241

[102] Kiran S. Kedlaya, *Étale covers of affine spaces in positive characteristic* (English, with English and French summaries), C. R. Math. Acad. Sci. Paris **335** (2002), no. 11, 921–926, DOI 10.1016/S1631-073X(02)02587-6. MR1952550

[103] Kiran S. Kedlaya, *A p-adic local monodromy theorem*, Ann. of Math. (2) **160** (2004), no. 1, 93–184, DOI 10.4007/annals.2004.160.93. MR2119719

[104] Kiran S. Kedlaya, *More étale covers of affine spaces in positive characteristic*, J. Algebraic Geom. **14** (2005), no. 1, 187–192, DOI 10.1090/S1056-3911-04-00381-9. MR2092132

[105] K.S. Kedlaya, Slope filtrations revisited, *Doc. Math.* **10** (2005), 447–525; errata, *ibid.* **12** (2007), 361–362; additional errata at http://kskedlaya.org/papers/.

[106] Kiran S. Kedlaya, *Slope filtrations for relative Frobenius* (English, with English and French summaries), Astérisque **319** (2008), 259–301. Représentations p-adiques de groupes p-adiques. I. Représentations galoisiennes et (ϕ, Γ)-modules. MR2493220

[107] Kiran Sridhara Kedlaya, *Relative p-adic Hodge theory and Rapoport-Zink period domains*, Proceedings of the International Congress of Mathematicians. Volume II, Hindustan Book Agency, New Delhi, 2010, pp. 258–279. MR2827795

[108] Kiran S. Kedlaya, *p-adic differential equations*, Cambridge Studies in Advanced Mathematics, vol. 125, Cambridge University Press, Cambridge, 2010. MR2663480

[109] Kiran S. Kedlaya, *Nonarchimedean geometry of Witt vectors*, Nagoya Math. J. **209** (2013), 111–165, DOI 10.1017/S0027763000010692. MR3032139

[110] Kiran S. Kedlaya, *New methods for (φ, Γ)-modules*, Res. Math. Sci. **2** (2015), Art. 20, 31, DOI 10.1186/s40687-015-0031-z. MR3412585

[111] Kiran S. Kedlaya, *Noetherian properties of Fargues-Fontaine curves*, Int. Math. Res. Not. IMRN **8** (2016), 2544–2567, DOI 10.1093/imrn/rnv227. MR3519123

[112] Kiran S. Kedlaya, *Reified valuations and adic spectra*, Res. Number Theory **1** (2015), Art. 20, 42, DOI 10.1007/s40993-015-0021-7. MR3501004

[113] Kiran S. Kedlaya, *On commutative nonarchimedean Banach fields*, Doc. Math. **23** (2018), 171–188. MR3846060

[114] K.S. Kedlaya, Some ring-theoretic properties of $\mathbf{A}_{\inf}$, arXiv:1602.09016v5 (2019).

[115] K.S. Kedlaya, The Nonarchimedean Scottish Book, http://scripts.mit.edu/~kedlaya/wiki/index.php?title=The_Nonarchimedean_Scottish_Book (retrieved February 2017).

[116] Kiran S. Kedlaya, *Noetherian properties of Fargues-Fontaine curves*, Int. Math. Res. Not. IMRN **8** (2016), 2544–2567, DOI 10.1093/imrn/rnv227. MR3519123

[117] K.S. Kedlaya and R. Liu, Relative p-adic Hodge theory: Foundations, *Astérisque* **371** (2015), 239 pages, MR3379653; errata, [**118**, Appendix A].

[118] K.S. Kedlaya and R. Liu, Relative p-adic Hodge theory, II: Imperfect period rings, arXiv:1602.06899v3 (2019).

[119] K.S. Kedlaya and Ruochuan Liu, Finiteness of cohomology of local systems on rigid analytic spaces, arXiv:1611.06930v2 (2019).

[120] Kiran S. Kedlaya and Jonathan Pottharst, *On categories of (φ, Γ)-modules*, Algebraic geometry: Salt Lake City 2015, Proc. Sympos. Pure Math., vol. 97, Amer. Math. Soc., Providence, RI, 2018, pp. 281–304. MR3821175

[121] Reinhardt Kiehl, *Theorem A und Theorem B in der nichtarchimedischen Funktionentheorie* (German), Invent. Math. **2** (1967), 256–273, DOI 10.1007/BF01425404. MR0210949

[122] Mark Kisin, *Crystalline representations and F-crystals*, Algebraic geometry and number theory, Progr. Math., vol. 253, Birkhäuser Boston, Boston, MA, 2006, pp. 459–496, DOI 10.1007/978-0-8176-4532-8_7. MR2263197

[123] Martin Kreidl, *On p-adic lattices and Grassmannians*, Math. Z. **276** (2014), no. 3-4, 859–888, DOI 10.1007/s00209-013-1225-y. MR3175163

[124] Robert E. Kottwitz, *Isocrystals with additional structure*, Compositio Math. **56** (1985), no. 2, 201–220. MR809866

[125] Robert E. Kottwitz, *Isocrystals with additional structure. II*, Compositio Math. **109** (1997), no. 3, 255–339, DOI 10.1023/A:1000102604688. MR1485921

[126] Robert A. Kucharczyk and Peter Scholze, *Topological realisations of absolute Galois groups*, Cohomology of arithmetic groups, Springer Proc. Math. Stat., vol. 245, Springer, Cham, 2018, pp. 201–288. MR3848820

[127] Laurent Lafforgue, *Chtoucas de Drinfeld et conjecture de Ramanujan-Petersson* (French), Astérisque **243** (1997), ii+329. MR1600006

[128] Laurent Lafforgue, *Chtoucas de Drinfeld et correspondance de Langlands* (French, with English and French summaries), Invent. Math. **147** (2002), no. 1, 1–241, DOI 10.1007/s002220100174. MR1875184

[129] V. Lafforgue, Introduction to chtoucas for reductive groups and to the global Langlands parameterization, arXiv:1404.6416v2 (2015).

[130] Vincent Lafforgue, *Chtoucas pour les groupes réductifs et paramétrisation de Langlands globale* (French), J. Amer. Math. Soc. **31** (2018), no. 3, 719–891, DOI 10.1090/jams/897. MR3787407

[131] E. Lau, On generalized $\mathcal{D}$-shtukas, PhD thesis, Universität Bonn, 2004; available at https://www.math.uni-bielefeld.de/~lau/ (retrieved February 2017).

[132] Michel Lazard, *Les zéros des fonctions analytiques d'une variable sur un corps valué complet* (French), Inst. Hautes Études Sci. Publ. Math. **14** (1962), 47–75. MR0152519

[133] Arthur-César Le Bras, *Espaces de Banach-Colmez et faisceaux cohérents sur la courbe de Fargues-Fontaine* (French, with English and French summaries), Duke Math. J. **167** (2018), no. 18, 3455–3532, DOI 10.1215/00127094-2018-0034. MR3881201

[134] Qing Liu, *Un contre-exemple au "critère cohomologique d'affinoïdicité"* (French, with English summary), C. R. Acad. Sci. Paris Sér. I Math. **307** (1988), no. 2, 83–86. MR954265

[135] Qing Liu, *Sur les espaces de Stein quasi-compacts en géométrie rigide* (French, with English summary), Sém. Théor. Nombres Bordeaux (2) **1** (1989), no. 1, 51–58. MR1050264

[136] Ju. I. Manin, *Theory of commutative formal groups over fields of finite characteristic* (Russian), Uspehi Mat. Nauk **18** (1963), no. 6 (114), 3–90. MR0157972

[137] Michel Matignon and Marc Reversat, *Sous-corps fermés d'un corps valué* (French), J. Algebra **90** (1984), no. 2, 491–515, DOI 10.1016/0021-8693(84)90186-8. MR760025

[138] Tomoki Mihara, *On Tate's acyclicity and uniformity of Berkovich spectra and adic spectra*, Israel J. Math. **216** (2016), no. 1, 61–105, DOI 10.1007/s11856-016-1404-8. MR3556963

[139] M. Morrow, Notes on the $\mathbb{A}_{\inf}$-cohomology of Integral p-adic Hodge theory, arXiv:1608.00922v1 (2016).

[140] David Mumford, *Curves and their Jacobians*, The University of Michigan Press, Ann Arbor, Mich., 1975. MR0419430

[141] D. Mumford, *An algebro-geometric construction of commuting operators and of solutions to the Toda lattice equation, Korteweg deVries equation and related nonlinear equation*, Proceedings of the International Symposium on Algebraic Geometry (Kyoto Univ., Kyoto, 1977), Kinokuniya Book Store, Tokyo, 1978, pp. 115–153. MR578857

[142] J. P. Murre, *Lectures on an introduction to Grothendieck's theory of the fundamental group*, Tata Institute of Fundamental Research, Bombay, 1967. Notes by S. Anantharaman; Tata Institute of Fundamental Research Lectures on Mathematics, No 40. MR0302650

[143] M. S. Narasimhan and C. S. Seshadri, *Stable and unitary vector bundles on a compact Riemann surface*, Ann. of Math. (2) **82** (1965), 540–567, DOI 10.2307/1970710. MR0184252

[144] Behrang Noohi, *Fundamental groups of topological stacks with the slice property*, Algebr. Geom. Topol. **8** (2008), no. 3, 1333–1370, DOI 10.2140/agt.2008.8.1333. MR2443246

[145] Madhav V. Nori, *On the representations of the fundamental group*, Compositio Math. **33** (1976), no. 1, 29–41. MR0417179

[146] Madhav V. Nori, *The fundamental group-scheme*, Proc. Indian Acad. Sci. Math. Sci. **91** (1982), no. 2, 73–122, DOI 10.1007/BF02967978. MR682517

[147] Martin C. Olsson, *On Faltings' method of almost étale extensions*, Algebraic geometry—Seattle 2005. Part 2, Proc. Sympos. Pure Math., vol. 80, Amer. Math. Soc., Providence, RI, 2009, pp. 811–936, DOI 10.1090/pspum/080.2/2483956. MR2483956

[148] Brian Osserman, *Frobenius-unstable bundles and p-curvature*, Trans. Amer. Math. Soc. **360** (2008), no. 1, 273–305, DOI 10.1090/S0002-9947-07-04218-3. MR2342003

[149] A. Pál and G. Zábrádi, *Cohomology and overconvergence for representations of powers of Galois groups*, arXiv:1705.03786v2 (2018).

[150] J. Pottharst, Harder–Narasimhan theory, available at https://vbrt.org/writings/HN.pdf (retrieved February 2017).

[151] M. Rapoport and M. Richartz, *On the classification and specialization of F-isocrystals with additional structure*, Compositio Math. **103** (1996), no. 2, 153–181. MR1411570

[152] Michael Rapoport and Eva Viehmann, *Towards a theory of local Shimura varieties*, Münster J. Math. **7** (2014), no. 1, 273–326. MR3271247

[153] M. Rapoport and Th. Zink, *Period spaces for p-divisible groups*, Annals of Mathematics Studies, vol. 141, Princeton University Press, Princeton, NJ, 1996. MR1393439

[154] David Rydh, *Submersions and effective descent of étale morphisms* (English, with English and French summaries), Bull. Soc. Math. France **138** (2010), no. 2, 181–230, DOI 10.24033/bsmf.2588. MR2679038

[155] Peter Schneider, *Nonarchimedean functional analysis*, Springer Monographs in Mathematics, Springer-Verlag, Berlin, 2002. MR1869547

[156] Peter Scholze, *Perfectoid spaces*, Publ. Math. Inst. Hautes Études Sci. **116** (2012), 245–313, DOI 10.1007/s10240-012-0042-x. MR3090258

[157] P. Scholze, *p-adic Hodge theory for rigid analytic varieties*, Forum of Math. Pi **1** (2013), doi:10.1017/fmp.2013.1; errata available at http://www.math.uni-bonn.de/people/scholze/pAdicHodgeErratum.pdf.

[158] Peter Scholze, *Perfectoid spaces: a survey*, Current developments in mathematics 2012, Int. Press, Somerville, MA, 2013, pp. 193–227. MR3204346

[159] Peter Scholze, *On torsion in the cohomology of locally symmetric varieties*, Ann. of Math. (2) **182** (2015), no. 3, 945–1066, DOI 10.4007/annals.2015.182.3.3. MR3418533

[160] P. Scholze, Étale cohomology of diamonds, preprint (2018) available at http://www.math.uni-bonn.de/people/scholze/ (retrieved December 2018; see also arXiv:1709.07343).

[161] Peter Scholze and Jared Weinstein, *Moduli of p-divisible groups*, Camb. J. Math. **1** (2013), no. 2, 145–237, DOI 10.4310/CJM.2013.v1.n2.a1. MR3272049

[162] P. Scholze and J. Weinstein, Berkeley lectures on *p*-adic geometry, http://www.math.uni-bonn.de/people/scholze/Berkeley.pdf (retrieved October 2018).

[163] Shankar Sen, *Ramification in p-adic Lie extensions*, Invent. Math. **17** (1972), 44–50, DOI 10.1007/BF01390022. MR0319949

[164] Jean-Pierre Serre, *Géométrie algébrique et géométrie analytique* (French), Ann. Inst. Fourier, Grenoble **6** (1955), 1–42. MR0082175

[165] Jean-Pierre Serre, *Exemples de variétés projectives conjuguées non homéomorphes* (French), C. R. Acad. Sci. Paris **258** (1964), 4194–4196. MR0166197

[166] The Stacks Project Authors, *Stacks Project*, http://stacks.math.columbia.edu, retrieved January 2017.

[167] Lynn Arthur Steen and J. Arthur Seebach Jr., *Counterexamples in topology*, Dover Publications, Inc., Mineola, NY, 1995. Reprint of the second (1978) edition. MR1382863

[168] Richard G. Swan, *On seminormality*, J. Algebra **67** (1980), no. 1, 210–229, DOI 10.1016/0021-8693(80)90318-X. MR595029

[169] John Tate, *Rigid analytic spaces*, Invent. Math. **12** (1971), 257–289, DOI 10.1007/BF01403307. MR0306196

[170] J. Tate, *Number theoretic background*, Automorphic forms, representations and L-functions (Proc. Sympos. Pure Math., Oregon State Univ., Corvallis, Ore., 1977), Proc. Sympos. Pure Math., XXXIII, Amer. Math. Soc., Providence, R.I., 1979, pp. 3–26. MR546607

[171] R. W. Thomason and Thomas Trobaugh, *Higher algebraic K-theory of schemes and of derived categories*, The Grothendieck Festschrift, Vol. III, Progr. Math., vol. 88, Birkhäuser Boston, Boston, MA, 1990, pp. 247–435, DOI 10.1007/978-0-8176-4576-2_10. MR1106918

[172] Burt Totaro, *Tensor products in p-adic Hodge theory*, Duke Math. J. **83** (1996), no. 1, 79–104, DOI 10.1215/S0012-7094-96-08304-0. MR1388844

[173] Yakov Varshavsky, *Moduli spaces of principal F-bundles*, Selecta Math. (N.S.) **10** (2004), no. 1, 131–166, DOI 10.1007/s00029-004-0343-0. MR2061225

[174] V. Voevodsky, *Homology of schemes*, Selecta Math. (N.S.) **2** (1996), no. 1, 111–153, DOI 10.1007/BF01587941. MR1403354

[175] T. Wedhorn, Lecture notes on adic spaces available at http://math.stanford.edu/~conrad/Perfseminar/refs/wedhornadic.pdf (retrieved February 2017).

[176] Jared Weinstein, $\mathrm{Gal}(\overline{\mathbf{Q}}_p/\mathbf{Q}_p)$ *as a geometric fundamental group*, Int. Math. Res. Not. IMRN **10** (2017), 2964–2997, DOI 10.1093/imrn/rnw072. MR3658130

[177] J. Weinstein, *Adic spaces*, Perfectoid Spaces: Lectures from the 2017 Arizona Winter School, Mathematical Surveys and Monographs, vol. 242, 1–43, American Mathematical Society, 2019.

[178] Jean-Pierre Wintenberger, *Extensions de Lie et groupes d'automorphismes des corps locaux de caractéristique p* (French, with English summary), C. R. Acad. Sci. Paris Sér. A-B **288** (1979), no. 9, A477–A479. MR529480

[179] Jean-Pierre Wintenberger, *Le corps des normes de certaines extensions infinies de corps locaux; applications* (French), Ann. Sci. École Norm. Sup. (4) **16** (1983), no. 1, 59–89. MR719763

[180] Gergely Zábrádi, *Multivariable (φ, Γ)-modules and smooth o-torsion representations*, Selecta Math. (N.S.) **24** (2018), no. 2, 935–995, DOI 10.1007/s00029-016-0259-5. MR3782415

[181] Gergely Zábrádi, *Multivariate (φ, Γ)-modules and products of Galois groups*, Math. Res. Lett. **25** (2018), no. 2, 687–721, DOI 10.4310/MRL.2018.v25.n2.a18. MR3826842

[182] Xinwen Zhu, *An introduction to affine Grassmannians and the geometric Satake equivalence*, Geometry of moduli spaces and representation theory, IAS/Park City Math. Ser., vol. 24, Amer. Math. Soc., Providence, RI, 2017, pp. 59–154. MR3752460

[183] Xinwen Zhu, *Affine Grassmannians and the geometric Satake in mixed characteristic*, Ann. of Math. (2) **185** (2017), no. 2, 403–492, DOI 10.4007/annals.2017.185.2.2. MR3612002

Email address: kedlaya@ucsd.edu

The Hodge-Tate decomposition via perfectoid spaces

Bhargav Bhatt

Contents

1. Lecture 1: Introduction
2. Lecture 2: The Hodge-Tate decomposition for abelian schemes
3. Lecture 3: The Hodge-Tate decomposition in general
4. Lecture 4: Integral aspects
5. Exercises
6. Projects

References

1. Lecture 1: Introduction

1.1. Statement and consequences of the Hodge-Tate decomposition.

Fix a prime number p. The goal of this series is to explain the p-adic analog of the following classical result, which forms the starting point of Hodge theory.

THEOREM 1.1 (Hodge decomposition). *Let $X/\mathbf{C}$ be a smooth proper variety. Then there exists a natural isomorphism*

$$H^n(X^{an}, \mathbf{C}) \simeq \bigoplus_{i+j=n} H^i(X, \Omega^j_{X/\mathbf{C}}).$$

A similar result holds true when X is a compact Kähler manifold.

Theorem 1.1 has many immediate consequences. For example, the "naturality" assertion above implies that the Hodge numbers are topological invariants in the following sense:

COROLLARY 1.2. *If $f : X \to Y$ is a map of smooth proper varieties that induces an isomorphism $H^n(Y^{an}, \mathbf{C}) \simeq H^n(X^{an}, \mathbf{C})$ for some $n \geq 0$, then one also has $H^i(Y, \Omega^j_{Y/\mathbf{C}}) \simeq H^i(X, \Omega^j_{X/\mathbf{C}})$ for each i, j with $i + j = n$.*

To move towards the p-adic analog, recall that the theory of étale cohomology provides an algebraic substitute for singular cohomology that works over any field k: the two roughly coincide when $k = \mathbf{C}$, but the former is constructed directly from algebraic geometry, and thus witnesses the action of algebraic symmetries, including those that might not be holomorphic when working over $k = \mathbf{C}$. As a concrete consequence, we have the following vaguely formulated statement:

THEOREM 1.3 (Grothendieck, Artin, ...). *Let $X/\mathbf{C}$ be an algebraic variety that is defined over $\mathbf{Q}$. Then the absolute Galois $G_\mathbf{Q}$ of $\mathbf{Q}$ acts canonically on $H^i(X^{an}, \mathbf{Z}/n)$ for any integer $n > 0$. Letting n vary through powers of a prime p, we obtain a continuous $G_\mathbf{Q}$-action[1] on the $\mathbf{Z}_p$-module $H^i(X^{an}, \mathbf{Z}_p)$, and thus on the $\mathbf{Q}_p$-vector space $H^i(X^{an}, \mathbf{Q}_p)$.*

Some important examples of this action are:

EXAMPLE 1.4 (Elliptic curves). Let $X = E$ be an elliptic curve over $\mathbf{C}$ which is defined over $\mathbf{Q}$. Then
$$H^1(X^{an}, \mathbf{Z}/n) \simeq H_1(X^{an}, \mathbf{Z}/n)^\vee \simeq \mathrm{Hom}(\pi_1(E), \mathbf{Z}/n) \simeq E[n]^\vee$$
is the $\mathbf{Z}/n$-linear dual of the n-torsion of E. In this case, Theorem 1.3 reflects the fact that all n-torsion points on E are defined over $\overline{\mathbf{Q}} \subset \mathbf{C}$, and are permuted by the Galois group $G_\mathbf{Q}$ as E has $\mathbf{Q}$-coefficients. Passing to the inverse limit, this endows the p-adic Tate module $T_p(E) := \lim_n E[n]$ and its $\mathbf{Z}_p$-linear dual $H^1(X^{an}, \mathbf{Z}_p)$ with canonical $G_\mathbf{Q}$-actions. As $T_p(E) \simeq \mathbf{Z}_p^2$ as a topological group, this discussion provides a continuous 2-dimensional representation $G_\mathbf{Q} \to \mathrm{GL}_2(\mathbf{Z}_p)$. More generally, the same discussion applies to any abelian variety of dimension g to yield a continuous representation $G_\mathbf{Q} \to \mathrm{GL}_{2g}(\mathbf{Z}_p)$.

EXAMPLE 1.5 (The torus and Tate twists). Another important example is the case of $X = \mathbf{G}_m$. In this case, by the same reasoning above, we have $H^1(X^{an}, \mathbf{Z}/n) \simeq \mu_n^\vee$ (where $\mu_n \subset \overline{\mathbf{Q}}^*$ denotes the set of n-th roots of 1) and $H^1(X^{an}, \mathbf{Z}_p) \simeq (\lim_n \mu_n)^\vee =: \mathbf{Z}_p(-1)$. It is easy to see that $\mathbf{Z}_p(-1)$ is a rank 1 free module over $\mathbf{Z}_p$, so we can make sense of $\mathbf{Z}_p(j)$ for any integer j: we simply set $\mathbf{Z}_p(j) := (\mathbf{Z}_p(-1)^{\otimes j})^\vee$. Moreover, the resulting represenation $G_\mathbf{Q} \to \mathrm{GL}_1(\mathbf{Z}_p)$ is highly non-trivial by class field theory. In general, for a $\mathbf{Z}_p$-algebra R, we shall write $R(i) := R \otimes_{\mathbf{Z}_p} \mathbf{Z}_p(i)$, and refer to this as the i-th Tate twist of R.

EXAMPLE 1.6 (Projective line and abelian varieties). Standard computations in algebraic topology are compatible with the Galois action from Theorem 1.3. Thus, for example, if $A/\mathbf{C}$ is an abelian variety of dimension g, then we know from topology that $H^*(A^{an}, \mathbf{Z}_p)$ is an exterior algebra on $H^1(A, \mathbf{Z}_p)$ under the cup product: we have $A \simeq (S^1)^{2g}$ as a topological space, so the claim follows by Künneth. It follows that the same description also applies in the world of $G_\mathbf{Q}$-modules. Likewise, via the Mayer-Vietoris sequence, we have a canonical isomorphism $H^2(\mathbf{P}^{1,an}, \mathbf{Z}_p) \simeq H^1(\mathbf{G}_m^{an}, \mathbf{Z}_p) \simeq \mathbf{Z}_p(-1)$ in the world of $G_\mathbf{Q}$-modules. More generally, if $X/\mathbf{C}$ is a smooth (or merely irreducible) projective variety of dimension d defined over $\mathbf{Q}$, then one can show $H^{2d}(X^{an}, \mathbf{Z}_p) \simeq \mathbf{Z}_p(-d)$ as a $G_\mathbf{Q}$-module.

From here on, we assume that the reader is familiar with the basics of étale cohomology theory[2]. Via Theorem 1.3 (and variants), this theory provides perhaps

---

[1]Slightly more precisely, for any prime p, there is a functor $Y \mapsto H^i(Y_{et}, \mathbf{Z}_p)$ defined on the category of all schemes. For $Y = X_{\overline{\mathbf{Q}}}$ (or even $X_\mathbf{C}$), this theory is canonically identified with $H^i(X^{an}, \mathbf{Z}_p)$; the Galois action on Y then induces (by functoriality) the action mentioned in Theorem 1.3; see also the next footnote.

[2]For a scheme X, a prime number p and a coefficient ring $\Lambda \in \{\mathbf{Z}/p, \mathbf{Z}/p^n, \mathbf{Z}_p, \mathbf{Q}_p\}$, we write $H^*(X_{et}, \Lambda)$ for the étale cohomology X with Λ-coefficients; we indulge here in the standard abuse

the most important examples of $G_{\mathbf{Q}}$-representations on p-adic vector spaces. To understand these objects, at a first approximation, one must understand the action of the local Galois groups (or decomposition groups) $D_\ell \subset G_{\mathbf{Q}}$ for a rational prime ℓ. When $\ell \neq p$, these actions can be understood[3] in terms of algebraic geometry over the finite field $\mathbf{F}_\ell$; in effect, due to the incompatibility of the ℓ-adic nature of D_ℓ with the p-adic topology, these actions are classified by the action of a single endomorphism (the Frobenius), and one has powerful tools coming from the solution of Weil conjectures at our disposal to analyze this endomorphism. However, if $\ell = p$, the resulting representations are much too rich to be understood in terms of a single endomorphism. Instead, these representations are best viewed as p-adic analogs of Hodge structures, explaining the name "p-adic Hodge theory" given to the study of these representations. Perhaps the first general result justifying this choice of name is the following, which gives the p-adic analog of the Hodge decomposition in Theorem 1.1 and forms the focus of this lecture series:

THEOREM 1.7 (Hodge-Tate decomposition). *Let $K/\mathbf{Q}_p$ be a finite extension, and let $\mathbf{C}_p$ be a completion of an algebraic closure $\overline{K}$ of K. Let X/K be a smooth proper variety. Then there exists a Galois equivariant decomposition*

$$(1) \qquad H^n(X_{\overline{K},et}, \mathbf{Q}_p) \otimes_{\mathbf{Q}_p} \mathbf{C}_p \simeq \bigoplus_{i+j=n} H^i(X, \Omega^j_{X/K}) \otimes_K \mathbf{C}_p(-j),$$

where $\mathbf{C}_p(-j)$ denotes the $(-j)$-th Tate twist of $\mathbf{C}_p$. This isomorphism is functorial in X. In particular, it respects the natural graded algebra structures on either side as n varies.

We take a moment to unravel this statement. The object $H^n(X_{\overline{K},et}, \mathbf{Q}_p)$ is the étale cohomology of $X_{\overline{K}} := X \otimes_K \overline{K}$, and hence admits a $G_K := \mathrm{Gal}(\overline{K}/K)$-action by transport of structure. The G_K-action on $\overline{K}$ is continuous, and hence extends to one on the completion $\mathbf{C}_p$. In particular, G_K acts on the left side of (1) via the tensor product action. On the right side, the only nontrivial G_K-action exists on Tate twists $\mathbf{C}_p(-j) := \mathbf{C}_p \otimes_{\mathbf{Z}_p} \mathbf{Z}_p(1)^{\otimes -j}$, where it is defined as the tensor product of G_K-actions on the two pieces. In particular, $\mathbf{C}_p(-j)$ is *not* a linear representation of G_K on a $\mathbf{C}_p$-vector space; instead, it is semilinear with respect to the standard G_K-action on $\mathbf{C}_p$.

To extract tangible consequences from Theorem 1.7, it is important to know that the Tate twists $\mathbf{C}_p(j)$ are distinct for different values of j. In fact, one has the much stronger statement that these Tate twists do not talk to each other for different values of j (see [**Ta**]):

THEOREM 1.8 (Tate). *Fix notation as in Theorem 1.7. Then, for $i \neq 0$, we have*

$$H^0(G_K, \mathbf{C}_p(i)) = H^1(G_K, \mathbf{C}_p(i)) = 0,$$

---

of notation where, for $\Lambda \in \{\mathbf{Z}_p, \mathbf{Q}_p\}$, the groups $H^n(X_{et}, \Lambda)$ are not the cohomology groups of a sheaf on the étale site X_{et}, but rather are defined by an inverse limit procedure.

[3] We are implicitly assuming in this paragraph that the prime p is a prime of good reduction for the variety under consideration. If X has bad reduction at p, then the resulting representations of D_ℓ are much more subtle: already when $\ell \neq p$, there is an extremely interesting additional piece of structure, called the monodromy operator, that is still not completely understood.

where the Galois cohomology groups are interpreted as continuous cohomology groups. For $i = 0$, each of these groups is a copy of K. In particular, we have

$$\mathrm{Hom}_{G_K, \mathbf{C}_p}(\mathbf{C}_p(i), \mathbf{C}_p(j)) = 0$$

for $i \neq j$.

In words, Theorem 1.8 states that there are no continuous G_K-equivariant maps between different Tate twists of $\mathbf{C}_p$, and that there are no non-trivial G_K-equivariant extensions of $\mathbf{C}_p(i)$ by $\mathbf{C}_p(j)$ for $i \neq j$. We now revisit the preceding examples armed with this understanding.

EXAMPLE 1.9. Consider $X := \mathbf{P}^1$ and $n = 2$. In this case, we have $H^2(X_{\overline{K}, et}, \mathbf{Q}_p) \simeq \mathbf{Q}_p(-1)$ by Example 1.6 (see also Example 1.5). Using Theorem 1.8, we see that Theorem 1.7 captures the statement that $H^0(X, \Omega^2_{X/K}) = H^2(X, \mathcal{O}_X) = 0$, while $H^1(X, \Omega^1_{X/K})$ is 1-dimensional.

EXAMPLE 1.10. Let $X = A$ be an abelian variety over K. By combining Example 1.4 and Theorem 1.7, we learn that

$$T_p(A) \otimes_{\mathbf{Z}_p} \mathbf{C}_p \simeq \left(H^1(A, \mathcal{O}_A)^\vee \otimes_K \mathbf{C}_p\right) \oplus \left(H^0(A, \Omega^1_{A/K})^\vee \otimes_K \mathbf{C}_p(1)\right).$$

One can identify the right side in more classical terms:

$$H^0(A, \Omega^1_{A/K})^\vee \simeq \mathrm{Lie}(A) \quad \text{and} \quad H^1(A, \mathcal{O}_A) \simeq \mathrm{Lie}(A^\vee),$$

where $A^\vee$ is the dual of A. Thus, we can rewrite the above decomposition as

$$T_p(A) \otimes_{\mathbf{Z}_p} \mathbf{C}_p \simeq \left(\mathrm{Lie}(A^\vee)^\vee \otimes_K \mathbf{C}_p\right) \oplus \left(\mathrm{Lie}(A) \otimes_K \mathbf{C}_p(1)\right).$$

As we shall see later, if A is merely defined over $\mathbf{C}_p$ instead of over a finite extension K as above, then we always have a short exact sequence

$$0 \to \mathrm{Lie}(A)(1) \to T_p(A) \otimes_{\mathbf{Z}_p} \mathbf{C}_p \to \mathrm{Lie}(A^\vee)^\vee \to 0,$$

but this sequence may not split in a canonical way: there is no Galois action present when A is defined merely over $\mathbf{C}_p$, so one cannot invoke Theorem 1.8 to obtain a (necessarily unique!) splitting of the previous sequence.

In number theory, one of the main applications of these ideas is in understanding the Galois representations of G_K arising as $H^n(X_{\overline{K}, et}, \mathbf{Q}_p)$. For example, Theorem 1.7 implies these representations are Hodge-Tate, which forms the first in a series of increasingly stronger restrictions placed on the representations arising in this fashion from algebraic geometry; upgrading this structure, one can even give a completely "linear algebraic" description of these Galois representations (see Remark 1.18), which is very useful for computations.

Theorem 1.7 also has applications to purely geometric statements. For example, applying Theorem 1.8 leads to the following concrete consequence concerning the recovery of the algebro-geometric invariants $H^i(X, \Omega^j_{X/K})$ from the topological/arithmetic invariant $H^n(X_{\overline{K}, et}, \mathbf{Q}_p)$:

COROLLARY 1.11 (Recovery of Hodge numbers). *With notation as in Theorem 1.7, we have*

$$H^i(X, \Omega^j_{X/K}) \simeq \left(H^{i+j}(X_{\overline{K}, et}, \mathbf{Q}_p) \otimes_{\mathbf{Q}_p} \mathbf{C}_p(j)\right)^{G_K}.$$

PROOF. Set $n = i + j$. Tensoring both sides of (1) (and replacing j with k in that formula)

$$H^n_{et}(X_{\overline{K}}, \mathbf{Q}_p) \otimes_{\mathbf{Q}_p} \mathbf{C}_p(j) \simeq \bigoplus_{i+k=n} H^i(X, \Omega^k_{X/K}) \otimes_K \mathbf{C}_p(j-k).$$

Applying $(-)^{G_K}$ then gives the claim as $K \simeq \mathbf{C}_p^{G_K}$ and $\mathbf{C}_p(j-k)^{G_K} = 0$ when $j \neq k$ by Theorem 1.8. □

In particular, Corollary 1.11 gives an analog of Corollary 1.2 in this setting. In fact, Corollary 1.11 is one of the key steps in Ito's alternative proof [**It**] of the following purely geometric result; the first proof of the latter gave birth to the theory of motivic integration [**Ko**, **DL**], and both proofs rely on Batyrev's [**Ba**] proving the analogous claim for Betti numbers via p-adic integration.

THEOREM 1.12 (Kontsevich, Denef, Loeser, Ito). *Let X and Y be smooth projective varieties over $\mathbf{C}$. Assume that both X and Y are Calabi-Yau (i.e., K_X and K_Y are trivial), and that X is birational to Y. Then*

$$\dim(H^i(X, \Omega^j_{X/\mathbf{C}})) = \dim(H^i(Y, \Omega^j_{Y/\mathbf{C}})).$$

for all i, j.

One may view Theorem 1.7 as relating the Galois representation on $H^n(X_{\overline{K},et}, \mathbf{Q}_p)$ to the algebraic geometry of X. An obvious question that then arises, and one we have essentially skirted in the discussion so far, is whether one can understand the p-torsion in $H^n(X_{\overline{K},et}, \mathbf{Z}_p)$ in terms of the geometry of X; in particular, we may ask for a geometric description of $H^n(X_{\overline{K},et}, \mathbf{F}_p)$. This integral story is much less understood than the rational theory above. Nevertheless, one has the following recent result [**BMS2**]:

THEOREM 1.13. *Fix notation as in Theorem 1.7. Assume that X extends to a proper smooth $\mathcal{O}_K$-scheme $\mathfrak{X}$. Write $\mathfrak{X}_k$ for the fiber of $\mathfrak{X}$ over the residue field k of K. Then we have*

$$\dim_{\mathbf{F}_p}(H^n(X_{\overline{K},et}, \mathbf{F}_p)) \leq \sum_{i+j=n} \dim_k H^i(\mathfrak{X}_k, \Omega^j_{\mathfrak{X}_k/k}).$$

Moreover, there exist examples where the inequality is strict.

In other words, the mod-p cohomology of $X_{\overline{K}}$ is related to the geometry of $\mathfrak{X}_k$. We shall sketch a proof of Theorem 1.13 towards the end of the lecture series.

Outline of proof and lectures. The main goal of this lecture series is to explain a proof of Theorem 1.7; towards the end, we shall also sketch some ideas going into Theorem 1.13. Our plan is to prove Theorem 1.7 following the perfectoid approach of Scholze [**Sc2**], which itself is inspired by the work of Faltings [**Fa1**, **Fa2**, **Fa3**, **Fa4**]. In broad strokes, there are two main steps:

(1) *Local study of Hodge cohomology via perfectoid spaces:* Construct a proétale cover $X_\infty \to X$ which is "infinitely ramified in characteristic p", and study the cohomology of X_∞. In fact, X_∞ shall be an example of a perfectoid space [**Sc1**], so the perfectoid theory gives a lot of control on the cohomology of X_∞. In particular, suitably interpreted, X_∞ carries no differential forms, so the full Hodge cohomology comes from the structure sheaf.

(2) *Descent:* Descend the preceding understanding of the Hodge cohomology of X_∞ down to X. In this step, we shall see that the differential forms on X, which vanished after pullback to X_∞, reappear in the descent procedure.

In fact, to illustrate this process in practice, we work out explicitly the case of abelian varieties with good reduction in §2. The general case is then treated in §3, while the integral theory is surveyed in §4.

1.2. Complementary remarks. We end this section by some remarks of a historical nature, complementing the theory discussed above.

REMARK 1.14. Theorem 1.7 was conjectured by Tate [**Ta**]. In the same paper, Tate also settled the case of abelian varieties (and, more generally, p-divisible groups) with good reduction; the case of general abelian varieties was then settled by Raynaud using the semistable reduction theory. The abelian variety case was revisited by Fontaine in [**Fo1**], who also provided a natural "differential" definition of the Tate twist. The general statement mentioned above was established by Faltings [**Fa1**], as a consequence of his machinery of almost étale extensions.

REMARK 1.15. (Hodge-Tate decomposition for rigid spaces) In [**Ta**, §4, Remark], Tate wondered if Theorem 1.7 should be valid more generally for any proper smooth rigid-analytic[4] space. This question was answered affirmatively by Scholze [**Sc2**, Corollary 1.8]. In fact, Scholze proves the following more general assertion (see [**Sc3**, Theorem 3.20]):

THEOREM 1.16 (Hodge-Tate filtration). *Let C be a complete and algebraically closed nonarchimedean extension of $\mathbf{Q}_p$. Let X/C be a proper smooth rigid-analytic space. Then there exists an E_2-spectral sequence, called the "Hodge-Tate spectral sequence",*

$$E_2^{i,j}: H^i(X, \Omega^j_{X/C})(-j) \Rightarrow H^{i+j}(X_{et}, \mathbf{Q}_p) \otimes C.$$

It is this more general result that is most naturally accessible to perfectoid techniques, and thus forms the focus of this lecture series.

REMARK 1.17 (Degeneration of the Hodge-Tate spectral sequence). When X is defined over a discretely valued subfield of C (such as a finite extension of $\mathbf{Q}_p$), the Hodge-Tate spectral sequence degenerates thanks to due to Theorem 1.8 (which holds true for over any such field). Indeed, the differentials are Galois equivariant and go between vector spaces with different Tate twists, so they must be 0 by the H^0 statements from Theorem 1.8. In fact, a little bit more care using Theorem 1.8 for H^* shows that one can even promote this degeneration to a canonical one, i.e., one can produce the Hodge-Tate decomposition for proper smooth rigid-analytic spaces, as inquired by Tate.

In the general case (i.e., if X is merely defined over C), one can still show that this spectral sequence always degenerates (see [**BMS2**, §13]), but there is no canonical Hodge-Tate decomposition.

---

[4]At first glance, this is very surprising: in complex geometry, the Hodge decomposition in Theorem 1.1 only applies to compact complex manifolds which are (not far from) Kähler, so one would also expect an analog of the Kähler condition in p-adic geometry. However, if one accepts that Kähler metrics are somewhat analogous to formal models (for example, the latter provides a well-behaved metric on the space of analytic functions), then the analogy with complex geometry is restored: as every rigid space admits a formal model by Raynaud [**BL2**, Theorem 4.1].

REMARK 1.18 (*p*-adic comparison theorems). The Hodge-Tate decomposition forms the first in a hierarchy of increasingly stronger statements (conjectured by Fontaine, and proven by various authors) describing the Galois representations of G_K on $H^n(X_{\overline{K},et}, \mathbf{Q}_p)$ in terms of the geometry of X. For example, the *p*-adic de Rham comparison isomorphism, which is formulated in terms of a certain filtered G_K-equivariant $\overline{K}$-algebra B_{dR} constructed by Fontaine, asserts:

THEOREM 1.19 (de Rham comparison). *There exists a canonical isomorphism*

$$H^n(X_{\overline{K},et}, \mathbf{Q}_p) \otimes_{\mathbf{Q}_p} B_{dR} \simeq H^n_{dR}(X/K) \otimes_K B_{dR}.$$

This isomorphism respects the Galois action and filtrations.

Theorem 1.19, together with some knowledge of B_{dR}, allows one to recover the de Rham cohomology $H^n_{dR}(X/K)$ as a *filtered vector space* from the G_K-representation $H^n(X_{\overline{K},et}, \mathbf{Q}_p)$. In fact, passage to the associated graded in Theorem 1.19 recovers Theorem 1.7, so one may view the de Rham comparison isomorphism as a non-trivial deformation of the Hodge-Tate decomposition. Continuing further, in the setting of good or semistable reduction, one can endow $H^n_{dR}(X/K)$ with some extra structure (namely, a Frobenius endomorphism, as well a monodromy endomorphism in the semistable case); the crystalline/semistable comparison theorems give an analogous comparison relating the G_K-representation $H^n(X_{\overline{K},et}, \mathbf{Q}_p)$ with the de Rham cohomology $H^n_{dR}(X/K)$ equipped with the aforementioned additional structure. A major advantage of these latter theorems is that the recovery process works in both directions. In particular, one can completely describe the G_K-representation $H^n(X_{\overline{K},et}, \mathbf{Q}_p)$ in terms of the linear algebra data on the de Rham side, thus facilitating calculations. We will not be discussing any of these comparison theorems in this lecture series, and refer the reader to [**BMS2**, §1.1] for more information.

REMARK 1.20 (Open and singular varieties). Theorem 1.7 has a natural extension to arbitrary varieties X/K. In this case, the correct statement of the decomposition is:

$$H^n(X_{\overline{K},et}, \mathbf{Q}_p) \otimes_{\mathbf{Q}_p} \mathbf{C}_p \simeq \bigoplus_{i+j=n} \mathrm{gr}_F^j H^{i+j}_{dR}(X/K) \otimes_K \mathbf{C}_p(-j),$$

where gr_F^j denotes the *j*-th graded piece for the Hodge filtration on $H^n_{dR}(X/K)$ constructed by Deligne's theory of mixed Hodge structures [**De1, De2**]. This result was proven in [**Ki1**] using de Jong's alterations theorem, and found a natural home in the recent approach of Beilinson [**Be**] to the *p*-adic comparison theorems based on vanishing theorems for the *h*-topology; we shall not discuss such extensions further in these notes.

Acknowledgements. I'm thankful to Kestutis Cesnavicus, David Hansen, Kiran Kedlaya, Marius Leonhardt, Daniel Litt, Jacob Lurie, Akhil Mathew, Matthew Morrow, Darya Schedrina, Peter Scholze, Koji Shimizu, Peter Wear for conversations related to and feedback on the material in these notes. I am especially grateful to the referee for a thorough reading and many suggestions that improved the exposition.

2. Lecture 2: The Hodge-Tate decomposition for abelian schemes

The main goal for this section is to introduce, in the case of abelian varieties with good reduction, certain "large" constructions that are generally useful in Hodge-Tate theory. Our hope is that encountering these "large" objects in a relatively simple setting will help demystify them.

2.1. The statement. Fix a finite extension $K/\mathbf{Q}_p$, a completed algebraic closure $K \hookrightarrow C$, and an abelian scheme $\mathcal{A}/\mathcal{O}_K$ with generic fiber A. Our goal is to sketch a proof of the following result:

THEOREM 2.1. *There exists a canonical isomorphism*

$$H^1(A_C, C) := H^1(A_{C,et}, \mathbf{Z}_p) \otimes_{\mathbf{Z}_p} C \simeq \left(H^1(A, \mathcal{O}_A) \otimes_K C\right) \oplus \left(H^0(A, \Omega^1_{A/K}) \otimes_K C(-1)\right).$$

In fact, we shall not construct the complete decomposition. Instead, we shall construct a map

$$\alpha_A : H^1(A, \mathcal{O}_A) \otimes_K C \to H^1(A_C, C)$$

using inspiration from the perfectoid theory (as well as [**Be**]), and a map

$$\beta_A : H^0(A, \Omega^1_{A/K}) \otimes_K C(-1) \to H^1(A_C, C)$$

exploiting the arithmetic of the base field K, following an idea of Fontaine [**Fo1**]. We can then put these together to get the map

$$\gamma_A = \alpha_A \oplus \beta_A : \left(H^1(A, \mathcal{O}_A) \otimes_K C\right) \oplus \left(H^0(A, \Omega^1_{A/K}) \otimes_K C(-1)\right) \to H^1(A_C, C),$$

that induces the Hodge-Tate decomposition.

REMARK 2.2 (Reminders on abelian varieties). The following facts about the cohomology of abelian varieties will be used below.
(1) Write $T_p(A) := \lim A[p^n](C)$ for the p-adic Tate module of A. Then there is a natural identification of $H^*(A, C) \simeq H^*(T_p(A), C)$, where $H^*(T_p(A), C)$ denotes the continuous group cohomology of the profinite group $T_p(A)$ with coefficients in the topological ring C; this essentially comes down to the assertion that an abelian variety of dimension g over $\mathbf{C}$ is homeomorphic to $(S^1)^{2g}$. In particular, since $T_p(A) \cong \mathbf{Z}_p^{2g}$, one calculates that $H^*(A, C)$ is an exterior algebra on $H^1(A, C) \simeq T_p(A)^\vee \otimes C$.
(2) The $\mathcal{O}_K$-module $H^1(\mathcal{A}, \mathcal{O}_\mathcal{A})$ is free of rank $g = \dim(A)$. In fact, this module is canonically identified with the Lie algebra of the dual abelian scheme $\mathcal{A}^\vee$. Moreover, the cohomology ring $H^*(\mathcal{A}, \mathcal{O}_\mathcal{A})$ is an exterior algebra on $H^1(\mathcal{A}, \mathcal{O}_\mathcal{A})$ via cup products. In particular, all cohomology groups of $\mathcal{O}_\mathcal{A}$ are torsionfree.

2.2. The perfectoid construction of the map α_A. The discussion in this section is geometric, so we work with the abelian $\mathcal{O}_C$-scheme $\mathcal{A}_{\mathcal{O}_C}$ directly. Write $\mathcal{A}_n = \mathcal{A}_{\mathcal{O}_C}$ for each $n \geq 0$, and consider the tower

$$\dots \to \mathcal{A}_{n+1} \xrightarrow{[p]} \mathcal{A}_n \xrightarrow{[p]} \dots \xrightarrow{[p]} \mathcal{A}_0 := \mathcal{A}_{\mathcal{O}_C}$$

of multiplication by p maps on the abelian scheme $\mathcal{A}_{\mathcal{O}_C}$. Write $\mathcal{A}_\infty := \lim \mathcal{A}_n$ for the inverse limit of this tower, and $\pi : \mathcal{A}_\infty \to \mathcal{A}_0$ for the resulting map to the bottom of the tower; this inverse limit exists as multiplication by p is a finite map on $\mathcal{A}_{\mathcal{O}_C}$ (see [**SP**, Tag 01YX]), and its cohomology with reasonable coefficients (such as the structure sheaf) can be calculated as the direct limit of the cohomologies of the $\mathcal{A}_n$'s.

Now obseve that translating by p^n-torsion points gives an action of $\mathcal{A}[p^n](\mathcal{O}_C) \simeq A[p^n](C)$ on the map $\mathcal{A}_n \to \mathcal{A}_0$; here we use the valuative criterion of properness for the identification $\mathcal{A}[p^n](\mathcal{O}_C) \simeq A[p^n](C)$. Taking inverse limits in n, we obtain an action of $T_p(A)$ on the map π. Taking pullbacks, we obtain a map

$$H^*(\mathcal{A}, \mathcal{O}_\mathcal{A}) \to H^*(\mathcal{A}_\infty, \mathcal{O}_{\mathcal{A}_\infty}).$$

Due to the presence of the group action, the image of this map is contained in the $T_p(A)$-invariants of the target. Thus, we can view preceding map as a map

$$H^*(\mathcal{A}, \mathcal{O}_\mathcal{A}) \to H^0(T_p(A), H^*(\mathcal{A}_\infty, \mathcal{O}_{\mathcal{A}_\infty})),$$

where we use the notation $H^0(G, -)$ for the functor of taking G-invariants for a group G. Deriving this story[5], we obtain a map

(2) $$\Psi : R\Gamma(\mathcal{A}, \mathcal{O}_\mathcal{A}) \to R\Gamma_{conts}(T_p(A), R\Gamma(\mathcal{A}_\infty, \mathcal{O}_{\mathcal{A}_\infty})),$$

where $R\Gamma(\mathcal{A}, \mathcal{O}_\mathcal{A})$ denotes the cohomology of the structure sheaf on $\mathcal{A}$ and $R\Gamma_{conts}(T_p(A), -)$ denotes continuous group cohomology theory for the profinite group $T_p(A)$, both in the sense of derived categories (see [**We**, §10] for a quick introduction). To proceed further, we observe the following vanishing theorem:

PROPOSITION 2.3. *The canonical map $\mathcal{O}_C \to R\Gamma(\mathcal{A}_\infty, \mathcal{O}_{\mathcal{A}_\infty})$ induces an isomorphism modulo any power of p, and hence after p-adic completion.*

PROOF. As $\mathcal{A}$ is an abelian scheme, its cohomology ring $H^*(\mathcal{A}, \mathcal{O}_\mathcal{A})$ is an exterior algebra on $H^1(\mathcal{A}, \mathcal{O}_\mathcal{A})$. Moreover, multiplication by an integer N on $\mathcal{A}$ induces multiplication by N on $H^1(\mathcal{A}, \mathcal{O}_\mathcal{A})$. By combining these observations with the formula

$$H^i(\mathcal{A}_\infty, \mathcal{O}_{\mathcal{A}_\infty}) \simeq \operatorname*{colim}_n H^i(\mathcal{A}_n, \mathcal{O}_{\mathcal{A}_n}) \simeq \operatorname*{colim}_{[p]^*} H^i(\mathcal{A}_{\mathcal{O}_C}, \mathcal{O}_{\mathcal{A}_{\mathcal{O}_C}}),$$

we learn that $H^i(\mathcal{A}_\infty, \mathcal{O}_{\mathcal{A}_\infty})$ is the constants $\mathcal{O}_C$ if $i = 0$, and $H^i(\mathcal{A}_{\mathcal{O}_C}, \mathcal{O}_{\mathcal{A}_{\mathcal{O}_C}})[\frac{1}{p}]$ for $i > 0$. In particular, working modulo any power of p, the latter vanishes, so we get the claim. □

Thus, after p-adic completion, the map Ψ gives a map

$$\widehat{\Psi} : R\Gamma(\mathcal{A}, \mathcal{O}_\mathcal{A}) \to R\Gamma_{conts}(T_p(A), \mathcal{O}_C).$$

On the other hand, as abelian varieties are $K(\pi, 1)$'s, we can interpret the preceding map as a map

$$\widehat{\Psi} : R\Gamma(\mathcal{A}, \mathcal{O}_\mathcal{A}) \to R\Gamma(A_C, \mathcal{O}_C).$$

In particular, applying H^1 and inverting p, we get a map

$$H^1(A, \mathcal{O}_A) \to H^1(A_C, C),$$

---

[5] Let us give a slightly more precisely explanation (see also Exercise 13). Say $f : X \to Y$ is a map of quasi-compact and quasi-separated schemes, and G is a profinite group (regarded as a profinite scheme over **Z**) acting on X equivariantly with respect to the trivial action on Y. Then there is a canonical factorization $R\Gamma(Y, \mathcal{O}_Y) \to R\Gamma_{conts}(G, R\Gamma(X, \mathcal{O}_X))$ of the pullback map $R\Gamma(Y, \mathcal{O}_Y) \to R\Gamma(X, \mathcal{O}_X)$. One way to see this is to observe that the map f acts as $X \to [X/G] \to Y$, where $[X/G]$ is the quotient of X by G in the category of (not necessarily algebraic) stacks for the pro-étale topology. Taking $R\Gamma$ of the structure sheaf for the preceding factorization gives the desired factorization as $R\Gamma([X/G], \mathcal{O}_{[X/G]}) \simeq R\Gamma_{conts}(G, R\Gamma(X, \mathcal{O}_X))$ since $X \to [X/G]$ is a pro-étale G-torsor.

which then linearizes to the promised map
$$\alpha_A : H^1(A, \mathcal{O}_A) \otimes C \to H^1(A_C, C).$$

REMARK 2.4 (Perfectoid abelian varieties). Given any affine open $U \subset \mathcal{A}_{\mathcal{O}_C}$, write $U_\infty \subset \mathcal{A}_\infty$ for its inverse image. Then the p-adic completion R of $\mathcal{O}(U_\infty)$ is an *integral perfectoid* $\mathcal{O}_C$-*algebra*, i.e., R is p-adically complete and p-torsionfree, and the Frobenius induces an isomorphism $R/p^{\frac{1}{p}} \simeq R/p$. In particular, the generic fiber A_∞ of $\mathcal{A}_\infty$ gives a perfectoid space. In fact, the map $A_\infty \to A_C$ is a pro-étale $T_p(A)$-torsor. This construction may be viewed as the analog for abelian varieties of the perfectoid torus from Example 3.27 below.

2.3. Fontaine's construction of the map β_A. For Fontaine's construction, we need the following fact about the arithmetic of p-adic fields:

THEOREM 2.5 (Differential forms on $\mathcal{O}_C$). *Write Ω for the Tate module of $\Omega^1_{\mathcal{O}_C/\mathcal{O}_K}$. This $\mathcal{O}_C$-module is free of rank 1. Moreover, there is a Galois equivariant isomorphism $C(1) \simeq \Omega[\frac{1}{p}]$.*

CONSTRUCTION OF THE MAP GIVING THE ISOMORPHISM. Consider the $d\log$ map
$$\mu_{p^\infty}(\mathcal{O}_C) \subset \mathcal{O}_C^* \to \Omega^1_{\mathcal{O}_C/\mathcal{O}_K}$$
given by $f \mapsto \frac{df}{f}$. On passage to Tate modules and linearizations, this gives a map
$$T_p(\mu_{p^\infty}(\mathcal{O}_C)) \otimes_{\mathbf{Z}_p} \mathcal{O}_C = \mathbf{Z}_p(1) \otimes_{\mathbf{Z}_p} \mathcal{O}_C = \mathcal{O}_C(1) \to \Omega.$$
Fontaine proves this map is injective with torsion cokernel, giving $C(1) \simeq \Omega[\frac{1}{p}]$; see [**Fo1**, §1] for Fontaine's proof, and [**Be**, §1.3] for a slicker (but terse) argument using the cotangent complex. □

REMARK 2.6 (The cotangent complex of $\mathcal{O}_C$). Theorem 2.5 also extends to the cotangent complex after a shift: one has $\widehat{L_{\mathcal{O}_C/\mathbf{Z}_p}} \simeq \Omega[1]$, where the completion on the left side is the derived p-adic completion. Although this can be deduced directly from Theorem 2.5, we do not explain this here; instead, we refer to Remark 3.19 where a more general statement is proven. This assertion will be useful later in constructing the Hodge-Tate filtration.

In particular, this result helps connect the Tate twist $C(1)$ (which lives on the Galois side of the story) to differential forms (which lie on the de Rham side). Using this, Fontaine's idea for constructing the map β_A is to pullback differential forms on $\mathcal{A}$ to those on $\mathcal{O}_C$ using points in $\mathcal{A}(\mathcal{O}_C)$. More precisely, this pullback gives a pairing
$$H^0(\mathcal{A}, \Omega^1_{\mathcal{A}/\mathcal{O}_K}) \otimes \mathcal{A}(\mathcal{O}_C) \to \Omega^1_{\mathcal{O}_C/\mathcal{O}_K}.$$
Passing to p-adic Tate modules, this gives a pairing
$$H^0(\mathcal{A}, \Omega^1_{\mathcal{A}/\mathcal{O}_K}) \otimes T_p(A) \to \Omega.$$
Using the identification $T_p(A) \simeq H^1(A_{C,et}, \mathbf{Z}_p)^\vee$, this gives a map
$$H^0(\mathcal{A}, \Omega^1_{\mathcal{A}/\mathcal{O}_K}) \to H^1(A_{C,et}, \mathbf{Z}_p) \otimes \Omega.$$
Inverting p and using Theorem 2.5, we get the map
$$H^0(A, \Omega^1_{A/K}) \to H^1(A_C, C)(1).$$

Linearizing and twisting gives the desired map
$$\beta_A : H^0(A, \Omega^1_{A/K}) \otimes C(-1) \to H^1(A_C, C).$$

2.4. Conclusion. Taking direct sums of the previous two constructions gives the map $\gamma_A = \alpha_A \oplus \beta_A$
$$\gamma_A : \bigl(H^1(A, \mathcal{O}_A) \otimes C\bigr) \oplus \bigl(H^0(A, \Omega^1_{A/K})(-1) \otimes C\bigr) \to H^1(A_C, C).$$

Each of the parenthesized summands on the left has dimension g, while the target has dimension on $2g$. Thus, to show γ_A is an isomorphism, it is enough to show injectivity. Moreover, by Tate's calculations in Theorem 1.8, it is enough to show that α_A and β_A are separately injective: the map γ_A is Galois equivariant, and the two summands have different Galois actions, so they cannot talk to each other. For Fontaine's map β_A, this follows by a formal group argument as it is enough to check the corresponding assertion for the formal group of $\mathcal{A}$. For the map α_A, we are not aware of a direct argument that does not go through one of the proofs of the p-adic comparison theorems. For lack of space, we do not give either argument here.

3. Lecture 3: The Hodge-Tate decomposition in general

Let C be a complete and algebraically closed extension of $\mathbf{Q}_p$. Let X/C be a smooth rigid-analytic space[6]. Our goal is to relate the étale cohomology of X to differential forms. More precisely, setting
$$H^n(X, C) := H^n(X_{et}, \mathbf{Z}_p) \otimes_{\mathbf{Z}_p} C,$$
we want to prove the following result (contained in Theorem 1.16 and Remark 1.17):

THEOREM 3.1 (Hodge-Tate spectral sequence). *Assume X is proper. Then there exists a (degenerate) E_2-spectral sequence*
$$E_2^{i,j} : H^i(X, \Omega^j_{X/C})(-j) \Rightarrow H^{i+j}(X, C).$$
The resulting filtration on $H^n(X, C)$ is called the Hodge-Tate filtration.

In this lecture, we shall explain the construction of this spectral sequence. The starting point of this relation between étale cohomology and differential forms is the completed structure sheaf $\widehat{\mathcal{O}_X}$ on the pro-étale site X_{proet} of X; these objects are defined in §3.2. To a first approximation, objects of X_{proet} may be viewed as towers $\{U_i\}$ of finite étale covers with $U_0 \to X$ étale, and $\widehat{\mathcal{O}_X}$ is the sheaf which assigns to such a tower the completion of the direct limit of the rings of analytic functions on the U_i's. In particular, this is a sheaf of C-algebras. The following comparison theorem [**Sc2**, Theorem 5.1] relates the cohomology of $\widehat{\mathcal{O}_X}$ to more topological invariants:

THEOREM 3.2 (Primitive comparison theorem). *If X is proper, then the inclusion $C \subset \widehat{\mathcal{O}_X}$ of the constants gives an isomorphism*
$$H^*(X, C) \simeq H^*(X_{proet}, \widehat{\mathcal{O}_X}).$$

---

[6]All results in this section are due to Scholze unless otherwise specified. When X is a smooth proper variety, some of the results were proven by Faltings [**Fa1, Fa4**] in a different language. When discussing the étale cohomology of adic spaces, we are implicitly using Huber's theory [**Hu2**]. When X arises as the analytification of an algebraic variety Y, Huber's étale cohomology groups agree with those of Y, so we can draw consequences for the algebraic theory as well.

Thus, to prove Theorem 3.1, it suffices to work with $H^n(X_{proet}, \widehat{\mathcal{O}_X})$ instead of $H^n(X, C)$, thus putting both sides of Theorem 3.1 into the realm of coherent cohomology. To proceed further, we recall that there is a canonical projection map

$$\nu : X_{proet} \to X_{et}$$

from the pro-étale site of X to the étale site of X; recall that a morphism of sites goes in the other direction from the underlying functor of categories, and the latter for ν simply captures the fact that an étale morphism is pro-étale. Theorem 3.1 arises from the Leray spectral sequence for ν using the following:

THEOREM 3.3 (Hodge-Tate filtration: local version). *There is a canonical isomorphism* $\Omega^1_{X/C}(-1) \simeq R^1\nu_*\widehat{\mathcal{O}_X}$. *Taking products, this gives isomorphisms* $\Omega^j_{X/C}(-j) \simeq R^j\nu_*\widehat{\mathcal{O}_X}$.

In the rest of this lecture, we will sketch a proof of this result. More precisely, §3.2 contains some reminders on the pro-étale site, especially its locally perfectoid nature. This is then used in §3.4 to construct the map giving the isomorphism of Theorem 3.3; this construction relies on the cotangent complex (whose basic theory is reviewed in §3.1), and differs from that in [**Sc2**]. Once the map has been constructed, we check that it is an isomorphism in §3.5 using the almost acyclicity of the structure sheaf for affinoid perfectoid spaces.

REMARK 3.4 (Hodge and Hodge-Tate filtrations). As mentioned in Remark 1.17, the differentials in Theorem 3.1 are, in fact, always 0, and thus one always has *some* Hodge-Tate decomposition as in Theorem 1.7. This result is explained in [**BMS2**, §13] and relies on the work of Conrad-Gabber [**CG**] on spreading out rigid-analytic families to reduce to the corresponding assertion over discretely valued fields. However, one cannot construct a Hodge-Tate decomposition as in Theorem 1.7 that varies p-adic analytically in X. Concretely, when X is an abelian variety, one has a canonical map $H^1(X, \mathcal{O}_X) \to H^1(X, C)$ as explained in §2, giving a piece of the Hodge-Tate filtration on $H^1(X, C)$; however, one cannot choose a splitting $H^1(X, C) \to H^1(X, \mathcal{O}_X)$ in a manner that is compatible in families of abelian varieties. Instead, the variation of the Hodge-Tate filtration in a family of abelian varieties provides a highly non-trivial and interesting invariant of the family: the Hodge-Tate period map from [**Sc4**, §III.3].

The above discussion is analogous to the following (perhaps more familiar and) more classical story over **C** (see [**Vo**, §10]): even though the Hodge-to-de Rham spectral sequence always degenerates for a smooth projective variety, one cannot choose a Hodge decomposition for smooth projective varieties that varies holomorphically in a family. Instead, it is the Hodge *filtration* on de Rham cohomology that varies holomorphically. In fact, the variation of this filtration in a family of smooth projective varieties provides an extremely important invariant of the family: the period map to the classifying space for Hodge structures.

REMARK 3.5 (The first obstruction to splitting the Hodge-Tate filtration). Under the primitive comparison theorem (Theorem 3.2), the Hodge-Tate filtration on $H^*(X, C)$ is induced by the canonical filtration on the complex $R\nu_*\widehat{\mathcal{O}_X}$. In this remark, we discuss the first obstruction to splitting this filtration locally on X. Consider the complex $K := \tau^{\leq 1} R\nu_*\widehat{\mathcal{O}_X}$. This complex has 2 nonzero cohomology sheaves (identified by Theorem 3.3), and thus sits in an exact triangle

$$\mathcal{O}_X \to K \to \Omega^1_{X/C}(-1)[-1].$$

The boundary map for this exact triangle is a map

$$\Omega^1_{X/C}(-1)[-1] \to \mathcal{O}_X[1],$$

and thus gives an element of $\mathrm{ob}_X \in \mathrm{Ext}^2_X(\Omega^1_{X/C}, \mathcal{O}_X(1))$. Unwinding definitions, the element ob_X measures the obstruction to finding a map $K \to \mathcal{O}_X$ splitting the map $\mathcal{O}_X \simeq R^0\nu_*\widehat{\mathcal{O}_X} \to K$. In the rest of this remark, we describe this element slightly more geometrically using some formalism of rigid geometry and p-adic Hodge theory; the reader unfamiliar with these notions should feel free to skip this remark.

Recall that Fontaine has constructed [**Fo2**, §1.2] a canonical surjection $\theta : A_{\mathrm{inf}}(\mathcal{O}_C) \to \mathcal{O}_C$ whose kernel is generated by a nonzerodivisor (see Remark 3.14 for a description adapted to these notes). This gives an exact sequence

$$0 \to \ker(\theta)/\ker(\theta)^2 \to A_{\mathrm{inf}}(\mathcal{O}_C)/\ker(\theta)^2 \xrightarrow{\overline{\theta}} A_{\mathrm{inf}}(\mathcal{O}_C)/\ker(\theta) \simeq \mathcal{O}_C \to 0.$$

The map $\overline{\theta}$ is a non-trivial square-zero extension of the commutative ring $\mathcal{O}_C$ by the invertible $\mathcal{O}_C$-module $\ker(\theta)/\ker(\theta)^2$. In fact, one can also show that this extension does not split over any subring $\mathcal{O}' \subset \mathcal{O}_C$ with $\mathcal{O}_C/\mathcal{O}'$ killed by a fixed power of p. It follows from the formalism of "passing to the generic fibre" that the map $\overline{\theta}[1/p]$ is a non-trivial square-zero extension of the Tate algebra C. One can then show (using the method of §3.4) that ob_X is precisely the obstruction to lifting X across the thickening $\overline{\theta}[\frac{1}{p}]$, thus giving some geometric meaning to the non-canonicity of the Hodge-Tate decomposition mentioned in Remark 3.4[7]. The characteristic p analog of this is the Deligne-Illusie result, recalled next.

REMARK 3.6 (Deligne-Illusie obstructions to liftability). Remark 3.5 is analogous to a more classical picture from [**DI**], which we recall. Let k be a perfect field of characteristic p, and let X/k be a smooth k-scheme. Consider the truncated de Rham complex $K := \tau^{\leq 1}\Omega^\bullet_{X/k}$. Note that the differentials in the de Rham complex $\Omega^\bullet_{X/k}$ are linear over k and the p-th powers $\mathcal{O}^p_X$ of functions on X. Thus, we may view $\Omega^\bullet_{X/k}$ (and thus K) as a complex of coherent sheaves on the Frobenius twist $X^{(1)}$ of X relative to k. By a theorem of Cartier, one has $\mathcal{H}^i(\Omega^\bullet_{X/k}) \simeq \Omega^i_{X^{(1)}/k}$. Thus, the complex K sits in an exact triangle

$$\mathcal{O}_{X^{(1)}} \to K \to \Omega^1_{X^{(1)}/k}[-1].$$

The boundary map for this triangle is a map

$$\Omega^1_{X^{(1)}/k}[-1] \to \mathcal{O}_{X^{(1)}}[1],$$

and can thus be viewed as an element $\mathrm{ob}_X \in \mathrm{Ext}^2_{X^{(1)}}(\Omega^1_{X^{(1)}/k}, \mathcal{O}_{X^{(1)}})$. One of the main observations of [**DI**] is that ob_X is precisely the obstruction to lifting the k-scheme $X^{(1)}$ along the square-zero extension $W_2(k) \to k$ of k.

---

[7]Forthcoming work of Conrad-Gabber [**CG**] shows that this obstruction class is always 0, at least when X is assumed to be proper. Nevertheless, this class admits an integral analog, which can be nonzero; see Remark 3.22.

3.1. The cotangent complex and perfectoid rings. We recall the construction and basic properties of the cotangent complex; much more thorough accounts can be found in [**Qu2**, **Ill1**, **Ill2**] and [**SP**, Tag 08P5]. Once the basics have been introduced, we shall explain some applications to the perfectoid theory; the key point is that maps between perfectoids are formally étale in a strong sense, and this perspective helps conceptualize certain results about them (such as the tilting correspondence and Fontaine's calculation of differential forms in Theorem 2.5) better. We begin with the following construction from non-abelian homological algebra:

CONSTRUCTION 3.7 (Quillen). For any ring A and a set S, we write $A[S]$ for the polynomial algebra over A on a set of variables x_s indexed by $s \in S$. The functor $S \mapsto A[S]$ is left adjoint to the forgetful functor from A-algebras to sets. In particular, for any A-algebra B, we have a canonical map $\eta_B : A[B] \to B$, which is evidently surjective. Repeating the construction, we obtain two natural A-algebra maps $\eta_{A[B]}, A[\eta_B] : A[A[B]] \to A[B]$. Iterating this process allows one to define a simplicial A-algebra $P^\bullet_{B/A}$ augmented over B that looks like

$$P^\bullet_{B/A} := \Big(...A[A[A[B]]] \Rrightarrow A[A[B]] \rightrightarrows A[B]\Big) \longrightarrow B.$$

This map is a resolution of B in the category of simplicial A-algebras, and is called the canonical simplicial A-algebra resolution of B; concretely, this implies that the chain complex underlying $P^\bullet_{B/A}$ (obtained by taking an alternating sum of the face maps as differentials) is a free resolution of B over A. Slightly more precisely, there is a model category of simplicial A-algebras, and the factorization $A \to P^\bullet_{B/A} \to B$ provides a functorial cofibrant replacement of B, and can thus be used to calculate non-abelian derived functors. We do not discuss this theory here, and will take certain results (such as the fact that such polynomial A-algebra resolutions are unique up to a suitable notion of homotopy) as blackboxes; a thorough discussion, in the language of model categories, can be found in [**Qu1**].

Using the previous construction, the main definition is:

DEFINITION 3.8 (Quillen). For any map $A \to B$ of commutative rings, we define its cotangent complex $L_{B/A}$, which is a complex of B-modules and viewed as an object of the derived category $D(B)$ of all B-modules, as follows: set $L_{B/A} := \Omega^1_{P^\bullet/A} \otimes_{P^\bullet} B$, where $P^\bullet \to B$ is a simplicial resolution of B by polynomial A-algebras. Here we view the simplicial B-module $\Omega^1_{P^\bullet/A} \otimes_{P^\bullet} B$ as a B-complex by taking an alternating sum of the face maps as a differential.

For concreteness and to obtain a strictly functorial theory, one may choose the canonical resolution $P^\bullet_{B/A}$ in the definition above. However, in practice, just like in homological algebra, it is important to allow the flexibility of changing resolutions without changing $L_{B/A}$ (up to quasi-isomorphism). The following properties can be checked in a routine fashion, and we indicate a brief sketch of the argument:

(1) *Polynomial algebras.* If B is a polynomial A-algebra, then $L_{B/A} \simeq \Omega^1_{B/A}[0]$: this follows because any two polynomial A-algebra resolutions of B are homotopic to each other, so we may use the constant simplicial A-algebra with value B to compute $L_{B/A}$.

(2) *Künneth formula.* If B and C are flat A-algebras, then $L_{B \otimes_A C/A} \simeq L_{B/A} \otimes_A C \oplus B \otimes_A L_{C/A}$: this reduces to the case of polynomial algebras by passage to resolutions. The flatness hypothesis gets used in concluding that if $P^\bullet \to B$ and $Q^\bullet \to C$ are polynomial A-algebra resolutions, then $P^\bullet \otimes_A Q^\bullet \to B \otimes_A C$ is also a polynomial A-algebra resolution. (In fact, this reasoning shows that the flatness hypotheses can be relaxed to the assumption $\mathrm{Tor}^A_{>0}(B,C) = 0$ provided one uses derived tensor products of chain complexes in the formula above.)

(3) *Transitivity triangle.* Given a composite $A \to B \to C$ of maps, we have a canonical exact triangle

$$L_{B/A} \otimes^L_B C \to L_{C/A} \to L_{C/B}$$

in $D(C)$. To prove this, one first settles the case where $A \to B$ and $B \to C$ are polynomial maps (which reduces to a classical fact in commutative algebra). The general case then follows by passage to the canonical resolutions as the exact sequences constructed in the previous case were functorial.

(4) *Base change.* Given a flat map $A \to C$ and an arbitrary map $A \to B$, we have $L_{B/A} \otimes_A C \simeq L_{B \otimes_A C/C}$. Again, one first settles the case of polynomial rings, and then reduces to this by resolutions, using flatness to reduce a derived base change to a classical one. (Again, this reasoning shows that the flatness hypothesis can be relaxed to the assumption $\mathrm{Tor}^A_{>0}(B,C) = 0$ provided one uses derived tensor products of chain complexes in the formula above. More generally, all assumptions can be dropped if one replaces $B \otimes_A C$ with $B \otimes^L_A C$, interpreted as a simplicial commutative ring.)

(5) *Vanishing for étale maps.* We claim that if $A \to B$ is étale, then $L_{B/A} \simeq 0$. For this, assume first that $A \to B$ is a Zariski localization. Then $B \otimes_A B \simeq B$, so (2) implies that $L_{B/A} \oplus L_{B/A} \simeq L_{B/A}$ via the sum map. This immediately gives $L_{B/A} = 0$ for such maps. In general, as $A \to B$ is étale, the multiplication map $B \otimes_A B \to B$ is a Zariski localization, and thus $L_{B/B \otimes_A B} \simeq 0$. By the transitivity triangle for $B \xrightarrow{i_1} B \otimes_A B \to B$, this yields $L_{B \otimes_A B/B} \otimes_{B \otimes_A B} B \simeq 0$. But, by (4), we have $L_{B \otimes_A B/B} \simeq L_{B/A} \otimes_A B$, so the base change of $L_{B/A}$ along $A \to B \to B \otimes_A B \to B$ vanishes. The latter is just the structure map $A \to B$, so $L_{B/A} \otimes_A B \simeq 0$. The standard map $L_{B/A} \to L_{B/A} \otimes_A B$ has a section coming from the B-action on $L_{B/A}$, so $L_{B/A} \simeq 0$.

(6) *Étale localization.* If $B \to C$ is an étale map of A-algebras, then $L_{B/A} \otimes_B C \simeq L_{C/A}$: this follows from (3) and (5) as $L_{C/B} \simeq 0$.

(7) *Relation to Kähler differentials.* For any map $A \to B$, we have $H^0(L_{B/A}) \simeq \Omega^1_{B/A}$. This can be shown directly from the definition.

(8) *Smooth algebras.* If $A \to B$ is smooth, then $L_{B/A} \simeq \Omega^1_{B/A}[0]$. By (6), there is a natural map $L_{B/A} \to \Omega^1_{B/A}[0]$. To show this is an isomorphism, we may work locally on A by (6). In this case, there is an étale map $B' := A[x_1, ..., x_n] \to B$. We know that $L_{B'/A} \simeq \Omega^1_{B'/A}[0]$ by (1) and $L_{B/B'} \simeq 0$ by (6). By (3), it follows that $L_{B/A} \simeq L_{B'/A} \otimes_{B'} B \simeq \Omega^1_{B/A}[0]$.

We give an example of the use of these properties in a computation.

EXAMPLE 3.9 (Cotangent complex for a complete intersection). Let R be a ring, let $I \subset R$ be an ideal generated by a regular sequence, and let $S = R/I$. Then we claim that $L_{S/R} \simeq I/I^2[1]$. In particular, this is a *perfect complex*, i.e., quasi-isomorphic to a finite complex of finite projective modules. To see this isomorphism, consider first the case $R = \mathbf{Z}[x_1, ..., x_r]$ and $I = (x_i)$. In this case, $S = \mathbf{Z}$, and the transitivity triangle for $\mathbf{Z} \to R \to S$ collapses to give $L_{S/R} \simeq \Omega^1_{R/\mathbf{Z}} \otimes_R S[1] \simeq I/I^2[1]$, where the isomorphism $I/I^2 \to \Omega^1_{R/\mathbf{Z}} \otimes_R S$ is defined by $f \mapsto df$. For general R, once we choose a regular sequence $f_1, ... f_r$ generating I, we have a pushout square of commutative rings

$$\begin{array}{ccc} \mathbf{Z}[x_1,...,x_r] & \xrightarrow{x_i \mapsto f_i} & R \\ \downarrow {\scriptstyle x_i \mapsto 0} & & \downarrow \\ \mathbf{Z} & \longrightarrow & S. \end{array}$$

As the f_i's form a regular sequence, this is also a derived pushout square, i.e., $\mathrm{Tor}^{\mathbf{Z}[x_1,...,x_r]}_{>0}(R, \mathbf{Z}) = 0$. Base change for the cotangent complex implies that $L_{S/R} \simeq L_{\mathbf{Z}/\mathbf{Z}[x_1,...,x_r]} \otimes_\mathbf{Z} S \simeq I/I^2[1]$.

Assume now that with R, I, S as above, the ring R is smooth over a base ring k. Then $L_{R/k} \simeq \Omega^1_{R/k}$ is locally free. The transitivity triangle for $k \to R \to S$ then tells us that $L_{S/k}$ is computed by the following 2-term complex of locally free S-modules:

$$I/I^2 \xrightarrow{f \mapsto df} \Omega^1_{R/k} \otimes_R S.$$

Here the identification of the differential involves unraveling some of the identifications above. In particular, $L_{S/k}$ is also a perfect complex. Conversely, it is a deep theorem of Avramov (conjectured by Quillen) that if k is a field and $L_{S/k}$ is perfect for a finite type k-algebra S, then S is a complete intersection.

REMARK 3.10 (Naive cotangent complex). For most applications in algebraic geometry and number theory (including all that come up in these notes), it suffices to work with the truncation $\tau^{\geq -1} L_{B/A}$. This is a complex of B-modules with (at most) two non-zero cohomology groups in degrees -1 and 0. It can be constructed explicitly using a presentation: if $A \to B$ factors as $A \to P \to B$ with $A \to P$ a polynomial algebra and $P \to B$ surjective with kernel I, then we have

$$\tau^{\geq -1} L_{B/A} := \left(I/I^2 \xrightarrow{f \mapsto df} \Omega^1_{P/A} \otimes_P B \right).$$

This object is sometimes called the *naive cotangent complex*, and its basic theory is developed in [**SP**, 00S0]. Despite the elementary definition, it is sometimes awkward to work with the truncated object, so we stick to the non-truncated version in these notes.

The main reason to introduce the cotangent complex is that it controls deformation theory in complete generality, analogous to how the tangent bundle controls deformations of smooth varieties; roughly speaking, the functors $\mathrm{Ext}^i_B(L_{B/A}, -)$ control square-zero extensions of B in the category of A-algebras. We do not discuss this formalism in depth here, and instead simply record the following consequence.

THEOREM 3.11 (Deformation invariance of the category of formally étale algebras). *For any ring A, write $\mathcal{C}_A$ for the category of flat A-algebras B such that $L_{B/A} \simeq 0$. Then for any surjective map $\tilde{A} \to A$ with nilpotent kernel, base change* induces an equivalence $\mathcal{C}_{\tilde{A}} \simeq \mathcal{C}_A$. In other words, every $A \to B$ in $\mathcal{C}_A$ lifts uniquely (up to unique isomorphism) to $\tilde{A} \to \tilde{B}$ in $\mathcal{C}_{\tilde{A}}$.

Any étale A-algebra B is an object of $\mathcal{C}_A$; conversely, every finitely presented A-algebra B in $\mathcal{C}_A$ is étale over A (see [**SP**, Tag 0D12] for a more general assertion). Thus, for such maps, Theorem 3.11 captures the topological invariance of the étale site (see [**SP**, Tag 04DZ]). However, the finite presentation hypothesis is too restrictive for applications in the perfectoid theory; instead, the following class of examples is crucial:

PROPOSITION 3.12. *Assume A has characteristic p. Let $A \to B$ be a flat map that is relatively perfect, i.e., the relative Frobenius $F_{B/A} : B^{(1)} := B \otimes_{A, F_A} A \to B$ is an isomorphism. Then $L_{B/A} \simeq 0$.*

PROOF. We first claim that for any A-algebra B, the relative Frobenius induces the 0 map $L_{F_{B/A}} : L_{B^{(1)}/A} \to L_{B/A}$: this is clear when B is a polynomial A-algebra (as $d(x^p) = 0$), and thus follows in general by passage to the canonical resolutions. Now if $A \to B$ is relatively perfect, then $L_{F_{B/A}}$ is also an isomorphism by functoriality. Thus, the 0 map $L_{B^{(1)}/A} \to L_{B/A}$ is an isomorphism, so $L_{B/A} \simeq 0$. □

This leads to the following conceptual description of the Witt vector functor:

EXAMPLE 3.13 (Witt vectors via deformation theory). Let R be a perfect ring of characteristic p. Then R is relatively perfect over $\mathbf{Z}/p$. Proposition 3.12 tells us that $L_{R/\mathbf{F}_p} \simeq 0$, so Theorem 3.11 implies that R has a flat lift R_n to $\mathbf{Z}/p^n$ for any $n \geq 1$, and that this lift is unique up to unique isomorphism. In fact, this lift is simply given by the Witt vector construction $W_n(R)$. Setting $W(R) = \lim_n W_n(R)$ gives the Witt vectors of R, which can also be seen as the unique p-adically complete p-torsionfree $\mathbf{Z}_p$-algebra lifting R. This perspective also allows one to see some additional structures on $W(R)$. For example, the map $R \to R$ of multiplicative monoids lifts uniquely across any map $W_n(R) \to R$: the monoid R is uniquely p-divisible, while the fiber over $1 \in R$ of $W_n(R) \to R$ is p-power torsion. Explicitly, one simply sends $r \in R$ to $\widetilde{r_n}^{p^n}$, where $\widetilde{r_n} \in W_n(R)$ denotes some lift of $r_n := r^{\frac{1}{p^n}}$. The resulting multiplicative maps $R \to W_n(R)$ and $R \to W(R)$ are called the Teichmüller lifts, and denoted by $r \mapsto [r]$.

REMARK 3.14 (Fontaine's A_{inf} and the map θ). Let us use the above discussion to introduce Fontaine's A_{inf}-construction (from [**Fo2**] and mentioned earlier in Remak 3.5). Fix a ring A and a map $A \to B$ in $\mathcal{C}_A$. With a bit more care in analyzing deformation theory via the cotangent complex (see [**SP**, Tag 0D11]), one can show the following lifting feature: if $C' \to C$ is a surjective A-algebra map with a nilpotent kernel, then every A-algebra map $B \to C$ lifts uniquely to an A-algebra map $B \to C'$. In particular, given a p-adically complete $\mathbf{Z}_p$-algebra C, a perfect ring D, and a map $D \to C/p$, we obtain a unique lift $W_n(D) \to C/p^n$ of the composition $W_n(D) \to D \to C/p$ for each n. Taking limits, we obtain unique map $W(D) \to C$ lifting the map $W(D) \to C/p$ arising via $W(D) \to D \to C/p$. Applying this in a universal example of such a D for a given C, we obtain Fontaine's map θ from [**Fo2**] via abstract nonsense:

PROPOSITION 3.15. *Given any p-adically complete ring R, the canonical projection map $\overline{\theta} : R^{\flat} := \lim_{\phi} R/p \to R/p$ lifts to a unique map $\theta : W(R^{\flat}) \to R$. The ring $W(R^{\flat})$ is also called $A_{\text{inf}}(R)$.*

Note that the map $\overline{\theta} : R^\flat \to R/p$ is surjective exactly when R/p is *semiperfect*, i.e., has a surjective Frobenius. In this case, the map θ is also surjective by p-adic completeness.

We next explain the relevance of these ideas to the perfectoid theory. As the definition of perfectoid algebras varies somewhat depending on context, we define the notion we need (see [**BMS2**, §3.2] for more on such rings), using the map θ introduced above:

DEFINITION 3.16. A ring R is *integral perfectoid* if it satisfies the following conditions:
(1) R is π-adically complete for some element π with $\pi^p \mid p$.
(2) The ring R/p has a surjective Frobenius.
(3) The kernel of $A_{inf}(R) := W(R^\flat) \to R$ is principal.

Note that being integral perfectoid is a property of the ring R as an abstract ring (as opposed to a topological ring, or an algebra over some other fixed ring). Important examples include the rings of integers of perfectoid fields (in the sense of [**Sc1**, Definition 3.1]), and any perfect ring of characteristic p. In fact, if C is a perfectoid field of characteristic 0, then a p-adically complete and p-torsionfree $\mathcal{O}_C$-algebra R is integral perfectoid exactly when the map $\mathcal{O}_C/p \to R/p$ is relatively perfect in the sense of Proposition 3.12.

REMARK 3.17. It takes a little practice to appreciate the strength of the conditions appearing in Definition 3.16. For example, condition (1) rules out the ring $\mathbf{Z}_p$, conditions (1) and (2) together rule out finite extensions of $\mathbf{Z}_p$, and condition (3) ensures that the perfectoid property does not trivially pass to (π-adically complete) quotients. We refer the reader to other references on perfectoid rings (such as [**BMS2**, §3], [**Mor**, §3], [**BIM**, §3], [**La**, §8]) for more thorough discussions.

REMARK 3.18 (Tilting). For an integral perfectoid ring R, the map $\theta : A_{inf}(R) \to R$ from Proposition 3.15 fits into the following commutative diagram

$$\begin{array}{ccc} A_{inf}(R) & \xrightarrow{\theta} & R \\ \downarrow & & \downarrow \\ R^\flat & \xrightarrow{\overline{\theta}} & R/p, \end{array}$$

where each map can be regarded as a pro-infinitesimal thickening of the target by the perfectoidness assumption. In particular, all 4 rings are pro-infinitesimal thickenings of R/p. Theorem 3.11 and Proposition 3.12 may then be used to prove half of the tilting correspondence from [**Sc1**, Theorem 5.2].

We make some remarks on the differential aspects of perfectoid rings. For this, we use the language of derived p-adic completions (see [**SP**, Tag 091N]) as well as Exercise 7 in §5.

REMARK 3.19 (Formally étale nature of A_{inf} and differential forms). Let A be a perfect ring of characteristic p. By Example 3.13, the map $\mathbf{Z}_p \to W(A)$ satisfies the following crucial feature: the cotangent complex $L_{A/\mathbf{F}_p} \simeq 0$, so the p-adic completion $\widehat{L_{W(A)/\mathbf{Z}_p}}$ vanishes by base change for cotangent complexes and Nakayama's lemma for p-adically complete complexes. By the transitivity triangle, for any $W(A)$-algebra R, we have $\widehat{L_{R/\mathbf{Z}_p}} \simeq \widehat{L_{R/W(A)}}$. Now specialize to the case where R is an integral perfectoid ring and $A = R^\flat$, with R viewed as an algebra over $W(A) = A_{inf}(R^\flat)$ via θ. Then we learn that

$$\widehat{L_{R/\mathbf{Z}_p}} \simeq \widehat{L_{R/A_{inf}(R^\flat)}}.$$

But the map $\theta : A_{inf}(R^\flat) \to R$ is a quotient by a nonzerodivisor in $A_{inf}(R^\flat)$ (see [**BMS2**, Lemma 3.10 (i)] for a proof). Using Example 3.9, this tells us that

$$\widehat{L_{R/\mathbf{Z}_p}} \simeq \ker(\theta)/\ker(\theta)^2[1].$$

In particular, this is a free R-module of rank 1.

Let us specialize the above discussion to the ring $R := \mathcal{O}_C$ of Theorem 2.5. In the notation there, the above discussion recovers the freeness assertion from Theorem 2.5 as one can show that $\widehat{L_{R/\mathbf{Z}_p}} \simeq \Omega[1]$ using Exercises 7 and 12. To recover the precise Galois module structure given in Theorem 2.5, one proves the following compatibility (which we leave here as an instructive exercise):

(∗) If $\underline{\epsilon} := (1, \epsilon_p, \epsilon_{p^2}, ...) \in \mathcal{O}_C^\flat$ is a compatible sequence of p-power roots of 1 with $\epsilon_{p^n} \in \mu_{p^n}(\mathcal{O}_C)$ and $\epsilon_p \neq 1$, then the element $\mu := [\underline{\epsilon}] - 1$ lies in $\ker(\theta)$, and its image under the isomorphism $\ker(\theta)/\ker(\theta)^2 \simeq \Omega$ mentioned above is exactly the compatible system $\{d\log(\epsilon_{p^n})\}_{n \geq 1} \in \Omega$ appearing in the discussion following Theorem 2.5.

In particular, it follows that μ generates (Galois equivariantly) a copy of $\mathcal{O}_C(1)$ in the invertible $\mathcal{O}_C$-module Ω. Inverting p then gives the isomorphism $\Omega[1/p] \simeq C(1)$ predicted by Theorem 2.5.

REMARK 3.20 (Breuil-Kisin twists). For future reference, we remark that the $\mathcal{O}_C$-module

$$\Omega := T_p(\Omega^1_{\mathcal{O}_C/\mathbf{Z}_p}) \simeq \widehat{L_{\mathcal{O}_C/\mathbf{Z}_p}}[-1] \simeq \ker(\theta)/\ker(\theta)^2$$

is a canonically defined invertible $\mathcal{O}_C$-module (as it is abstractly free of rank 1), and we shall write

$$M \mapsto M\{i\} := M \otimes_{\mathcal{O}_C} \Omega^{\otimes i}$$

for the corresponding twisting operation on $\mathcal{O}_C$-modules; when M carries a Galois action, so does the twist. These objects are called the *Breuil-Kisin twists* of M, and are related to the Tate twist via an inclusion $M(i) \subset M\{i\}$ for $i \geq 0$ with a torsion cokernel (see the construction following Theorem 2.5 for an explanation of the origin of this inclusion). Slightly more generally, the same discussion applies when $\mathcal{O}_C$ is replaced by an integral perfectoid ring R to define a twisting operation $M \mapsto M\{1\} := M \otimes_R \ker(\theta)/\ker(\theta)^2$ on R-modules (but one loses the analog of the inclusion $M(1) \subset M\{1\}$ available for $R = \mathcal{O}_C$).

One fruitful viewpoint on integral perfectoid rings is to view them as integral analogs of perfect rings: they share some of the miraculous properties of perfect characteristic p rings without themselves having characteristic p. This perspective leads one to predict certain results in mixed characteristic, and we explain how this plays out for the Deligne-Illusie theorem in the next two remarks.

REMARK 3.21 (Deligne-Illusie, revisited). Let k be a perfect field of characteristic p. Then $k \simeq W(k)/p$, so $L_{k/W(k)} \simeq k[1]$ by Example 3.9. Now consider a smooth k-algebra R. The transitivity triangle for $W(k) \to k \to R$ is

$$L_{k/W(k)} \otimes_k R \to L_{R/W(k)} \to L_{R/k}.$$

Using the smoothness of R and the previous computation of $L_{k/W(k)}$, this simplifies to
$$R[1] \to L_{R/W(k)} \to \Omega^1_{R/k}.$$
As this construction is functorial in R, we may sheafify it to obtain the following: for smooth k-scheme X, we have a functorial exact triangle
$$\mathcal{O}_X[1] \to L_{\mathcal{O}_X/W(k)} \to \Omega^1_{X/k}.$$
In particular, the boundary map for this triangle is
$$\Omega^1_{X/k} \to \mathcal{O}_X[2],$$
and can thus be identified as a class
$$\mathrm{ob}_{X/W(k)} \in \mathrm{Ext}^2(\Omega^1_{X/k}, \mathcal{O}_X).$$
Using the deformation-theoretic interpretation of the cotangent complex (and unravelling Example 3.9), one can show that $\mathrm{ob}_{X/W(k)}$ is precisely the obstruction to lifting X to $W_2(k)$. The main theorem of Deligne-Illusie [**DI**] is that the obstruction class ob_X constructed in Remark 3.6 via the de Rham complex coincides with $\mathrm{ob}_{X^{(1)}/W(k)}$; equivalently, the complex $L_{X^{(1)}/W}[-1]$ identifies with $\tau^{\leq 1}\Omega^\bullet_{X/k}$.

REMARK 3.22 (The integral analog of Deligne-Illusie). The analogy between perfect rings and integral perfectoid rings is strong enough that we can directly repeat the discussion of Remark 3.21 when the ring k is only assumed to be an integral perfectoid ring. In this case, we must replace $W(k)$ with $A_{inf}(k) = W(k^\flat)$ and the map $W(k) \to k$ with Fontaine's map $\theta : A_{inf}(k) \to k$. Given a smooth k-scheme X, the discussion in Remark 3.21 goes through (using Remark 3.19) to construct a class $\mathrm{ob}_{X/A_{inf}(k)} \in \mathrm{Ext}^1_X(\Omega^1_{X/k}, \mathcal{O}_X\{1\})$ from the complex $L_{X/A_{inf}(k)}$: it measures the failure to lift X across $\overline{\theta} : A_{inf}(k)/\ker(\theta)^2 \to k$. In this setting, the analog of the Deligne-Illusie theorem is then the subject of [**BMS2**, §8]; the rational version for $k = \mathcal{O}_C$ with C an algebraically closed perfectoid field of characteristic 0 is the identification of $L_{X/A_{inf}(k)}[-1][\frac{1}{p}]$ with $\tau^{\leq 1}R\nu_*\widehat{\mathcal{O}_{X_C}}$, as alluded to in Remark 3.5.

The notation A_{inf} (and its cousin A_{crys}) were adopted for geometric reasons, as we briefly recall.

REMARK 3.23 (Nomenclature of A_{inf} and A_{crys}). Let R be an integral perfectoid ring. By definition, the map $\overline{\theta} : R^\flat \to R/p$ is a projective limit of the maps $R/p \xrightarrow{\phi^n} R/p$; by the perfectoidness of R, each of these latter maps is a infinitesimal thickening (i.e., is surjective with nilpotent kernel). Thus, we may regard $R^\flat$ as a projective limit of infinitesimal thickenings on R. Moreover, by perfectness, $R^\flat \to R/p$ is the universal such object in characteristic p rings: for any other infinitesimal thickening $S \to R/p$ with S an $\mathbf{F}_p$-algebra, there is a unique map $R^\flat \to S$ factoring $\overline{\theta}$. As in Proposition 3.15, one then checks that $\theta : A_{inf}(R) \to R/p$ is also a projective limit of infinitesimal thickenings of R/p, and is the universal such object amongst all thickenings. Stated differently, $A_{inf}(R)$ is the global sections of the structure sheaf of the infinitesimal site for $\mathrm{Spec}(R/p)$ (see [**Gro**] for the infinitesimal and crystalline site); this is the origin of Fontaine's notation $A_{inf}(R)$ (which arose in the example $R = \mathcal{O}_{\mathbf{C}_p}$ first). Likewise, in this case, adjoining divided powers along the kernel of θ and p-adically completing produces Fontaine's period ring $A_{crys}(R)$, which comes equipped with a factorization $A_{inf}(R) \to A_{crys}(R) \to R$; one can then show that the map $A_{crys}(R) \to R$ realizes $A_{crys}(R)$ as the global sections of the structure sheaf on the crystalline site of $\mathrm{Spec}(R/p)$, once again explaining the notation.

3.2. Recollections on the pro-étale site. We now return to the setup at the start of the section: C is a complete and algebraically closed extension of $\mathbf{Q}_p$, and X/C is a smooth rigid-analytic space. Viewing X as an adic space, Scholze has attached its pro-étale site X_{proet} in [**Sc2**, §3] (see also lectures by Kedlaya and Weinstein[8]). A typical object here is a pro-object $U := \{U_i\}$ of X_{et} such that all transition maps $U_i \to U_j$ are finite étale covers for i, j large. Heuristically, one wishes to allow towers of finite étale covers of an open in X. The following class of objects in X_{proet} plays a crucial role:

DEFINITION 3.24. An object $U := \{U_i\} \in X_{proet}$ is called *affinoid perfectoid* if it satisfies the following:

(1) Each $U_i = \mathrm{Spa}(R_i, R_i^+)$ is affinoid.
(2) Setting $R^+ := \widehat{\mathrm{colim}_i R_i^+}$ (where the completion is p-adic) and $R = R^+[\frac{1}{p}]$, the pair (R, R^+) is a perfectoid affinoid algebra.

For such an object U, we write $\widehat{U} := \mathrm{Spa}(R, R^+)$ for the corresponding perfectoid space.

For our purposes, the main reason to enlarge the étale site X_{et} to the pro-étale site X_{proet} is that the following theorem, stating roughly that there are enough affinoid perfectoid objects to cover any object, becomes true (see [**Sc2**, Corollary 4.7]):

THEOREM 3.25 (Locally perfectoid nature of X_{proet}). *The collection of $U \in X_{proet}$ which are affinoid perfectoid form a basis for the topology.*

REMARK 3.26. The construction of pro-étale site makes sense any noetherian adic space X over $\mathrm{Spa}(\mathbf{Q}_p, \mathbf{Z}_p)$. Moreover, Theorem 3.25 is true in this generality; this is due to Colmez, see [**Sc2**, Proposition 4.8].

Theorem 3.25 is a remarkable assertion: it allows us to reduce statements about (pro-)étale sheaves on rigid-analytic spaces to those for perfectoid spaces. In practice, this means that affinoid perfectoids play a role in p-adic geometry that is somewhat analogous to the role of unit polydisks in complex analytic geometry. We do not prove Theorem 3.25 in these notes. Instead, we content ourselves by describing the key construction that goes into its proof, which is analogous to the one in §2.2.

EXAMPLE 3.27 (The perfectoid torus). Let $X := \mathbb{T}^1 := \mathrm{Spa}(C\langle T^{\pm 1}\rangle, \mathcal{O}_C\langle T^{\pm 1}\rangle)$ be the torus. Consider the object $U := \{U_i\}_{i \in \mathbf{N}} \in X_{proet}$ given by setting $U_n = X$ for all n, with the transition map $U_{n+1} \to U_n$ being given by the p-power map on the torus. To avoid confusion, choose co-ordinates so as to write

---

[8]The definition of the pro-étale site has evolved a bit over time. For our purposes, the original one from [**Sc2**, §3], which is perhaps the most intuitive, suffices. Other variants that are technically much more useful were discovered later, and are discussed in the lectures of Kedlaya and Weinstein. In particular, the notion discussed in these notes is called the *flattening pro-étale topology* in Kedlaya's lectures. The reader may freely use any of these variants whilst reading these notes.

$U_n = \mathrm{Spa}(C\langle T^{\pm \frac{1}{p^n}}\rangle, \mathcal{O}_C\langle T^{\pm \frac{1}{p^n}}\rangle)$. Then U is indeed affinoid perfectoid: the corresponding perfectoid affinoid algebra is simply $(C\langle T^{\pm \frac{1}{p^\infty}}\rangle, \mathcal{O}_C\langle T^{\pm \frac{1}{p^\infty}}\rangle)$. Note that each map $U_{n+1} \to U_n$ is a $\mu_p(C)$-torsor, and hence $U \to X$ is a pro-étale $\mathbf{Z}_p(1)$-torsor. Explicitly, we have a (continuous) direct sum decomposition

$$C\langle T^{\pm \frac{1}{p^\infty}}\rangle \simeq \widehat{\bigoplus_{i \in \mathbf{Z}[\frac{1}{p}]}} C \cdot T^i.$$

This decomposition is equivariant for the $\mathbf{Z}_p(1)$-action, and an element $\underline{\epsilon} := (\epsilon_n) \in \lim \mu_{p^n}(C) =: \mathbf{Z}_p(1)$ acts on the summands via

$$T^{\frac{a}{p^m}} \mapsto \epsilon_m^a T^{\frac{a}{p^m}}.$$

For convenience, we often abbreviate this action as

$$T^i \mapsto \underline{\epsilon}^i T^i.$$

In particular, in this case, we have a profinite étale cover of X by an affinoid perfectoid in X_{proet}. More generally, a similar construction applies when X admits an étale map to the n-dimensional torus $\mathbb{T}^n$ that factors as a composition of rational subsets and finite étale maps (see [**Sc2**, Lemma 4.6]). In general, one can always cover X by affinoid opens that admit such maps.

We now recall some "vanishing theorems" on X_{proet}. Recall that we have already discussed the morphism $\nu : X_{proet} \to X_{et}$ of sites; precisely, this is the morphism of topoi inverse to the operation of observing that any $U \in X_{et}$ trivially gives an object of X_{proet} as the constant pro-system. Using this morphism, we obtain the sheaves $\mathcal{O}_X^+ := \nu^* \mathcal{O}_{X_{et}}^+$ and $\mathcal{O}_X := \nu^* \mathcal{O}_{X_{et}}$ on X_{proet}; here $\mathcal{O}_{X_{et}}$ and $\mathcal{O}_{X_{et}}^+$ are the usual structure sheaves on the étale site X_{et}. The completed structure sheaves are then defined as $\widehat{\mathcal{O}_X^+} = \lim \mathcal{O}_X^+/p^n$ and $\widehat{\mathcal{O}_X} = \widehat{\mathcal{O}_X^+}[\frac{1}{p}]$. Given an affinoid perfectoid $U := \{\mathrm{Spa}(R_i, R_i^+)\} \in X_{proet}$ as in Definition 3.24 with limit $\widehat{U} := \mathrm{Spa}(R, R^+)$, one has the expected formulae

$$\widehat{\mathcal{O}_X^+}(U) = R^+ \quad \text{and} \quad \widehat{\mathcal{O}_X}(U) = R,$$

see [**Sc2**, Lemma 4.10]. The first vanishing theorem concerns the cotangent complex:

COROLLARY 3.28. *The cotangent complex* $L_{\widehat{\mathcal{O}_X^+}/\mathcal{O}_C}$ *vanishes modulo p on X_{proet}. Hence, the p-adic completion of* $L_{\widehat{\mathcal{O}_X^+}/\mathcal{O}_C}$ *vanishes.*

PROOF. By Theorem 3.25, it is enough to show that the presheaf $U \mapsto L_{\widehat{\mathcal{O}_X^+}(U)/\mathcal{O}_C} \otimes_{\mathbf{Z}_p}^L \mathbf{Z}/p$ vanishes on affinoid perfectoid $U \in X_{proet}$. But $\widehat{\mathcal{O}_X^+}(U) = R^+$ for a perfectoid affinoid (R, R^+). We are then reduced to the vanishing modulo p of the cotangent complex for perfectoids, which may be deduced from Proposition 3.12 as $\mathcal{O}_C/p \to R^+/p$ is flat and relatively perfect. $\square$

In other words, there is no differential geometric information available when working on the ringed site $(X_{proet}, \widehat{\mathcal{O}_X})$. We shall see later that the differential forms on X can nevertheless be recovered from $(X_{proet}, \widehat{\mathcal{O}_X})$ via pushforward down to X_{et}. The second vanishing theorem concerns the cohomology of $\widehat{\mathcal{O}_X}$ on affinoid perfectoids (see [**Sc2**, Lemma 4.10]):

THEOREM 3.29 (Acyclicity of the structure sheaf on affinoid perfectoids). *Let $U \in X_{proet}$ be an affinoid perfectoid. Then $H^i(U, \widehat{\mathcal{O}_X^+})$ is almost zero[9] for $i > 0$, and thus $H^i(U, \widehat{\mathcal{O}_X}) = 0$ for $i > 0$.*

In particular, this theorem gives us a technique for calculating the cohomology of $\widehat{\mathcal{O}_X}$ for any affinoid $U \in X_{proet}$: if we choose a pro-étale cover $V \to U$ with V affinoid perfectoid as provided by Theorem 3.25, then Cech theory (see [**SP**, Tag 01ET]) gives an identification

$$H^i(U, \widehat{\mathcal{O}_X}) \simeq H^i\Big(\widehat{\mathcal{O}_X}(V) \to \widehat{\mathcal{O}_X}(V \times_U V) \to \widehat{\mathcal{O}_X}(V \times_U V \times_U V) \to \dots\Big)$$

as $V, V \times_U V, V \times_U V \times_U V$, etc. are all affinoid perfectoid; here the differentials are the alternating sums of the pullbacks along the various projections. If we can further ensure that $V \to U$ is a G-torsor for a profinite group (see Example 3.27 for an example), then the above complex takes on a particularly simple form related to group cohomology (as in [**SP**, Tag 09B3]). Indeed, we can identify $V \times_U V \simeq V \times \underline{G}$ (where $\underline{G}$ is the "topologically constant" sheaf on X_{proet} defined by the association $W \mapsto \text{Map}_{conts}(|W|, G)$, where $|W|$ is the natural topological space attached to $W \in X_{proet}$), so the above formula simplifies to

$$H^i(U, \widehat{\mathcal{O}_X}) = H^i_{conts}(G, \widehat{\mathcal{O}_X}(V)).$$

In other words, we can calculate the cohomology of $\widehat{\mathcal{O}_X}$ in terms of the continuous group cohomology. The same strategy also applies for the integral sheaf $\widehat{\mathcal{O}_X^+}$ in the almost category, and will be used repeatedly in the sequel.

3.3. The key calculation. Continuing the notation from §3.2, we record the main calculation describing $R\nu_*\widehat{\mathcal{O}_X}$.

LEMMA 3.30. *The $\mathcal{O}_X$-module $R^1\nu_*\widehat{\mathcal{O}_X}$ is locally free of rank n, and taking cup products gives an isomorphism $\wedge^i R^1\nu_*\widehat{\mathcal{O}_X} \simeq R^i\nu_*\widehat{\mathcal{O}_X}$.*

The strategy of the proof is roughly analogous to that employed in §2.2: we find a map $X_\infty \to X$ which is a torsor for a profinite group with the property that X_∞ is perfectoid, and then proceed via descent.

PROOF. This is a local assertion, so we may assume that X is affinoid, and that there exists an étale map $X \to \mathbb{T}^n$ that factors as a composition of rational subsets and finite étale covers. By the vanishing of higher coherent sheaf cohomology on

---

[9]The phrase "almost zero" refers to a notion introduced by Faltings [**Fa1**]: an $\mathcal{O}_C$-module is almost zero if it is killed by the maximal ideal of $\mathcal{O}_C$. Intuitively, such a module is "very small" and can often be safely ignored when performing computations. Faltings theory of "almost mathematics" (expounded in [**GR**]) is based on the idea of systematically developing various notions of commutative algebra and algebraic geometry up to almost zero error terms (as in Theorem 3.29), i.e., one works with rings, modules, etc. in the ⊗-category of almost $\mathcal{O}_C$-modules, defined as the quotient of the category of all $\mathcal{O}_C$-modules by almost zero ones. Whilst we have avoided any discussion of this notion in these notes, it is important to note that almost mathematics lurks in the background when working with perfectoid spaces, and is most directly visible in the integral aspects of theory.

affinoids, it is enough to show the following:

(1) The $\mathcal{O}_X(X)$-module $H^1(X_{proet}, \widehat{\mathcal{O}_X})$ is free of rank n.
(2) Taking cup products gives an isomorphism
$$\wedge^i H^1(X_{proet}, \widehat{\mathcal{O}_X}) \simeq H^i(X_{proet}, \widehat{\mathcal{O}_X})$$
for each i.
(3) The preceding two properties are compatible with étale localization on X.

We shall explain the first two in the key example of a torus, leaving the rest to the references.

Consider first the case $X = \mathbb{T}^1 := \mathrm{Spa}(C\langle T^{\pm 1}\rangle, \mathcal{O}_C\langle T^{\pm 1}\rangle)$ of a 1-dimensional torus with co-ordinate T. Write $X_\infty \in X_{proet}$ for the affinoid perfectoid object constructed in Example 3.27. Then Theorem 3.29 shows that
$$R\Gamma(X_{\infty, proet}, \widehat{\mathcal{O}_X}) \simeq C\langle T^{\pm \frac{1}{p^\infty}}\rangle.$$

As $X_\infty \to X$ is a $\mathbf{Z}_p(1)$-torsor, this implies (see discussion following Theorem 3.29) that
$$R\Gamma(X_{proet}, \widehat{\mathcal{O}_X}) \simeq R\Gamma_{conts}(\mathbf{Z}_p(1), C\langle T^{\pm \frac{1}{p^\infty}}\rangle).$$

Now that canonical presentation
$$C\langle T^{\pm \frac{1}{p^\infty}}\rangle \simeq \widehat{\bigoplus_{i \in \mathbf{Z}[\frac{1}{p}]}} C \cdot T^i$$
is equivariant for the action of $\mathbf{Z}_p(1)$ described in Example 3.27. In particular, if $\underline{\epsilon} = (\epsilon_n) \in \lim \mu_{p^n}(C) = \mathbf{Z}_p(1)$ is a generator, then, by standard facts about the continuous group cohomology of pro-cyclic groups, we have
$$R\Gamma(X_{proet}, \widehat{\mathcal{O}_X}) \simeq \widehat{\bigoplus_{i \in \mathbf{Z}[\frac{1}{p}]}} \left(C \cdot T^i \xrightarrow{T^i \mapsto (\underline{\epsilon}^i - 1)T^i} C \cdot T^i \right);$$

here we follow the convention that if $i = \frac{a}{p^m}$, $a \in \mathbf{Z}$, then $\underline{\epsilon}^i = \epsilon_m^a$. In particular, the differential is trivial on the summands indexed by $i \in \mathbf{Z}$ (as $\underline{\epsilon}^i = 1$ for such i) and an isomorphism for non-integral $i \in \mathbf{Z}[\frac{1}{p}]$ (as $\underline{\epsilon}^i - 1 \neq 0$ for such i). Thus, up to quasi-isomorphism, we can ignore the non-integral summands to get
$$R\Gamma(X_{proet}, \widehat{\mathcal{O}_X}) \simeq \widehat{\bigoplus_{i \in \mathbf{Z}}} \left(C \cdot T^i \xrightarrow{0} C \cdot T^i \right).$$

This presentation (and some unraveling of isomorphisms) shows that $H^*(X_{proet}, \widehat{\mathcal{O}_X})$ is the exterior algebra on its H^1, and that $H^1(X_{proet}, \widehat{\mathcal{O}_X})$ is free of rank 1, as wanted.

The preceding analysis applies equally well (modulo bookkeeping) when $X = \mathbb{T}^n$ is an n-dimensional torus for any $n \geq 1$. The general case is then deduced from this one by the almost purity theorem and base change properties of group cohomology, as explained in [**Sc3**, Proposition 3.23] and [**Sc2**, Lemma 4.5, 5.5]. □

3.4. Construction of the map.

In this section, we give a global construction of the map
$$\Phi^i : \Omega^i_{X/C}(-i) \to R^i\nu_*\widehat{\mathcal{O}_X}$$
that will eventually give the isomorphism in Theorem 3.3. This construction is analogous to the one in [**BMS2**, §8.2] and differs from that in [**Sc3**, §3.3].

We choose a formal model $\mathfrak{X}/\mathcal{O}_C$ of X, and write $\mathfrak{X}_{aff}$ for the category of affine opens in $\mathfrak{X}$ with the indiscrete topology (i.e., only isomorphisms are covers, so all presheaves are sheaves). Then we have evident morphisms
$$(X_{proet}, \widehat{\mathcal{O}_X}) \xrightarrow{\nu} (X_{et}, \mathcal{O}_X) \xrightarrow{\pi} (\mathfrak{X}_{aff}, \mathcal{O}_{\mathfrak{X}})$$
of ringed sites, and write $\mu = \pi \circ \nu$ for the composite. We shall construct[10] a natural morphism
$$\Phi^{1,'} : \Omega^1_{\mathfrak{X}/\mathcal{O}_C} \to R^1\mu_*\widehat{\mathcal{O}_X}(1).$$
By formal properties of adjoints, this defines a map
$$\pi^*\Omega^1_{\mathfrak{X}/\mathcal{O}_C} = \pi^{-1}\Omega^1_{\mathfrak{X}/\mathcal{O}_C} \otimes_{\pi^{-1}\mathcal{O}_{\mathfrak{X}}} \mathcal{O}_X \to R^1\nu_*\widehat{\mathcal{O}_X}(1).$$
The left side identifies with $\Omega^1_{X/C}$, so untwisting defines the desired map Φ^1. The remaining Φ^i's are obtained by passage to exterior powers using the anticommutative cup product on $\oplus_i R^i\nu_*\widehat{\mathcal{O}_X}$.

Consider the maps
$$\mathbf{Z}_p \to \mathcal{O}_C \to \widehat{\mathcal{O}_X^+}$$
of sheaves of rings on X_{proet}. Attached to this, there is a standard exact triangle
$$L_{\mathcal{O}_C/\mathbf{Z}_p} \otimes_{\mathcal{O}_C} \widehat{\mathcal{O}_X^+} \to L_{\widehat{\mathcal{O}_X^+}/\mathbf{Z}_p} \to L_{\widehat{\mathcal{O}_X^+}/\mathcal{O}_C}$$
of cotangent complexes. Corollary 3.28 shows that the last term vanishes after a p-adic completion. Hence, we obtain an isomorphism
$$L_{\mathcal{O}_C/\mathbf{Z}_p} \otimes_{\mathcal{O}_C} \widehat{\mathcal{O}_X^+} \simeq \widehat{L_{\widehat{\mathcal{O}_X^+}/\mathbf{Z}_p}}.$$
By Theorem 2.5, the first term identifies with $\Omega \otimes_{\mathcal{O}_C} \widehat{\mathcal{O}_X^+}[1]$, where Ω is a free $\mathcal{O}_C$-module of rank 1 that Galois equivariantly looks like $\mathcal{O}_C(1)$ up to torsion. In particular, inverting p gives

(3) $$\widehat{\mathcal{O}_X}(1)[1] \simeq \widehat{L_{\widehat{\mathcal{O}_X^+}/\mathbf{Z}_p}}[\frac{1}{p}].$$

Now consider the map μ of ringed sites. Via pullback, this yields a map

(4) $$\widehat{L_{\mathfrak{X}/\mathbf{Z}_p}} \to R\mu_*\widehat{L_{\widehat{\mathcal{O}_X^+}/\mathbf{Z}_p}} \to R\mu_*\widehat{L_{\widehat{\mathcal{O}_X^+}/\mathbf{Z}_p}}[\frac{1}{p}] \simeq R\mu_*\widehat{\mathcal{O}_X}(1)[1].$$

To proceed further, we claim that there is a natural identification

(5) $$\mathcal{H}^0(\widehat{L_{\mathfrak{X}/\mathbf{Z}_p}}) \simeq \Omega^1_{\mathfrak{X}/\mathcal{O}_C}.$$

---

[10] Here $\Omega^1_{\mathfrak{X}/\mathcal{O}_C}$ denotes the sheaf of Kähler differentials on the formal scheme, and is computed as follows: if $\mathfrak{X} = \mathrm{Spf}(R)$ for flat $\mathcal{O}_C$-algebra R that is topologically of finite presentation, then $\Omega^1_{\mathfrak{X}/\mathcal{O}_C}$ is the coherent $\mathcal{O}_{\mathfrak{X}}$-sheaf associated to the finitely presented R-module of continuous Kähler differentials on R, see [**EGA**, §0.20.1] and [**GR**, 7.1.23]. This module is computed as the p-adic completion of module $\Omega^1_{R/\mathcal{O}_C}$ in the algebraic sense. In particular, the sheaf $\Omega^1_{\mathfrak{X}/\mathcal{O}_C}$ has the following key feature for our purposes: its values on affines are p-adically complete.

Granting this claim, passage to $\mathcal{H}^0$ in (4) yields the map
$$\Phi^{1,'} : \Omega^1_{\mathfrak{X}/\mathcal{O}_C} \to R^1\mu_*\widehat{\mathcal{O}_X}(1),$$
and hence the maps Φ^i, as explained earlier. To prove (3.4), consider the sequence
$$\mathbf{Z}_p \to \mathcal{O}_C \to \mathcal{O}_{\mathfrak{X}}$$
of rings on $\mathfrak{X}_{aff}$. The transitivity triangle then takes the form
$$L_{\mathcal{O}_C/\mathbf{Z}_p} \otimes_{\mathcal{O}_C} \mathcal{O}_{\mathfrak{X}} \to L_{\mathfrak{X}/\mathbf{Z}_p} \to L_{\mathfrak{X}/\mathcal{O}_C}.$$
On applying the derived p-adic completion functor, we obtain an exact triangle where the term on the left has no $\mathcal{H}^0$, so we obtain an identification
$$\mathcal{H}^0(\widehat{L_{\mathfrak{X}/\mathbf{Z}_p}}) \simeq \mathcal{H}^0(\widehat{L_{\mathfrak{X}/\mathcal{O}_C}}).$$
The upshot of this reduction is that the right side is a geometric object: $\mathfrak{X}$ is a topologically finitely presented formal scheme over $\mathcal{O}_C$. Using this fact, one can check that
$$\mathcal{H}^0(\widehat{L_{\mathfrak{X}/\mathcal{O}_C}}) = \Omega^1_{\mathfrak{X}/\mathcal{O}_C},$$
which then gives the desired (3.4); we refer to the discussion surrounding [**GR**, Lemma 7.1.25] for more on the relationship between the cotangent complex and continuous Kähler differentials, and [**GR**, Proposition 7.1.27] for the proof of the above equality.

3.5. Conclusion: the isomorphy of Φ^i. Combining the material in §3.4 with the calculation §3.3, we learn that both the source and the target of
$$\oplus_i \Phi^i : \bigoplus_i \wedge^i(\Omega^1_{X/C}(-1)) \to \bigoplus_i R^i\nu_*\widehat{\mathcal{O}_X}$$
are exterior algebras on the $i = 1$ terms. Thus, to prove that Φ^i is an isomorphism for all i, it suffices to do so for $i = 1$. Moreover, note that both sides are coherent sheaves of an étale local nature on X. Thus, we may assume $X = \mathbb{T}^n$, and may pass to global sections. Thus, we need to show that the
$$\Phi^1(X) : \Omega^1_{X/C}(-1) \to H^1(X_{proet}, \widehat{\mathcal{O}_X})$$
of free rank n $\mathcal{O}_X(X)$-modules is an isomorphism. Both sides are compatible with taking products of adic spaces, so one reduces to the case $n = 1$. Choose coordinates to write $X := \mathbb{T}^1 := \mathrm{Spa}(C\langle T^{\pm 1}\rangle, \mathcal{O}_C\langle T^{\pm 1}\rangle)$. Then $d\log(T) \in \Omega^1_{X/C}$ is a generator, and it suffices to show that $\Psi^1(d\log(T))$ is also a generator. This can be checked by making explicit the construction of §3.4 as in [**BMS2**, §8.3].

4. Lecture 4: Integral aspects

Let $C/\mathbf{Q}_p$ be a complete and algebraically closed field with residue field k. Let $\mathfrak{X}$ be a smooth and proper formal scheme[11] over $\mathcal{O}_C$. Write $X = \mathfrak{X}_C$ for the

---

[11] Not much will be lost if one assumes that $C = \mathbf{C}_p$ and that $\mathfrak{X}$ arises as the p-adic completion of a proper smooth $\mathcal{O}_C$-scheme $\mathcal{X}$. For our constructions, though, it will nevertheless be convenient to work with formal schemes. The added generality is also useful in some geometric applications, see [**CLL**] for a recent concrete example arising from the following phenomenon: it is possible for a K3 surface X/C to only extend to a smooth and proper algebraic space $\mathcal{X}/\mathcal{O}_C$ (and not a smooth and proper scheme); nevertheless, the special fibre of $\mathcal{X}/\mathcal{O}_C$ is a scheme (as any proper smooth 2-dimensional algebraic space over a field is a scheme), and hence the formal completion $\mathfrak{X}/\mathcal{O}_C$ of $\mathcal{X}/\mathcal{O}_C$ is actually a proper smooth formal scheme.

generic fibre, and $\mathfrak{X}_k$ for the special fibre. We then have the degenerate Hodge-Tate spectral sequence

$$E_2^{i,j} : H^i(X, \Omega^j_{X/C})(-j) \Rightarrow H^{i+j}(X, C).$$

leading to the equality

(6) $$\dim_{\mathbf{Q}_p} H^n(X_{et}, \mathbf{Q}_p) = \dim_C H^n(X, C) = \sum_{i+j=n} \dim_C H^i(X, \Omega^j_{X/C})$$

relating étale and Hodge-cohomology for the generic fiber. As X admits a good model $\mathfrak{X}$, the groups appearing on either side of the equality above admit good integral and mod-p variants: we have

$$H^i(\mathfrak{X}_k, \Omega^j_{\mathfrak{X}/k}) \quad \text{and} \quad H^n(X_{et}, \mathbf{F}_p).$$

It is thus natural to ask if (6) admits a mod-p variant. The following theorem was proven recently in [**BMS2**]

THEOREM 4.1. *One has inequalities*

(7) $$\dim_{\mathbf{F}_p} H^n(X_{et}, \mathbf{F}_p) \leq \sum_{i+j=n} \dim_k H^i(\mathfrak{X}_k, \Omega^j_{\mathfrak{X}/k}).$$

The primary goal of this section, accomplished in §4.2 and §4.3, is to present a broader context for Theorem 4.1 and to sketch a proof. But first, in §4.1, we record some examples showing that the inequality in Theorem 4.1 (as well as a mod p^2 variant) can be strict.

4.1. Examples indicating Theorem 4.1 cannot be strengthened.

4.1.1. *An example where* (7) *is strict.* The goal of this subsection is to record Example 4.3, showing that (7) can be strict. Our strategy is to construct certain interesting degenerations of group schemes, and then to approximate their classifying stacks. To motivate this idea and subsequent constructions, we begin with a purely topological calculation; the reader not interested in this motivation may skip ahead to Example 4.3 for the actual example relevant to Theorem 4.1.

EXAMPLE 4.2. Let $G = \mathbf{Z}/p$. Consider the classifying space BG of G-torsors; this space can be defined as EG/G, where EG is a contractible space with a free G-action. The cohomology of BG agrees with the group cohomology of G. We claim that there exist G-torsors $f_i : X_i \to BG$ for $i \in \{0,1\}$ such that $H^1(X_0, \mathbf{F}_p) \simeq 0$, but $H^1(X_1, \mathbf{F}_p) \neq 0$. In fact, for X_0, we take $X_0 = EG$, with $f_0 : X_0 \to BG$ being the universal G-torsor: as EG is contractible, we have $H^{>0}(X_0, \mathbf{F}_p) \simeq 0$. For X_1, we simply take $X_1 = BG \times G$ with $f_1 : X_1 \to BG$ being the projection, realizing X_1 as the trivial G-torsor over BG. Then $H^1(X_1, \mathbf{F}_p)$ contains $H^1(BG, \mathbf{F}_p)$ as a summand, and is thus nonzero since $H^1(BG, \mathbf{F}_p) \simeq \mathrm{Hom}(G, \mathbf{F}_p) \neq 0$.

As a thought experiment, imagine that one can construct a family degenerating f_0 to f_1, i.e., a continuous one parameter family $f_t : X_t \to BG$ of G-torsors indexed by $t \in [0, 1]$ coinciding with the construction above for $t = 0, 1$. The total space $\mathcal{X}$ of this degeneration would then admit a fibration $\mathcal{X} \to [0, 1]$ whose fibers have varying $\mathbf{F}_p$-cohomologies. Unfortunately, it is impossible to find such a degeneration in topology. Indeed, any such family would correspond to a non-constant path in the "space of G-torsors on BG" that degenerates the non-trivial torsor X_0 to the trivial torsor X_1. The space of such torsors is tautologically $\mathrm{Map}(BG, BG)$; as G is discrete and abelian, this space admits no non-trivial paths[12], so no such families exist. (Even more directly, the fibers of a fibration over $[0,1]$ are homotopy-equivalent, and hence can't have distinct cohomologies.)

However, we *can* produce such a degeneration in algebraic geometry in positive or mixed characteristic, essentially because morphisms between finite group schemes can vary in families in this setting; for example, $\mathrm{Hom}(\mathbf{Z}/p, \mu_p) \simeq \mu_p$ is not discrete in characteristic p. Using this idea, one can rather readily find the phenomenon described in the previous paragraph in the world of algebraic stacks (see [**BMS1**, Example 4.1]). To stay within the world of schemes, one needs an additional approximation argument. The example recorded next (from [**BMS2**, §2.1]) accomplishes both of these tasks, albeit in a hidden fashion.

EXAMPLE 4.3. Assume $p = 2$. Let $S/\mathcal{O}_C$ be a proper smooth morphism with $\pi_1(S_C) \xrightarrow{\simeq} \pi_1(S) \xleftarrow{\simeq} \pi_1(S_k) \simeq \mathbf{Z}/2$; one may construct an Enriques surface with such properties. Let $E/\mathcal{O}_C$ be an elliptic curve with good ordinary reduction. Hence, there is a canonical subgroup $\mu_2 \subset E$ (see Caraiani's lectures). Choosing the element $-1 \in \mu_2(\mathcal{O}_C)$ defines a map
$$\alpha : \mathbf{Z}/2 \to \mu_2 \subset E$$
of group schemes over $\mathcal{O}_C$. If $\tilde{S} \to S$ denotes the universal $\mathbf{Z}/2$-cover of S, then we may push out $\tilde{S} \to S$ along α to obtain an E-torsor $f : Y \to S$ by setting $Y := \tilde{S} \times_{\mathbf{Z}/2} E$ (where $\mathbf{Z}/2$ acts via the covering involution on $\tilde{S}$, and by translation using α on E) with $f : Y \to S$ the map induced by projection onto $\tilde{S}/(\mathbf{Z}/2) \simeq S$. This E-torsor has the following properties:

(1) The special fiber $Y_k \to S_k$ is identified with the split torsor $E_k \times S_k \to S_k$: the construction of Y is compatible with restriction to the special fibre, and α_k is the 0 map as $-1 = 1$ over k.

(2) The generic fibre $Y_C \to S_C$ is a non-split E_C-torsor (i.e., it has no section): using the exact sequence

(8) $$0 \to \mathbf{Z}/2_C \xrightarrow{\alpha_C} E_C \xrightarrow{\beta} E'_C \to 0$$

(where E'_C is defined as the quotient of the elliptic curve E_C by the non-trivial 2-torsion point coming from α, and is thus also an elliptic curve over C), the triviality of the torsor $Y_C \to S_C$ would give a non-constant map $S_C \to E'_C$. But one can show that there are no non-constant maps from a smooth proper variety over C with finite étale fundamental group into an abelian variety[13], so we are done.

---

[12]More precisely, for any pair of discrete groups H and G, the space $\mathrm{Map}(BH, BG)$ can be modeled by a groupoid whose objects are group homomorphisms $f : H \to G$, and morphisms $f \to f'$ are given by group elements $g \in G$ that conjugate f to f'. When G is abelian (as above), this description collapses to identify $\mathrm{Map}(BH, BG)$ as the product of the discrete set $\mathrm{Hom}(H, G)$ with the groupoid BG. In particular, a path in $[0, 1] \to \mathrm{Map}(BH, BG)$ is "trivial," i.e., the corresponding map $BH \times [0, 1] \to BG$ factors through the second projection.

[13]Fix a map $g : Z \to A$ over C, where Z is a smooth proper variety, A is an abelian variety, and $\pi_1^{et}(Z)$ is finite. Then the induced map $g_* : \pi_1^{et}(Z) \to \pi_1^{et}(A)$ is constant as $\pi_1^{et}(A)$ is topologically a free abelian group, and thus the pullback $g^* : H^1(A, C) \to H^1(Z, C)$ is the 0 map. As $H^*(A, C) \simeq \wedge^* H^1(A, C)$ via cup products, it follows that $g^* : H^n(A, C) \to H^n(Z, C)$ is 0 for all $n > 0$. In particular, if $L \in \mathrm{Pic}(A)$ is an ample line bundle, then $c_1(g^*L) = g^*c_1(L)$ is 0. On the other hand, if g was non-constant, then there would exist a curve $i : C \hookrightarrow Z$ such that $g \circ i : C \to A$ is finite, and thus i^*g^*L is ample, so $\deg(i^*g^*L) = c_1(i^*g^*L) = i^*c_1(g^*L)$ is

Using these properties, we shall calculate both sides of (7) in this example in degree 1 and arrive at the following

(9) $\dim_{\mathbf{F}_2} H^1(Y_{C,et}, \mathbf{F}_2) = 2$ and $\dim_k H^1(Y_k, \mathcal{O}_{Y_k}) + \dim_k H^0(Y_k, \Omega^1_{Y_k/k}) \geq 3$

showing that (7) can indeed be strict. The rest of this example may be viewed as an exercise, and we encourage the reader to try it on their own.

PROVING THE INEQUALITIES IN (9). Let us begin on the étale side. We give a topological argument after choosing an isomorphism $C \simeq \mathbf{C}$; alternately, a purely algebraic version of the same set of ideas can be found in [**BMS2**, §2.1]. The map $Y_C \to S_C$ is an E_C-torsor, so fixing a (suppressed) base point on S_C gives an exact sequence of homotopy groups

$$\pi_1(E_C) \xrightarrow{\mu} \pi_1(Y_C) \xrightarrow{\nu} \pi_1(S_C) \to 0,$$

where the surjectivity on the right comes from the connectedness of the fibers. We shall show that μ is injective, and identify the resulting extension. Consider the map $Y \to E'_C := E_C/(\mathbf{Z}/2)$ coming from the definition of Y. The composite $E_C \to Y \to E'_C$ is clearly injective on π_1 (as it is a non-constant map of smooth proper curves of genus 1), and thus μ must be injective. This data fits into a map of short exact sequences:

$$\begin{array}{ccccccccc} 0 & \longrightarrow & \pi_1(E_C) & \xrightarrow{\mu} & \pi_1(Y_C) & \xrightarrow{\nu} & \pi_1(S_C) & \longrightarrow & 0 \\ & & \| & & \downarrow & & \downarrow{\eta} & & \\ 0 & \longrightarrow & \pi_1(E_C) & \longrightarrow & \pi_1(E'_C) & \xrightarrow{\tau} & \mathbf{Z}/2 & \longrightarrow & 0. \end{array}$$

Here the target of τ is identified via the boundary map induced by the fibration coming from the short exact sequence (8). Unraveling definitions shows that η is the identity; in fact, slightly more canonically, the target of τ is naturally $\mu_2(C)$ viewed as the canonical subgroup on $E(C)$, and the map η arises from our choice of $-1 \in \mu_2(\mathcal{O}_C)$ at the start of the construction defining α. Putting these together, we see that η is an isomorphism, and hence $\pi_1(Y_C) \simeq \pi_1(E'_C) \simeq \mathbf{Z}^{\oplus 2}$. In particular, we get

(10) $$\dim_{\mathbf{F}_2} H^1(Y_{C,et}, \mathbf{F}_2) = 2.$$

We now move to the Hodge side. Here, we have $Y_k \simeq S_k \times E_k$. In particular, one has

$$h^{0,1}(Y_k) = h^{0,1}(S_k) + h^{0,1}(E_k) \quad \text{and} \quad h^{1,0}(Y_k) = h^{1,0}(S_k) + h^{1,0}(E_k)$$

by the Künneth formula for the cohomology of the structure sheaf and differential forms. Now $h^{0,1}(E_k) = h^{1,0}(E_k) = 1$ by general facts about elliptic curves. Also, we claim that $H^1(S_k, \mathcal{O}_{S_k}) \neq 0$, and hence $h^{0,1}(S_k) > 0$: as $\pi_1(S_k) \simeq \mathbf{Z}/2$, there is a non-trivial element in $H^1(S_{k,et}, \mathbf{F}_2)$, which contributes a non-trivial element to $H^1(S_k, \mathcal{O}_{S_k})$ from the Artin-Schreier exact sequence

$$0 \to \mathbf{F}_2 \to \mathcal{O}_{S_k} \xrightarrow{F-1} \mathcal{O}_{S_k} \to 0.$$

---

positive. This contradicts the triviality of $c_1(g^*L)$, so there are no such curves, and hence g must be constant.

Putting these together, we learn that
$$\dim_k H^1(Y_k, \mathcal{O}_{Y_k}) + \dim_k H^0(Y_k, \Omega^1_{Y_k/k}) \geq 3, \tag{11}$$
as promised. In particular, comparing (10) and (11) shows that (7) can be strict. □

4.1.2. *An example where the exponent of p-primary torsion can change.* The inequality (7) is a consequence of the following stronger inequality
$$\dim_{\mathbf{F}_p} H^n(X_{et}, \mathbf{F}_p) \leq \dim_k H^n_{dR}(\mathfrak{X}_k/k),$$
proven in (12) below. Now both sides have a canonical mixed characteristic deformation: étale cohomology with $\mathbf{Z}_p$-coefficients on the left, and crystalline cohomology on the right. In fact, as explained in (13), the previous inequality may be improved to compare the torsion in the two lifts: one has
$$\ell_{\mathbf{Z}_p}(H^i(X_{et}, \mathbf{Z}_p)_{tors}) \leq \ell_{W(k)}(H^i_{crys}(\mathfrak{X}_k/W(k))_{tors}).$$
It is natural to ask if this last inequality is actually a reflection of an inclusion of groups. For example, if $H^i(X_{et}, \mathbf{Z}_p)$ contains an element of order p^2, is the same true for $H^i_{crys}(\mathfrak{X}_k/W(k))$? We shall answer this question negatively. The crucial idea going into the construction of the example is again a phenomenon exhibited by finite flat group schemes away from equicharacteristic 0: one can degenerate a finite group scheme of order exactly p^2 in characteristic 0 into a finite group scheme that is killed by p in characteristic p. An explicit construction of such a degeneration is recorded next.

CONSTRUCTION 4.4 (Degenerating a p^2-torsion group scheme to a p-torsion one). Let $\mathcal{E}/\mathcal{O}_C$ be an elliptic curve with supersingular reduction. Choose a point $x \in \mathcal{E}(C)$ of order exactly p^2, so x defines an inclusion $\mathbf{Z}/p^2 \hookrightarrow \mathcal{E}_C$ of group schemes. Taking the closure, we obtain a finite flat group subscheme $G \hookrightarrow \mathcal{E}$ with $G|_C \simeq \mathbf{Z}/p^2$. The special fibre $G_k \subset \mathcal{E}_k$ is a subgroup of order p^2 on the elliptic curve $\mathcal{E}_k$. As $\mathcal{E}_k$ is supersingular, all p-power torsion subgroups of $\mathcal{E}_k$ are connected (i.e., infinitesimal neighbourhoods of 0). As the collection of all infinitesimal neighbourhoods of 0 is totally ordered under inclusion (as E is 1-dimensional), it follows that E carries a unique subgroup of order p^2, given (as a scheme) by the (p^2-1)-th infinitesimal neighbourhood of $0 \in \mathcal{E}_k$. But $\mathcal{E}_k[p]$ is a subgroup of order p^2, so we must have $G_k = \mathcal{E}_k[p]$. In particular, the group scheme G_k is killed by p, while G_C is a cyclic group scheme of exact order p^2.

Passing from the above construction of group schemes to their classifying stacks yields the sought-for examples in the world of algebraic stacks; the example below approximates this construction using smooth projective varieties.

EXAMPLE 4.5. Choose G as in Construction 4.4. Then we may choose a smooth projective $\mathcal{O}_C$-scheme $\mathcal{Y}$ that has relative dimension 2 and comes equipped with a free G-action. In fact, one may (and we do) choose[14] $\mathcal{Y}$ to be a general complete intersection surface in $\mathbf{P}^n$ for $n \gg 0$. Set $\mathfrak{X} = \mathcal{Y}/G$ to be the quotient, so $\mathfrak{X}$ is a smooth projective $\mathcal{O}_C$-scheme of relative dimension 2 equipped with a G-torsor $\pi: \mathcal{Y} \to \mathfrak{X}$.

---

[14]The existence of such complete intersections is a general fact that is valid for all finite flat group schemes; this fact goes back to the work of Serre [Se] and Atiyah-Hirzeburch [AH]. More recent accounts of this construction include [To, §1], [MV, §4.2], and [Ill3, §6], and the details necessary for our purposes can be found in [BMS2, §2.2].

On the étale side, the Hochschild-Serre spectral sequence for the G-torsor π shows that $H^2(X_{C,et}, \mathbf{Z}_p)_{tors} \simeq \mathbf{Z}/p^2$. Indeed, as $\mathcal{Y}$ is a complete intersection surface, the groups $H^i(\mathcal{Y}_{C,et}, \mathbf{Z}_p)$ are torsionfree for $i \in \{0, 2\}$ and 0 for $i = 1$ by the Lefschetz theorems; the desired claim immediately falls out of the low degree terms for the spectral sequence. Slightly more conceptually, the G-torsor π is classified by a map $\mathfrak{X} \to BG$; we have $H^2(BG_{C,et}, \mathbf{Z}_p)_{tors} = H^2(\mathbf{Z}/p^2, \mathbf{Z}_p)_{tors} = \mathbf{Z}/p^2$, and this group maps isomorphically to $H^2(X_{et}, \mathbf{Z}_p)_{tors}$.

On the crystalline side, we claim that $H^2_{crys}(\mathfrak{X}_k/W(k))_{tors}$ is killed by multiplication by p. By repeating the reasoning used above, we are reduced to showing that $H^i_{crys}(BG_k/W(k))$ is killed by multiplication by p. But G_k itself is killed by multiplication by p, and hence so is its cohomology. (The argument given in the last sentence is meant to convey intuition, and is not a rigorous one as the relevant technology to analyze the crystalline cohomology of stacks has not been documented (to the best of the author's knowledge); a more indirect but precise argument can be found in [**BMS2**, §2.2].)

Putting the conclusions of the previous paragraphs together, we learn that $H^2(X_{et}, \mathbf{Z}_p)_{tors}$ contains an element of order p^2, while $H^2_{crys}(\mathfrak{X}_k/W(k))_{tors}$ is killed by p. In particular, the length inequality

$$\ell_{\mathbf{Z}_p}(H^i(X_{et}, \mathbf{Z}_p)_{tors}) \leq \ell_{W(k)}(H^i_{crys}(\mathfrak{X}_k/W(k))_{tors})$$

cannot be upgraded to an inclusion of groups.

4.2. The main theorem. Theorem 4.1 is proven by constructing a new cohomology theory – the A_{inf}-cohomology – that witnesses a specialization from étale cohomology to de Rham cohomology. In this subsection, we summarize the structure of this theory, and indicate how to deduce some consequences (such as Theorem 4.1). Our discussion here emphasizes certain aspects relevant to this paper, and is not complete; more complete discussions can be found in [**BMS2**, **Mor**, **Bh**].

Fix a complete and algebraically closed field $C/\mathbf{Q}_p$ with residue field k. As C is a perfectoid field, its valuation ring $\mathcal{O}_C$ is integral perfectoid, giving rise to its deformation $A_{inf} := A_{inf}(\mathcal{O}_C)$ as in Proposition 3.15; write $\phi: A_{inf} \to A_{inf}$ for the automorphism deduced by functoriality from Frobenius on $\mathcal{O}/p$, and write $\widetilde{\theta} := \theta \circ \phi^{-1} : A_{inf} \to \mathcal{O}_C$. Writing $C^\flat$ for the fraction field of $\mathcal{O}_C^\flat$, we also have the maps $A_{inf} \to W(C^\flat)$ and $A_{inf} \to W(k)$ arising from the functoriality of $W(-)$, and the map $A_{inf} \to \mathcal{O}_C^\flat$ arising by setting $p = 0$. The scheme $\mathrm{Spec}(A_{inf})$ together with the points and divisors arising from all these maps is depicted in Figure 1 (which is borrowed from [**Bh**]).

Fix a proper smooth formal scheme $\mathfrak{X}/\mathcal{O}_C$ with generic fibre X of dimension d. Theorem 4.1 asserts the existence of a numerical inequality between two mod-p cohomology theories: one is topological in nature and is attached to the generic fibre X, while the other is algebro-geometric and is attached to the special fibre $\mathfrak{X}_k$. This inequality is deduced by constructing a specialization from one cohomology theory to the other over the base A_{inf}, as follows:

THEOREM 4.6 (The A_{inf}-cohomology theory). *There exists a functorial perfect complex $R\Gamma_A(\mathfrak{X}) \in D(A_{inf})$ together with a Frobenius action $\phi_\mathfrak{X} : \phi^* R\Gamma_A(\mathfrak{X}) \to R\Gamma_A(\mathfrak{X})$ that is an isomorphism outside the divisor $\mathrm{Spec}(\mathcal{O}_C) \xrightarrow{\widetilde{\theta}} \mathrm{Spec}(A_{inf})$ defined* by $\widetilde{\theta}$. Moreover, one has the following comparison isomorphisms[15]:

(1) *Étale cohomology:* there exists a canonical ϕ-equivariant identification
$$R\Gamma_A(\mathfrak{X}) \otimes_{A_{inf}} W(C^\flat) \simeq R\Gamma(X_{et}, \mathbf{Z}_p) \otimes W(C^\flat).$$
In fact, such an isomorphism already exists after base change to $A_{inf}[\frac{1}{\mu}]$, where $\mu \in A_{inf}$ is the element from Remark 3.19.

(2) *de Rham cohomology:* there exists a canonical isomorphism
$$R\Gamma_A(\mathfrak{X}) \otimes^L_{A_{inf},\theta} \mathcal{O}_C \simeq R\Gamma_{dR}(\mathfrak{X}/\mathcal{O}_C).$$

(3) *Hodge-Tate cohomology:* there exists an E_2-spectral sequence
$$E_2^{i,j} : H^i(\mathfrak{X}, \Omega^j_{\mathfrak{X}/\mathcal{O}_C})\{-j\} \Rightarrow H^{i+j}(\widetilde{\theta}^* R\Gamma_A(\mathfrak{X})).$$
Here the twist $\{-j\}$ refers to the Breuil-Kisin twist from Remark 3.20.

(4) *Crystalline cohomology of the special fibre:* there exists a canonical ϕ-equivariant identification
$$R\Gamma_A(\mathfrak{X}) \otimes^L_{A_{inf}} W(k) \simeq R\Gamma_{crys}(\mathfrak{X}_k/W(k)).$$

In fact, the properness assumption on $\mathfrak{X}$ is only necessary for Theorem 4.6 (1): the de Rham, Hodge-Tate and crystalline comparisons hold true for any smooth formal scheme $\mathfrak{X}$. Applications of Theorem 4.6 include the following:

(1) *Recovering the Hodge-Tate decomposition.* The element $\mu \in A_{inf}$ is invertible at the generic point of the divisor $\text{Spec}(\mathcal{O}_C) \xhookrightarrow{\widetilde{\theta}} \text{Spec}(A_{inf})$ (marked as the Hodge-Tate specialization in Figure 1). Thus, the base change of $R\Gamma_A(\mathfrak{X})$ along $A_{inf} \xrightarrow{\widetilde{\theta}} \mathcal{O}_C \subset C$ is described by both Theorem 4.6 (1) and (3). Combining these gives the Hodge-Tate spectral sequence from Theorem 3.1.

(2) *Recovering the inequality in Theorem* 4.1. Consider the perfect complex $K := R\Gamma_A(\mathfrak{X}) \otimes_{A_{inf}} \mathcal{O}^\flat_C$ over the valuation ring $\mathcal{O}^\flat_C$ (which is labelled as the modular specialization in Figure 1). By Theorem 4.6 (1), we have
$$K \otimes C^\flat \simeq R\Gamma(X_{et}, \mathbf{F}_p) \otimes C^\flat.$$
By Theorem 4.6 (2) or (3), we have
$$K \otimes k \simeq R\Gamma_{dR}(\mathfrak{X}_k/k).$$
By semicontinuity for the ranks of the cohomology groups of a perfect complex, we learn that

(12) $$\dim_{\mathbf{F}_p} H^n(X_{et}, \mathbf{F}_p) \leq \dim_k H^n_{dR}(\mathfrak{X}_k/k).$$

On the other hand, the existence of the Hodge-to-de Rham spectral sequence shows that
$$\dim_k H^n_{dR}(\mathfrak{X}_k/k) \leq \sum_{i+j=n} \dim_k H^i(\mathfrak{X}_k, \Omega^j_{\mathfrak{X}_k/k}).$$
Combining these, we obtain (7).

---

[15] See Figure 1 for a depiction of the loci in $\text{Spec}(A_{inf})$ where this comparison isomorphisms take place.

(3) *Relating torsion in étale to crystalline or de Rham cohomology.* The reasoning used above can be upgraded to show the following inequality

$$\ell_{\mathbf{Z}_p}(H^i(X_{et}, \mathbf{Z}_p)_{tors}/p^n) \leq \ell_{W(k)}(H^i_{crys}(\mathfrak{X}_k/W(k))_{tors}/p^n)$$

for all $n \geq 0$, and thus

(13) $$\ell_{\mathbf{Z}_p}(H^i(X_{et}, \mathbf{Z}_p)_{tors}) \leq \ell_{W(k)}(H^i_{crys}(\mathfrak{X}_k/W(k))_{tors}).$$

In particular, if $H^i_{crys}(\mathfrak{X}_k/W(k))$ is torsion free, so is $H^i(X_{et}, \mathbf{Z}_p)$. Once one defines a suitable normalized length[16] for finitely presented torsion $\mathcal{O}_C$-modules, the de Rham analogs of the previous two inequalities also hold true, as observed by Cesnavicus [**Ce**, Theorem 4.12]: one has

$$\ell_{\mathbf{Z}_p}(H^i(X_{et}, \mathbf{Z}_p)_{tors}/p^n) \leq \ell_{\mathcal{O}_C}(H^i_{dR}(\mathfrak{X}/\mathcal{O}_C)_{tors}/p^n)$$

for all $n \geq 0$, and thus

(14) $$\ell_{\mathbf{Z}_p}(H^i(X_{et}, \mathbf{Z}_p)_{tors}) \leq \ell_{\mathcal{O}_C}(H^i_{dR}(\mathfrak{X}/\mathcal{O}_C)_{tors}).$$

Example 4.5 shows that this inequalities cannot be upgraded to an inclusion of groups in general.

(4) *The zero locus of $\phi_{\mathfrak{X}}$.* Theorem 4.6 asserts that the map $\phi_{\mathfrak{X}} : \phi^* R\Gamma_A(\mathfrak{X}) \to R\Gamma_A(\mathfrak{X})$ is an isomorphism outside the divisor $\mathrm{Spec}(\mathcal{O}_C) \stackrel{\tilde{\theta}}{\hookrightarrow} \mathrm{Spec}(A_{inf})$ defined by $\tilde{\theta}$. Specializing this picture along $A_{inf} \to W(k)$ and using the crystalline comparison recovers the Berthelot-Ogus theorem [**BO1**, Theorem 1.3] that $\phi_{\mathfrak{X}_k}$ is an isogeny on $R\Gamma_{crys}(\mathfrak{X}_k/W(k))$.

(5) *The absolute crystalline comparison theorem.* Recall from Remark 3.23 that Fontaine's period ring A_{crys} is defined as the p-adic completion of the divided power envelope of the map $\theta : A_{inf} \to \mathcal{O}_C$; concretely, we choose a generator $\xi \in \ker(\theta)$ and define A_{crys} as the p-adic completion of $A_{inf}[\{\frac{\xi^n}{n!}\}_{n \geq 1}] \subset A_{inf}[\frac{1}{p}]$. The Frobenius automorphism ϕ of A_{inf} induces a Frobenius endomorphism ϕ of A_{crys}. More conceptually, the ring A_{crys} may be regarded as the absolute crystalline cohomology of $\mathrm{Spec}(\mathcal{O}_C/p)$, with ϕ corresponding to Frobenius. The image of the map $\mathrm{Spec}(A_{crys}) \to \mathrm{Spec}(A_{inf})$ is depicted in Figure 1.

The absolute crystalline cohomology $R\Gamma_{crys}(\mathfrak{X}_{\mathcal{O}/p})$ of $\mathfrak{X}_{\mathcal{O}_C/p}$ is naturally an A_{crys}-complex. One may show that this A_{crys}-complex lifts the de Rham cohomology $R\Gamma_{dR}(\mathfrak{X}/\mathcal{O}_C)$ of $\mathfrak{X}$ along the map $A_{crys} \to \mathcal{O}_C$ arising from θ, and lifts the crystalline cohomology $R\Gamma_{crys}(\mathfrak{X}_k/W(k))$ along the map $A_{crys} \to W(k)$ factoring the canonical map $A_{inf} \to W(k)$. For this object, one has the following comparison isomorphism, which unifies and generalizes Theorem 4.6 (2) and (4): there exists a canonical ϕ-equivariant isomorphism

(15) $$R\Gamma_A(\mathfrak{X}) \otimes^L_{A_{inf}} A_{crys} \simeq R\Gamma_{crys}(\mathfrak{X}_{\mathcal{O}_C/p}),$$

---

[16] More precisely, given a finitely presented torsion $\mathcal{O}_C$-module M, there is a unique way to define a number $\ell_{\mathcal{O}_C}(M) \in \mathbf{R}_{\geq 0}$ that behaves additively under short exact sequences, and carries $\mathcal{O}_C/p$ to 1. A high-brow perspective on this length arises from algebraic K-theory: by the excision sequence for $\mathcal{O}_C \to C$, one may identify K_0 of the category of finitely presented torsion $\mathcal{O}_C$-modules with $K_1(C)/K_1(\mathcal{O}_C) \simeq C^*/\mathcal{O}_C^*$. Postcomposing with the p-adic valuation map $C^*/\mathcal{O}_C^* \to \mathbf{R}$ (normalized to send p to 1) gives the desired normalized length function; see also [**Ce**, §4.10].

which is the absolute crystalline comparison theorem. This isomorphism can be then used to prove the crystalline comparison theorem, see [**BMS2**, Theorem 14.5].

(6) *Bounding the failure of integral comparison maps to be isomorphisms.* Consider the element

$$\xi = \mu/\phi^{-1}(\mu) = \frac{[\epsilon]-1}{[\epsilon^{\frac{1}{p}}]-1} = \sum_{i=0}^{p-1}[\epsilon^{\frac{i}{p}}].$$

This element can be checked to be a generator for $\ker(\theta)$, and thus $\phi(\xi)$ generates $\ker(\widetilde{\theta})$. We also have the formula $\mu = \xi \cdot \phi^{-1}(\mu)$ which provides justification for the heuristic formula "$\mu = \prod_{n \geq 0} \phi^{-n}(\xi)$." The zero locus of μ is depicted in orange in Figure 1.

The construction of $R\Gamma_A(\mathfrak{X})$ shows that there is a naturally defined map

$$R\Gamma_A(\mathfrak{X}) \to R\Gamma(X_{et}, \mathbf{Z}_p) \otimes_{\mathbf{Z}_p} A_{inf}$$

in the almost category[17] that has an inverse up to μ^d, where $d = \dim(X)$. Specializing along the natural map $A_{inf} \to A_{crys}$ and using (5), we obtain a naturally defined almost map

$$R\Gamma_{crys}(\mathfrak{X}_{\mathcal{O}_C/p}) \to R\Gamma(X_{et}, \mathbf{Z}_p) \otimes_{\mathbf{Z}_p} A_{crys}$$

which is also invertible up to μ^d. In particular, one has reasonable control on the failure of the integral comparison maps to be isomorphism, as in the work of Faltings [**Fa3**, **Fa4**].

(7) *Recovering crystalline cohomology of the special fibre from the generic fibre, integrally.* Each cohomology group M of $R\Gamma_A(\mathfrak{X})$ can be shown to be a finitely presented A_{inf}-module equipped with a map $\phi_M : \phi^*M \to M$ that is an isomorphism outside $\text{Spec}(\mathcal{O}_C) \xrightarrow{\widetilde{\theta}} \text{Spec}(A_{inf})$, and is free after inverting p; such pairs (M, ϕ_M) are analogues over C of the Breuil-Kisin modules from [**Ki2**], were introduced and studied by Fargues, and were called Breuil-Kisin-Fargues modules in [**BMS2**, §4.3]. Using some abstract properties of such modules and Theorem 4.6, one can show the following: if $H^i_{crys}(\mathfrak{X}_k/W(k))$ and $H^{i+1}_{crys}(\mathfrak{X}_k/W(k))$ are torsionfree, then $H^i_{crys}(\mathfrak{X}_k/W(k))$ is determined functorially from the generic fibre X (see [**BMS2**, Theorem 1.4]). In particular, in naturally arising geometric situations (such as K3 surfaces), this implies that for different good models for the same generic fibre X, the integral crystalline cohomology of the special fibres is independent of the choice of good model.

4.3. Strategy of the proof. Theorem 4.6 posits the existence of an A_{inf}-valued cohomology theory attached to $\mathfrak{X}$. A natural way to construct such a theory is to work *locally* on $\mathfrak{X}$, i.e., construct a complex $A\Omega_{\mathfrak{X}}$ of sheaves of A_{inf}-modules on the formal scheme $\mathfrak{X}$, and try to prove all the comparisons in Theorem 4.6 at the level of sheaves. With one caveat, this is essentially how the construction goes.

---

[17]The footnote to Theorem 3.29 described almost mathematics over $\mathcal{O}_C$. In the present remark, one needs an extension of this notion to A_{inf}-modules. Roughly, one declares an A_{inf}-module to be *almost zero* if it is annihilated by the kernel of $A_{\text{inf}} = W(\mathcal{O}_C^{\flat}) \to W(k)$. Care must be exercised in using this notion uue to certain subtleties with completions; we refer to [**Bh**, §3] precise definitions.

The necessary tools are:

(1) *The nearby cycles map.* Breaking from the notation used in §3, we write $\nu : X_{proet} \to \mathfrak{X}$ for the *nearby cycles map*; this it the map on topoi whose pullback is induced by the observation that if $\mathfrak{U} \subset \mathfrak{X}$ is an open subset, then we get a rational open subset $U \subset X$ on passage to generic fibres. The reason behind calling this map the "nearby cycles map" name is a theorem of Huber [**Hu2**, Theorem 0.7.7]: for any integer n, the stalk of $R\nu_*\mathbf{Z}/n$ at a point $x \in \mathfrak{X}$ is given by the cohomology of the "nearby fiber", or the "Milnor fiber", i.e., by $R\Gamma(\mathrm{Spec}(\mathcal{O}_{\mathfrak{X},x}^{sh}[\frac{1}{p}])_{et}, \mathbf{Z}/n)$.

(2) *The pro-étale sheaf $A_{inf,X}$.* Fontaine's construction of $A_{inf}(R) := W(R^\flat)$ and the map $\theta : A_{inf}(R) \to R$ makes sense for any ring p-adically complete ring R (see Remark 3.14). In particular, this yields a presheaf $A_{inf,X} := A_{inf}(\mathcal{O}_X^+)$ of A_{inf}-modules on the pro-étale site of X. Using the locally perfectoid nature of X_{proet} from Theorem 3.25, this presheaf can be checked to be a sheaf. By a variant of the primitive comparison theorem (see Theorem 3.2), the cohomology of $A_{inf,X}$ is almost isomorphic to $H^*(X_{et}, \mathbf{Z}_p) \otimes_{\mathbf{Z}_p} A_{inf}$; as we shall see, this is the only place where properness enters the proof of Theorem 4.6.

(3) *Killing torsion in the derived category.* Given a ring A and a nonzerodivisor $f \in A$, we need a systematic technique for killing the f-torsion in the homology of a chain complex K of A-modules; the adjective "systematic" means roughly that the construction should only depend on the class of K in the derived category $D(A)$. While this is impossible to achieve with an *exact* functor $D(A) \to D(A)$, the following non-exact functor on chain complexes does the job: given a chain complex $K^\bullet$ of f-torsionfree A-modules, define a new chain complex $\eta_f K^\bullet$ as a subcomplex of $K^\bullet[\frac{1}{f}]$ with the following terms:

$$(\eta_f K^\bullet)^i = \{\alpha \in f^i K^i \mid d(\alpha) \in f^{i+1} K^{i+1}\}.$$

One easily checks that $H^i(\eta_f K^\bullet)$ identifies with $H^i(K^\bullet)/(f\text{-torsion})$, and thus the association $K^\bullet \to \eta_f K^\bullet$ derives to give a functor $L\eta_f : D(A) \to D(A)$. This construction is motivated by ideas of Berthelot-Ogus in crystalline cohomology [**BO2**, §8], can be thought of as a "decalage" of the f-adic filtraton on K in the sense of Deligne [**De1**], and discussed in much more depth in [**BMS2**, §6], [**Bh**, §6], [**Mor**, §2].

With these tools in play, here are the two main steps in the construction:

(1) *The first approximation.* Consider the complex $A\Omega_{\mathfrak{X}}^{pre} := R\nu_* A_{inf,X}$ as an object of the derived category $D(\mathfrak{X}, A_{inf})$ of A_{inf}-modules on the formal scheme $\mathfrak{X}$. As explained above, we have

$$R\Gamma(\mathfrak{X}, R\nu_* A_{inf,X}) = R\Gamma(X_{proet}, A_{inf,X}) \stackrel{a}{\simeq} R\Gamma(X_{et}, \mathbf{Z}_p) \otimes_{\mathbf{Z}_p} A_{inf}.$$

As almost zero modules die[18] after base change along $A_{inf} \to W(C^\flat)$, this tells us that the complex $R\Gamma(\mathfrak{X}, A\Omega_{\mathfrak{X}}^{pre})$ satisfies Theorem 4.6 (1). Now let's instead consider the Hodge-Tate specialization $\widetilde{\theta}^* R\Gamma(\mathfrak{X}, A\Omega_{\mathfrak{X}}^{pre})$,

---

[18]In terms of Figure 1, the locus where almost zero modules lives is the crystalline specialization, which does not intersect the locus defined by $\mathrm{Spec}(W(C^\flat)) \to \mathrm{Spec}(A_{inf})$ the étale specialization.

where $\tilde{\theta} = \theta \circ \phi^{-1} : A_{inf} \to \mathcal{O}_C$. By formal nonsense with the projection formula, this complex identifies with the $\mathcal{O}_C$-complex $R\Gamma(X_{proet}, \widehat{\mathcal{O}_X^+})$, viewed as an A_{inf}-complex via $\tilde{\theta}$. To compute this explicitly, assume further that $\mathfrak{X} = \mathrm{Spf}(\mathcal{O}_C\langle t^{\pm 1}\rangle)$ is the formal torus. One can then essentially repeat the calculation given in Lemma 3.30 to obtain that

$$
\begin{aligned}
R\Gamma(X_{proet}, \widehat{\mathcal{O}_X^+}) &\simeq \widehat{\bigoplus_{i \in \mathbf{Z}[\frac{1}{p}]}} \left(\mathcal{O}_C \cdot T^i \xrightarrow{T^i \mapsto (\underline{\epsilon}^i - 1)T^i} \mathcal{O}_C \cdot T^i\right) \\
&\simeq \widehat{\bigoplus_{i \in \mathbf{Z}}} \left(\mathcal{O}_C \cdot T^i \xrightarrow{T^i \mapsto (\underline{\epsilon}^i - 1)T^i} \mathcal{O}_C \cdot T^i\right) \\
&\oplus \widehat{\bigoplus_{i \in \mathbf{Z}[\frac{1}{p}] - \mathbf{Z}}} \left(\mathcal{O}_C \cdot T^i \xrightarrow{T^i \mapsto (\underline{\epsilon}^i - 1)T^i} \mathcal{O}_C \cdot T^i\right) \\
&\simeq \widehat{\bigoplus_{i \in \mathbf{Z}}} \left(\mathcal{O}_C \cdot T^i \xrightarrow{0} \mathcal{O}_C \cdot T^i\right) \oplus \mathrm{Err},
\end{aligned}
$$
(16)

where Err is an $\mathcal{O}_C$-complex whose homology is killed by $\epsilon^{\frac{1}{p}} - 1$ (since $\epsilon^i - 1 \mid \epsilon^{\frac{1}{p}} - 1$ for any $i \in \mathbf{Z}[\frac{1}{p}] - \mathbf{Z}$). Thus, when viewed as an A_{inf}-complex via $\tilde{\theta}$, this tells us that $R\Gamma(X_{proet}, \widehat{\mathcal{O}_X^+})$ looks like it has the right size for the Hodge-Tate comparison, up to an error term Err whose homology is killed by $\mu := [\underline{\epsilon}] - 1$. One can also repeat the same calculation without specializing to compute $R\Gamma(\mathfrak{X}, A\Omega_{\mathfrak{X}}^{pre})$ directly in this case[19] to see that the error term Err above comes from an analogous summand of $R\Gamma(\mathfrak{X}, A\Omega_{\mathfrak{X}}^{pre})$ whose homology is also μ-torsion. Thus, we want to modify $A\Omega_{\mathfrak{X}}^{pre}$ in a manner that functorially kill the μ-torsion in its homology.

(2) *The main construction.* The preceding analysis suggests defining

$$A\Omega_{\mathfrak{X}} := L\eta_\mu A\Omega_{\mathfrak{X}}^{pret} := L\eta_\mu R\nu_* A_{inf, X} \quad \text{and} \quad R\Gamma_A(\mathfrak{X}) := R\Gamma(\mathfrak{X}, A\Omega_{\mathfrak{X}})$$

In this definition, the Frobenius $\phi_{\mathfrak{X}}$ is induced by the sequence

$$\phi^*(A\Omega_{\mathfrak{X}}) \simeq L\eta_{\phi(\mu)} \phi^* R\nu_* A_{inf, X} \simeq L\eta_{\phi(\underline{\epsilon})} L\eta_\mu R\nu_* A_{inf, X} \to L\eta_\mu R\nu_* A_{inf, X} =: A\Omega_{\mathfrak{X}},$$

where the first isomorphism is by "transport of structure", the second isomorphism relies on a transitivity property of the $L\eta$-functor (namely, $L\eta_f \circ L\eta_g \simeq L\eta_{fg}$ with obvious notation), the third map exists because of the structure of $R\nu_* A_{inf, X}$ (namely, the construction of $L\eta_f$ shows that if K can be represented by a chain complex $K^\bullet$ of f-torsionfree modules with $K^i = 0$ for $i < 0$, then there is an evident map $L\eta_f(K) \to K$) and the fact that $\phi^* A_{inf, X} \simeq A_{inf, X}$, and the last isomorphism is a definition.

This definition does indeed work, and we only briefly indicate what goes into proving the required comparison isomorphisms:

- Étale cohomology. We have already explained in (1) above why $R\Gamma(\mathfrak{X}, A\Omega_{\mathfrak{X}}^{pre})$ satisfies the requisite comparison isomorphism with

---

[19]The entire calculation remains the same: one simply replaces $\mathcal{O}_C$ with A_{inf} in the formulas above, and one is not allowed to simplify the differential on the first summand to 0 as $[\underline{\epsilon}]^i - 1$ is not zero on A_{inf} for $i \in \mathbf{Z}$.

étale cohomology after base change to $W(C^\flat)$. The rest follows immediately $L\eta_\mu(K)$ and K are naturally isomorphic after inverting μ for any complex K.
- Hodge-Tate cohomology. This comparison was essentially forced to be true by the calculation in (1) above. More precisely, one defines a map a $\Omega^1_{\mathfrak{X}/\mathcal{O}_C}\{-1\} \to \mathcal{H}^1(\widetilde{\theta}^* A\Omega_{\mathfrak{X}})$ via a variant of the construction in §3.4, and then checks that it yields isomorphisms
$$\Omega^i_{\mathfrak{X}/\mathcal{O}_C}\{-i\} \simeq \mathcal{H}^i(\widetilde{\theta}^* A\Omega_{\mathfrak{X}})$$
by unraveling the preceding map and matching it with the computation in (4). The Hodge-Tate spectral sequence is then simply the standard spectral sequence expressing the hypercohomology of a complex of sheaves in terms of the hypercohomology of its cohomology sheaves. We refer to [**BMS2**, §8], [**Bh**, §6] for more details.
- de Rham cohomology. This comparison results from the previous one using the following observation:

PROPOSITION 4.7. *For any ring A with a nonzerodivisor $f \in A$ and a complex $K \in D(A)$, the complex $L\eta_f K/f$ is naturally represented by the chain complex*

$$(H^*(K/f), \mathrm{Bock}_f)$$
$$:= \left(\cdots \to H^i(f^i K/f^{i+1} K) \xrightarrow{\mathrm{Bock}_f} H^{i+1}(f^{i+1} K/f^{i+2} K) \to \cdots\right),$$

where Bock_f is the boundary map "Bockstein" on cohomology associated to the exact triangle

$$f^{i+1}K/f^{i+2}K \xrightarrow{\mu} f^i K/f^{i+2} K \xrightarrow{\mathrm{std}} f^i K/f^{i+1} K$$

in $D(A)$. Moreover, when K admits the structure of a commutative algebra in $D(A)$, the preceding identification naturally makes $L\eta_f(K)/f$ into a differential graded algebra via cup products.

We apply this observation to $K = L\eta_{\phi^{-1}\mu} R\nu_* A_{\inf,X}$ and $f = \xi = \mu/\phi^{-1}(\mu)$ is the displayed generator of $\ker(\theta)$. Note that $A\Omega_{\mathfrak{X}} \simeq L\eta_\xi(K)$. Applying the previous observation tells us that $\theta^* A\Omega_{\mathfrak{X}} \simeq A\Omega_{\mathfrak{X}}/\xi$ is naturally represented by the differential graded algebra $(H^*(K/\xi), \mathrm{Bock}_\xi)$. The complex K is a Frobenius twist of $A\Omega_{\mathfrak{X}}$; keeping track of the twists, one learns that K/ξ is the Hodge-Tate specialization $\widetilde{\theta}^* A\Omega_{\mathfrak{X}}$. Thus, by the previous comparison, the i-th term of $H^*(K/\xi)$ is thus given by

$$\Omega^i_{\mathfrak{X}/\mathcal{O}_C}\{-i\} \otimes_{\mathcal{O}_C} \xi^i/\xi^{i+1} \simeq \Omega^i_{\mathfrak{X}/\mathcal{O}_C},$$

i.e., by differential forms. Unraveling these isomorphisms, the Bockstein differential Bock_ξ can then be checked to coincide with the de Rham differential, thus proving that $A\Omega_{\mathfrak{X}}/\xi \simeq \Omega^\bullet_{\mathfrak{X}/\mathcal{O}_C}$. We refer to [**Mor**, Theorem 5.9] and [**Bh**, Proposition 7.9] for more details on the implementation of this approach.
- Crystalline cohomology. There are two possible approaches here: one either repeats the arguments given for the de Rham comparison above using de Rham-Witt complexes to identify $A\Omega_{\mathfrak{X}}/\mu$ with the

relative de Rham-Witt complex of $\mathfrak{X}/\mathcal{O}$, or one directly proves that $A\Omega_{\mathfrak{X}} \widehat{\otimes}^L_{A_{inf}} A_{crys}$ identifies with the absolute crystalline cohomology of $\mathfrak{X}$ over A_{crys}. Both approaches yield strictly finer statements than Theorem 4.6 (4). We refer to [**BMS2**, §11], [**Mor**] for the first approach, and [**BMS2**, §12] for the second approach.

5. Exercises

This section was written jointly with Daniel Litt.

Using the Hodge-Tate decomposition.

(1) Calculate $h^{i,j}(X)$ (in the sense of Deligne's mixed Hodge theory) for the following varieties X by using the Hodge-Tate decomposition and calculating the corresponding étale cohomology groups (as Galois modules) first.
   (a) $X = \mathbf{Gr}(k,n)$ is a Grassmannian.
   (b) X is a smooth affine curve.
   (c) $X = \mathbf{P}^1/\{0,\infty\}$ is nodal rational curve.
   (d) $X \subset \mathbf{P}^2$ is a cubic curve with 1 cusp.

(2) Let R be a finitely-generated integral **Z**-algebra with fraction field K, and let X, Y be smooth proper R-schemes. Suppose that if $\mathfrak{p}$ is any closed point of $\mathrm{Spec}(R)$, and $k/\kappa(\mathfrak{p})$ is any finite extension, then $\#|X(k)| = \#|Y(k)|$.
   (a) Use the Hodge-Tate decomposition to show that $h^{i,j}(X_K) = h^{i,j}(Y_K)$ for all X, Y. (Hint: Use the Lefschetz fixed-point formula to figure out how Frobenii act; use Chebotarev to conclude that the Galois representations on the cohomology of X and Y are the same. Use the Hodge-Tate decomposition to finish the proof.)
   (b) * Let X, Y be birational Calabi-Yau varieties over the complex numbers (i.e. varieties with trivial canonical bundle). Show that they have the same Hodge numbers. (Hint: Use p-adic integration to count points of reductions.)

(3) The goal of this exercise is to use the Hodge-Tate decomposition to translate a point-counting statement to a geometric one[20]. Let $X/\mathbf{C}$ be a smooth projective variety that is defined over $\mathbf{Q}$. For a prime p, write X_p for a reduction of X to $\overline{\mathbf{F}_p}$; this makes sense for all but finitely many p's once an integral model of X has been chosen. Assume that there exists a polynomial P_X such that for all but finitely many p, we have $P_X(p) = \#X(\mathbf{F}_p)$. We shall
   (a) Show that for each n, the $G_{\mathbf{Q}}$-representation $H^n(X, \mathbf{Q}_\ell)$ is isomorphic to a direct sum of copies of $\mathbf{Q}_\ell(-i)$ up to semisimplification. (Hint: use the Weil conjectures and Chebotarev.)
   (b) Show that $h^{i,j}(X) = 0$ for $i \neq j$. (Hint: use the Hodge-Tate decomposition.)
   We also encourage the reader to think about the converse assertion: if $h^{i,j}(X) = 0$ for $i \neq j$, then is the function $p \mapsto \#X(\mathbf{F}_p)$ given by a polynomial, at least on a large set of primes? (Hint: try to use the "Newton-lies-above-Hodge" theorem.)

---

[20]We restrict ourselves to working over $\mathbf{Q}$ to avoid notational complications. The general version of the result in this exercise is due Katz, see [**HR**, Appendix].

THE HODGE-TATE DECOMPOSITION VIA PERFECTOID SPACES 231

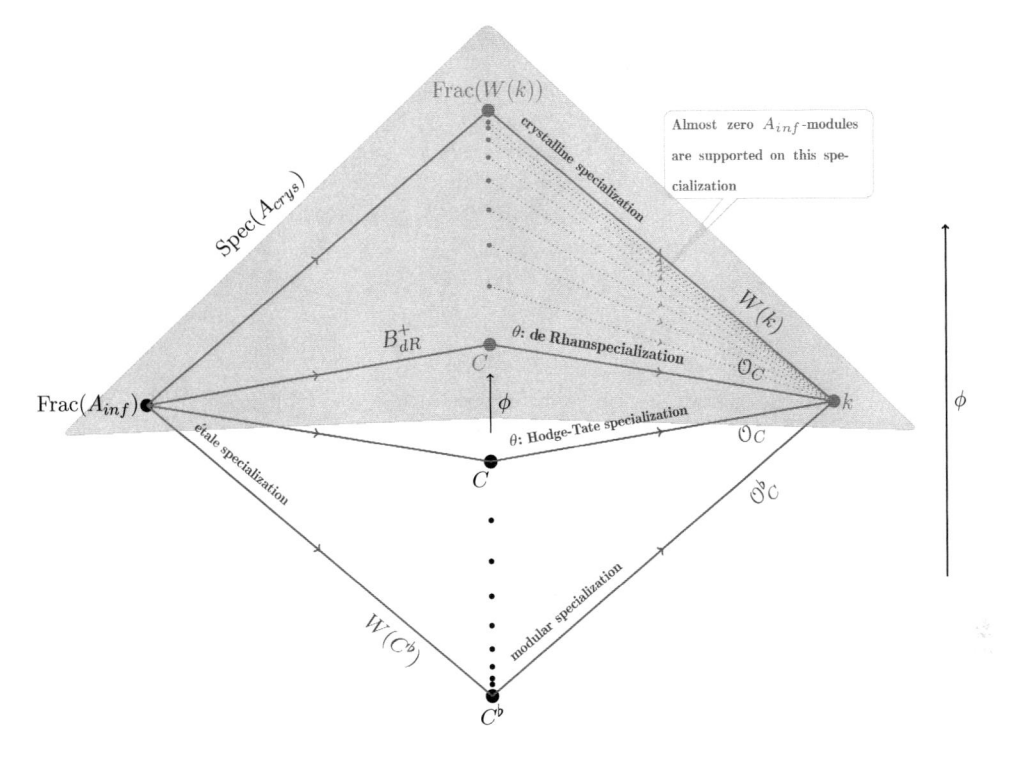

FIGURE 1. A cartoon of $\mathrm{Spec}(A_{inf})$. This depiction of the poset of prime ideals in A_{inf} emphasizes certain vertices and edges that are relevant to p-adic cohomology theories.

- The darkened vertices (labelled '●' or '●') indicate (certain) points of $\mathrm{Spec}(A_{inf})$ and are labelled by the corresponding residue field.
- The gray/orange arrows indicate specializations in the spectrum, while the blue label indicates the completed local ring along the specialization.
- The arrow labelled ϕ on the far right indicates the Frobenius action on $\mathrm{Spec}(A_{inf})$, which fixes the 4 vertices of the outer diamond in the above picture.
- The labels in purple match the arrows to one of the specializations that are important for p-adic comparison theorems.
- The smaller bullets (labelled '·' or '·') down the middle are meant to denote the $\phi^{\mathbf{Z}}$-translates to the two drawn points labelled C (with $\phi^{\mathbf{Z}_{\geq 0}}$ translates of the generic point of the de Rham specialization in orange, and the rest in black), and are there to remind the reader that not all points/specialization in $\mathrm{Spec}(A_{inf})$ have been drawn.
- The vertices/labels/arrows in orange mark the points and specializations that lie in $\mathrm{Spec}(A_{inf}/\mu) \subset \mathrm{Spec}(A_{inf})$.
- The triangular region covered in teal identifies the image of $\mathrm{Spec}(A_{crys}) \to \mathrm{Spec}(A_{inf})$.

Inverse limits of schemes and perfectoid abelian varieties.

(4) Let $\{X_i\}$ be a cofiltered system of quasi-compact and quasi-separated schemes with affine transition maps $f_{ij} : X_i \to X_j$.
  (a) Show that the inverse limit $X_\infty := \lim_i X_i$ exists in the category of schemes, and coincides with the inverse limit in the category of locally ringed spaces. Write $f_i : X_\infty \to X_i$ for the projection map.
  (b) For any quasi-coherent sheaf $\mathcal{F}$ on some X_0, show that the natural pullback induces a isomorphisms
  $$H^*(X_\infty, f_i^*\mathcal{F}) \simeq \operatorname*{colim}_{f_{i0}:X_i \to X_0} H^*(X_i, f_{i0}^*\mathcal{F}).$$

Much more material on such limits can be gleaned from [**SP**, Tag 01YT].

(5) Let k be an algebraically closed field and A/k an abelian variety of dimension g. The purpose of this problem is to show that A is a $K(\pi, 1)$.
  (a) Show that any connected finite étale cover of A is also an abeliann variety (note that this is not true for commutative group schemes which are not proper – find a commutative group scheme with a connected finite étale cover which does not admit the structure of a group scheme).
  (b) Deduce from the previous part that the étale fundamental group of A is canonically isomorphic to its Tate module.
  (c) Let B be an abelian group. Show that any class in $H^1(A_{\text{ét}}, B)$ is killed by some finite étale cover of A.
  (d) Observe that if $R = \mathbf{F}_q$ is a finite field, the ring $H^*(A_{\text{ét}}, R)$ is given a Hopf-algebra structure by the multiplication on A. Conclude that if the characteristic of R is different from that of k, then $H^*(A_{\text{ét}}, R)$ is an exterior algebra on $2g$ generators in degree 1. What happens if the characteristic of R equals that of k?
  (e) Deduce from the previous part the following fact: for any finite abeliann group B, the natural map
  $$H^*(\pi_1^{\text{ét}}(A), B) \to H^*(A_{\text{ét}}, B)$$
  is an isomorphism.

(6) The goal of this exercise is to sketch why the inverse limit of multiplication by p on an abelian scheme over $\mathcal{O}_C$ gives a perfectoid space. For this exercise, we shall need the relative Frobenius map: if S is a scheme of characteristic p, and $f : X \to S$ is a map, then we define the Frobenius twist $X^{(1)} := X \times_{\operatorname{Frob}_S, S} S$ as the base change of f along the Frobenius on S, and write $F_{X/S} : X \to X^{(1)}$ for map induced by the Frobenius on X. This fits into the following diagram:

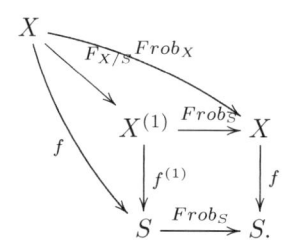

$$\begin{array}{ccc}
X & & \\
& \searrow^{\operatorname{Frob}_X} & \\
\downarrow^{F_{X/S}} & & \\
& X^{(1)} \xrightarrow{\operatorname{Frob}_S} X & \\
\downarrow^{f} & \downarrow^{f^{(1)}} & \downarrow^{f} \\
& S \xrightarrow{\operatorname{Frob}_S} S. &
\end{array}$$

Given a flat[21] map $f : X \to S$, we shall say that f is *relatively perfect* if $F_{X/S}$ is an isomorphism. Note that the functor $X \mapsto X^{(1)}$ on S-schemes preserves finite limits, and thus carries (commutative) group schemes to (commutative) group schemes.

(a) Let R be a p-adically complete and p-torsionfree $\mathcal{O}_C$-algebra such that the map $\mathrm{Spec}(R/p) \to \mathrm{Spec}(\mathcal{O}_C/p)$ relatively perfect. Show that $R[\frac{1}{p}]$ is naturally a perfectoid algebra.

(b) Let A be a ring of characteristic p, and let G be a finite flat group scheme over A. Assume that the relative Frobenius map $G \to G^{(1)}$ is the trivial map. Using Verschiebung, show that G is killed by p. Deduce the following: if H is a smooth group scheme over A, then the relative Frobenius map $H \to H^{(1)}$ factors multiplication by p on H.

(c) Let A be a ring of characteristic p. Let $\mathcal{A}$ be an abelian scheme over A. Show that the inverse limit of multiplication by p on $\mathcal{A}$ is relatively perfect over A.

(d) Let $\mathcal{A}/\mathcal{O}_C$ be a smooth abelian group scheme with generic fiber A. Show that the inverse limit $\lim_p A$ of multiplication by p on A is naturally a perfectoid space.

(e) Let $\mathcal{A}/\mathcal{O}_C$ be a smooth abelian group scheme. Show that the p-adic completion of the inverse limit $\lim_p \mathcal{A}$ depends only on the abelian $\mathcal{O}_C/p$-scheme $\mathcal{A} \otimes_{\mathcal{O}_C} \mathcal{O}_C/p$.

Derived completions of complexes.

(7) For any complex K of torsionfree abelian groups, define $\widehat{K} := \lim K/p^n K$.

(a) Show that the operation $K \mapsto \widehat{K}$ passes to the derived category $D(\mathrm{Ab})$ of abelian groups, i.e., it carries quasi-isomorphisms of chain complexes to quasi-isomorphisms. We write the resulting functor $D(\mathrm{Ab}) \to D(\mathrm{Ab})$ also by $K \mapsto \widehat{K}$, and call it the p-adic completion functor.

(b) Show that the p-adic completion functor is given by the formula
$$K \mapsto R\lim_n (K \otimes_{\mathbf{Z}}^L \mathbf{Z}/p^n).$$

(c) Show that the p-adic completion functor is exact, i.e., preserves exact triangles.

(d) Show that $\widehat{\widehat{K}} \simeq \widehat{K}$, i.e., the completion is complete.

(e) Show that $K \in D(\mathrm{Ab})$ is complete (i.e., $K \simeq \widehat{K}$) if and only if $\mathrm{RHom}(\mathbf{Z}[\frac{1}{p}], K) \simeq 0$.

(f) Prove Nakayama's lemma: for $K \in D(\mathrm{Ab})$, if $K \otimes_{\mathbf{Z}}^L \mathbf{Z}/p \simeq 0$, then $\widehat{K} \simeq 0$.

(g) If A is a p-divisible abelian group, show that $\widehat{A} \simeq T_p(A)[1]$, where $T_p(A)$ is the Tate module.

The cotangent complex, perfect rings, perfectoid rings.

---

[21]More generally, it is convenient to adopt the same terminology if f and Frob_S are Tor-independent.

(8) Let $A \to B$ be an lci map of rings, i.e., after Zariski localization on both rings, the map factors as $A \xrightarrow{a} P \xrightarrow{b} B$, where a is a polynomial extension, and b is a quotient defined by a regular sequence.
   (a) Show that $H^1(L_{B/A})$ is torsionfree (i.e., not killed by a nonzerodivisor on B).
   (b) Show that if $A \to B$ is flat and $f \in A$ is a nonzerodivisor with $A[\frac{1}{f}] \to B[\frac{1}{f}]$ smooth, then $L_{B/A} \simeq \Omega^1_{B/A}$.
   (c) Let $K/\mathbf{Q}_p$ be a nonarchimedean extension, and let L/K be an algebraic extension. Show that $L_{\mathcal{O}_L/\mathcal{O}_K} \simeq \Omega^1_{\mathcal{O}_L/\mathcal{O}_K}$.
(9) Let A be a perfect $\mathbf{F}_p$-algebra.
   (a) Use the "transitivity triangle" to show that $L_{A/\mathbf{F}_p} = 0$.
   (b) Deduce that A admits a unique flat deformation over $\mathbf{Z}/p^n$ for any n.
   (c) Using (a), show that the derived p-adic completion of $L_{W(A)/\mathbf{Z}_p}$ vanishes. Convince yourself that it is necessary to take a completion here.
   (d) Using the transitivity triangle, show that for any map $A \to B$ of perfect $\mathbf{F}_p$-algebras, the derived p-adic completion of $L_{W(B)/W(A)}$ vanishes.
   (e) More generally, if $R \to S$ is a map of p-torsionfree $\mathbf{Z}_p$-algebras such that $R/p \to S/p$ is relatively perfect, show that the p-adic completion of $L_{S/R}$ vanishes.
(10) Let $A \to B$ be a map of integral perfectoid rings.
   (a) Show that the square
   $$\begin{array}{ccc} W(A) & \longrightarrow & W(B) \\ \downarrow \theta & & \downarrow \theta \\ A & \longrightarrow & B \end{array}$$
   is a pushout square of commutative rings. (Hint: use [**BMS2**, Remark 3.11]).
   (b) Show that the derived p-adic completion of $L_{B/A}$ vanishes.
(11) Give examples of:
   (a) Give an example of a map $A \to B$ of finite type $\mathbf{C}$-algebras where $L_{B/A} \in D^{\leq -2}(B)$, i.e., $H^i(L_{B/A}) = 0$ for $i \geq -1$.
   (b) A p-adically complete ring A such that A/p is semiperfect, but $W(A) \xrightarrow{\theta} A$ does not have a principal kernel.
   (c) A semiperfect $\mathbf{F}_p$-algebra A such that $L_{A/\mathbf{F}_p}$ is nonzero.
   (d) (*) An $\mathbf{F}_p$-algebra A such that $L_{A/\mathbf{F}_p} = 0$, but A is not perfect.
(12) This exercise is meant to illustrate a general feature of certain valuation rings, and is not relevant to the rest of these notes. Let $\mathbf{Z}_p \to V$ be a faithfully flat map with V a valuation ring. Assume that $\mathrm{Frac}(V)$ is algebraically closed.
   (a) (*) Show that V can be written as a filtered colimit of regular $\mathbf{Z}_p$-algebras. (Hint: use de Jong's alterations theorem from [**dJ**]).
   (b) Deduce that $V[\frac{1}{p}]$ is ind-smooth over $\mathbf{Q}_p$. (This can be proven without using (a)).

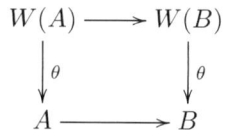

(c) Show that any regular $\mathbf{Z}_p$-algebra is lci over $\mathbf{Z}_p$.
(d) Deduce that $L_{V/\mathbf{Z}_p} \simeq \Omega^1_{V/\mathbf{Z}_p}$.

Group cohomology and the pro-étale site.

(13) Fix a finite group G. Let X be a topological space equipped with an action of G, and let $f : X \to Y$ be a G-equivariant map (for the trivial G-action on Y). Let A be a coefficient ring.
   (a) Show that the natural pullback $H^0(Y, A) \to H^0(X, A)$ has image contained inside the G-invariants $H^0(X, A)^G$. Using the spectral sequence for a composition of derived functors, deduce that there is a natural map $H^i(Y, A)$ to groups $H^i_G(X, A)$ which are computed by a Hochschild-Serre spectral sequence
   $$E_2^{i,j} : H^i(G, H^j(X, A)) \Rightarrow H^{i+j}_G(X, A).$$
   (b) Lift the preceding assertion to construct a natural map
   $$R\Gamma(Y, A) \to R\Gamma(G, R\Gamma(X, A))$$
   in the derived category $D(A)$.
   (c) If f is a G-torsor (i.e., f realizes Y as the quotient of X by G, and the G-action has no non-trivial stabilizers on X), then show that the maps above are isomorphisms, i.e., we have
   $$H^i(Y, A) \simeq H^i_G(X, A) \quad \text{and} \quad R\Gamma(Y, A) \simeq R\Gamma(G, R\Gamma(X, A)).$$
   (d) Assume that X is contractible, and that f is a G-torsor. Show that the above maps identify $H^*(X, A)$ with the group cohomology $H^*(G, A)$ of G.

(14) The goal of this exercise is to show that the ideas going into the construction of the pro-étale site lead to a sheaf-theoretic perspective on continuous cohomology, at least with a large class of coefficients; see [**BS**, §4.3], [**Sc2**, §3, erratum] for more. Let G be a profinite group. Let $\mathcal{C}_G$ be the category of sets equipped with a continuous G-action. Equip $\mathcal{C}_G$ with the structure of a site by declaring all continuous surjective maps to be covers. Write $H^*(\mathcal{C}_G, -)$ for the derived functors of $\mathcal{F} \mapsto \mathcal{F}(*)$, where $*$ is the 1 point set with the trivial G-action.
   (a) Let X be a topological space equipped with a continuous G-action. Show that $\mathrm{Hom}_G(-, X)$ defines a sheaf on $\mathcal{C}_G$. We write $\mathcal{F}_X$ for this sheaf; if X is a G-module, then $\mathcal{F}_X$ is naturally a sheaf of abelian groups, likewise for rings, etc..
   (b) Let A be a topological abelian group equipped with a continuous G-action. By considering the Cech nerve of the continuous G-equivariant map $G \to *$, show that there is a canonical map
   $$c_A : H^*_{cts}(G, A) \to H^*(\mathcal{C}_G, \mathcal{F}_A).$$
   Write $\mathcal{D}$ for the category of all A such that c_A is an isomorphism.
   (c) Show that any discrete G-module lies in $\mathcal{D}$. (Hint: first show the analogous assertion for the category $\mathcal{C}_G^f$ of finite G-sets with a continuous G-action, and then analyze the natural morphism $\mathrm{Sh}(\mathcal{C}_G) \to \mathrm{Sh}(\mathcal{C}_G^f)$ on the categories of sheaves.)

(d) Fix a sequence
$$M_1 \to M_2 \to \ldots \to M_n \xrightarrow{f_n} M_{n+1}\ldots.$$
in $\mathcal{D}$ with M_i being Hausdorff and the f_n's being closed immersions. Show that the colimit $\mathrm{colim}_i M_i$ also belongs to $\mathcal{D}$.

(e) Fix a sequence
$$\ldots M_{n+1} \xrightarrow{f_n} M_n \to \ldots \to M_2 \to M_1 \to M_0 = 0$$
Assume that each f_n has sections after base change along a continuous map $K \to M_i$ with K a profinite set, and that $\ker(f_n) \in \mathcal{D}$ for all $n \geq 1$. Then $\lim_n M_n \in \mathcal{D}$.

(f) Fix a finite extension $K/\mathbf{Q}_p$. Let $G = \mathrm{Gal}(\overline{K}/K)$. Fix a completed algebraic closure $\mathbf{C}_p$ of K, and let V be a finite dimensional $\mathbf{C}_p$-vector space with a continuous semilinear G-action. Show that $V \in \mathcal{D}$.

(15) Let $G = \oplus_{i=1}^n \mathbf{Z}_p \cdot \gamma_i$ be a finite free $\mathbf{Z}_p$-module with generators γ_i; we view G as a profinite group. Let M be a discrete G-module. Show that $H^*_{cts}(G, M)$ is computed as the cohomology of the complex
$$\otimes_{i=1}^n \left(M \xrightarrow{\gamma_i - 1} M \right).$$

Étale and de Rham cohomology in equicharacteristic p.

(16) Let k be a field of characteristic $p > 0$ and X a k-variety. Compute $H^1(X_{\mathrm{\acute{e}t}}, \mathbf{F}_p)$ if
   (a) $X = \mathbf{A}^1_k$ (Hint: Use the Artin-Schreier exact sequence).
   (b) X is a smooth, proper, geometrically connected curve of genus 1 (Hint: The answer depends on the curve).

(17) Let X be a smooth variety over a perfect field k of characteristic $p > 0$.
   (a) Suppose X admits a flat lift X' to $W_2(k)$, and that Frobenius lifts to X'. Show that the Cartier isomorphism lifts to a map of complexes
   $$\Omega^1_{X^{(p)}/k} \to F_* \Omega^\bullet_{X/k}.$$
   (b) In the situation above, let F_1, F_2 be two different lifts of Frobenius. Show that the maps constructed in (a) using these two lifts are homotopic.
   (c) Now suppose that X lifts to $W_2(k)$, but do not assume that Frobenius lifts. Show that the Cartier isomorphism lifts to a map
   $$\Omega^1_{X^{(p)}/k} \to F_* \Omega^\bullet_{X/k}$$
   in $D^b(X)$. (Hint: Cover X by affines and use a Cech complex.)

p-adic Hodge theory.

(18) Let X be a commutative group scheme over $\mathcal{O}_{\mathbf{C}_p}$.
   (a) Use the construction of the Hodge-Tate comparison map to define a pairing
   $$\int : T_p(X) \times H^1_{dR}(X) \to \mathbf{C}_p(1).$$
   (b) One can think of the above pairing as "integrating a form along a (closed) cycle." What is the analogue of a path integral?
   (c) In the case $X = \mathbf{G}_m$, make everything as explicit as you can.

(19) Let C be a complete and algebraically closed extension of $\mathbf{Q}_p$. Let $K/\mathbf{Q}_p$ be a finite extension that is contained in C. Recall that there is a natural surjective map $A_{inf} \xrightarrow{\theta} \mathcal{O}_C$. Write $B_{dR}^+ \to C$ for map obtained from the previous one by inverting p and completing, i.e., B_{dR}^+ is the completion of $A_{inf}[\frac{1}{p}]$ along $\ker(\theta[\frac{1}{p}])$.
  (a) Show that the map $\mathcal{O}_K \to \mathcal{O}_C$ lifts across $A_{inf} \to \mathcal{O}_C$ if and only if $K/\mathbf{Q}_p$ is unramified.
  (b) Show that the map $K \to C$ always lifts uniquely across $B_{dR}^+ \to C$.

  Now let X_0/K be a smooth rigid space, and let X/C denote its base change.
  (c) Using the deformation theoretic interpretation from the notes, show that the complex $\tau^{\leq 1} R\nu_* \widehat{\mathcal{O}}_X$ on X_{proet} splits for X as above.

6. Projects

This section was written jointly with Matthew Morrow. Let C be a complete and algebraically closed extension of $\mathbf{Q}_p$.

(1) **Understand the Hodge-Tate filtration for singularities**[22]. The primitive comparison theorem holds true for non-smooth spaces X as well. Thus, for X proper, we still have a "Hodge-Tate" spectral sequence
$$E_2^{i,j} : H^i(X, R^j \nu_* \widehat{\mathcal{O}}_X) \Rightarrow H^{i+j}(X, C).$$
It is thus of interest to understand the sheaves $R^j \nu_* \widehat{\mathcal{O}}_X$. This problem turns out to be closely related to the singularities of X. Recall first that a ring R is called *semi-normal* if and only if, for any $y, z \in R$ satisfying $y^3 = z^2$, there exists a unique $x \in R$ satisfying $x^2 = y$, $x^3 = z$. A relevant source for the basic theory of semi-normal rings, schemes, and rigid analytic spaces is [**KL**, §1.4, §3.7]. In particular, perfectoid rings are semi-normal and so, for any rigid analytic space X, the pro-étale sheaf $\widehat{\mathcal{O}}_X$ takes values in semi-normal rings; in fact, the pro-étale site of X and of its semi-normalisation are equivalent (as ringed topoi).
  (a) Deduce that if X is not semi-normal, then $\mathcal{O}_X \to R^0 \nu_* \widehat{\mathcal{O}}_X$ cannot be an isomorphism. See this explicitly in the case of a cusp $X = \mathrm{Sp}(C\langle X, Y\rangle/(X^2 - Y^3))$ by computing $H^0(X_{proet}, \widehat{\mathcal{O}}_X)$. In fact, [**KL**, Theorem 8.23] proves that $\mathcal{O}_X \to R^0 \nu_* \widehat{\mathcal{O}}_X$ is an isomorphism if and only if X is semi-normal; their proof shows how resolutions of singularities enters the picture.
  (b) Are the sheaves $R^j \nu_* \widehat{\mathcal{O}}_X$ coherent? A first attempt might be to try and reduce to the smooth case using resolution of singularities.
  (c) The construction given in the notes still produces a map
$$\Omega^1_{X/C}(-1) \to R^1 \nu_* \widehat{\mathcal{O}}_X.$$
When is this map an isomorphism? Moreover, when is the induced map $\Omega^i_{X/C}(-i) \to R^i \nu_* \widehat{\mathcal{O}}_X$ an isomorphism? For example, is it true with mild control on the singularities of X, such as quotient singularities? Note that if X has quotient singularities (say $X = Y/G$) then the "h-differential forms on X" equal the G-stable forms on Y,

---

[22] This question comes from David Hansen via Kedlaya.

by [**HJ**, Proposition 4.10]. For general X, the case $j = \dim X$ may be most accessible.

(d) Combining the isomorphism of (a) with [**HJ**, Proposition 4.5] shows that $R^0\nu_*\widehat{\mathcal{O}_X}$ is related to the h-sheafification of $\mathcal{O}_X$ (here we implicitly assume that X is an algebraic variety, and we abusively also write X for the associated rigid analytic space). Is there a similar relation between $R^j\nu_*\widehat{\mathcal{O}_X}$ and the h-sheaves $\Omega^j_{-/C,h}$ obtained by sheafifying $U \mapsto H^0(U, \Omega^j_{U/C})$ for the h-topology on varieties over C. For example, do we have
$$\dim H^i(X, R^j\nu_*\widehat{\mathcal{O}_X}) = \dim H^i_h(X, \Omega^j_{-/C,h})$$
when X is proper? Note that $H^i_h(X, \Omega^j_{-/C,h})$ is gr^j for Deligne's Hodge filtration on $H^{i+j}_{dR}(X)$.

An alternative approach to some of these questions may come from the notion of sousperfectoid rings. An affinoid algebra R over C is said to be *sousperfectoid* if and only if there exists a perfectoid Tate algebra R_∞ and a continuous algebra homomorphism $R \to R_\infty$ which admits an R-module splitting. It seems to be true that this is equivalent to R being semi-normal[23]. If $R \to R_\infty$ is flat, then sousperfectoid implies semi-normal by [**KL**, Lemma 1.4.13].

(2) **Understanding torsion discrepancies.** Let $\mathfrak{X}$ be a proper smooth formal scheme over $\mathcal{O}_C$ with generic fibre X. In this situation, we have several natural integral cohomology theories:
- Étale cohomology $H^n(X_{et}, \mathbf{Z}_p)$.
- Hodge-Tate cohomology $H^n(\widetilde{\theta}^* R\Gamma_A(\mathfrak{X}))$.
- de Rham cohomology $H^n_{dR}(\mathfrak{X}/\mathcal{O}_C)$.
- Crystalline cohomology $H^n_{crys}(\mathfrak{X}_k/W(k))$.
- Hodge cohomology $\oplus_{i+j=n} H^i(\mathfrak{X}, \Omega^j_{\mathfrak{X}/\mathcal{O}_C})$.

Each of these is a finitely presented module over a p-adic valuation ring, and they all have the same rank by fundamental results of p-adic Hodge theory. The first four of these are essentially specializations of $R\Gamma_A(\mathfrak{X})$; the order in which they appear above is roughly inverse to the order in which the corresponding specializations are described in A_{inf}-picture in the notes.

The main theorems of [**BMS2**], as explained in the notes, imply that the torsion in étale cohomology is bounded above by the torsion in the de Rham and crystalline cohomology. One expects the same relation to hold for Hodge and Hodge-Tate cohomology as well:

(a) Does one have
$$\ell_{\mathbf{Z_p}}(H^n(X_{et}, \mathbf{Z}_p)_{tors}) \leq \ell_{\mathcal{O}_C}(H^i(\widetilde{\theta}^* R\Gamma_A(\mathfrak{X}))_{tors}) \leq \sum_{i+j=n} \ell_{\mathcal{O}_C}(H^i(\mathfrak{X}, \Omega^j_{\mathfrak{X}/\mathcal{O}_C})_{tors}),$$
where $\ell_{\mathcal{O}_C}$ is the normalized length, as explained in the notes?

As we have seen in the notes, such inequalities can sometimes be strict, and cannot in general be upgraded to an inclusion of torsion subgroups.

---

[23]This is asserted in problem 6 of http://scripts.mit.edu/~kedlaya/wiki/index.php?title=The_Nonarchimedean_Scottish_Book. It might be worthwhile to rediscover the proof.

The goal of this project is to investigate relationships between the torsion subgroups occurring in these cohomology theories, both theoretically as well as through examples. Two natural unanswered questions here are:

(b) By [**BMS2**], de Rham and Hodge-Tate cohomologies occur as specializations of $R\Gamma_A(\mathfrak{X})$ along θ and $\widetilde{\theta}$. Is there a relation between the torsion subgroups of these cohomology theories? For example, is it always the case that $\ell_{\mathcal{O}_C}(H^n_{dR}(\mathfrak{X}/\mathcal{O}_C)_{tors}) \geq \ell_{\mathcal{O}_C}(H^n(\widetilde{\theta}^* R\Gamma_A(\mathfrak{X}))_{tors})$? In a search for counterexamples, a natural starting point, as in [**BMS2**, §2], is to construct "interesting" finite flat group schemes over $\mathcal{O}_C$, and to consider cohomology of quotients of smooth projective schemes by free actions of such groups.

(c) Does there exist an example of an $\mathfrak{X}$ as above where the étale and de Rham cohomologies are torsionfree, but the Hodge cohomology is not? What about an example where Hodge cohomology has more torsion than de Rham cohomology?

Notice that we did not include crystalline cohomology above. The reason is that [**BMS2**, Lemma 4.18] asserts: for a fixed n, $H^n_{crys}(\mathfrak{X}_k/W(k))$ is torsionfree if and only if $H^n_{dR}(\mathfrak{X}/\mathcal{O}_C)$. This is a statement entirely on the "de Rham" side and requires no knowledge of étale cohomology; however, the proof passes through the A_{inf}-cohomology theory and étale cohomology of the generic fibre.

(d) Find a direct proof of the preceding assertion without passing through étale cohomology or the generic fibre.

We end by briefly discussing spectral sequences. The construction of the Hodge-Tate spectral sequence also works integrally to give a spectral sequence

$$E_2^{i,j} : H^i(\mathfrak{X}, \Omega^j_{\mathfrak{X}/\mathcal{O}_C})\{-j\} \Rightarrow H^{i+j}(\widetilde{\theta}^* R\Gamma_A(\mathfrak{X}))$$

converging to the Hodge-Tate cohomology introduced above.

(e) Show that by reduction modulo the maximal ideal of $\mathcal{O}_C$, the integral Hodge-Tate spectral sequence admits a natural map to the conjugate spectral sequence

$$E_2^{i,j} : H^i(\mathfrak{X}_k^{(1)}, \Omega^j_{\mathfrak{X}^{(1)}/\mathcal{O}_C}) \Rightarrow H^{i+j}_{dR}(\mathfrak{X}_k/k),$$

where $\mathfrak{X}_k^{(1)}$ denotes the Frobenius twist relative to k of $\mathfrak{X}$. (This exercise entails understanding the construction of $R\Gamma_A(\mathfrak{X})$.)

(f) Find an $\mathfrak{X}$ as above for which integral Hodge-Tate spectral sequence does not degenerate. In view of the preceding compatibility, a natural starting point would be to find a smooth variety Y/k for which the conjugate spectral sequence does not degenerate, and then find a lift $\mathfrak{X}$ of Y to $\mathcal{O}_C$. Note that the non-degeneration of the conjugate spectral sequence is closely related to the non-liftability of Y to $W_2(k)$ (and thus the non-liftability of $\mathfrak{X}$ to $A_{inf}/\ker(\theta)^2$); this suggests that a suitable Y might be constructed by approximating a finite flat group scheme over k that lifts to $\mathcal{O}_C$ but not to $W_2(k)$.

(3) **Perfectoid universal covers for abelian varieties:** Let A/C be an abelian variety. Consider the tower
$$A_\infty := \left(... \to A \xrightarrow{p} A \xrightarrow{p} A\right)$$
of multiplication by p maps on A. This tower is an object $A_\infty \in A_{proet}$, and the structure map $f : A_\infty \to A$ is a pro-étale $T_p(A)$-torsor. The question we want to explore is: is A_∞ representable by a perfectoid space? More precisely, is there a perfectoid space that is $\sim$ to (in the sense of [**Sc3**, Definition 2.20]) the limit of the above tower? In some ways, this question appears to be the p-adic analog of the fact that the universal cover of a complex abelian variety is a Stein space[24].

Why this should be true. When A has good reduction, the arguments sketched in the exercises explain why A_∞ is naturally a perfectoid space. More generally, an affirmative answer in general can likely be extracted in general from a careful reading of [**Sc4**, §III]. However, any such argument would be necessarily indirect (as it would entail invoking the structure of the boundary in the minimal compactification of $\mathcal{A}_g$, as well as using the Hodge-Tate period map to move the moduli point of A to a "sufficiently close to ordinary" one), and it would be better to come up with a direct argument that is intrinsic to A.

Possible strategy via p-adic uniformization. One might try to construct A_∞ as a perfectoid space by mimicing the construction that works in the good reduction case using the Neron model to replace the non-existent good model, i.e., by contemplating the generic fibre of the p-adically completed inverse limit of multiplication by p on the identity component $\mathcal{A}$ of the Neron model of A. However, this does not quite work: when A has bad reduction, the generic fibre of the p-adic completion of $\mathcal{A}$ is not all of A, but rather just an open subgroup of A (as adic spaces), so at best this approach would construct an open subspace of A_∞ as a perfectoid space. But this suggests an obvious strategy: using p-adic uniformization of abelian varieties, we may write $A = E/M$ in rigid geometry, where E is an extension of an abelian variety B with good reduction by a torus T (and is constructed as an enlargement of the generic fibre of $\mathcal{A}$), and $M \subset T \subset E$ is a lattice of "periods" defining A. In fact, the covering map $\pi : E \to A$ can be constructed from $\mathcal{A}$ (see [**BL1**, §1] for a summary, and [**Hu1**, §5] for the adic geometry variant) and has sections locally on A. Thus, one may attempt the following:

(a) Try to show that the inverse limit of multiplication by p on E is naturally a perfectoid space by putting together the analogous assertions for B and T.

(b) If (a) works, then try to conclude that A_∞ is perfectoid using the fact that π has local sections.

Assuming the preceding strategy to represent A_∞ by a perfectoid space works, we would learn:

---

[24]For example, the perfectoidness of A_∞ implies the following, which can also be seen using the Stein property of the universal cover in complex geometry: for any constructible sheaf F of $\mathbf{F}_p$-vector spaces A, the direct limit $\varinjlim_n H^i(A, [p^n]^*F)$ vanishes for $i > \dim(\operatorname{Supp}(F))$. In other words, the cohomology of constructible $\mathbf{F}_p$-sheaves on A_∞ behaves like that on a Stein space.

- Unlike the approach via the Hodge-Tate period map, the approach via p-adic uniformization also potentially applies to "abeloid spaces" A that are not necessarily algebraic (see [**Lu**]), i.e., the rigid-geometry analog of complex tori; this appears to be the correct generality, at least in analogy with the universal cover from complex geometry.
- The perfectoidness of A_∞ should yield, via [**Sc4**, II.2] and almost purity theorem, the following: for any subvariety $X \subset A$, the "universal cover" $X_\infty \to X$ is naturally a perfectoid space. For the more geometrically inclined, it might be fun to try to prove this last statement directly when X is a hyperbolic curve.

(4) $L\eta$ **and pro-complexes.** This is essentially a question in homological algebra, but it is motivated by integral p-adic Hodge theory. Fix a complex $K \in D(\mathbf{Z}_p)$ that is derived p-adically complete together with an isomorphism $\phi : L\eta_p(K) \simeq K$. The Berthelot-Ogus theorem [**BO2**, §8] tells us that the crystalline cohomology complex of any smooth affine scheme in characteristic p carries this structure. What can be said about such K's in general?

   (a) Iterating ϕ gives an isomorphism $L\eta_{p^n}(K) \simeq K$. Proposition 4.7 then tells us that K/p^n can be represented by the chain complex $(H^*(K/p^n), \text{Bock}_{p^n})$ for all n. As K is derived p-adically complete, it is tempting to guess that the pair (K, ϕ) carries no homotopical information. More precisely, say $\mathcal{C}$ is the ∞-category of all K as above (suitably defined). Is $\mathcal{C}$ discrete?

   (b) One has the standard restriction map $K/p^{n+1} \to K/p^n$. Via the identification of K/p^n as $(H^*(K/p^n), \text{Bock}_{p^n})$, one can check that this gives a map $R : H^i(K/p^{n+1}) \to H^i(K/p^n)$ on the i-th term that is compatible with the Bockstein differential. On the other hand, there is also a standard map $F : H^i(K/p^{n+1}) \to H^i(K/p^n)$. How are these related? Is there a connection to the F-V-pro-complexes appearing in the work of Langer-Zink (see [**LZ**], [**BMS2**, §10.2])?

REMARK 6.1 (Update on the projects). Between the time of the winter school and the publication of these proceedings, several of the problems listed above have been solved:

(1) The paper [**BGHSSWY**] solves problem (3).
(2) The paper [**BLM**] solves problem (4).
(3) The forthcoming PhD thesis of Haoyang Guo addresses problem (1).

References

[AH] M. F. Atiyah and F. Hirzebruch, *Analytic cycles on complex manifolds*, Topology **1** (1962), 25–45, DOI 10.1016/0040-9383(62)90094-0. MR0145560

[Be] A. Beilinson, *p-adic periods and derived de Rham cohomology*, J. Amer. Math. Soc. **25** (2012), no. 3, 715–738, DOI 10.1090/S0894-0347-2012-00729-2. MR2904571

[BO1] P. Berthelot and A. Ogus, *F-isocrystals and de Rham cohomology. I*, Invent. Math. **72** (1983), no. 2, 159–199, DOI 10.1007/BF01389319. MR700767

[BO2] Pierre Berthelot and Arthur Ogus, *Notes on crystalline cohomology*, Princeton University Press, Princeton, N.J.; University of Tokyo Press, Tokyo, 1978. MR0491705

[Ba] Victor V. Batyrev, *Birational Calabi-Yau n-folds have equal Betti numbers*, New trends in algebraic geometry (Warwick, 1996), London Math. Soc. Lecture Note Ser., vol. 264, Cambridge Univ. Press, Cambridge, 1999, pp. 1–11, DOI 10.1017/CBO9780511721540.002. MR1714818

[BS] Bhargav Bhatt and Peter Scholze, *The pro-étale topology for schemes* (English, with English and French summaries), Astérisque **369** (2015), 99–201. MR3379634

[BMS1] B. Bhatt, M. Morrow, and P. Scholze, *Integral p-adic Hodge theory—announcement*, Math. Res. Lett. **22** (2015), no. 6, 1601–1612, DOI 10.4310/MRL.2015.v22.n6.a3. MR3507252

[BMS2] Bhargav Bhatt, Matthew Morrow, and Peter Scholze, *Integral p-adic Hodge theory*, Publ. Math. Inst. Hautes Études Sci. **128** (2018), 219–397, DOI 10.1007/s10240-019-00102-z. MR3905467

[Bh] Bhargav Bhatt, *Specializing varieties and their cohomology from characteristic 0 to characteristic p*, Algebraic geometry: Salt Lake City 2015, Proc. Sympos. Pure Math., vol. 97, Amer. Math. Soc., Providence, RI, 2018, pp. 43–88. MR3821167

[BIM] B. Bhatt, S. Iyengar, L. Ma, *Regular rings and perfectoid rings*, available at https://arxiv.org/abs/1803.03229

[BLM] B. Bhatt, J. Lurie, A. Mathew, *Revisiting the de Rham-Witt complex*, available at https://arxiv.org/abs/1805.05501

[BGHSSWY] C. Blakestad, D. Gvirtz, B. Heuer, D. Shchedrina, K. Shimizu, P. Wear, Z. Yao, *Perfectoid covers of abelian varieties*, available at https://arxiv.org/abs/1804.04455

[BL1] Siegfried Bosch and Werner Lütkebohmert, *Degenerating abelian varieties*, Topology **30** (1991), no. 4, 653–698, DOI 10.1016/0040-9383(91)90045-6. MR1133878

[BL2] Siegfried Bosch and Werner Lütkebohmert, *Formal and rigid geometry. I. Rigid spaces*, Math. Ann. **295** (1993), no. 2, 291–317, DOI 10.1007/BF01444889. MR1202394

[Ce] K. Cesnavicus, *The A_{inf}-cohomology theory in the semistable case*, available at https://math.berkeley.edu/~kestutis/Ainf-coho.pdf.

[CLL] B. Chiarellotto, C. Lazda, C. Liedtke, *Crystalline Galois Representations arising from K3 Surfaces*, available at https://arxiv.org/abs/1701.02945.

[CG] B. Conrad, O. Gabber *Spreading out rigid-analytic families*, to appear.

[dJ] A. J. de Jong, *Smoothness, semi-stability and alterations*, Inst. Hautes Études Sci. Publ. Math. **83** (1996), 51–93. MR1423020

[De1] Pierre Deligne, *Théorie de Hodge. II* (French), Inst. Hautes Études Sci. Publ. Math. **40** (1971), 5–57. MR0498551

[De2] Pierre Deligne, *Théorie de Hodge. III* (French), Inst. Hautes Études Sci. Publ. Math. **44** (1974), 5–77. MR0498552

[DI] Pierre Deligne and Luc Illusie, *Relèvements modulo p^2 et décomposition du complexe de de Rham* (French), Invent. Math. **89** (1987), no. 2, 247–270, DOI 10.1007/BF01389078. MR894379

[DL] Jan Denef and François Loeser, *Definable sets, motives and p-adic integrals*, J. Amer. Math. Soc. **14** (2001), no. 2, 429–469, DOI 10.1090/S0894-0347-00-00360-X. MR1815218

[Fa1] Gerd Faltings, *p-adic Hodge theory*, J. Amer. Math. Soc. **1** (1988), no. 1, 255–299, DOI 10.2307/1990970. MR924705

[Fa2] Gerd Faltings, *Crystalline cohomology and p-adic Galois-representations*, Algebraic analysis, geometry, and number theory (Baltimore, MD, 1988), Johns Hopkins Univ. Press, Baltimore, MD, 1989, pp. 25–80. MR1463696

[Fa3] Gerd Faltings, *Integral crystalline cohomology over very ramified valuation rings*, J. Amer. Math. Soc. **12** (1999), no. 1, 117–144, DOI 10.1090/S0894-0347-99-00273-8. MR1618483

[Fo1] Jean-Marc Fontaine, *Formes différentielles et modules de Tate des variétés abéliennes sur les corps locaux* (French), Invent. Math. **65** (1981/82), no. 3, 379–409, DOI 10.1007/BF01396625. MR643559

[Fo2] Jean-Marc Fontaine, *Le corps des périodes p-adiques* (French), Astérisque **223** (1994), 59–111. With an appendix by Pierre Colmez; Périodes p-adiques (Bures-sur-Yvette, 1988). MR1293971

[Fa4] Gerd Faltings, *Almost étale extensions*, Astérisque **279** (2002), 185–270. Cohomologies p-adiques et applications arithmétiques, II. MR1922831

[EGA] A. Grothendieck, *Éléments de géométrie algébrique. I. Le langage des schémas*, Inst. Hautes Études Sci. Publ. Math. **4** (1960), 228. MR0217083

[Gro]　A. Grothendieck, *Crystals and the de Rham cohomology of schemes*, Dix exposés sur la cohomologie des schémas, Adv. Stud. Pure Math., vol. 3, North-Holland, Amsterdam, 1968, pp. 306–358. Notes by I. Coates and O. Jussila. MR269663

[GR]　Ofer Gabber and Lorenzo Ramero, *Almost ring theory*, Lecture Notes in Mathematics, vol. 1800, Springer-Verlag, Berlin, 2003. MR2004652

[HR]　Tamás Hausel and Fernando Rodriguez-Villegas, *Mixed Hodge polynomials of character varieties*, Invent. Math. **174** (2008), no. 3, 555–624, DOI 10.1007/s00222-008-0142-x. With an appendix by Nicholas M. Katz. MR2453601

[Hu1]　R. Huber, *A generalization of formal schemes and rigid analytic varieties*, Math. Z. **217** (1994), no. 4, 513–551, DOI 10.1007/BF02571959. MR1306024

[Hu2]　Roland Huber, *Étale cohomology of rigid analytic varieties and adic spaces*, Aspects of Mathematics, E30, Friedr. Vieweg & Sohn, Braunschweig, 1996. MR1734903

[HJ]　Annette Huber and Clemens Jörder, *Differential forms in the h-topology*, Algebr. Geom. **1** (2014), no. 4, 449–478, DOI 10.14231/AG-2014-020. MR3272910

[Ill1]　Luc Illusie, *Complexe cotangent et déformations. I* (French), Lecture Notes in Mathematics, Vol. 239, Springer-Verlag, Berlin-New York, 1971. MR0491680

[Ill2]　Luc Illusie, *Complexe cotangent et déformations. II* (French), Lecture Notes in Mathematics, Vol. 283, Springer-Verlag, Berlin-New York, 1972. MR0491681

[Ill3]　Luc Illusie, *Grothendieck's existence theorem in formal geometry*, Fundamental algebraic geometry, Math. Surveys Monogr., vol. 123, Amer. Math. Soc., Providence, RI, 2005, pp. 179–233. With a letter (in French) of Jean-Pierre Serre. MR2223409

[It]　Tetsushi Ito, *Stringy Hodge numbers and p-adic Hodge theory*, Compos. Math. **140** (2004), no. 6, 1499–1517, DOI 10.1112/S0010437X04001095. MR2098399

[Ke]　Kiran S. Kedlaya, *More étale covers of affine spaces in positive characteristic*, J. Algebraic Geom. **14** (2005), no. 1, 187–192, DOI 10.1090/S1056-3911-04-00381-9. MR2092132

[KL]　Kiran S. Kedlaya and Ruochuan Liu, *Relative p-adic Hodge theory: foundations* (English, with English and French summaries), Astérisque **371** (2015), 239. MR3379653

[KL]　Kiran S. Kedlaya and Ruochuan Liu, *Relative p-adic Hodge theory: foundations* (English, with English and French summaries), Astérisque **371** (2015), 239. MR3379653

[Ki1]　Mark Kisin, *Potential semi-stability of p-adic étale cohomology*, Israel J. Math. **129** (2002), 157–173, DOI 10.1007/BF02773161. MR1910940

[Ki2]　Mark Kisin, *Crystalline representations and F-crystals*, Algebraic geometry and number theory, Progr. Math., vol. 253, Birkhäuser Boston, Boston, MA, 2006, pp. 459–496, DOI 10.1007/978-0-8176-4532-8_7. MR2263197

[Ko]　M. Kontsevich, *Lecture at Orsay*, 1995.

[LZ]　Andreas Langer and Thomas Zink, *De Rham-Witt cohomology for a proper and smooth morphism*, J. Inst. Math. Jussieu **3** (2004), no. 2, 231–314, DOI 10.1017/S1474748004000088. MR2055710

[La]　Eike Lau, *Dieudonné theory over semiperfect rings and perfectoid rings*, Compos. Math. **154** (2018), no. 9, 1974–2004, DOI 10.1112/s0010437x18007352. MR3867290

[Lu]　Werner Lütkebohmert, *The structure of proper rigid groups*, J. Reine Angew. Math. **468** (1995), 167–219, DOI 10.1515/crll.1995.468.167. MR1361790

[MV]　Fabien Morel and Vladimir Voevodsky, $\mathbf{A}^1$-*homotopy theory of schemes*, Inst. Hautes Études Sci. Publ. Math. **90** (1999), 45–143 (2001). MR1813224

[Mor]　M. Morrow, *Notes on the A_{inf}-cohomology of Integral p-adic Hodge theory*, preprint (2016). Available at http://arxiv.org/abs/1608.00922.

[Qu1]　Daniel G. Quillen, *Homotopical algebra*, Lecture Notes in Mathematics, No. 43, Springer-Verlag, Berlin-New York, 1967. MR0223432

[Qu2]　Daniel Quillen, *On the (co-) homology of commutative rings*, Applications of Categorical Algebra (Proc. Sympos. Pure Math., Vol. XVII, New York, 1968), Amer. Math. Soc., Providence, R.I., 1970, pp. 65–87. MR0257068

[Sc1]　Peter Scholze, *Perfectoid spaces*, Publ. Math. Inst. Hautes Études Sci. **116** (2012), 245–313, DOI 10.1007/s10240-012-0042-x. MR3090258

[Sc2]　Peter Scholze, *p-adic Hodge theory for rigid-analytic varieties*, Forum Math. Pi **1** (2013), e1, 77, DOI 10.1017/fmp.2013.1. MR3090230

[Sc3] Peter Scholze, *Perfectoid spaces: a survey*, Current developments in mathematics 2012, Int. Press, Somerville, MA, 2013, pp. 193–227. MR3204346

[Sc4] Peter Scholze, *On torsion in the cohomology of locally symmetric varieties*, Ann. of Math. (2) **182** (2015), no. 3, 945–1066, DOI 10.4007/annals.2015.182.3.3. MR3418533

[SW] P. Scholze, J. Weinstein, *Lectures on p-adic geometry*, notes for Scholze's course at Berkeley in Fall 2014, available at http://math.bu.edu/people/jsweinst/Math274/ScholzeLectures.pdf

[Se] Jean-Pierre Serre, *Sur la topologie des variétés algébriques en caractéristique p* (French), Symposium internacional de topología algebraica International symposium on algebraic topology, Universidad Nacional Autónoma de México and UNESCO, Mexico City, 1958, pp. 24–53. MR0098097

[SP] *The Stacks Project*. Available at http://stacks.math.columbia.edu.

[Ta] J. T. Tate, *p-divisible groups*, Proc. Conf. Local Fields (Driebergen, 1966), Springer, Berlin, 1967, pp. 158–183. MR0231827

[To] Burt Totaro, *The Chow ring of a classifying space*, Algebraic K-theory (Seattle, WA, 1997), Proc. Sympos. Pure Math., vol. 67, Amer. Math. Soc., Providence, RI, 1999, pp. 249–281, DOI 10.1090/pspum/067/1743244. MR1743244

[Vo] Claire Voisin, *Hodge theory and complex algebraic geometry. I*, Cambridge Studies in Advanced Mathematics, vol. 76, Cambridge University Press, Cambridge, 2002. Translated from the French original by Leila Schneps. MR1967689

[We] Charles A. Weibel, *An introduction to homological algebra*, Cambridge Studies in Advanced Mathematics, vol. 38, Cambridge University Press, Cambridge, 1994. MR1269324

Perfectoid Shimura varieties

Ana Caraiani

ABSTRACT. This is an expanded version of the lecture notes for the minicourse I gave at the 2017 Arizona Winter School. In these notes, I discuss Scholze's construction of Galois representations for torsion classes in the cohomology of locally symmetric spaces for GL_n, with a focus on his proof that Shimura varieties of Hodge type with infinite level at p acquire the structure of perfectoid spaces. I also briefly discuss some recent vanishing results for the cohomology of Shimura varieties with infinite level at p.

Contents

1. Introduction
2. Locally symmetric spaces and Shimura varieties
3. Background from p-adic Hodge theory
4. The canonical subgroup and the anticanonical tower
5. Perfectoid Shimura varieties and the Hodge–Tate period morphism
6. The cohomology of locally symmetric spaces: conjectures and results
References

1. Introduction

One of the famous consequences of the Langlands program is the theorem that all elliptic curves over $\mathbb{Q}$ are modular [**Wil95, TW95, BCDT01**]. The proof of this theorem for semistable elliptic curves led to Wiles's proof of Fermat's last theorem [**Wil95**] and had an enormous impact on number theory over the decades since.

What does it mean to say that an elliptic curve is modular? It roughly means that the elliptic curve corresponds to a modular form. For example, the elliptic curve $E/\mathbb{Q}$ defined by the equation

$$y^2 + y = x^3 - x^2$$

corresponds to the modular form $f(z)$ with Fourier expansion

$$f(z) = q \cdot \prod_{n=1}^{\infty} (1-q^n)^2 (1-q^{11n})^2 = \sum_{n=1}^{\infty} a_n q^n,$$

This work was supported in part by a Royal Society University Research Fellowship.

l	2	3	5	7	13	17	19	23	29
$\#E(\mathbb{F}_l)$	5	5	5	10	10	20	20	25	30
a_l	-2	-1	1	-2	4	-2	0	-1	0

FIGURE 1. The number of points on the elliptic curve $E : y^2 + y = x^3 - x^2$, and the coefficients of the modular form $\sum a_i q^i = q \prod_{n=1}^{\infty}(1-q^n)^2(1-q^{11n})^2$. These satisfy $l + 1 - \#E(\mathbb{F}_l) = a_l$ for primes $l \neq 11$.

where $q = e^{2\pi i z}$. The connection between E and f can be made explicit, by relating the number of points of E over finite fields to the Fourier coefficients of f. Concretely, we have

$$\ell + 1 - \#E(\mathbb{F}_\ell) = a_\ell$$

for every prime number ℓ.

The more sophisticated statement that encodes the relationship between E and f says that the p-adic *Galois representations* attached to each of these two objects are isomorphic

$$\rho_E \simeq \rho_f : G_{\mathbb{Q}} := \mathrm{Gal}(\overline{\mathbb{Q}}/\mathbb{Q}) \to \mathrm{GL}_2(\mathbb{Q}_p),$$

for every prime number p.

We recall that the p-adic Galois representation attached to E arises from the Tate module of E, using the natural $G_{\mathbb{Q}}$-action on the p^n-torsion points of E for every integer $n \geq 1$:

$$\rho_E : G_{\mathbb{Q}} \to \mathrm{GL}(\varprojlim_n E[p^n]) \simeq \mathrm{GL}_2(\varprojlim_n \mathbb{Z}/p^n\mathbb{Z}) \simeq \mathrm{GL}_2(\mathbb{Z}_p).$$

We can rephrase this by saying that the Galois representation arises from the first étale homology of the elliptic curve $E/\mathbb{Q}$. The Galois representation ρ_f satisfies the Eichler–Shimura relation

$$\mathrm{tr}(\rho_f(\mathrm{Frob}_\ell)) = a_\ell,$$

where Frob_ℓ is the geometric Frobenius at the prime number $\ell \neq p, 11$, which determines a conjugacy class in $G_{\mathbb{Q}}$.

The equalities

$$\ell + 1 - \#E(\mathbb{F}_\ell) = a_\ell$$

can be recovered from

$$\rho_E \simeq \rho_f$$

when $\ell \neq p, 11$ by taking the traces of Frob_ℓ on either side, applying the Lefschetz trace formula for the action of Frob_ℓ on the p-adic étale homology of $E/\mathbb{F}_\ell$, and applying the Eichler-Shimura relation for f.

EXERCISE 1.0.1. *Convince yourself that $\rho_E \simeq \rho_f$ really does recover the relation $\ell + 1 - \#E(\mathbb{F}_\ell) = a_\ell$ for every prime $\ell \neq p, 11$. Of course, we can vary p. What happens for $\ell = 11$?*

These notes are meant to explain how to vastly generalize the construction of the Galois representation ρ_f, so we start by recalling the key elements involved in the construction of ρ_f, going back to Eichler and Shimura. Recall that, under a first approximation, modular forms are holomorphic functions on the upper-half plane

$$\mathbb{H}^2 = \{z \in \mathbb{C} | \; \mathrm{Im} \; z > 0\}$$

which satisfy many symmetries. These symmetries are defined in terms of certain discrete subgroups of $SL_2(\mathbb{R})$. The upper-half plane has a transitive action of $SL_2(\mathbb{R})$ by Möbius transformations

$$\gamma = \begin{pmatrix} a & b \\ c & d \end{pmatrix}, \gamma : z \mapsto \frac{az+b}{cz+d}.$$

The modular form f is a cusp form of weight 2 and level

$$\Gamma_0(11) := \{\gamma \in SL_2(\mathbb{Z}) | \gamma \equiv \begin{pmatrix} * & * \\ 0 & * \end{pmatrix} \pmod{11}\},$$

a subgroup of $SL_2(\mathbb{Z})$ defined by congruence conditions. The weight and the level of f specify the symmetries that f must satisfy:

$$f\left(\frac{az+b}{cz+d}\right) = (cz+d)^2 f(z).$$

REMARK 1.0.2. The Möbius transformations are actually all the holomorphic isometries of $\mathbb{H}^2$ when we endow $\mathbb{H}^2$ with the hyperbolic metric $\frac{(dx)^2+(dy)^2}{y^2}$, where $z = x + iy$. The stabilizer of the point $i \in \mathbb{H}^2$ in $SL_2(\mathbb{R})$ is $SO_2(\mathbb{R})$, so we can identify

$$\mathbb{H}^2 \simeq SL_2(\mathbb{R})/SO_2(\mathbb{R}),$$

as smooth real manifolds together with a Riemannian metric. The subgroup $SO_2(\mathbb{R}) \subset SL_2(\mathbb{R})$ is maximal compact and SL_2 is semisimple, so we can identify $\mathbb{H}^2$ with the symmetric space for the group SL_2, as defined in Section 2.

In the case of the group SL_2, the symmetric space $\mathbb{H}^2$ has a natural complex structure and, as a result, one can prove that its quotients by congruence subgroups such as $\Gamma_0(11)$ are Riemann surfaces. It turns out that the symmetries that f satisfies allow us to consider instead of f the holomorphic differential $\omega_f := f(z)dz$ on the (non-compact) Riemann surface $\Gamma_0(11) \backslash \mathbb{H}^2$.

EXERCISE 1.0.3. *Prove that f indeed descends to a well-defined holomorphic differential on the quotient $\Gamma_0(11) \backslash \mathbb{H}^2$.*

The Riemann surface $\Gamma_0(11) \backslash \mathbb{H}^2$ is an example of a *locally symmetric space* for the group SL_2, in the sense of the definition we give in section 2.

Moreover, f is a simultaneous eigenvector for all Hecke operators T_ℓ (with $\ell \neq 11$), i.e. a Hecke eigenform. The ℓth Fourier coefficient a_ℓ can in fact be identified with the eigenvalue of T_ℓ acting on f.[1] (This can be seen by computing the dimension of the space of cusp forms of weight 2 and level $\Gamma_0(11)$, e.g. by computing the dimension of the space of holomorphic differentials on (the compactification of) $\Gamma_0(11) \backslash \mathbb{H}^2$. The space turns out to be one-dimensional and thus generated by f.)

Set $\Gamma := \Gamma_0(11)$. In the special case of the group SL_2, it turns out that the quotients $\Gamma \backslash \mathbb{H}^2$ have even more structure: there exists an algebraic curve Y_Γ defined over $\mathbb{Q}$ such that $\Gamma \backslash \mathbb{H}^2$ can be identified with $Y_\Gamma(\mathbb{C})$. This follows from the fact that $\mathbb{H}^2$ can be interpreted as a moduli of Hodge structures of elliptic curves[2], and, as a result, the quotients $\Gamma \backslash \mathbb{H}^2$ are (coarse) moduli spaces of elliptic curves over $\mathbb{C}$ equipped with certain extra structures. The particular moduli problem for $\Gamma = \Gamma_0(11)$ gives rise to a canonical model Y_Γ over $\mathbb{Q}$. Y_Γ is a smooth, quasi-projective but not projective curve, known as the *modular curve of level Γ*.

---

[1] In these notes, we will only be concerned with Hecke eigenforms, not with all modular forms and, more generally, we will be interested in *systems of Hecke eigenvalues*.

[2] We make this precise in section 2, when we discuss Shimura varieties. See Example 2.3.2.

The modular form f determines the holomorphic differential $\omega_f \in H^1_{\mathrm{dR}}(\Gamma\backslash\mathbb{H}^2)$. A refinement of Hodge theory for the non-compact Riemann surface $Y_\Gamma(\mathbb{C}) \simeq \Gamma\backslash\mathbb{H}^2$ shows that ω_f determines a system of Hecke eigenvalues in
$$H^1_{\mathrm{Betti}}(Y_\Gamma(\mathbb{C}),\mathbb{C}).$$
This system of Hecke eigenvalues is actually defined over $\mathbb{Q}$ (in this case, the T_ℓ eigenvalues for $\ell \neq 11$ match the Fourier coefficients of f; the system of Hecke eigenvalues will be defined over a number field in general). Now the comparison between the Betti and the étale cohomology of Y_Γ shows that it determines a system of Hecke eigenvalues in
$$H^1_{\mathrm{ét}}(Y_\Gamma \times_\mathbb{Q} \bar{\mathbb{Q}}, \mathbb{Q}_p).$$
Eichler and Shimura show that the corresponding eigenspace is two-dimensional (this follows from a refinement of the Hodge decomposition) and the natural Galois action on it is *the Galois representation* ρ_f. By the Cebotarev density theorem, the Galois representation ρ_f (which is absolutely irreducible) is determined by $\rho_f(\mathrm{Frob}_\ell)$ for $\ell \neq 11, p$ and the relationship between ρ_f and f is encoded in the Eichler–Shimura relation
$$\mathrm{tr}(\rho_f(\mathrm{Frob}_\ell)) = a_\ell$$
for all such primes ℓ.

Higher-dimensional analogues of modular forms are *automorphic representations* and they can be associated to any connected reductive group $G/\mathbb{Q}$ (or over a more general number field). Modular forms correspond to the group SL_2 (or GL_2). [3] In order to associate Galois representations to more general automorphic representations, one first relates automorphic representations to systems of Hecke eigenvalues occurring in the Betti cohomology of locally symmetric spaces, as we did above. If the corresponding locally symmetric spaces have the structure of algebraic varieties defined over number fields, as modular curves do, then one can sometimes find the desired Galois representations in their étale cohomology. If the locally symmetric spaces do not have an algebraic structure, the question of constructing Galois representations is much more difficult than in the algebraic case. The question is even more difficult if we are interested in understanding *torsion classes* occurring in the Betti cohomology of locally symmetric spaces rather than characteristic 0 classes. Nevertheless, there has been a spectacular amount of progress recently due to Scholze [**Sch15**], building on work of Harris–Lan–Taylor–Thorne [**HLTT16**] for the characteristic 0 case.

The goal of these lecture notes is to describe some recent progress in the Langlands program, namely the construction of Galois representations associated to torsion classes in the Betti cohomology of locally symmetric spaces for GL_n/F, where F is a totally real or imaginary CM field. This gives as a corollary the existence of Galois representations for a certain class of automorphic representations of GL_n/F, namely those which are regular and L-algebraic, cf. [**BG14**]. We will focus on the ingredients coming from the theory of *Shimura varieties*, which are higher-dimensional analogues of modular curves, and from the theory of *perfectoid spaces*, as recently introduced by Scholze [**Sch12a**]. A central part of these notes concerns Scholze's theorem that the tower of Shimura varieties with increasing level

---
[3] From the representation-theoretic perspective, a modular form is actually a vector inside an automorphic representation of SL_2.

at p has the structure of a perfectoid space and that it admits a period morphism to a flag variety, the *Hodge–Tate period domain*.

REMARK 1.0.4. While the focus of these notes is the geometry of Shimura varieties and the construction of Galois representations (thus understanding the automorphic to Galois direction), we started the introduction by mentioning a *modularity* result. The modularity result is proved by the so-called *Taylor–Wiles patching method*, which relies on working in p-adic families, both on the side of the Galois representations (coming from elliptic curves) and on the side of modular forms. The existence of the automorphic to Galois direction, $f \mapsto \rho_f$, is a prerequisite to applying the Taylor–Wiles method. Indeed, modularity is not proved by directly matching ρ_E with ρ_f, but rather by considering a universal Galois deformation ring for the residual representation $\bar{\rho}_E$ and comparing this ring to the Hecke algebra acting on a space of modular forms that contains f. The map from the Galois deformation ring to the Hecke algebra is obtained by interpolating the correspondence $f \mapsto \rho_f$.

In order to prove such modularity results in higher dimensions (or even over imaginary quadratic fields), one needs to understand the automorphic to Galois direction first. Moreover, as the insight of Calegari–Geraghty shows [**CG18**], one needs to understand Galois representations attached not just to characteristic 0 automorphic representations, but also to classes in the cohomology of locally symmetric spaces with torsion coefficients, which are a reasonable substitute for p-adic and mod p automorphic forms.[4]

1.1. Organization. In Section 2, we first introduce locally symmetric spaces, then we specialize to the case of Shimura varieties. We discuss examples and counter-examples. In Section 3, we recall the necessary background from p-adic Hodge theory on the (relative) Hodge–Tate filtration.[5] In Section 4, we recall the theory of the canonical subgroup and construct the anticanonical tower, which has a perfectoid structure. In Section 5, we show that (many) Shimura varieties with infinite level at p are perfectoid and describe the geometry of the Hodge–Tate period morphism. In Section 6, we discuss some conjectures and results about the cohomology of locally symmetric spaces and about the corresponding Galois representations.

1.2. Notation. If F is a local or global field, we let G_F denote the absolute Galois group of F. If S is a finite set of places of the global field F, we let $G_{F,S}$ denote the Galois group of the maximal extension of F which is unramified at all primes of F not in S.

If F is a number field, we let $\mathbb{A}_F$ denote the adèles of F, $\mathbb{A}_{F,f}$ the finite adèles, $\mathbb{A}_{F,f}^{\mathfrak{p}}$ the finite adèles away from some prime $\mathfrak{p}$ of F, and $\mathbb{A}_{F,f}^{S}$ the finite adèles of F away from some finite set of primes S. If $\mathfrak{p}$ is a prime of F, we let $\mathrm{Frob}_{\mathfrak{p}}$ denote a choice of geometric Frobenius at the prime $\mathfrak{p}$.

We let $\mathbb{Q}_p^{\mathrm{cycl}}$ be the p-adic completion of the field $\mathbb{Q}_p(\mu_{p^\infty})$ obtained by adjoining all the pth power roots of unity to $\mathbb{Q}_p$. We let $\mathbb{Z}_p^{\mathrm{cycl}}$ be the ring of integers inside $\mathbb{Q}_p^{\mathrm{cycl}}$.

---

[4] In Section 6.2, we explain why torsion classes give a reasonable notion of mod p and p-adic automorphic forms for a general reductive group, by discussing Emerton's notion of *completed cohomology*.

[5] See also the lecture notes of Bhatt for more details on the Hodge–Tate filtration.

If G is a Lie group, we let G° denote the connected component of the identity in G.

If $R \subseteq S$ are rings and V is an R-module, we write $V_S := V \otimes_R S$.

1.3. Acknowledgements. This article was developed from lecture notes for a minicourse taught at the 2017 Arizona Winter School. We thank the organizers of the A.W.S. for a wonderful experience. We are grateful to Peter Scholze for generously sharing his ideas on perfectoid Shimura varieties over several years. We are also grateful to Dan Gulotta, Christian Johansson, and James Newton for many useful conversations. Finally, we thank Johannes Anschütz, Pol van Hoften, Christian Johansson, Judith Ludwig, Mafalda Santos, Peter Scholze, Romyar Sharifi, and the anonymous referee for reading an earlier version of this article, for catching and correcting mistakes, and for many useful suggestions.

2. Locally symmetric spaces and Shimura varieties

In this section, we discuss locally symmetric spaces and Shimura varieties. In § 2.1, we introduce locally symmetric spaces for a general connected reductive group over $\mathbb{Q}$ and we give examples of locally symmetric spaces which admit the structure of complex algebraic varieties and which do not. In § 2.2, in preparation for discussing Shimura varieties, we review some necessary background from Hodge theory, in particular the notion of a variation of polarizable Hodge structures. Finally, in § 2.3 we discuss the axioms of a Shimura datum, the corresponding Shimura variety, and give many examples of Shimura varieties.

2.1. Locally symmetric spaces. Let $G/\mathbb{Q}$ be a connected reductive algebraic group. Let A_G denote the maximal $\mathbb{Q}$-split torus in the center of G. Let $K_\infty \subset G(\mathbb{R})$ denote a maximal compact subgroup and let $A_\infty = A_G(\mathbb{R})$. To G, we can attach a *symmetric space* as follows:
$$X = G(\mathbb{R})/K_\infty^\circ A_\infty^\circ.[6]$$
This is a disjoint union of smooth real manifolds of some dimension d, it has an induced action of $G(\mathbb{R})$, and it can be endowed with a $G(\mathbb{R})$-invariant Riemannian metric.

Two subgroups Γ_1, Γ_2 of the same group are *commensurable* if the intersection $\Gamma_1 \cap \Gamma_2$ has finite index in both Γ_1 and Γ_2. A subgroup Γ of $G(\mathbb{Q})$ is *arithmetic* if it is commensurable with $G(\mathbb{Q}) \cap \mathrm{GL}_N(\mathbb{Z})$, for some embedding $G \hookrightarrow \mathrm{GL}_N$ of algebraic groups over $\mathbb{Q}$.[7] For an arithmetic subgroup $\Gamma \subset G(\mathbb{Q})$, we can define the *locally symmetric space*
$$X_\Gamma := \Gamma \backslash X.$$
If Γ is torsion-free, the space X_Γ is a smooth real manifold of dimension d endowed with an induced Riemannian metric. (If Γ is not torsion-free, then X_Γ is an *orbifold*.)

Suppose we have a model $\mathcal{G}/\mathbb{Z}$ of G which is a flat affine group scheme of finite type over $\mathbb{Z}$.

---

[6]The term A_∞° is included to ensure that the locally symmetric spaces we obtain have finite volume.

[7]More generally, one can define a *lattice* $\Gamma \subset G(\mathbb{R})$ as a discrete subgroup with finite covolume with respect to the Haar measure on $G(\mathbb{R})$. A remarkable theorem of Margulis shows that, if $G(\mathbb{R})$ is a semisimple Lie group with no factor isogenous to $\mathrm{SO}(n,1)$ or $\mathrm{SU}(n,1)$, any lattice $\Gamma \subset G(\mathbb{R})$ is an arithmetic subgroup. See Section 3.3 of [Mil04] for more details on arithmetic subgroups.

EXERCISE 2.1.1. *Show that a finite index subgroup $\Gamma \subset \mathcal{G}(\mathbb{Z})$ is an arithmetic subgroup of $G(\mathbb{Q})$.*

From now on, we will only consider locally symmetric spaces X_Γ, where $\Gamma \subset \mathcal{G}(\mathbb{Z})$ is a finite-index subgroup. In fact, we will only consider arithmetic subgroups which are *congruence subgroups* of $\mathcal{G}(\mathbb{Z})$, i.e. subgroups which contain

$$\Gamma(N) := \ker\left(\mathcal{G}(\mathbb{Z}) \to \mathcal{G}(\mathbb{Z}/N\mathbb{Z})\right)$$

for some $N \in \mathbb{Z}_{\geq 1}$.[8]

If Γ is a congruence subgroup, the cohomology $H^*_{\text{Betti}}(X_\Gamma, \mathbb{C})$ can be computed in terms of automorphic representations of G [**BW00**, **Fra98**]. This is easier to see in the case when the locally symmetric space X_Γ is compact. Then Matsushima's formula expresses $H^*_{\text{Betti}}(X_\Gamma, \mathbb{C})$ in terms of the relative Lie algebra cohomology $H^*(\mathfrak{g}, K_\infty, \pi_\infty)$, where $\pi = \pi_f \otimes \pi_\infty$ runs over automorphic representations of G. The fact that one can express $H^*_{\text{Betti}}(X_\Gamma, \mathbb{C})$ in terms of $(\mathfrak{g}, K_\infty)$-cohomology uses the induced Riemannian structure on X_Γ and Hodge theory for Riemannian manifolds.

We will mostly be interested in the converse direction: realizing certain automorphic representations of G as classes occurring in the Betti cohomology of locally symmetric spaces. Results of Franke guarantee that we can do this, at least for so-called *cohomological* automorphic representations. For GL_n/F, cohomological automorphic representations match regular L-algebraic ones up to a twist, see [**BG14**] for more details.

EXAMPLE 2.1.2.
(1) If $G = \text{SL}_2$ (and we can take $\mathcal{G} = \text{SL}_2/\mathbb{Z}$), the corresponding symmetric space is the upper-half plane $\mathbb{H}^2$. The locally symmetric spaces are the Riemann surfaces corresponding to modular curves, which are discussed in the introduction. These locally symmetric spaces are non-compact Riemann surfaces.

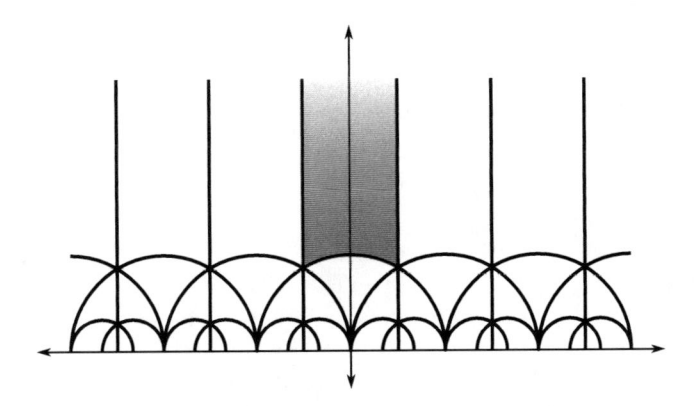

If $G = \text{D}^\times$, where $\text{D}/\mathbb{Q}$ is a quaternion algebra which is split at infinity, i.e. $\text{D}(\mathbb{R}) \simeq \text{M}_2(\mathbb{R})$, then the corresponding symmetric domain is again $\mathbb{H}^2$ but the locally symmetric spaces are now compact Riemann surfaces.

---

[8]It can be shown that $\text{SL}_2(\mathbb{Z})$ contains infinitely many conjugacy classes of finite-index subgroups which are non-congruence, but for $n \geq 3$, every finite-index subgroup of $\text{SL}_n(\mathbb{Z})$ is a congruence subgroup.

These correspond to certain so-called Shimura curves, which give another example of Shimura varieties.

(2) If $G = \mathrm{Res}_{\mathbb{Q}[i]/\mathbb{Q}}\mathrm{SL}_2$ (and we can take $\mathcal{G} = \mathrm{Res}_{\mathbb{Z}[i]/\mathbb{Z}}\mathrm{SL}_2$), the corresponding symmetric space can be identified with 3-dimensional hyperbolic space

$$\mathrm{SL}_2(\mathbb{C})/\mathrm{SU}_2(\mathbb{C}) \simeq \mathbb{H}^3$$

and the locally symmetric spaces are called *Bianchi manifolds*. They are examples of arithmetic hyperbolic 3-manifolds and, since their real dimension is odd, they have no chance of having the structure of algebraic varieties.

(3) If F is a totally real or imaginary CM field with ring of integers $\mathcal{O}$, set $G = \mathrm{Res}_{F/\mathbb{Q}}\mathrm{GL}_n$. In some cases, the corresponding locally symmetric spaces match ones we have already studied. For example, the symmetric space for $\mathrm{GL}_2/\mathbb{Q}$ is

$$\mathrm{GL}_2(\mathbb{R})/\mathrm{SO}_2(\mathbb{R})\mathbb{R}^\times_{>0} \simeq \mathbb{H}^{2,\pm},$$

the disjoint union of the upper and lower half complex planes. The corresponding locally symmetric spaces are disjoint unions of finitely many copies of modular curves.

If F is totally real with $[F:\mathbb{Q}] \geq 2$, the locally symmetric spaces for $\mathrm{Res}_{F/\mathbb{Q}}\mathrm{GL}_2$ are not complex algebraic varieties. The obstruction comes from the center $\mathrm{Res}_{F/\mathbb{Q}}\mathrm{GL}_1$, which is no longer a $\mathbb{Q}$-split torus; by considering a variant of the group we obtain *Hilbert modular varieties*.

If F is totally real and $n \geq 3$, the locally symmetric spaces do not have the structure of complex algebraic varieties. If F is an imaginary CM field and $n \geq 2$, the locally symmetric spaces also do not have the structure of complex algebraic varieties. One way to see this is as follows. Set

$$l_0 := \mathrm{rank}\, G(\mathbb{R}) - \mathrm{rank}\, K_\infty A_\infty.$$

(This is the so-called "defect" of the group G, see [**BW00, CG18**] for a discussion.) The axioms for a Shimura variety introduced in Section 2.3.1 below imply that $l_0 = 0$ whenever the group G admits a Shimura variety. However, when F is a general number field with r_1 real places and r_2 complex places, one can compute l_0 for $\mathrm{Res}_{F/\mathbb{Q}}\mathrm{GL}_n$ to be

$$l_0 = \begin{cases} r_1\left(\frac{n-2}{2}\right) + r_2(n-1) & n \text{ even,} \\ r_1\left(\frac{n-1}{2}\right) + r_2(n-1) & n \text{ odd.} \end{cases}$$

(4) If $G(\mathbb{R})$ is compact (or more generally, if $G(\mathbb{R})/A_\infty$ is compact), then $G(\mathbb{R})/K_\infty^\circ A_\infty^\circ$ is a finite set of points and the locally symmetric spaces attached to G are also just finite sets of points, called Shimura sets. This situation is very favorable for setting up the Taylor–Wiles method, because the cohomology of the locally symmetric space is then concentrated in only one degree, namely in degree 0. This happens, for example, in the case of a *definite* unitary group defined over a totally real field (whose signature at each infinite place is $(0, n)$). However, this situation is not very interesting from the point of view of geometry!

In these notes, we will mostly use the adelic perspective on locally symmetric spaces. Recall that we have chosen a model $\mathcal{G}/\mathbb{Z}$ of $G/\mathbb{Q}$. Let $K \subset G(\mathbb{A}_f)$ be a

compact open subgroup of the form $\prod_v K_v$, where v runs over primes of $\mathbb{Q}$ and $K_v \subseteq \mathcal{G}(\mathbb{Z}_v)$, and such that $K_v = \mathcal{G}(\mathbb{Z}_v)$ for all but finitely many primes v. Define the double quotient
$$X_K := G(\mathbb{Q}) \backslash (X \times G(\mathbb{A}_f)/K),$$
where the action of $G(\mathbb{Q})$ on the two factors is via the diagonal embedding. The set $G(\mathbb{Q}) \backslash G(\mathbb{A}_f)/K$ is finite; this follows from [**PR94**][Thm 5.1].

EXERCISE 2.1.3. *When $G = \operatorname{Res}_{F/\mathbb{Q}} \operatorname{GL}_n$, prove directly that $G(\mathbb{Q}) \backslash G(\mathbb{A}_f)/K$ is finite.*

Let $g_1, \ldots, g_r$ be a set of double coset representatives. For $i = 1, \ldots, r$, let $\Gamma_i := G(\mathbb{Q}) \cap g_i K g_i^{-1}$. This is a discrete subgroup of $G(\mathbb{Q})$ and it is in fact a congruence subgroup of $\mathcal{G}(\mathbb{Z})$. Then we have
$$X_K = G(\mathbb{Q}) \backslash (X \times G(\mathbb{A}_f)/K) = \sqcup_{i=1}^r \Gamma_i \backslash X = \sqcup_{i=1}^r X_{\Gamma_i},$$
so the adelic version of a locally symmetric space is a finite disjoint union of the locally symmetric spaces introduced above.

We say that K is *neat* if $G(\mathbb{Q}) \cap gKg^{-1}$ is torsion-free for any $g \in G(\mathbb{A}_f)$, in which case X_K is a smooth real manifold of dimension d. If K is sufficiently small, then it is neat.

As seen in Example 2.1.2 (1) above, the locally symmetric spaces X_K can be non-compact. Borel and Serre [**BS73**] constructed a compactification of X_K (or rather, of the individual spaces X_Γ), which is a smooth real manifold with corners. If X_K^{BS} denotes the Borel-Serre compactification of X_K, the inclusion
$$X_K \hookrightarrow X_K^{\mathrm{BS}}$$
is a homotopy equivalence. This shows that X_K has the same homotopy type as that of a finite CW complex, so in particular the vector spaces $H^i_{\mathrm{Betti}}(X_K, \mathbb{C})$ are finite-dimensional. Similarly, the cohomology groups $H^i_{\mathrm{Betti}}(X_K, \mathbb{Z}/p^N\mathbb{Z})$ are finite $\mathbb{Z}/p^N\mathbb{Z}$-modules and the groups $H^i_{\mathrm{Betti}}(X_K, \mathbb{Q}_p)$ are finite-dimensional for every prime p.

As K varies, we have a tower of locally symmetric spaces $(X_K)_K$. If K, K' are two compact-open subgroups of $G(\mathbb{A}_f)$ and if $g \in G(\mathbb{A}_f)$ is such that $g^{-1}K'g \subseteq K$, we have a finite étale morphism $c_g : X_{K'} \to X_K$ induced by $hK' \mapsto hgK$ for $h \in G(\mathbb{A}_f)$. If one takes $K' := K \cap gKg^{-1}$, one obtains a correspondence
$$(c_g, c_1) : X_{K'} \to X_K \times X_K,$$
called a *Hecke correspondence*. This correspondence induces an endomorphism of $H^i_{(c)}(X_K)$, where we take the Betti cohomology of the locally symmetric space with coefficients in either $\mathbb{C}, \mathbb{Q}_p, \mathbb{Z}/p^N\mathbb{Z}$ for $N \in \mathbb{Z}_{\geq 1}$ and this endomorphism only depends on the double coset KgK.

2.2. Review of Hodge structures. Roughly speaking, a Shimura variety is an algebraic variety defined over a number field whose underlying complex manifold is a locally symmetric space corresponding to some connected reductive group $G/\mathbb{Q}$. As we have seen in Example 1, this can exist only in special circumstances, for certain groups G. In this section, we recall some notions related to Hodge structures and variations of Hodge structures, which will be useful for explaining the axioms defining a Shimura datum in Section 2.3. For a more in-depth discussion of these notions, see Chapter II of [**Mil04**].

2.2.1. *Hodge structures.* Recall that a (pure) *Hodge structure* on a finite-dimensional real vector space V is a direct sum decomposition of the complexification $V_{\mathbb{C}}$ of V of the form
$$V_{\mathbb{C}} = \oplus_{(i,j) \in \mathbb{Z}^2} V^{i,j}$$
such that the following relation, known as *Hodge symmetry*, holds: for every $(i,j) \in \mathbb{Z}^2$, the complex conjugate of $V^{i,j}$ is $V^{j,i}$. The direct sum decomposition is called the Hodge decomposition. If $V_{\mathbb{C}} = \oplus_{k \in I} V^{i_k, j_k}$, we say that V has a Hodge structure of *type* $(i_k, j_k)_{k \in I}$. If, moreover, $i_k + j_k = n$ for every $k \in I$ then we say that the Hodge structure on V is pure of *weight* n. The weight decomposition is the direct sum decomposition of V indexed by weight and it is already defined over $\mathbb{R}$. A morphism of Hodge structures is a morphism of real vector spaces which respects the Hodge decomposition of their complexifications.

More generally, one can define *rational* and *integral* Hodge structures. An integral (resp. rational) Hodge structure is a free $\mathbb{Z}$-module of finite rank (resp. finite-dimensional $\mathbb{Q}$-vector space) together with a Hodge decomposition of $V_{\mathbb{R}}$ such that the weight decomposition is defined over $\mathbb{Q}$.

EXAMPLE 2.2.2. If $X/\mathbb{C}$ is an smooth projective variety[9], then the Betti cohomology groups $H^n(X(\mathbb{C}), \mathbb{Z})$ are endowed with integral Hodge structures coming from the Hodge decomposition
$$H^n(X(\mathbb{C}), \mathbb{Z}) \otimes_{\mathbb{Z}} \mathbb{C} = \bigoplus_{i+j=n} H^j(X, \Omega^i_{X/\mathbb{C}});$$
we set $V^{i,j} := H^j(X, \Omega^i_{X/\mathbb{C}})$.

If $X = A$ is an abelian variety over $\mathbb{C}$, the Hodge decomposition is
$$H^1(A(\mathbb{C}), \mathbb{Z}) \otimes_{\mathbb{Z}} \mathbb{C} = H^0(A, \Omega^1_{A/\mathbb{C}}) \oplus H^1(A, \mathcal{O}_A);$$
then $H^1(A(\mathbb{C}), \mathbb{Z})$ has an integral Hodge structure of type $(1,0), (0,1)$. The dual $H_1(A(\mathbb{C}), \mathbb{Z})$ has a Hodge structure of type $(-1, 0), (0, -1)$. Giving a Hodge structure of this type on $H_1(A(\mathbb{C}), \mathbb{Z})$ is equivalent to giving a complex structure on $H_1(A(\mathbb{C}), \mathbb{Z}) \otimes_{\mathbb{Z}} \mathbb{R}$.

The category of integral Hodge structures of type $(-1, 0), (0, -1)$ is equivalent to the category of complex tori. (If A is an abelian variety, then $A(\mathbb{C})$ is a complex torus, though not every complex torus arises from an abelian variety.)

EXAMPLE 2.2.3. If $n \in \mathbb{Z}$, we define the Hodge structure $\mathbb{R}(n)$ to be the unique Hodge structure on $\mathbb{R}$ of type $(-n, -n)$. We define $\mathbb{Q}(n)$ and $\mathbb{Z}(n)$ analogously.

Let $\mathbb{S} := \operatorname{Res}_{\mathbb{C}/\mathbb{R}} \mathbb{G}_m$; this is a real algebraic group such that $\mathbb{S}(\mathbb{R}) = \mathbb{C}^\times$. The group $\mathbb{S}$ is the *Tannakian group* for the category of Hodge structures on real vector spaces.[10] This implies that there is an equivalence of categories between the category of Hodge structures on finite-dimensional real vector spaces and the category of finite-dimensional representations of $\mathbb{S}$ on real vector spaces. We describe the functor in one direction: a representation of $\mathbb{S}$ on a real vector space V determines an action of $\mathbb{C}^\times$ on the complexification $V_{\mathbb{C}}$. Then $V_{\mathbb{C}}$ decomposes as a direct sum

---

[9]We could take, more generally, X to be a compact Kähler manifold, in which case the Betti cohomology decomposes as $H^n(X, \mathbb{C}) = \oplus_{i+j=n} H^{i,j}(X)$, where $H^{i,j}(X)$ denotes the space of cohomology classes of type (i, j).

[10]See, for example, Chapter I of [**Mil90**] for a discussion of Tannakian categories and the corresponding Tannakian groups as relevant to Shimura varieties.

of subspaces $V^{i,j}$ with $i,j \in \mathbb{Z}$, such that the action of $\mathbb{C}^\times$ on $V^{i,j}$ is through the cocharacter $z \mapsto z^{-i}\bar{z}^{-j}$. This direct sum decomposition defines a Hodge structure on V. Thus, we can think of a Hodge structure on a real vector space V as a pair (V,h), where $h : \mathbb{S} \to \mathrm{GL}(V)$ is a homomorphism.

A *polarizable* Hodge structure is a Hodge structure which can be equipped with a polarization. A *polarization* on a real Hodge structure (V,h) of weight n is a morphism of Hodge structures

$$\Psi : V \times V \to \mathbb{R}(-n)$$

such that the bilinear form $(v,w) \mapsto \Psi(v, h(i)w)$ is symmetric and positive definite. (One can similarly define polarizable integral and rational Hodge structures.)

Hodge structures coming from algebraic geometry are polarizable.[11] For example, recall Riemann's classification result for abelian varieties over $\mathbb{C}$.

THEOREM 2.2.4. *The functor $A \mapsto H_1(A, \mathbb{Z})$ defines an equivalence of categories between the category of abelian varieties over $\mathbb{C}$ and the category of polarizable integral Hodge structures of type $(-1, 0), (0, -1)$.*

2.2.5. *Variations of polarizable Hodge structures.* In order to have a Shimura variety, the symmetric space X should be interpreted as a "moduli space" of polarizable Hodge structures. The precise notion of "moduli space" we will use is that of a *variation of Hodge structures*.

For a Hodge structure on V of weight n, we define the associated *Hodge–de Rham filtration*[12] to be the descending filtration given by

$$F^i V := \bigoplus_{i' \geq i} V^{i',j'} \subset V_\mathbb{C}.$$

EXAMPLE 2.2.6. If $X/\mathbb{C}$ is a smooth projective variety, the Hodge structure on the Betti cohomology $H^*(X(\mathbb{C}), \mathbb{Z})$ has the Hodge–de Rham filtration

$$F^i\left(H^*(X(\mathbb{C}), \mathbb{Z}) \otimes_\mathbb{Z} \mathbb{C}\right) = \oplus_{i' \geq i} H^{j'}(X, \Omega^{i'}_{X/\mathbb{C}}).$$

Under the canonical comparison isomorphism between Betti and de Rham cohomology, the Hodge–de Rham filtration on $H^*(X(\mathbb{C}), \mathbb{Z}) \otimes_\mathbb{Z} \mathbb{C}$ matches the filtration on the algebraic de Rham cohomology $H^*_{\mathrm{dR}}(X)$ induced from the degeneration of the Hodge–de Rham spectral sequence

$$E_1^{i,j} = H^j(X, \Omega^i_{X/\mathbb{C}}) \Rightarrow H^{i+j}(X, \Omega^\bullet_{X/\mathbb{C}}) =: H^{i+j}_{\mathrm{dR}}(X).$$

If $X = A$ is an abelian variety over $\mathbb{C}$, the Hodge–de Rham filtration is determined by $F^1\left(H^1(A(\mathbb{C}), \mathbb{Z}) \otimes_\mathbb{Z} \mathbb{C}\right) = H^0(A, \Omega^1_{A/C})$ (F^0 is everything and F^2 is zero).

We remark that, if X is defined over a number field E, then the algebraic de Rham cohomology $H^{i+j}_{\mathrm{dR}}(X)$ is an E-vector space and the Hodge–de Rham filtration on algebraic de Rham cohomology is also defined over E. This observation, together with standard comparison results between the cohomology of schemes and of the corresponding rigid-analytic varieties, will be used in Section 3. The degeneration of the Hodge–de Rham spectral sequence, which is needed to obtain the Hodge

---

[11] More precisely, if $X/\mathbb{C}$ is a smooth projective variety, then its Betti cohomology carries a Hodge structure equipped with a rational polarization. The polarization comes from the hard Lefschetz theorem applied to a rational Kähler cohomology class.

[12] We prefer to refer to the Hodge filtration as the Hodge–de Rham filtration in order to avoid confusion with the Hodge–Tate filtration which will be discussed in Section 3.

filtration on de Rham cohomology, is a deep result, originally established using analytic techniques (Hodge theory), but it was later on proved purely algebraically in [**DI87**].

A variation of (pure) Hodge structures should model the Hodge structure on the local system coming from the Betti cohomology of a continuous family of smooth projective varieties over some base. We start with an elementary definition, which we will apply to the case of Shimura varieties, and which can be formulated very concretely. We then give the more general definition.

Let X^+ be a simply-connected connected complex manifold.[13] Fix a real vector space V and a positive integer n. Assume that for each $h \in X^+$ we have a Hodge structure on V of weight n. Let $V_h^{i,j} \subset V_{\mathbb{C}}$ be the subspace of type (i,j) corresponding to the Hodge structure attached to h, and let $F_h^i(V_{\mathbb{C}}) \subset V_C$ be the ith graded piece of the Hodge–de Rham filtration on $V_{\mathbb{C}}$ determined by h.

DEFINITION 2.2.7. *We say that the family of Hodge structures indexed by X^+ is a variation of Hodge structures of weight n if the following conditions are satisfied.*

(1) *Firstly, for each (i,j), the subspace $V_h^{i,j}$ varies continuously with $h \in X^+$. This means that the dimension of the subspace $V_h^{i,j}$ is equal to a constant $d(i,j) \in \mathbb{Z}_{\geq 0}$, so there is a natural map to the Grassmannian parametrizing $d(i,j)$-dimensional subspaces of $V_{\mathbb{C}}$*

$$X^+ \to \mathrm{Gr}^{d(i,j)}(V_{\mathbb{C}}).$$

Moreover, the above map is required to be continuous.

(2) *The Hodge filtration $F_h^\bullet$ varies holomorphically with $h \in X^+$. More precisely, let $\mathrm{Fl}^{\mathrm{std}}(V_{\mathbb{C}})$ be the flag variety parametrizing descending filtrations on $V_{\mathbb{C}}$ of type $(d(i))_{i \in \mathbb{Z}}$, where $d(i) = \sum_{i' \geq i} d(i', n - i')$. The first condition guarantees that there exists a map*

$$\pi_{\mathrm{HdR}}^+ : X^+ \to \mathrm{Fl}^{\mathrm{std}}(V_{\mathbb{C}}), h \mapsto F_h^\bullet$$

and we require this to be a map of complex manifolds (i.e. holomorphic).

(3) *(Griffiths transversality) The tangent space of $\mathrm{Fl}^{\mathrm{std}}(V_{\mathbb{C}})$ at a point corresponding to a filtration $F^\bullet$ on $V_{\mathbb{C}}$ is contained in $\oplus_{i \in \mathbb{Z}} \mathrm{Hom}(F^i, V_{\mathbb{C}}/F^i)$. Let $h \in X^+$. The final condition is that we require that the differential $d\pi_{\mathrm{HdR}}^+$, which is a map*

$$d\pi_{\mathrm{HdR}}^+ : T_h X^+ \to T_{F_h^\bullet} \mathrm{Fl}^{\mathrm{std}}(V_{\mathbb{C}}) \subset \bigoplus_{i \in \mathbb{Z}} \mathrm{Hom}(F^i, V_{\mathbb{C}}/F^i),$$

to satisfy the following transversality condition:

$$\mathrm{Im}(d\pi_{\mathrm{HdR}}^+) \subset \bigoplus_{i \in \mathbb{Z}} \mathrm{Hom}(F^i, F^{i-1}/F^i) \subset \bigoplus_{i \in \mathbb{Z}} \mathrm{Hom}(F^i, V_{\mathbb{C}}/F^i).$$

REMARK 2.2.8. In fixed weight n, the Hodge–de Rham filtration determines the Hodge decomposition via $V^{p,q} = F^p(V_{\mathbb{C}}) \cap \overline{F^q(V_{\mathbb{C}})}$. This implies that, if the Hodge structures parametrized by X^+ are all distinct, the holomorphic map

$$\pi_{\mathrm{HdR}}^+ : X^+ \hookrightarrow \mathrm{Fl}^{\mathrm{std}}(V_{\mathbb{C}})$$

is injective. We call such a map a *period morphism*. One of the protagonists of these lecture notes is the p-adic analogue of this morphism, called the *Hodge–Tate*

---

[13]For example, we could take $X^+ = \mathbb{H}^2$, the upper half-plane.

period morphism. This will not be injective, in general, but in many situations we will be able to understand its fibers. Note also that, while the Hodge–de Rham filtration varies holomorphically in families, the same is not true for the Hodge decomposition.

EXERCISE 2.2.9. *Check that the tangent space of* $\mathrm{Fl}^{\mathrm{std}}(V_\mathbb{C})$ *at a point corresponding to a filtration* $F^\bullet$ *on* $V_\mathbb{C}$ *is indeed contained in* $\oplus_{i\in\mathbb{Z}}\mathrm{Hom}(F^i, V_\mathbb{C}/F^i)$.

A variation of polarizable Hodge structures on X^+ is a variation of Hodge structures on X^+ together with a bilinear form
$$\Psi : V \times V \to \mathbb{R}$$
such that Ψ induces for any $h \in X^+$ a polarization on the Hodge structure determined by h.

More generally, let X be a complex manifold. A variation of Hodge structures of some weight $n \in \mathbb{Z}$ on X is a locally constant sheaf of finitely generated $\mathbb{Z}$-modules $\mathcal{V}_\mathbb{Z}$ on X (we call such an object a $\mathbb{Z}$-local system on X) together with the following additional structures. Define $\mathcal{E} := \mathcal{V}_\mathbb{Z} \otimes_\mathbb{Z} \mathcal{O}_X$, where $\mathcal{O}_X$ is the sheaf of holomorphic functions on X. Then $\mathcal{E}$ is a holomorphic vector bundle on X; this is equipped with a canonical flat connection
$$\nabla : \mathcal{E} \to \mathcal{E} \otimes_{\mathcal{O}_X} \Omega^1_X,$$
induced from $\partial : \mathcal{O}_X \to \Omega^1_X$ (here, Ω^1_X denotes the sheaf of holomorphic differentials on X). The connection ∇ is called the *Gauss-Manin connection*. The additional structure is a descending filtration $F^\bullet\mathcal{E}$ on $\mathcal{E}$ by holomorphic sub-bundles such that

(1) The filtration $F^\bullet\mathcal{E}$ induces Hodge structures of weight n on the fibers of $\mathcal{E}$.
(2) (Griffiths transversality) For all $i \in \mathbb{Z}$, the Gauss-Manin connection satisfies
$$\nabla : F^i\mathcal{E} \to F^{i-1}\mathcal{E} \otimes_{\mathcal{O}_X} \Omega^1_X \subset \mathcal{E} \otimes_{\mathcal{O}_X} \Omega^1_X.$$

If X is simply-connected, the local system $\mathcal{V}_\mathbb{Z}$ on X is trivial. By choosing a trivialization of $\mathcal{V}_\mathbb{R}$, we recover Definition 2.2.7. As above, we can also define variations of polarizable Hodge structures. With this more general definition, we have the following example.

EXAMPLE 2.2.10. Let $f : Y \to X$ be a smooth and projective morphism of complex varieties, such that X is smooth. Let $(R^n f_*\mathbb{Z})_{\mathrm{tf}}$ be the torsion-free part of $R^n f_*\mathbb{Z}$. Then the local system $(R^n f_*\mathbb{Z})_{\mathrm{tf}}$ on $X(\mathbb{C})$ is a variation of polarizable Hodge structures of weight n.

2.3. Shimura varieties.

2.3.1. *Definition of a Shimura variety.* Shimura varieties are described by *Shimura data*, which are certain pairs (G, X), consisting of a connected reductive group G defined over $\mathbb{Q}$, and a $G(\mathbb{R})$-conjugacy class X of homomorphisms
$$\mathbb{S} \to G_\mathbb{R}.$$

As we saw above that $\mathbb{S}$ is the Tannakian group for the category of real Hodge structures, for any finite-dimensional representation V of G on a real vector space, X parametrizes a family of Hodge structures with underlying vector space V. If we choose an element $h \in X$, we can identify X with $G(\mathbb{R})/K^h_\infty$, where K^h_∞ is the stabilizer of h in $G(\mathbb{R})$ under conjugacy. We will impose certain additional conditions on (G, X) which will ensure that X carries a unique complex structure making the family of Hodge structures that X parametrizes a variation of polarizable Hodge structures.

In order for a pair (G, X) as above to be a Shimura datum, it has to also satisfy the following axioms.

(1) Let $\mathfrak{g}$ denote the Lie algebra of $G(\mathbb{R})$. For any choice of $h \in X$, the composite
$$h : \mathbb{S} \to G_{\mathbb{R}} \to G_{\mathbb{R}}^{\mathrm{ad}} \to \mathrm{GL}(\mathfrak{g}),$$
i.e. the composite with the adjoint action of $G_{\mathbb{R}}$ on $\mathfrak{g}$, induces a Hodge structure of type $(-1, 1), (0, 0), (1, -1)$ on $\mathfrak{g}$.

(2) For any choice of $h \in X$, $h(i)$ is a Cartan involution on $G^{\mathrm{ad}}(\mathbb{R})$.

(3) G^{ad} has no factor defined over $\mathbb{Q}$ whose real points form a compact group.

Note that, while the first two conditions are formulated for any choice of $h \in X$, it is enough to check them for one choice of $h \in X$. We discuss the role that each of the three axioms plays below. Assume, for simplicity, that X is connected.

The first axiom implies, in particular, that the Hodge structure on $\mathfrak{g}$ induced by the adjoint representation has weight 0, which in turn implies that $h(\mathbb{R}^{\times})$ lies in the center of $G(\mathbb{R})$ for one $h \in X$ (equivalently, for all $h \in X$). Even though a given real representation V of G may not give rise to a family of Hodge structures which are homogeneous of a given weight, the fact that $h(\mathbb{R}^{\times})$ is central means that we can write V as a direct sum of G-invariant pieces which do give rise to Hodge structures that are homogeneous of a given weight, independent of the choice of $h \in X$. In other words, the weight decomposition on V is independent of $h \in X$.

We can now ask whether the family of Hodge structures parametrized by X can be made into a variation of polarizable Hodge structures, by endowing X with an appropriate complex structure. Choose V to be the direct sum of the representations in a faithful family of representations of G. The fact that the weight decomposition on V is independent of $h \in X$ is all that is needed to show that X carries a unique complex structure for which the family of Hodge structures varies holomorphically. Indeed, if we let $\mathrm{Fl}^{\mathrm{std}}(V_{\mathbb{C}})$ be the product of the flag varieties defined above for each homogenous piece of V, we have an injection
$$X \hookrightarrow \mathrm{Fl}^{\mathrm{std}}(V_{\mathbb{C}}).$$
The complex structure on X is induced from the natural complex structure on the flag variety $\mathrm{Fl}^{\mathrm{std}}(V_{\mathbb{C}})$. Furthermore, the requirement for the family of Hodge structures on X to satisfy Griffiths transversality is equivalent to $\mathfrak{g} = F^{-1}\mathfrak{g}$. Since the Hodge structure on $\mathfrak{g}$ has weight 0, this is in turn equivalent to asking that the Hodge structure on $\mathfrak{g}$ be of type $(-1, 1), (0, 0), (1, -1)$. See Section 1.1 of [**Del79**] for more details.

For the second axiom, note that $h(i)$ induces an involution of $G^{\mathrm{ad}}(\mathbb{R})$ because the adjoint action of $h(-1)$ is trivial. The fact that $h(i)$ is a Cartan involution of $G^{\mathrm{ad}}(\mathbb{R})$ means that the inner form over $\mathbb{R}$ of G^{ad} defined by the fixed points of the involution $g \mapsto h(i)\bar{g}h(i)^{-1}$ is compact. It is easy to see now that the second axiom is independent of the choice of conjugacy class of $h(i)$. The second axiom guarantees that the variation of Hodge structures on X (obtained by choosing any V as above) is a variation of polarizable Hodge structures. See Section 1.1 [**Del79**] for more details. We note that this axiom implies that the stabilizer $K_{\infty}^{h} \subset G(\mathbb{R})$ of any $h \in X$ is compact modulo center.

The third axiom is fairly harmless to assume (since we could replace G with its quotient by a connected normal subgroup whose group of real points is compact), and it allows us to use strong approximation when G is simply-connected.

When (G, X) is a Shimura datum, Deligne proves that X is a finite disjoint union of *Hermitian symmetric domains* in [**Del79**]. For a neat compact open subgroup $K \subset G(\mathbb{A}_f)$, the double quotient

$$G(\mathbb{Q}) \setminus (X \times G(\mathbb{A}_f)/K)$$

has the structure of an algebraic variety, called a *Shimura variety*. The Shimura variety has a canonical model which is a smooth, quasi-projective variety defined over a number field E, called the *reflex field* of the Shimura datum. Choose a representative $h \in X$. This gives rise to a cocharacter

$$\mu_h := h \times_{\mathbb{R}} \mathbb{C}|_{(1\text{st } \mathbb{G}_m \text{ factor})} : \mathbb{G}_{m,\mathbb{C}} \to G_{\mathbb{C}}.$$

The axioms in the definition of a Shimura datum imply that the cocharacter μ_h is *minuscule*, i.e. its pairing with any root of $G_{\mathbb{C}}$ is in the set $\{-1, 0, 1\}$. The $G(\mathbb{C})$-conjugacy class $\{\mu_h\}$ is independent of h. The reflex field E is the field of definition of the conjugacy class $\{\mu_h\}$ (this may be smaller than the field of definition of the cocharacter μ_h). From now on, we denote by X_K the canonical model of the Shimura variety over E.

EXAMPLE 2.3.2 (Modular curves). Let V be a 2-dimensional vector space over $\mathbb{Q}$. We consider the algebraic group over $\mathbb{Q}$ given by $G := \mathrm{GL}(V)$. Let X be the set of complex structures on $V \otimes_{\mathbb{Q}} \mathbb{R}$, i.e. of embeddings $\mathbb{C} \subset \mathrm{End}_{\mathbb{R}}(V \otimes_{\mathbb{Q}} \mathbb{R})$. Then X can be identified with a $G(\mathbb{R})$-conjugacy class of homomorphisms

$$h : \mathbb{S} \to G_{\mathbb{R}}$$

via $x \in X \mapsto h_x : \mathbb{S} \to G_{\mathbb{R}}$, where for every $z \in \mathbb{S}(\mathbb{R}) \simeq \mathbb{C}^{\times}$, $h_x(z) \in \mathrm{GL}(V_{\mathbb{R}})$ is identified with $z \in \mathbb{C}^{\times} \subset \mathrm{Aut}_{\mathbb{R}}(V \otimes_{\mathbb{Q}} \mathbb{R})$. One can check that the three axioms for (G, X) to be a Shimura datum are satisfied.

By choosing a basis of V, we can identify G with GL_2 and X with $\mathbb{H}^{\pm}$, the disjoint union of the upper and lower half planes. We see that the symmetric space for $\mathrm{GL}_2/\mathbb{Q}$ can be identified with the conjugacy class X. The corresponding Shimura varieties are disjoint unions of finitely many copies of connected modular curves.

Let Λ be a fixed $\mathbb{Z}$-lattice in V. By Example 2.2.2, we see that X can be identified with the set of integral Hodge structures of type $(-1, 0), (0, -1)$ on Λ. All such Hodge structures are polarizable, so X can be identified with a moduli of Hodge structures of elliptic curves over $\mathbb{C}$. This is the reason for the moduli interpretation of modular curves in terms of elliptic curves together with level structures.

The period morphism taking a Hodge structure to the corresponding Hodge-de Rham filtration can be identified with the natural embedding

$$\mathbb{H}^{\pm} \hookrightarrow \mathbb{P}^1(\mathbb{C})$$

Note that this is equivariant for the action of $\mathrm{GL}_2(\mathbb{R})$ on both sides: given by Möbius transformations on the left hand side and factoring through the usual action of $\mathrm{GL}_2(\mathbb{R})$ on $\mathbb{P}^1(\mathbb{C})$.

EXERCISE 2.3.3. *Write down the identification $\mathbb{H}^{\pm} \simeq X$ such that the usual action of $\mathrm{GL}_2(\mathbb{R})$ on $\mathbb{H}^{\pm}$ given by the Möbius transformations*

$$\gamma = \begin{pmatrix} a & b \\ c & d \end{pmatrix} \in \mathrm{GL}_2(\mathbb{R}), \gamma : z \mapsto \frac{az+b}{cz+d}$$

can be recovered from the conjugation action of $\mathrm{GL}_2(\mathbb{R})$ on the set of homomorphisms $\mathbb{S} \to \mathrm{GL}_{2,\mathbb{R}}$ from Example 2.3.2.

We end this section by giving examples of higher-dimensional Shimura varieties. The key examples that we will consider in these lecture notes will be Siegel modular varieties (which are the simplest Shimura varieties from the point of view of the moduli problem that they satisfy) and Shimura varieties for quasi-split unitary groups (which also have an explicit moduli interpretation).

EXAMPLE 2.3.4 (Siegel modular varieties). Let $n \geq 1$ and let

$$(V, \psi) = \left(\mathbb{Q}^{2n}, \psi((a_i), (b_i)) = \sum_{i=1}^{n} (a_i b_{n+i} - a_{n+i} b_i) \right)$$

be the split symplectic space of dimension $2n$ over $\mathbb{Q}$. Consider the symplectic similitude group $\mathrm{GSp}_{2n} := \mathrm{GSp}(V, \psi)$; this is the algebraic group over $\mathbb{Q}$ defined by

$$\mathrm{GSp}_{2n}(R) = \{(g, \lambda) \in \mathrm{GL}(V \otimes_{\mathbb{Q}} R) \times R^{\times} | \psi(gv, gw) = \lambda \cdot \psi(v, w), \forall v, w \in V \otimes_{\mathbb{Q}} R\}$$

for any $\mathbb{Q}$-algebra R. In other words, GSp_{2n} is the group of automorphisms of V preserving the symplectic form up to a scalar, called the similitude factor, which is a unit. One can identify the corresponding symmetric space X with the Siegel double space, which has the following explicit description

$$\{Z \in \mathrm{M}_n(\mathbb{C}) | Z = Z^t, \mathrm{Im}(Z) \text{ positive or negative definite}\},$$

where $\mathrm{Im}(Z)$ denotes the imaginary part of the matrix Z. The Siegel double space has an action of $\mathrm{GSp}_{2n}(\mathbb{R})$, via

$$\Gamma = \begin{pmatrix} A & B \\ C & D \end{pmatrix} \in \mathrm{GSp}_{2n}(\mathbb{R}), \Gamma : Z \mapsto (AZ+B)(CZ+D)^{-1},\,^{14}$$

which is transitive. The stabilizer in $\mathrm{GSp}_{2n}(\mathbb{R})$ of the matrix $i \cdot \mathrm{Id}_n$ can be identified with $\mathrm{U}(n) \times \mathbb{R}_{>0}$; the unitary group $U(n)$ is the identity component of a maximal compact subgroup of $\mathrm{GSp}_{2n}(\mathbb{R})$. This shows that we do have an identification of the Siegel double space with the symmetric space for GSp_{2n}.

EXERCISE 2.3.5. *Check that the action of $\mathrm{GSp}_{2n}(\mathbb{R})$ described above preserves the Siegel double space, that it is transitive and compute the stabilizer of $i \cdot \mathrm{Id}_n$.*

The Siegel double space X is a disjoint union of two copies of a Hermitian symmetric domain. Using the classification of Hermitian symmetric domains in [**Del79**], one sees that X can be identified with a conjugacy class of homomorphisms

$$h : \mathbb{S} \to \mathrm{GSp}_{2n,\mathbb{R}}$$

such that the pair (GSp_{2n}, X) satisfies the three axioms in the definition of a Shimura datum. The corresponding Shimura varieties are called *Siegel modular varieties*. When $n = 1$, we have an isomorphism $\mathrm{GSp}_2 \simeq \mathrm{GL}_2$ of algebraic groups over $\mathbb{Q}$, and in this case we recover the modular curves.

---

[14]These are $n \times n$-matrices, so for $n > 1$ the order of multiplication matters.

Fix the lattice $\Lambda = \mathbb{Z}^{2n}$ in V (which is self-dual under the symplectic form ψ). For every $h \in X$, let
$$\mu_h := h \times_{\mathbb{R}} \mathbb{C}|_{(\text{1st } \mathbb{G}_{m,\mathbb{C}}\text{-factor})};$$
this defines a cocharacter $\mu_h : \mathbb{G}_{m,\mathbb{C}} \to \mathrm{GSp}_{2n,\mathbb{C}}$. For every $h \in X$, the Hodge structure induced by μ_h on V has type $(-1,0),(0,-1)$ and is polarizable by the second axiom in the definition of a Shimura datum. This Hodge structure gives rise by Theorem 2.2.4 to the abelian variety over $\mathbb{C}$ with associated complex torus $V^{(-1,0)}/\Lambda$. This abelian variety has dimension n.

For $K \subset \mathrm{GSp}_{2n}(\mathbb{A}_f)$ a neat compact open subgroup, the corresponding Shimura variety X_K is a moduli space of polarized g-dimensional abelian varieties with level-K-structure. X_K has a canonical model, which is a smooth, quasi-projective variety over the reflex field $\mathbb{Q}$. It carries a universal abelian variety A^{univ} and a natural ample line bundle ω given by the determinant of the sheaf of invariant differentials on A^{univ}.

If p is a good prime for the level K (i.e.), X_K, A^{univ}, ω admit integral models over the localization $\mathbb{Z}_{(p)}$. The integral model X_K is a smooth, quasi-projective but not projective scheme over $\mathrm{Spec}\,\mathbb{Z}_p$. It admits a minimal (Baily–Borel–Satake) compactification $X_K \hookrightarrow X_K^*$, which is a projective but usually not smooth scheme over $\mathrm{Spec}\,\mathbb{Z}_{(p)}$. This was constructed by Faltings and Chai in [**FC90**]. The ample line bundle ω extends canonically to X_K^*.

EXAMPLE 2.3.6 (Shimura varieties of PEL type). Shimura varieties of PEL type are Shimura varieties which admit a moduli interpretation in terms of abelian varieties equipped with *p*olarizations, *e*ndomorphisms and *l*evel structure. Siegel modular varieties give examples of PEL-type Shimura varieties, since they parametrize abelian varieties equipped with polarizations and level structure. General PEL-type Shimura varieties admit closed embeddings into Siegel modular varieties and they can be studied via these closed embeddings, but they can also be studied directly via their moduli interpretation. One of the key examples of PEL type Shimura varieties that we will consider in these lecture notes will be that of unitary Shimura varieties (and, in particular, those for quasi-split unitary similitude groups).

Let F be an imaginary CM field, with $F^+ \subset F$ maximal totally real subfield. Let $x \mapsto x^*$ denote the non-trivial automorphism in $\mathrm{Gal}(F/F^+)$. Let V be a $2n$-dimensional F-vector space and let
$$\psi(\ ,\ ) : V \times V \to \mathbb{Q}$$
be a non-degenerate alternating $*$-Hermitian form on V. Let $G/\mathbb{Q}$ be the algebraic group of unitary similitudes of (V, ψ): if R is a $\mathbb{Q}$-algebra, then
$$G(R) := \left\{ (g, \lambda) \in \mathrm{GL}(V \otimes_{\mathbb{Q}} R) \times R^\times \mid \psi(gv, gw) = \lambda \cdot \psi(v, w), \forall v, w \in V \otimes_{\mathbb{Q}} R \right\}.$$
The group of real points $G(\mathbb{R})$ can be identified with
$$G\left(\prod_{i=1}^{[F^+:\mathbb{Q}]} U(p_i, q_i) \right),$$
where i indexes embeddings $F^+ \hookrightarrow \mathbb{R}$ and $U(p_i, q_i)$ is the real unitary group of signature (p_i, q_i) with $p_i + q_i = 2n$. (By the notation $G(\ )$, we mean that the similitude factors for all embeddings $F^+ \hookrightarrow \mathbb{R}$ match.)

If $F^+ = \mathbb{Q}$, then we only have one signature (p, q). The corresponding group of real points $G(\mathbb{R})$ can then be identified with $GU(p,q)$, the group of unitary similitudes which preserve up to a scalar the form

$$\langle (a_j), (b_j) \rangle = \sum_{j=1}^{p} a_j \bar{b}_j - \sum_{j=p+1}^{n} a_j \bar{b}_j.$$

We can always arrange that G is a quasi-split group over $\mathbb{Q}$ (this depends on the choice of ψ). Since $2n = \dim_F V$ is even, the quasi-split inner form will have signature (n, n) at every embedding $F^+ \hookrightarrow \mathbb{R}$. (If we had worked V with $\dim_F V = 2n+1$, then $U(n+1, n)$ and $U(n, n+1)$ are isomorphic quasi-split unitary groups over $\mathbb{R}$.)

REMARK 2.3.7.
  (1) For the purposes of studying the corresponding Shimura varieties, we can assume that the set of signatures $(p_i, q_i)_{i \in \{1, \ldots, [F^+:\mathbb{Q}]\}}$ is arbitrary. We do note that the Hasse principle for unitary groups gives a restriction on whether a unitary group with given signatures at real embeddings and with specific ramification conditions at finite places exists. See [**Clo91**] for more details; we will not dwell on this aspect since we will ultimately only need to work with the quasi-split group.
  (2) A criterion for PEL-type Shimura varieties to be compact can be found in [**Lan13a**, §5.3.3]. This satisfied, for example, if one works with a unitary similitude group for which one of the signatures is equal to $(0, n)$ or $(n, 0)$. The Shimura varieties attached to the quasi-split unitary similitude group are non-compact.

An example of a *rational PEL datum* is given by a tuple $(F, *, V, \psi, h)$, where $F, *, V, \psi$ are as above and h is an $\mathbb{R}$-algebra homomorphism

$$h : \mathbb{C} \to \mathrm{End}_{F \otimes_{\mathbb{Q}} \mathbb{R}}(V \otimes_{\mathbb{Q}} \mathbb{R}),$$

such that $\psi(h(z)v, w) = \psi(v, h(\bar{z})w)$ for all $z \in \mathbb{C}$ and such that the pairing

$$\langle v, w \rangle := \psi(v, h(i)w)$$

is symmetric and positive definite. Such a homomorphism puts a complex structure on $V \otimes_{\mathbb{Q}} \mathbb{R}$, which is the same as a Hodge structure of type $(-1, 0), (0, -1)$. By restricting h to $\mathbb{C}^\times$ and noticing that it then preserves ψ up to a scalar in $\mathbb{R}^\times$, we get a homomorphism of algebraic groups over $\mathbb{R}$:

$$h|_{\mathbb{C}^\times} : \mathbb{S} \to G_{\mathbb{R}}.$$

Let X be the $G(\mathbb{R})$-conjugacy class of $h|_{\mathbb{C}^\times}$.

EXERCISE 2.3.8. *Assume that the signatures of G at real embeddings are not all $(0,n)$ or $(n,0)$. Check that the pair (G, X) satisfies the axioms in the definition of a Shimura datum.*

Choose a rational PEL datum as above, giving rise to a Shimura datum (G, X). Let $K \subset G(\mathbb{A}_f)$ be a neat compact open subgroup. Let X_K be the corresponding Shimura variety; it is a smooth, quasi-projective scheme over the reflex E, of dimension $\sum_{i=1}^{[F^+:\mathbb{Q}]} p_i \cdot q_i$. It represents the following moduli problem over E. Let S be a connected, locally noetherian, Spec E-scheme and s a geometric point of S. The moduli problem represented by X_K sends the pair (S, s) to the set of isomorphism classes of tuples $(A, \lambda, \iota, \bar\eta)$, which is described as follows.

(1) A is an abelian scheme over S of dimension $n \cdot [F^+ : \mathbb{Q}]$.
(2) $\lambda : A \to A^\vee$ (where $A^\vee$ is a dual abelian variety) is a polarization.
(3) $\iota : F \hookrightarrow \mathrm{End}^0(A) := \mathrm{End}(A) \otimes_\mathbb{Z} \mathbb{Q}$ is an embedding of $\mathbb{Q}$-algebras giving an action of F on A by quasi-isogenies.[15] This action satisfies the following compatibility with λ: $\lambda \circ \iota(x^*) = \iota(x)^\vee \circ \lambda$ for all $x \in F$.
(4) $\bar\eta$ is a $\pi_1^{\text{ét}}(S, s)$-invariant K-orbit of $F \otimes_\mathbb{Q} \mathbb{A}_f$-equivariant isomorphisms
$$\eta : V \otimes_\mathbb{Q} \mathbb{A}_f \xrightarrow{\sim} V_f A_s,$$
where $V_f A_s$ is the rational adelic Tate module of the abelian variety A_s, such that η takes the pairing induced by ψ on $V \otimes_\mathbb{Q} \mathbb{A}_f$ to an $\mathbb{A}_f^\times$-multiple of the λ-Weil pairing on $V_f A_s$. [16]

Such a tuple is required to satisfy the following *determinant condition*: the complex structure on $V \otimes_\mathbb{Q} \mathbb{R}$ induced by h gives rise to the Hodge decomposition $V \otimes_\mathbb{Q} \mathbb{C} = V^{0,-1} \oplus V^{-1,0}$. Explicitly, we must have
$$\det(x | V^{-1,0}) = \det_{\mathcal{O}_S}(x | \mathrm{Lie}\, A), x \in F.$$
This should be understood as an equality of polynomials with $\mathcal{O}_S$-coefficients rather than as an equality of numbers, where we choose a basis for F over $\mathbb{Q}$ and write the indeterminate $x \in F$ in terms of the chosen basis. In characteristic 0, this is just a condition on ranks of $F \otimes_\mathbb{Q} \mathcal{O}_S$-modules, but it is more subtle in characteristic p. Intuitively, the determinant condition matches the Hodge structure of the abelian variety A, as it decomposes under the action of F, with the Hodge structures parametrized by the Hermitian symmetric domain X, which are also restricted by the action of F on (V, ψ).

Two such tuples $(A, \lambda, \iota, \bar\eta)$ and $(A', \lambda', \iota', \bar\eta')$, satisfying the determinant condition, are isomorphic if there exists an isogeny $A \to A'$ taking λ to a rational multiple of λ', and taking ι to ι', $\bar\eta$ to $\bar\eta'$.

If p is a good prime (for the PEL-type Shimura datum and for the level K) then one can also define an integral model of X_K, which is a smooth, quasi-projective scheme over the localization $\mathcal{O}_{E_{(p)}}$. This integral model is also constructed as the universal scheme representing a moduli problem, this time with integral data. For more details on integral models in the case of PEL-type Shimura varieties, see [**Kot92b**]. There exists a minimal (Baily–Borel) compactification $X_K \hookrightarrow X_K^*$, constructed in this case by Lan in [**Lan13b**], which is a projective, but not necessarily smooth scheme over $\mathrm{Spec}\, \mathcal{O}_{E_{(p)}}$.

2.3.9. *Shimura varieties of Hodge type.* Shimura varieties of Hodge type form a class of Shimura varieties which contain the ones of PEL type. To define them, we will first describe morphisms of Shimura data.

DEFINITION 2.3.10. *A morphism of Shimura data $(G, X) \to (G', X')$ is a homomorphism of algebraic groups $G \to G'$ inducing a map $X \to X'$. We call a morphism of Shimura data an embedding if the map $G \to G'$ is injective.*

---

[15] These are the "endomorphisms" in the PEL-type moduli problem.

[16] This definition can be shown to be independent of the choice of geometric point s and can be extended to non-connected schemes in the obvious way.

A Shimura datum of *Hodge type* is a Shimura datum (G, X) which admits an embedding $(G, X) \hookrightarrow (\widetilde{G}, \widetilde{X})$ into some Siegel datum $(\widetilde{G}, \widetilde{X})$. Given a Shimura datum of Hodge type and a neat compact open subgroup $K \subset G(\mathbb{A}_f)$, one can find a neat compact open subgroup $\widetilde{K} \subset \widetilde{G}(\mathbb{A}_f)$ such that we have a closed embedding of Shimura varieties (Proposition 1.15 of [**Del71**])

$$X_K \hookrightarrow \widetilde{X}_{\widetilde{K}}.$$

The Shimura variety X_K is said to be *of Hodge type*. The universal abelian variety $\widetilde{A}^{\mathrm{univ}}$ over $\widetilde{X}_{\widetilde{K}}$ restricts to an abelian variety A^{univ} over X_K.

Let (V, ψ) be the $2n$-dimensional split symplectic space over $\mathbb{Q}$ as defined above and set $\widetilde{G} = \mathrm{GSp}(V, \psi)$. If $(G, X) \hookrightarrow (\widetilde{G}, \widetilde{X})$ is an embedding of Shimura data, then there exists a finite collection of tensors

$$s_\alpha \subset V^\otimes := \oplus_{m,r \in \mathbb{Z}_{\geq 0}} V^{\otimes m} \otimes (V^\vee)^{\otimes r}, m, r \in \mathbb{Z}$$

such that $G = \mathrm{Stab}_{\widetilde{G}}(\{s_\alpha\})$. This holds by Proposition 3.1 of [**Del82**]. If we consider any choice of $h \in X$ we get an action of $\mathbb{S}$ on V by composing h with $G(\mathbb{R}) \hookrightarrow \widetilde{G}(\mathbb{R}) = \mathrm{GSp}(V_\mathbb{R}, \psi)$. Since G stabilizes the collection $\{s_\alpha\}$, we see that the tensors $s_\alpha \otimes 1 \in V_\mathbb{R}^\otimes$ are also stabilized by $\mathbb{S}$. This can be reformulated to say that the tensors s_α live in Hodge degree $(0,0)$, i.e. that they are *Hodge tensors*. Once we understand Siegel modular varieties, Shimura varieties of Hodge type can be studied by keeping track of Hodge tensors.

The symplectic form ψ gives rise to a Hodge tensor. In the case of Shimura varieties of PEL type, the additional Hodge tensors one needs to keep track of are particularly simple: they are given by the endomorphisms by the CM field F. Indeed, an endomorphism of a Hodge structure V respecting the Hodge decomposition can be thought of as a degree $(0, 0)$ element in $V \otimes V^\vee$. This explains why Shimura varieties of PEL type are a subclass of Shimura varieties of Hodge type.

EXAMPLE 2.3.11. A Shimura variety of Hodge type that is not of PEL type is obtained as follows: we consider the same setup as in Example 2.3.6, but rather than taking the group of unitary similitudes we consider the group of unitary isometries:

$$G(R) := \{g \in \mathrm{GL}(V \otimes_\mathbb{Q} R) \mid \psi(gv, gw) = \psi(v, w), \forall v, w \in V \otimes_\mathbb{Q} R\}.$$

It is not hard to see that, over $\overline{\mathbb{Q}}$, these Shimura varieties are connected components of the corresponding PEL-type Shimura varieties for the similitude group.

Let (G, X) be a Shimura datum of Hodge type and let μ denote the Hodge cocharacter determined by a choice of $h \in X$. Recall that the axioms for (G, X) to be a Shimura datum imply that μ is a minuscule cocharacter. The cocharacter μ determines two opposite parabolic subgroups of G:

$$P_\mu^{\mathrm{std}} := \{g \in G \mid \lim_{t \to \infty} \mathrm{ad}(\mu(t))g \text{ exists}\}, \text{ and}$$

$$P_\mu := \{g \in G \mid \lim_{t \to 0} \mathrm{ad}(\mu(t))g \text{ exists}\}.$$

REMARK 2.3.12. From the Tannakian point of view, the first parabolic can be thought of as the "stabilizer of the Hodge–de Rham filtration". Indeed, the Hodge cocharacter μ induces a *grading* on the Tannakian category $\mathrm{Rep}_\mathbb{C}(G)$, the category of finite-dimensional representations G on $\mathbb{C}$-vector spaces. This means that for

any $(V, \varphi) \in \mathrm{Rep}_{\mathbb{C}}(G)$, the composition $\varphi \circ \mu$ defines an action of $\mathbb{G}_{m,\mathbb{C}}$ on V, which is the same as a grading
$$V = \oplus_{p \in \mathbb{Z}} V^p$$
of the $\mathbb{C}$-vector space V. Note that this is *not* the same as defining a grading on V as a representation of G. The grading depends functorially on V and is compatible with tensor products in $\mathrm{Rep}_{\mathbb{C}}(G)$.

To the grading on $\mathrm{Rep}_{\mathbb{C}}(G)$ one can naturally associate two filtrations. We let $\mathrm{Fil}^\bullet(\mu)$ be the descending filtration on $\mathrm{Rep}_{\mathbb{C}}(G)$ defined by $\mathrm{Fil}^p(V) = \oplus_{p' \geq p} V^{p'}$ for each $(V, \varphi) \in \mathrm{Rep}_{\mathbb{C}}(G)$. The parabolic subgroup P_μ^{std} can be defined as the stabilizer of $\mathrm{Fil}^\bullet(\mu)$ in G. The other filtration is the ascending filtration $\mathrm{Fil}_\bullet(\mu)$ defined by $\mathrm{Fil}_p(V) = \oplus_{p' \leq p} V^{p'}$ for $(V, \varphi) \in \mathrm{Rep}_{\mathbb{C}}(G)$; the parabolic P_μ is the stabilizer of $\mathrm{Fil}_\bullet(\mu)$ in G.

Choose an embedding of Shimura data $(G, X) \hookrightarrow (\widetilde{G}, \widetilde{X})$ with $\widetilde{G} = \mathrm{GSp}(V, \psi)$, and compatible levels $K \subset G(\mathbb{A}_f)$, $\widetilde{K} \subset \widetilde{G}(\mathbb{A}_f)$. The representation V of $G(\mathbb{Q})$ determines a $\mathbb{Q}$-local system on the Shimura variety $X_K(\mathbb{C})$. This local system is the same as the relative rational Betti homology $\mathcal{V}_\mathrm{B}$ of abelian variety $\mathcal{A}(\mathbb{C})$ over $X_K(\mathbb{C})$, obtained by restriction from the universal abelian variety over $\widetilde{X}_{\widetilde{K}}(\mathbb{C})$. By considering the relative de Rham homology of $\mathcal{A}$, we also have a vector bundle $\mathcal{V}_\mathrm{dR}$ on X_K, equipped with an integrable connection. The filtration $\mathrm{Fil}^\bullet(V_\mathbb{C})$ gets identified, under the comparison between relative Betti and de Rham homologies, with the Hodge–de Rham filtration on $\mathcal{V}_\mathrm{dR}$. This is the sense in which we mean that P_μ^{std} is the "stabilizer of the Hodge–de Rham filtration".

The conjugacy classes of both parabolics are defined over the reflex field E of the Shimura datum. The two parabolics determine two flag varieties $\mathrm{Fl}_{G,\mu}^{\mathrm{std}}$ and $\mathrm{Fl}_{G,\mu}$ over E, which parametrize parabolic subgroups in the given conjugacy class, or equivalently, filtrations on $\mathrm{Rep}_{\mathbb{C}}(G)$ conjugate to $\mathrm{Fil}^\bullet(\mu)$. There is a holomorphic embedding
$$X \hookrightarrow \mathrm{Fl}_{G,\mu}^{\mathrm{std}}, h \mapsto \mathrm{Fil}^\bullet(\mu_h)^{17}.$$
The map π_HdR defined in Section 2.2 factors through the above embedding. The two flag varieties and the embedding π_HdR are functorial in the Shimura data.

3. Background from p-adic Hodge theory

In this section, we recall the relevant background from p-adic Hodge theory. Let L be a complete, discretely valued field of characteristic 0 with perfect residue field k of characteristic p.[18] Consider a proper smooth morphism $\pi : \mathcal{Y} \to \mathcal{X}$ of smooth rigid analytic spaces over L, considered as adic spaces over $\mathrm{Spa}(L, \mathcal{O}_L)$. In this section, we will:

(1) give a construction of the *relative Hodge–Tate filtration* for $\pi_C : \mathcal{Y}_C \to \mathcal{X}_C$, where C is an algebraically closed perfectoid field extension of L;
(2) explain its relationship to the relative p-adic-de Rham comparison isomorphism and to the relative Hodge–de Rham filtration;
(3) work out the specific example of the morphism
$$\pi : \mathcal{A} \to \mathcal{X}_K$$

---

[17]There is also an embedding $X \hookrightarrow \mathrm{Fl}_{G,\mu}$, but this is anti-holomorphic.

[18]Later on, L will be a finite extension of $\mathbb{Q}_p$, more precisely the completion $E_\mathfrak{p}$ of the reflex field E at a prime $\mathfrak{p}$ above a good prime p.

obtained by applying the adification functor

$$\{\text{Schemes}/\operatorname{Spec} E_\mathfrak{p}\} \to \{\text{Adic spaces}/\operatorname{Spa}(E_\mathfrak{p}, \mathcal{O}_{E,\mathfrak{p}})\}.$$

to $\pi : A^{\mathrm{univ}} \to X_K$, where X_K is a Shimura variety of Hodge type with reflex field E, A^{univ} is the universal abelian variety over X_K and $\mathfrak{p} \mid p$ is a prime of E. If one is merely interested in the form of the relative Hodge–Tate filtration rather than in its construction and relationship to the Hodge–de Rham filtration, one can skip to Example 3.2.6.

REMARK 3.0.1. For this section, we assume as prerequisites: adic spaces, perfectoid spaces, the *flattened pro-étale topology*, i.e the pro-étale topology as used in [**Sch13**], and the flattened pro-étale site $\mathcal{X}_{\mathrm{proét}}$ of a smooth adic space $\mathcal{X}$ over $\operatorname{Spa}(L, \mathcal{O}_L)$. The proofs of the statements from p-adic Hodge theory are given in Bhatt's lecture notes in this volume, so we contend ourselves to stating the precise results we will use in the study of Shimura varieties. The references we follow are [**Sch13**], Section 3 of [**Sch12b**], and Section 2.2 of [**CS17**].

3.1. The Hodge–Tate filtration. We will start with an extended example, where we discuss the Hodge–Tate filtration in the case over points, i.e when $\mathcal{X} = \operatorname{Spa}(L, \mathcal{O}_L)$. Let C be the p-adic completion of an algebraic closure $\overline{L}$ of L, with ring of integers $\mathcal{O}_C$. Let $\mathcal{X}_C := \operatorname{Spa}(C, \mathcal{O}_C)$, with tilt $\mathcal{X}_C^\flat := \operatorname{Spa}(C^\flat, \mathcal{O}_{C^\flat})$. We recall the construction of the ring $B_{\mathrm{dR},C}$, originally due to Fontaine: denote by $B_{\mathrm{dR},C}^+$ the completion of $W(\mathcal{O}_{C^\flat})[1/p]$ along the kernel of the map

$$\theta : W(\mathcal{O}_{C^\flat})[1/p] \to C$$

and then set $B_{\mathrm{dR},C} := B_{\mathrm{dR},C}^+[1/\xi]$ for a generator ξ of $\ker \theta$. The field $B_{\mathrm{dR},C}$ is the field of periods which shows up in the original comparison isomorphism between de Rham and p-adic étale cohomology, i.e. in the setting of schemes. The subring $B_{\mathrm{dR},C}^+ \subset B_{\mathrm{dR},C}$ is a complete discrete valuation ring with residue field C and with uniformizer ξ. There is a $\operatorname{Gal}(\overline{L}/L)$-action on ξ, which is via the cyclotomic character. There is a natural decreasing filtration on the ring $B_{\mathrm{dR},C}$ defined by $\operatorname{Fil}^i B_{\mathrm{dR},C} := \xi^i B_{\mathrm{dR},C}^+$ for $i \in \mathbb{Z}$. It has graded pieces $\operatorname{Gr}^i B_{\mathrm{dR},C} \simeq C(i)$.

We now let $\pi : \mathcal{Y} \to \mathcal{X}$ be a proper smooth rigid analytic variety. Because $\mathcal{Y}$ is defined over L, the Hodge–de Rham spectral sequence

$$E_1^{i-j,j} = H^j(\mathcal{Y}, \Omega_{\mathcal{Y}}^{i-j}) \Rightarrow H_{\mathrm{dR}}^i(\mathcal{Y})$$

degenerates on the first page (cf [**Sch13**, Thm. 1.6]). The induced filtration on $H_{\mathrm{dR}}^i(\mathcal{Y})$ is called the *Hodge–de Rham filtration*; it has graded pieces $H^j(\mathcal{Y}, \Omega_{\mathcal{Y}}^{i-j})$. There is a natural comparison isomorphism between the p-adic étale cohomology of $\mathcal{Y}$ and the de Rham cohomology of $\mathcal{Y}$

(3.1.1) $$H^i(\mathcal{Y}_{\overline{L},\mathrm{ét}}, \mathbb{Q}_p) \otimes_{\mathbb{Q}_p} B_{\mathrm{dR},C} \simeq H_{\mathrm{dR}}^i(\mathcal{Y}) \otimes_L B_{\mathrm{dR},C},$$

cf. [**Sch13**, Cor. 1.8]. This isomorphism is $\operatorname{Gal}(\overline{L}/L)$-equivariant and also compatible with the following filtrations on each side: on the LHS, we consider the filtration induced from the natural filtration on $B_{\mathrm{dR},C}$ and on the RHS we consider the convolution of the de Rham filtration on $H_{\mathrm{dR}}^i(\mathcal{Y})$ and the natural filtration on $B_{\mathrm{dR},C}$.

By applying Gr^0 on both sides of (3.1.1), we obtain the direct sum decomposition

$$H^i(\mathcal{Y}_{\overline{L},\mathrm{ét}}, \mathbb{Q}_p) \otimes_{\mathbb{Q}_p} C \simeq \oplus_{j=0}^i H^{i-j}(\mathcal{Y}, \Omega_{\mathcal{Y}}^j) \otimes_L C(-j),$$

known as the *Hodge–Tate decomposition*. When we consider the analogous p-adic comparison isomorphism in the relative setting, there is no longer a direct sum decomposition. This fact mirrors the complex phenomenon, where only the Hodge–de Rham filtration varies holomorphically, not the Hodge decomposition. It is therefore better to only remember the filtration

$$\mathrm{Fil}^j\left(H^i(\mathcal{Y}_{\bar{L},\mathrm{\acute{e}t}},\mathbb{Q}_p)\otimes_{\mathbb{Q}_p} C\right) = \oplus_{k=j}^i H^{i-k}(\mathcal{Y},\Omega_{\mathcal{Y}}^k)\otimes_L C(-k)$$

induced by this direct sum decomposition.

The following is the perspective that generalizes to the relative setting: we interpret the comparison isomorphism as saying that the p-adic étale cohomology of $\mathcal{Y}_{\bar{L}}$ and the de Rham cohomology of $\mathcal{Y}$ give rise to two $B_{\mathrm{dR},C}^+$-lattices contained in the same $B_{\mathrm{dR},C}$-vector space. Indeed, we define:

$$M = H^i(\mathcal{Y}_{\bar{L},\mathrm{\acute{e}t}},\mathbb{Z}_p)\otimes_{\mathbb{Z}_p} B_{\mathrm{dR},C}^+, \text{ and } M_0 = H^i_{\mathrm{dR}}(\mathcal{Y})\otimes_L B_{\mathrm{dR},C}^+.$$

We can define filtrations on both M, M_0 which measure the relative position of the two lattices. The filtration on $M_0/\xi M_0 = H^i_{\mathrm{dR}}(\mathcal{Y})\otimes_L C$ induced by the lattice M agrees with the Hodge–de Rham filtration.

On $M/\xi M = H^i(\mathcal{Y}_{C,\mathrm{\acute{e}t}},\mathbb{Q}_p)\otimes_{\mathbb{Q}_p} C$, we have the *Hodge–Tate filtration*, defined as follows in [**Sch13**]. Consider the morphism of sites

$$\nu : \mathcal{Y}_{C,\mathrm{pro\acute{e}t}} \to \mathcal{Y}_{C,\mathrm{\acute{e}t}};$$

this gives rise to a spectral sequence

$$E_2^{i-j,j} = H^{i-j}(\mathcal{Y}_{C,\mathrm{\acute{e}t}}, R^j\nu_*\widehat{\mathcal{O}}_{\mathcal{Y}_C}) \Rightarrow H^i(\mathcal{Y}_{C,\mathrm{pro\acute{e}t}},\widehat{\mathcal{O}}_{\mathcal{Y}_C}) \xrightarrow{\sim} H^i(\mathcal{Y}_{C,\mathrm{\acute{e}t}},\mathbb{Q}_p)\otimes_{\mathbb{Q}_p} C^{19}.$$

In [**Sch13**], Scholze shows that there are natural isomorphisms

$$\Omega_{\mathcal{Y}_C}^j(-j) \simeq R^j\nu_*\widehat{\mathcal{O}}_{\mathcal{Y}_C}$$

for all $j \geq 0$. The Hodge–Tate spectral sequence

$$E_2^{i-j,j} = H^{i-j}(\mathcal{Y},\Omega_{\mathcal{Y}}^j)\otimes_L C(-j) \Rightarrow H^i(\mathcal{Y}_{C,\mathrm{\acute{e}t}},\mathbb{Q}_p)\otimes_{\mathbb{Q}_p} C$$

then degenerates on the E_2 page because $\mathcal{Y}$ is defined over the subfield $L \subset C$ and the differentials are $\mathrm{Gal}(\bar{L}/L)$-equivariant. The corresponding filtration on $H^i(\mathcal{Y}_{C,\mathrm{\acute{e}t}},\mathbb{Q}_p)\otimes_{\mathbb{Q}_p} C$ is the Hodge–Tate filtration. This is the same as the filtration on $M/\xi M$ induced by the lattice M_0. (In the case when $\mathcal{X}$ is a point, the fact that these two filtrations on $H^i(\mathcal{Y}_{\bar{L},\mathrm{\acute{e}t}},\mathbb{Q}_p)\otimes_{\mathbb{Q}_p} C$ agree can be seen from the fact that the Hodge–Tate decomposition is canonical, because it is $\mathrm{Gal}(\bar{L}/L)$-equivariant.)

EXAMPLE 3.1.1. If we set $i = 1$, we have $\xi M \subset M_0 \subset M$, with $M_0/\xi M \simeq H^1(\mathcal{Y},\mathcal{O}_{\mathcal{Y}})\otimes_L C$ and $M/M_0 \simeq H^0(\mathcal{Y},\Omega_{\mathcal{Y}}^1)\otimes_L C(-1)$. The Hodge–de Rham filtration on $H^i_{\mathrm{dR}}(\mathcal{Y})\otimes_L C$ is given by

$$0 \to \xi M/\xi M_0 \to M_0/\xi M_0 \to M_0/\xi M \to 0,$$

which becomes

$$0 \to H^0(\mathcal{Y},\Omega_{\mathcal{Y}}^1)\otimes_L C \to H^1_{\mathrm{dR}}(\mathcal{Y})\otimes_L C \to H^1(\mathcal{Y},\mathcal{O}_{\mathcal{Y}})\otimes_L C \to 0.$$

The Hodge–Tate filtration on $H^i(\mathcal{Y}_{\bar{L},\mathrm{\acute{e}t}},\mathbb{Z}_p)\otimes_{\mathbb{Z}_p} C$ is given by

$$0 \to M_0/\xi M \to M/\xi M \to M/M_0 \to 0,$$

---

[19] This last isomorphism is the *primitive comparison theorem* in p-adic Hodge theory. For schemes, the primitive comparison theorem goes back to Faltings, cf. [**Fal02**], and for rigid-analytic varieties this is [**Sch13**, Thm 1.3]. The latter result underlies all other p-adic comparison theorems for rigid-analytic varieties.

which becomes
$$0 \to H^1(\mathcal{Y}, \mathcal{O}_\mathcal{Y}) \to H^1(\mathcal{Y}_{\bar{L},\text{ét}}, \mathbb{Z}_p) \otimes_{\mathbb{Z}_p} C \to H^0(\mathcal{Y}, \Omega^1_\mathcal{Y}) \otimes_L C(-1) \to 0.$$

Note that the graded pieces of these two filtration are isomorphic (up to Tate twists) but the filtrations themselves are not directly related.

3.2. The relative Hodge–Tate filtration. We now discuss a relative definition of the Hodge–Tate filtration, which will be crucial to our application to Shimura varieties.

For $\mathcal{X}$ a smooth adic space over $\text{Spa}(L, \mathcal{O}_L)$, we have the following sheaves on $\mathcal{X}_{\text{proét}}$, as defined in [**Sch13**]: the (integral) completed structure sheaf $\widehat{\mathcal{O}}_\mathcal{X}^{(+)}$, the (integral) tilted completed structure sheaf $\widehat{\mathcal{O}}_{\mathcal{X}^\flat}^{(+)}$, the relative period sheaves $\mathbb{B}_{\text{dR},\mathcal{X}}^{(+)}$, and the structural de Rham sheaves $\mathcal{O}\mathbb{B}_{\text{dR},\mathcal{X}}^{(+)}$. We recall the definitions of these sheaves.

DEFINITION 3.2.1.
(1) *The integral completed structure sheaf $\widehat{\mathcal{O}}_\mathcal{X}^+$ is the inverse limit of the sheaves $\mathcal{O}_\mathcal{X}^+/p^n$ on $\mathcal{X}_{\text{proét}}$. The titled integral structure sheaf $\widehat{\mathcal{O}}_{\mathcal{X}^\flat}^+$ is the inverse limit on $\mathcal{X}_{\text{proét}}$ of $\mathcal{O}_\mathcal{X}^+/p$ with respect to the Frobenius morphism.*
(2) *The relative period sheaf $\mathbb{B}_{\text{dR},\mathcal{X}}^+$ is the completion of $W(\widehat{\mathcal{O}}_{\mathcal{X}^\flat}^+)[1/p]$ along the kernel of the natural map*
$$\theta : W(\widehat{\mathcal{O}}_{\mathcal{X}^\flat}^+)[1/p] \to \widehat{\mathcal{O}}_\mathcal{X}.$$
The relative period sheaf $\mathbb{B}_{\text{dR},\mathcal{X}}$ is $\mathbb{B}_{\text{dR},\mathcal{X}}^+[\xi^{-1}]$, where ξ is any element that generates the kernel of θ. This is well-defined because such a ξ exists proétale locally on $\mathcal{X}$, is not a zero divisor, and is unique up to a unit.
(3) *We now define the sheaf $\mathcal{O}\mathbb{B}_{\text{dR},\mathcal{X}}^+$ as the sheafification of the following presheaf. If $U = \text{Spa}(R, R^+)$ is affinoid perfectoid, with (R, R^+) the completed direct limit of (R_i, R_i^+), the presheaf sends U to the direct limit over i of the completion of*
$$\left(R_i^+ \hat{\otimes}_{W(k)} W(R^{\flat+}) \right)[1/p]$$
along $\ker \theta$, where
$$\theta : \left(R_i^+ \hat{\otimes}_{W(k)} W(R^{\flat+}) \right)[1/p] \to R$$
is the natural map. We set $\mathcal{O}\mathbb{B}_{\text{dR},\mathcal{X}} := \mathcal{O}\mathbb{B}_{\text{dR},\mathcal{X}}^+[\xi^{-1}]$ as before.

These sheaves are equipped with the following structures. The relative period sheaves $\mathbb{B}_{\text{dR},\mathcal{X}}^{(+)}$ are equipped with compatible filtrations: $\text{Fil}^i \mathbb{B}_{\text{dR},\mathcal{X}} := \xi^i \mathbb{B}_{\text{dR},\mathcal{X}}^+$, with $\text{Gr}^0 \mathbb{B}_{\text{dR},\mathcal{X}} = \widehat{\mathcal{O}}_\mathcal{X}$. The structural de Rham sheaves $\mathcal{O}\mathbb{B}_{\text{dR},\mathcal{X}}^{(+)}$ are equipped with filtrations and connections
$$\nabla : \mathcal{O}\mathbb{B}_{\text{dR},\mathcal{X}}^{(+)} \to \mathcal{O}\mathbb{B}_{\text{dR},\mathcal{X}}^{(+)} \otimes_{\mathcal{O}_\mathcal{X}} \Omega^1_\mathcal{X}$$
We have a natural identification
$$\left(\mathcal{O}\mathbb{B}_{\text{dR},\mathcal{X}}^{(+)} \right)^{\nabla=0} = \mathbb{B}_{\text{dR},\mathcal{X}}^{(+)}.$$

The following theorem states the relative p-adic-de Rham comparison isomorphism for a proper smooth morphism $\pi : \mathcal{Y} \to \mathcal{X}$ of smooth adic spaces over L.

We consider the sheaf $R^i\pi_{\mathrm{dR}*}\mathcal{O}_{\mathcal{Y}}$ on $\mathcal{X}_{\mathrm{pro\acute{e}t}}$ obtained by taking the ith cohomology sheaf of the derived pushforward $R\pi_*$ applied to the complex of relative differentials $\Omega^{\cdot}_{\mathcal{Y}/\mathcal{X}}$ on $\mathcal{Y}_{\mathrm{pro\acute{e}t}}$. The sheaf $R^i\pi_{\mathrm{dR}*}\mathcal{O}_{\mathcal{Y}}$ is an $\mathcal{O}_{\mathcal{X}}$-module equipped with a filtration (the Hodge–de Rham filtration) and with an integrable connection (the Gauss–Manin connection ∇_{GM}). The Gauss–Manin connection satisfies Griffiths transversality with respect to the Hodge–de Rham filtration.

THEOREM 3.2.2. *(Thm 8.8 of [**Sch13**]) For all $i \geq 0$, there is a natural isomorphism of sheaves on $\mathcal{X}_{\mathrm{pro\acute{e}t}}$*

$$R^i\pi_*\widehat{\mathbb{Z}}_{p,\mathcal{Y}} \otimes_{\widehat{\mathbb{Z}}_{p,\mathcal{X}}} \mathcal{O}\mathbb{B}_{\mathrm{dR},\mathcal{X}} \simeq R^i\pi_{\mathrm{dR}*}\mathcal{O}_{\mathcal{Y}} \otimes_{\mathcal{O}_{\mathcal{X}}} \mathcal{O}\mathbb{B}_{\mathrm{dR},\mathcal{X}},$$

compatible with the filtrations and connections on both sides. [20]

REMARK 3.2.3. Note that [**Sch13**, Thm. 8.8] applies under the assumption that $R^i\pi_{\mathrm{pro\acute{e}t}*}\widehat{\mathbb{Z}}_p$ is a lisse $\widehat{\mathbb{Z}}_p$-sheaf on $\mathcal{X}$. This is now guaranteed by [**SW17**, Thm. 10.5.1].

We will see that the Hodge–de Rham filtration on $R^i\pi_{\mathrm{dR}*}\mathcal{O}_{\mathcal{Y}}$ induces, via the comparison isomorphism in Theorem 3.2.2, a filtration on $R^i\pi_*\widehat{\mathbb{Z}}_{p,\mathcal{Y}}\otimes_{\widehat{\mathbb{Z}}_{p,\mathcal{X}}}\widehat{\mathcal{O}}_{\mathcal{X}}$, which we will call the *relative Hodge–Tate filtration*. To make this precise, we construct two $\mathbb{B}^+_{\mathrm{dR},\mathcal{X}}$-local systems on $\mathcal{X}$. The first one, which is closely related to the relative étale cohomology of $\mathcal{Y}$ is given by

$$\mathbb{M} := R^i\pi_*\widehat{\mathbb{Z}}_{p,\mathcal{Y}} \otimes_{\mathbb{Z}_{p,\mathcal{X}}} \mathbb{B}^+_{\mathrm{dR},\mathcal{X}}.$$

The second one, which is closely related to the relative de Rham cohomology of $\mathcal{Y}$, is given by

$$\mathbb{M}_0 := \left(R^i\pi_{\mathrm{dR}*}\mathcal{O}_{\mathcal{Y}} \otimes_{\mathcal{O}_{\mathcal{X}}} \mathcal{O}\mathbb{B}^+_{\mathrm{dR},\mathcal{X}}\right)^{\nabla=0}.$$

A consequence of the comparison isomorphism is that $\mathbb{M}$ and $\mathbb{M}_0$ are two "lattices" contained in the same $\mathbb{B}_{\mathrm{dR},\mathcal{X}}$-local system on $\mathcal{X}$.

The following is (a reformulation of) Proposition 7.9 of [**Sch13**] and Proposition 2.2.3 of [**CS17**].

PROPOSITION 3.2.4. *There exists a canonical isomorphism*

$$\mathbb{M} \otimes_{\mathbb{B}^+_{\mathrm{dR},\mathcal{X}}} \mathbb{B}_{\mathrm{dR},\mathcal{X}} \simeq \mathbb{M}_0 \otimes_{\mathbb{B}^+_{\mathrm{dR},\mathcal{X}}} \mathbb{B}_{\mathrm{dR},\mathcal{X}}.$$

Consider the descending filtration $\mathrm{Fil}^j \mathbb{M}_{(0)}$ on $\mathbb{M}_{(0)}$ induced by the canonical filtration on $\mathbb{B}^+_{\mathrm{dR},\mathcal{X}}$. For any $j \in \mathbb{Z}$, there is an identification

$$(\mathbb{M} \cap \mathrm{Fil}^j \mathbb{M}_0)/(\mathbb{M} \cap \mathrm{Fil}^{j+1}\mathbb{M}_0) = (\mathrm{Fil}^{-j}R^i\pi_{\mathrm{dR}*}\mathcal{O}_{\mathcal{Y}}) \otimes_{\mathcal{O}_{\mathcal{X}}} \widehat{\mathcal{O}}_{\mathcal{X}}(j)$$

$$\subset \mathrm{Gr}^j \mathbb{M}_0 = R^i\pi_{\mathrm{dR}*}\mathcal{O}_{\mathcal{Y}} \otimes_{\mathcal{O}_{\mathcal{X}}} \widehat{\mathcal{O}}_{\mathcal{X}}(j).$$

In particular, we always have $\mathbb{M}_0 \subset \mathbb{M}$. Moreover, considering the relative position of $\mathbb{M}$ and $\mathbb{M}_0$ induces an ascending filtration on

$$\mathrm{Gr}^0 \mathbb{M} = R^i\pi_*\widehat{\mathbb{Z}}_{p,\mathcal{Y}} \otimes_{\widehat{\mathbb{Z}}_{p,\mathcal{X}}} \widehat{\mathcal{O}}_{\mathcal{X}}$$

---

[20] The filtration and connection on the left hand side are simply induced from the filtration and connection on $\mathcal{O}\mathbb{B}_{\mathrm{dR},\mathcal{X}}$. On the right hand side, one must take the convolution of the Hodge–de Rham filtration with the one on $\mathcal{O}\mathbb{B}_{\mathrm{dR},\mathcal{X}}$ and the connection is $\nabla_{\mathrm{GM}} \otimes 1 + 1 \otimes \nabla$.

given by
$$\mathrm{Fil}_{-j}(R^i\pi_*\widehat{\mathbb{Z}}_{p,\mathcal{Y}} \otimes_{\widehat{\mathbb{Z}}_{p,\mathcal{X}}} \widehat{\mathcal{O}}_{\mathcal{X}}) := (\mathbb{M} \cap \mathrm{Fil}^j\mathbb{M}_0)/(\mathrm{Fil}^1\mathbb{M} \cap \mathrm{Fil}^j\mathbb{M}_0).$$

We call this filtration the *relative Hodge–Tate filtration*.

REMARK 3.2.5. In this section, we gave the construction of the relative Hodge–Tate filtration via the comparison isomorphism rather than a version of the construction in § 3.1 via the morphism of sites from the proétale to the étale site. For applications to Shimura varieties of Hodge type, we will only need to use the first filtration step on $R^1\pi_*\widehat{\mathbb{Z}}_{p,\mathcal{Y}} \otimes_{\widehat{\mathbb{Z}}_{p,\mathcal{X}}} \widehat{\mathcal{O}}_{\mathcal{X}}$. Proposition 2.2.5 of [**CS17**], which works in the relative case, shows that the two constructions of the Hodge–Tate filtration agree on the first filtration step.

We made the choice of presenting the construction of the Hodge–Tate filtration via the p-adic comparison isomorphism because this perspective is the one used in constructing the Hodge–Tate period morphism for Shimura varieties of Hodge type in [**CS17**] (and, as a result, also for Shimura varieties of abelian type in [**She17**]). We explain this further in section 5. We also chose to present this construction in order to emphasize the close analogy between the period morphisms for Hermitian symmetric domains and the Hodge–Tate period morphism.

EXAMPLE 3.2.6 (The relative Hodge–Tate filtration for the universal abelian variety). Let (G, X) be a Shimura datum of Hodge type, $K \subset G(\mathbb{A}_f)$ a neat compact open subgroup, and X_K the corresponding Shimura variety over the reflex field E. Since X_K admits a closed embedding into some Siegel modular variety, there exists an abelian scheme $\pi : A^{\mathrm{univ}} \to X_K$.

We let $\mathfrak{p}|p$ be a prime of E, $L := E_\mathfrak{p}$, and consider the proper smooth morphism of adic spaces $\pi : \mathcal{A} \to \mathcal{X}_K$ over $\mathrm{Spa}(L, \mathcal{O}_L)$. The relative Hodge–Tate filtration on $R^1\pi_*\widehat{\mathbb{Z}}_p \otimes_{\widehat{\mathbb{Z}}_p} \widehat{\mathcal{O}}_{\mathcal{X}_K}$ is encoded in the short-exact sequence of sheaves on $\mathcal{X}_{K,\mathrm{proét}}$

$$0 \to R^1\pi_*\mathcal{O}_\mathcal{A} \otimes_{\mathcal{O}_{\mathcal{X}_K}} \widehat{\mathcal{O}}_{\mathcal{X}_K} \to R^1\pi_*\widehat{\mathbb{Z}}_p \otimes_{\widehat{\mathbb{Z}}_p} \widehat{\mathcal{O}}_{\mathcal{X}_K} \to \pi_*\Omega^1_\mathcal{A} \otimes_{\mathcal{O}_{\mathcal{X}_K}} \widehat{\mathcal{O}}_{\mathcal{X}_K}(-1) \to 0.$$

Proposition 2.2.5 of [**CS17**] shows that the first map in the short exact sequence can be identified with the natural injection

$$R^1\pi_*\mathcal{O}_\mathcal{A} \otimes_{\mathcal{O}_{\mathcal{X}_K}} \widehat{\mathcal{O}}_{\mathcal{X}_K} \hookrightarrow R^1\pi_*\widehat{\mathcal{O}}_\mathcal{A}$$

of sheaves on $\mathcal{X}_{K,\mathrm{proét}}$, where we have used the primitive relative comparison isomorphism

$$R^1\pi_*\widehat{\mathbb{Z}}_p \otimes_{\widehat{\mathbb{Z}}_p} \widehat{\mathcal{O}}_{\mathcal{X}_K} \simeq R^1\pi_*\widehat{\mathcal{O}}_\mathcal{A}.$$

4. The canonical subgroup and the anticanonical tower

In this section, we describe the theory of the canonical subgroup. We use this theory to explain the construction of the anticanonical tower of formal schemes over the ordinary locus of Siegel modular varieties, which has the following extremely useful properties

(1) it overconverges, i.e. it extends to an ε-neighborhood of the ordinary locus;
(2) its adic generic fiber gives rise to a perfectoid space.

These two properties, together with the Hodge–Tate period morphism, which is the focus of section 5, are the key ingredients in proving that Siegel modular varieties with infinite level at p are perfectoid. We follow Section 3 of [**Sch15**], but aim to give more background and fewer technical details.

4.1. The ordinary locus inside Siegel modular varieties.
In this section, we will only work with the Siegel modular varieties of Example 2.3.4. The same techniques could also be applied directly to the unitary Shimura varieties described in Example 2.3.6, if they are associated to a quasi-split unitary group over $\mathbb{Q}$. We leave this case as an exercise to the reader.[21]

Let $n \geq 1$ and let

$$(V, \psi) = \left(\mathbb{Q}^{2n}, \psi((a_i), (b_i)) = \sum_{i=1}^{n} (a_i b_{n+i} - a_{n+i} b_i) \right)$$

be the split symplectic space of dimension $2n$ over $\mathbb{Q}$. Let $\Lambda = \mathbb{Z}^{2n}$ be the standard lattice in V, which is self-dual under the symplectic form ψ. Consider the group of symplectic similitudes of Λ, $\mathrm{GSp}(\Lambda, \psi)$. This is an algebraic group over $\mathbb{Z}$. Fix a prime number p and a compact open subgroup $K^p \subset \mathrm{GSp}_{2n}(\mathbb{A}_f^p)$ contained in

$$\left\{ g \in \mathrm{GSp}_{2n}(\widehat{\mathbb{Z}}^p) \mid g \equiv 1 \pmod{N} \right\}$$

for some $N \geq 3$ such that $(N, p) = 1$. (This condition is enough to ensure that any level $K = K^p K_p$, with $K_p \subset G(\mathbb{Q}_p)$ compact open is neat.)

Set $K_p = \mathrm{GSp}_{2n}(\mathbb{Z}_p)$, $K := K^p K_p$ and let X_K be the model over $\mathbb{Z}_{(p)}$ of the corresponding Shimura variety. This is the moduli space of principally polarized n-dimensional abelian varieties with K^p-level structure. Since we will keep the tame level K^p fixed in this section, we denote X_K by X_{K_p}. Over Over X_{K_p}, we have a natural line bundle ω, given by the top exterior power of the sheaf of invariant differentials on the universal abelian scheme.

REMARK 4.1.1. As seen above, the case $n = 1$ corresponds to the group GL_2 and the case of modular curves; the constructions and techniques used in this section will be interesting (and relatively novel) even in this case. One may specialize to the case $n = 1$ on a first reading of this section.

On the level of generic fibers, we will also consider the versions with K_p-level structure for other compact open subgroups $K_p \subset G(\mathbb{Q}_p)$. We will be particularly interested in the case

$$\Gamma_0(p^m) := \{ g \in \mathrm{GSp}_{2n}(\mathbb{Z}_p) \mid g \equiv \begin{pmatrix} * & * \\ 0 & * \end{pmatrix} \pmod{p^m}, \lambda(g) \equiv 1 \pmod{p^m} \},$$

where $\lambda(g)$ is the symplectic similitude factor of g. For each $m \in \mathbb{Z}_{\geq 1}$, the Shimura variety $X_{\Gamma_0(p^m)}$ admits a morphism to $\mathrm{Spec}\, \mathbb{Q}_p(\mu_{p^m})$ corresponding to the symplectic similitude factor. We will consider the tower $(X_{\Gamma_0(p^m)})_m$ over the perfectoid field $\mathbb{Q}_p^{\mathrm{cycl}}$, by taking the base change at level m along the natural morphism $\mathrm{Spec}\, \mathbb{Q}_p(\mu_{p^m}) \to \mathrm{Spec}\, \mathbb{Q}_p^{\mathrm{cycl}}$.

---

[21] In fact, the same techniques should be applicable directly to any Shimura variety of PEL type where the *ordinary locus* is non-empty. The main theorem of [**Wed99**] shows that the ordinary locus inside the special fiber of the Shimura variety is non-empty if and only if p splits completely in the reflex field E of the Shimura datum.

We let $\mathfrak{X}_{K_p}$ be the p-adic completion of $X_{K_p} \times_{\mathbb{Z}_{(p)}} \mathbb{Z}_p^{\mathrm{cycl}}$ along its special fiber. This is a formal scheme over $\mathrm{Spf}\,\mathbb{Z}_p^{\mathrm{cycl}}$. We let $\mathcal{X}_{K_p}$ be its adic generic fiber, an analytic adic space over $\mathrm{Spa}(\mathbb{Q}_p^{\mathrm{cycl}}, \mathbb{Z}_p^{\mathrm{cycl}})$. Then $\mathcal{X}_{K_p}$ is a proper open subset of the analytic adic space $\left(X_{K_p} \times_{\mathbb{Z}_{(p)}} \mathbb{Q}_p^{\mathrm{cycl}}\right)^{\mathrm{ad}}$. The subset $\mathcal{X}_{K_p}$ is referred to in [**Sch15**] as the *good reduction locus*, i.e. the locus where the universal abelian scheme over $\left(X_{K_p} \times_{\mathbb{Z}_{(p)}} \mathbb{Q}_p^{\mathrm{cycl}}\right)^{\mathrm{ad}}$ has good reduction.

EXAMPLE 4.1.2. Let $\mathbb{A}^1_{\mathbb{Z}_p} := \mathrm{Spec}\,\mathbb{Z}_p[x]$ be one-dimensional affine space over $\mathbb{Z}_p$ and $\mathbb{P}^1_{\mathbb{Z}_p}$ be the one-dimensional projective space. The open immersion $\mathbb{A}^1_{\mathbb{Z}_p} \hookrightarrow \mathbb{P}^1_{\mathbb{Z}_p}$ is a toy model for the embedding $X_{K_p} \hookrightarrow X^*_{K_p}$ of the integral model X_{K_p} into its minimal compactification. The formal scheme corresponding to $\mathbb{A}^1_{\mathbb{Z}_p}$ is $\mathrm{Spf}\,\mathbb{Z}_p\langle x\rangle$, where

$$\mathbb{Z}_p\langle x\rangle = \left\{\sum_{i=0}^{\infty} a_i x^i \mid a_i \in \mathbb{Z}_p, \lim_{i\to\infty} |a_i|_p = 0\right\}$$

and its adic generic fiber is the closed unit disk $\mathrm{Spa}(\mathbb{Q}_p\langle x\rangle, \mathbb{Z}_p\langle x\rangle)$. On the other hand, the adic space $\mathbb{A}^{1,\mathrm{ad}}_{\mathbb{Q}_p}$ corresponding to the scheme $\mathbb{A}^1_{\mathbb{Q}_p}$ is the increasing union of closed disks

$$\bigcup_{m\geq 0} \mathrm{Spa}(\mathbb{Q}_p\langle p^m x\rangle, \mathbb{Z}_p\langle p^m x\rangle)$$

over $m \geq 0$.

EXERCISE 4.1.3. *Check that for $\mathbb{P}^1_{\mathbb{Z}_p}$, both constructions give rise to the same space $\mathbb{P}^{1,\mathrm{ad}}_{\mathbb{Q}_p}$.*

For any $m \in \mathbb{Z}_{\geq 1}$, we consider the adic space $\left(X_{\Gamma_0(p^m)} \times_{\mathbb{Q}_p(\mu_{p^m})} \mathbb{Q}_p^{\mathrm{cycl}}\right)^{\mathrm{ad}}$, equipped with the natural projection to $\left(X_{K_p} \times_{\mathbb{Z}_{(p)}} \mathbb{Q}_p^{\mathrm{cycl}}\right)^{\mathrm{ad}}$. We define $\mathcal{X}_{\Gamma_0(p^m)}$ to be the inverse image of the good reduction locus $\mathcal{X}_{K_p}$ under this projection.

REMARK 4.1.4. The adic space $\mathcal{X}_{\Gamma_0(p^m)}$ parametrizes pairs (A, D), where A is an abelian variety equipped with a principal polarization, K^p-level structure, and having "good reduction" and $D \subset A[p^m]$ is a totally isotropic subgroup scheme of rank p^{nm}.

The special fiber $\overline{X}_{K_p}$ of X_{K_p} (at least after base change to $\overline{\mathbb{F}}_p$) admits a stratification called the *Newton stratification*, which is defined in terms of the p-divisible groups (up to isogeny, and together with their extra structures) of the abelian varieties parametrized by $\overline{X}_{K_p}$. In these notes, we will only describe one Newton stratum, namely the *ordinary locus*. When it is non-empty, which holds for Siegel modular varieties, the ordinary locus is open and dense in $\overline{X}_{K_p}$.

We start by recalling the *Hasse invariant*. Let S be a scheme of characteristic p and let $\pi : A \to S$ be an abelian scheme of dimension n. The sheaf $\pi_*\Omega_{A/S}$ on S is locally free of rank n. We let $\omega_{A/S}$ be its top exterior power; this is a line bundle on S. Let $A^{(p)}$ denote the pullback of A along the absolute Frobenius of S. The Verschiebung isogeny $A^{(p)} \to A$ induces a morphism $\omega_{A/S} \to \omega_{A^{(p)}/S} \simeq \omega_{A/S}^{\otimes p}$ which can be identified with a section $\mathrm{Ha}(A/S) \in \omega_{A/S}^{\otimes (p-1)}$. This section is called the Hasse invariant of A/S.

DEFINITION 4.1.5. *We say that an abelian scheme A/S of dimension n is ordinary if for all geometric points $\bar{s}$ of S, the set $A[p](\bar{s})$ (obtained by evaluating the sheaf $A[p]$ on $S_{\text{ét}}$ on the geometric point $\bar{s}$) has p^n elements.*

This definition only depends on the p-divisible group $\mathcal{G} := A[p^\infty]$.

EXERCISE 4.1.6. *Prove that A is ordinary if and only if the p-divisible group $\mathcal{G}_{\bar{s}}$ is isomorphic to $(\mu_{p^\infty})^n \times (\mathbb{Q}_p/\mathbb{Z}_p)^n$ for all geometric points $\bar{s}$ of S.*

The following is a well-known result, in the formulation of Lemma 3.2.5 of [**Sch15**].

LEMMA 4.1.7. *The section $\mathrm{Ha}(A/S) \in \omega_{A/S}^{\otimes (p-1)}$ is invertible if and only if A/S is ordinary.*

PROOF. The Hasse invariant is the determinant of the map on co-tangent spaces induced by the Verschiebung morphism. Thus, the Hasse invariant is invertible if and only if the Verschiebung $V : A^{(p)} \to A$ induces an isomorphism on tangent spaces. This is equivalent to asking that Verschiebung be finite étale, which is in turn equivalent to asking that $\ker V$ has p^n (the degree of V) distinct geometric points above any geometric point $\bar{s}$ of S. If we let $F : A \to A^{(p)}$ be the Frobenius isogeny (i.e. the relative Frobenius of A) then $VF := p : A \to A$ and F is a purely inseparable map. Thus $A[p](\bar{s}) = (\ker V)(\bar{s})$ and we get the desired equivalence. $\square$

Now consider $\overline{A}^{\mathrm{univ}}/\overline{X}_{K_p}$. The complement of the vanishing locus of the Hasse invariant $\mathrm{Ha} := \mathrm{Ha}(\overline{A}^{\mathrm{univ}}/\overline{X}_{K_p})$ is called the ordinary locus $\overline{X}_{K_p}^{\mathrm{ord}} \subset \overline{X}_{K_p}$. We also let $\mathfrak{X}_{K_p}(0) \subset \mathfrak{X}_{K_p}$ be the open formal subscheme where Ha is invertible. If we let $\mathcal{X}_{K_p}(0)$ be the adic generic fiber of $\mathfrak{X}_{K_p}(0)$, then $\mathcal{X}_{K_p}(0) \subset \mathcal{X}_{K_p}$ is the open subset cut out by the condition $|\mathrm{Ha}| \geq 1$.

Let $0 \leq \varepsilon < 1/2$ be such that there exists an element $p^\varepsilon \in \mathbb{Z}_p^{\mathrm{cycl}}$ of p-adic valuation ε. Our goal for the rest of this section is to define a tower of formal schemes $\mathfrak{X}_{K_p}(m,\varepsilon)$ over $\mathbb{Z}_p^{\mathrm{cycl}}$ indexed by $m \in \mathbb{Z}_{\geq 0}$ which has the following properties:

(1) For $m = 0$ and $\varepsilon = 0$ we recover the formal scheme $\mathfrak{X}_{K_p}(0)$, corresponding to the ordinary locus. For general ε, the formal scheme $\mathfrak{X}_{K_p}(0,\varepsilon)$ will be a neighborhood of the ordinary locus.
(2) The transition morphisms $\mathfrak{X}_{K_p}(m+1,\varepsilon) \to \mathfrak{X}_{K_p}(m,\varepsilon)$ reduce modulo $p^{1-\varepsilon}$ to the relative Frobenius morphism.
(3) For each $m \in \mathbb{Z}_{\geq 1}$, there is a compatible system maps
$$\mathcal{X}_{K_p}(m,\varepsilon) \xrightarrow{\sim} \mathcal{X}_{\Gamma_0(p^m)}(\varepsilon)_{\mathrm{anti}} \hookrightarrow \mathcal{X}_{\Gamma_0(p^m)},$$
where the first map is an isomorphism and the second is an open embedding of adic spaces. The adic space $\mathcal{X}_{\Gamma_0(p^m)}(\varepsilon)_{\mathrm{anti}}$ is an "ε-neighborhood" of the so-called anticanonical part of the ordinary locus in $\mathcal{X}_{\Gamma_0(p^m)}$. The inverse system $(\mathcal{X}_{\Gamma_0(p^m)}(\varepsilon)_{\mathrm{anti}})_{m \in \mathbb{Z}_{\geq 1}}$ of adic spaces gives rise to a perfectoid space $\mathcal{X}_{\Gamma_0(p^\infty)}(\varepsilon)_{\mathrm{anti}}$ over $\mathbb{Z}_p^{\mathrm{cycl}}$.

In § 4.2, we explain the construction of the tower $\mathfrak{X}_{K_p}(m,0)$ over $\mathfrak{X}_{K_p}(0)$; this is not a logically necessary step in the argument, but we think it helps clarify the construction of the anticanonical tower. In § 4.3, we use the theory of the canonical subgroup to construct an "ε-neighborhood" $\mathfrak{X}_{K_p}(m,\varepsilon)$ of $\mathfrak{X}_{K_p}(m,0)$. In

§ 4.4, we use Faltings's almost purity theorem to construct a perfectoid version of the anticanonical tower $\mathcal{X}_{\Gamma(p^\infty)}(\varepsilon)_{\mathrm{anti}}$ at full level $\Gamma(p^\infty)$. Finally, in Remark 4.4.4, we briefly indicate how this construction extends to the boundary of the minimal compactification.

4.2. The anticanonical tower over the ordinary locus. Let R be a p-adically complete, flat $\mathbb{Z}_p^{\mathrm{cycl}}$-algebra and let $A \to \mathrm{Spec}\ R$ be an abelian scheme with reduction $A_0 \to \mathrm{Spec}\ (R/p)$. A_0 is equipped with the Frobenius $F : A_0 \to A_0^{(p)}$ and the Verschiebung $V : A^{(p)} \to A$ isogenies. For any $m \in \mathbb{Z}_{\geq 1}$, the p^m-torsion $A_0[p^m]$ fits into a short exact sequence of finite locally free group schemes over R/p

$$0 \to \ker\ F^m \to A_0[p^m] \to \mathcal{G}_0 \to 0,$$

where $\mathcal{G}_0 := \ker\ V^m : A_0^{(p^m)} \to A_0$. If A_0 is ordinary, i.e. if $\mathrm{Ha}(A_0/\mathrm{Spec}\ (R/p))$ is invertible, then $\mathcal{G}_0$ is a finite étale group scheme, which therefore lifts uniquely to a group scheme $\mathcal{G}$ over $\mathrm{Spec}\ R$. We get a short exact sequence

$$0 \to C_m \to A[p^m] \to \mathcal{G} \to 0.$$

The subgroup $C_m \subset A[p^m]$ is called the *canonical subgroup* of A of level m.

EXERCISE 4.2.1. *Let R be a p-adically complete, flat $\mathbb{Z}_p^{\mathrm{cycl}}$-algebra and let $A/\mathrm{Spec}\ R$ be an ordinary abelian variety. Take $C_m \subset A[p^m]$ to be the canonical subgroup of A of level m.*

(1) *Prove that*
$$A' := A/C_m$$
is also an ordinary abelian variety over $\mathrm{Spec}\ R$.

(2) *Understand the relationship between the canonical subgroup C_1' of A' and the subgroup $A[p]/C_1 \subset (A/C_1)[p] = A'[p]$.*

For $m = 0$, we take $\mathfrak{X}_{K_p}(0,0) := \mathfrak{X}_{K_p}(0)$. Note that we have an abelian variety $\mathfrak{A}_{K_p}(0,0)$ over $\mathfrak{X}_{K_p}(0,0)$, which is principally polarized, carries level K^p-structure and whose reduction is ordinary.

For $m \in \mathbb{Z}_{\geq 1}$, we define $\mathfrak{X}_{K_p}(m,0)$ to be abstractly isomorphic to $\mathfrak{X}_{K_p}(0)$, but the map to the base of the tower $\mathfrak{X}_{K_p}(m,0) \to \mathfrak{X}_{K_p}(0,0)$ is the canonical lift to characteristic 0 of the mth relative Frobenius morphism

$$F_m : \mathfrak{X}_{K_p}(m,0) \to (\mathfrak{X}_{K_p}(0)/p)^{(p^m)} \simeq \mathfrak{X}_{K_p}(0)/p.$$

We explain how to construct such a characteristic 0 lift: let $\mathfrak{C}_m$ be the canonical subgroup of the abelian variety $\mathfrak{A}_{K_p}(0,0)$. The abelian variety $\mathfrak{A}' := \mathfrak{A}_{K_p}(0,0)/\mathfrak{C}_m$ is also principally polarized and carries a level K^p-structure. By the universal property of $\mathfrak{X}_{K_p}$, $\mathfrak{A}'$ comes by pullback from a morphism

$$\mathfrak{X}_{K_p}(m,0) \to \mathfrak{X}_{K_p},$$

and, since $\mathfrak{A}'$ is ordinary, this morphism lifts uniquely to a morphism

$$\widetilde{F}_m : \mathfrak{X}_{K_p}(m,0) \to \mathfrak{X}_{K_p}(0,0).$$

We call the morphism $\widetilde{F}_m$ a *canonical Frobenius lift*. Modulo p, $\widetilde{F}_m$ agrees with the mth relative Frobenius, up to the isomorphism $(\mathfrak{X}_{K_p}(0)/p)^{(p^m)} \simeq \mathfrak{X}_{K_p}(0)/p$.

For $m' \in \mathbb{Z}$, $m' \geq m$, we obtain in the same way a morphism

$$\mathfrak{X}_{K_p}(m',0) \to \mathfrak{X}_{K_p}(m,0)$$

which is a canonical lift of the $(m - m')$th relative Frobenius, thus we have an inverse system of formal schemes $(\mathfrak{X}_{K_p}(m,0))_{m \in \mathbb{Z}_{\geq 0}}$.

This tower satisfies the first two desired properties by construction. We are left to identify the adic generic fibers $\mathcal{X}_{K_p}(m,0)$ of the formal schemes $\mathfrak{X}_{K_p}(m,0)$ with open adic subspaces of $\mathcal{X}_{\Gamma_0(p^m)}$.

Let $\mathcal{X}_{\Gamma_0(p^m)}(0)_{\text{anti}}$ be the open and closed locus inside the ordinary locus

$$\mathcal{X}_{\Gamma_0(p^m)}(0) \subset \mathcal{X}_{\Gamma_0(p^m)}$$

which parametrizes pairs (A, D) such that

(1) A is an ordinary abelian variety equipped with a principal polarization and a K^p-level structure (and with good reduction);
(2) $D \subset A[p^m]$ is a totally isotropic subgroup scheme of order p^{mn} such that $D[p] \cap C_1 = \{0\}$, where C_1 is the canonical subgroup of level 1 of A.

We see from the moduli interpretation in 4.1.4 that $\mathcal{X}_{\Gamma_0(p^m)}(0)_{\text{anti}}$ is indeed an open subspace of $\mathcal{X}_{\Gamma_0(p^m)}$. We call $\mathcal{X}_{\Gamma_0(p^m)}(0)_{\text{anti}}$ the *anticanonical* part of the ordinary locus at level m.

LEMMA 4.2.2. *For every $m \in \mathbb{Z}_{\geq 1}$, we have a natural isomorphism of adic spaces*

$$\mathcal{X}_{K_p}(m,0) \xrightarrow{\sim} \mathcal{X}_{\Gamma_0(p^m)}(0)_{\text{anti}}.$$

PROOF. Over $\mathcal{X}_{K_p}(m,0)$ we have an ordinary abelian variety $\mathcal{A}_{K_p}(m,0)$ together with a canonical subgroup $\mathcal{C}_m$ of level m, which is totally isotropic. The morphism

$$\mathcal{X}_{K_p}(m,0) \to \mathcal{X}_{\Gamma_0(p^m)}$$

is defined to be the one giving rise to the pair $(\mathcal{A}_{K_p}(m,0)/\mathcal{C}_m, \mathcal{A}_{K_p}(m,0)[p^m]/\mathcal{C}_m)$ over $\mathcal{X}_{K_p}(m,0)$, by pullback from the universal objects over $\mathcal{X}_{\Gamma_0(p^m)}$. Using Exercise 4.2.1, we identify the image of this map with $\mathcal{X}_{\Gamma_0(p^m)}(0)_{\text{anti}}$.

Consider also the morphism

$$\mathcal{X}_{\Gamma_0(p^m)} \to \mathcal{X}_{K_p}$$

defined by $(A, D) \mapsto A/D$ (with the canonical principal polarization and level K^p-structure). [22]

The composition of the two morphisms above

$$\mathcal{X}_{K_p}(m,0) \to \mathcal{X}_{\Gamma_0(p^m)} \to \mathcal{X}_{K_p}$$

is an open embedding: it corresponds to pulling back the universal abelian variety over $\mathcal{X}_{K_p}$ to $\mathcal{A}_{K_p}(m,0)$. Furthermore, the second map is étale. With the same proof as in the case of schemes, one deduces that the first map is an open embedding of adic spaces. □

The tower $(\mathcal{X}_{\Gamma_0(p^m)}(0)_{\text{anti}})_m$ is called *the anticanonical tower* over the ordinary locus. It gives rise to a perfectoid space $\mathcal{X}_{\Gamma_0(p^\infty)}(0)_{\text{anti}}$ which lives over the ordinary locus.

---

[22] Over $\mathcal{X}_{\Gamma_0(p^m)}(0)_{\text{anti}}$ this has no direct relation to the morphism $\widetilde{F}_m$, we are quotienting out precisely by subgroups $D \subset A[p^m]$ such that $D[p] \cap C_1 = \{0\}$ rather than by the canonical subgroup of level m.

4.3. The overconvergent anti-canonical tower.
We start by showing the existence of a canonical subgroup (of some level m) of an abelian scheme, as long as the valuation of the Hasse invariant of that abelian scheme is not too large (with respect to m). This will generalize the existence of the canonical subgroup in the case where the abelian scheme is ordinary, i.e when the Hasse invariant is invertible, and will follow roughly the same line of argument.

Let $0 < \varepsilon < 1/2$. Let R be a p-adically complete flat $\mathbb{Z}_p^{\mathrm{cycl}}$-algebra, and let $A \to \mathrm{Spec}\, R$ be an abelian scheme, with reduction $A_0 \to \mathrm{Spec}\, (R/p)$. Let $m \in \mathbb{Z}_{\geq 1}$. The following is Corollary III.2.6 of [**Sch15**].

PROPOSITION 4.3.1. *Assume that*
$$(\mathrm{Ha}(A_0/\mathrm{Spec}\,(R/p)))^{\frac{p^m-1}{p-1}} \mid p^\varepsilon.$$
Then there exists a unique closed subgroup $C_m \subset A[p^m]$ such that
$$C_m \equiv \ker F^m \subset A_0[p^m] \pmod{p^{1-\varepsilon}}.$$

PROOF. We sketch the argument in [**Sch15**]. As in the ordinary case, the key is to consider the group scheme $\mathcal{G}_0 := A_0[p^m]/\ker F^m$. The assumption on the Hasse invariant is made such that p^ε kills the Lie complex of $\mathcal{G}_0$. The results of Illusie's thesis on deformation theory imply that there exists a finite flat group scheme $\mathcal{G}$ over R such that $\mathcal{G}$ and $\mathcal{G}_0$ agree modulo $p^{1-\varepsilon}$. Furthermore, the map $A_0[p^m] \to \mathcal{G}_0$ modulo $p^{1-\varepsilon}$ gives rise to a map $A[p^m] \to \mathcal{G}$ that agrees with the original map modulo $p^{1-2\varepsilon}$. The canonical subgroup C_m is defined as $\ker(A[p^m] \to \mathcal{G})$. □

Now we make the analogous constructions to the ones in Section 4.2 using the fact that the canonical subgroup of any given level m overconverges (as shown above).[23]

We now define a formal scheme $\mathfrak{X}_{K_p}(\varepsilon) \to \mathfrak{X}_{K_p}$ that recovers $\mathfrak{X}_{K_p}(0)$ for $\varepsilon = 0$. First, define the functor $\mathfrak{X}_{K_p}(\varepsilon) \to \mathfrak{X}_{K_p}$ over $\mathbb{Z}_p^{\mathrm{cycl}}$ which sends any p-adically complete flat $\mathbb{Z}_p^{\mathrm{cycl}}$-algebra S to the set of pairs (f, u) where $f : \mathrm{Spf}\, S \to \mathfrak{X}_{K_p}$ is a map and $u \in H^0\left(\mathrm{Spf}\, S, f^*\omega^{\otimes(1-p)}\right)$ is a section such that
$$u \cdot \mathrm{Ha}(\bar{f}) = p^\varepsilon \in S/p,$$
up to the equivalence $(f, u) \simeq (f', u')$ if $f = f'$ and there exists some $h \in S$ with $u' = u(1 + p^{1-\varepsilon}h)$. Lemma 3.2.12 of [**Sch15**] shows that the functor $\mathfrak{X}_{K_p}(\varepsilon)$ is representable by a formal scheme which is flat over $\mathbb{Z}_p^{\mathrm{cycl}}$ and we have an explicit description of this formal scheme over affines $\mathrm{Spf}(R\hat{\otimes}_{\mathbb{Z}_p}\mathbb{Z}_p^{\mathrm{cycl}}) \subset \mathfrak{X}_{K_p}$ given by
$$\mathrm{Spf}(R\hat{\otimes}_{\mathbb{Z}_p}\mathbb{Z}_p^{\mathrm{cycl}})\langle u\rangle/(u\widetilde{\mathrm{Ha}} - p^\varepsilon)$$
for a lift $\widetilde{\mathrm{Ha}}$ of Ha. The adic generic fiber $\mathcal{X}_{K_p}(\varepsilon) \subset \mathcal{X}_{K_p}$ is the open subset defined by $|\mathrm{Ha}| \geq |p^\varepsilon|$.

For $m \in \mathbb{Z}_{\geq 1}$, we let the formal scheme at level m in the tower be $\mathfrak{X}_{K_p}(m, \varepsilon) := \mathfrak{X}_{K_p}(p^{-m}\varepsilon)$, with the morphism to the base of the tower $\mathfrak{X}_{K_p}(\varepsilon)$ given by a canonical lift $\widetilde{F}_m$ of the mth relative Frobenius modulo $p^{1-\varepsilon}$. We explain how to do this for

---
[23]While the canonical subgroup of any given level m overconverges, i.e. can be extended to an $\varepsilon = \varepsilon(m)$ neighborhood of the ordinary locus, if we let $m \to \infty$, we get $\varepsilon(m) \to 0$. The *canonical tower* does not overconverge.

$m = 1$. We need to construct a canonical lift of the relative Frobenius, i.e. a map of formal schemes
$$\widetilde{F}_1 : \mathfrak{X}_{K_p}(p^{-1}\varepsilon) \to \mathfrak{X}_{K_p}(\varepsilon)$$
which reduces to the relative Frobenius modulo $p^{1-\varepsilon}$. For this, we simply need to show that the natural map
$$\mathfrak{X}_{K_p}(p^{-1}\varepsilon) \to \mathfrak{X}_{K_p}$$
induced by quotienting out the universal abelian variety by the level 1 canonical subgroup factors through $\mathfrak{X}_{K_p}(\varepsilon)$. The key point is now to observe that quotienting out by the canonical subgroup of level 1 raises Ha to the pth power. Thus, from the initial condition $u \cdot \mathrm{Ha}(A) = p^{\frac{1}{p}\varepsilon}$ on $\mathfrak{X}_{K_p}(p^{-1}\varepsilon)$, we get $u^p \cdot \mathrm{Ha}(A/C_1) = p^\varepsilon$, which gives the desired factorization through $\mathfrak{X}_{K_p}(\varepsilon)$.

Using this argument at higher levels, we obtain the tower of formal schemes $(\mathfrak{X}_{K_p}(p^{-m}\varepsilon))_m$, where the transition map at level m is given by the relative Frobenius modulo $p^{1-\varepsilon}$.[24] From this property of the transition morphisms, we can see that the tower of adic generic fibers $(\mathcal{X}_{K_p}(p^{-m}\varepsilon))_m$ gives rise to a perfectoid space.

We are left with one question, namely identify the adic generic fiber $\mathcal{X}_{K_p}(p^{-m}\varepsilon)$ as an open subspace of the Shimura variety $\mathcal{X}_{\Gamma_0(p^m)}$. Let $\mathcal{X}_{\Gamma_0(p^m)}(\varepsilon)$ be the inverse image of $\mathcal{X}_{K_p}(\varepsilon)$ under the map $\mathcal{X}_{\Gamma_0(p^m)} \to \mathcal{X}_{K_p}$.

LEMMA 4.3.2. *$\mathcal{X}_{K_p}(p^{-m}\varepsilon)$ is isomorphic to the open and closed locus $\mathcal{X}_{\Gamma_0(p^m)}(\varepsilon)_{\mathrm{anti}}$ in $\mathcal{X}_{K_p}(\varepsilon)$ where the universal totally isotropic subgroup $\mathcal{D} \subset \mathcal{A}(\varepsilon)[p^m]$ satisfies $\mathcal{D}[p] \cap \mathcal{C}_1 = \{0\}$ for $\mathcal{C}_1 \subset \mathcal{A}(\varepsilon)[p]$ the canonical subgroup of level 1.*

We remark that in order to identify $\mathcal{X}_{\Gamma_0(p^m)}(\varepsilon)_{\mathrm{anti}}$ with $\mathcal{X}_{K_p}(p^{-m}\varepsilon)$, we use the map induced by
$$(A, D) \mapsto A/D.$$
When $D[p] \cap C_1 = \{0\}$, this decreases the valuation of the Hasse invariant, so it indeed defines a map
$$\mathcal{X}_{\Gamma_0(p^m)}(\varepsilon)_{\mathrm{anti}} \to \mathcal{X}_{K_p}(p^{-m}\varepsilon).$$

These maps are compatible as m varies. For each $m \in \mathbb{Z}_{\geq 1}$, we have a commutative diagram

$$\begin{array}{ccc} \mathcal{X}_{\Gamma_0(p^{m+1})}(\varepsilon)_{\mathrm{anti}} & \xrightarrow{\sim} & \mathcal{X}_{K_p}(p^{-m+1}\varepsilon) \\ \downarrow & & \downarrow \\ \mathcal{X}_{\Gamma_0(p^m)}(\varepsilon)_{\mathrm{anti}} & \xrightarrow{\sim} & \mathcal{X}_{K_p}(p^{-m}\varepsilon) \end{array}$$

where the horizontal maps are as described above, the left vertical map is the natural projection (i.e. the forgetful map from the moduli-theoretic point of view), and the right vertical map is the canonical lift of relative Frobenius.

REMARK 4.3.3. Unlike the canonical tower, the overconvergent anticanonical tower $(\mathcal{X}_{\Gamma_0(p^m)}(\varepsilon))_m$ inside the tower $(\mathcal{X}_{\Gamma_0(p^m)})_m$ has constant radius ε.

---

[24]More precisely, the map from level m to the base of the tower agrees with the mth relative Frobenius modulo $p^{1-\frac{\varepsilon}{p^m-1}}$, which is a multiple of $p^{1-\varepsilon}$.

4.4. An application of almost purity.

We have just seen that $\mathcal{X}_{\Gamma_0(p^\infty)}(\varepsilon)_{\mathrm{anti}}$ is perfectoid. For each $m \in \mathbb{Z}_{\geq 1}$, consider the congruence subgroups

$$\Gamma(p^m) := \{g \in \mathrm{GSp}_{2n}(\mathbb{Z}_p) \mid g \equiv \begin{pmatrix} \mathrm{Id}_n & 0 \\ 0 & \mathrm{Id}_n \end{pmatrix} \pmod{p^m}\}.$$

We will show that there exists a unique perfectoid space over $\mathbb{Q}_p^{\mathrm{cycl}}$ such that

$$\mathcal{X}_{\Gamma(p^\infty)}(\varepsilon)_{\mathrm{anti}} \sim \varprojlim_m \mathcal{X}_{\Gamma(p^m)}(\varepsilon)_{\mathrm{anti}}.$$

For this, the key input is Faltings's almost purity theorem, which we now recall.

THEOREM 4.4.1. *Let L be a perfectoid field and R a perfectoid L-algebra. Let S/R be finite étale. Then S is a perfectoid L-algebra and S° is almost finite étale over R°.*

For us, the perfectoid field L will be $\mathbb{Q}_p^{\mathrm{cycl}}$ throughout. Note that the projection maps

$$\mathcal{X}_{\Gamma(p^m)} \to \mathcal{X}_{\Gamma_0(p^m)}$$

are finite étale for every $m \in \mathbb{Z}_{\geq 1}$ and therefore the same thing holds true for their restriction to an ε-neighborhood of the anticanonical tower. By combining this observation with Theorem 4.4.1 and the fact that $\mathcal{X}_{\Gamma_0(p^\infty)}(\varepsilon)_{\mathrm{anti}}$ is a perfectoid space over $\mathbb{Q}_p^{\mathrm{cycl}}$, we conclude the following.

PROPOSITION 4.4.2. *For any $m \in \mathbb{Z}_{\geq 1}$, there exists a unique perfectoid space*

$$\mathcal{X}_{\Gamma(p^m) \cap \Gamma_0(p^\infty)}(\varepsilon)_{\mathrm{anti}}$$

over $\mathbb{Q}_p^{\mathrm{cycl}}$ such that

$$\mathcal{X}_{\Gamma(p^m) \cap \Gamma_0(p^\infty)}(\varepsilon)_{\mathrm{anti}} \sim \varprojlim_{m'} \mathcal{X}_{\Gamma(p^m) \cap \Gamma_0(p^{m'})}(\varepsilon)_{\mathrm{anti}}.$$

By varying m, we obtain an inverse system of perfectoid spaces with finite étale transition maps. Take an affinoid perfectoid cover of $\mathcal{X}_{\Gamma(p) \cap \Gamma_0(p^\infty)}(\varepsilon)_{\mathrm{anti}}$. The preimage of any affinoid perfectoid element of the cover in any $\mathcal{X}_{\Gamma(p^m) \cap \Gamma_0(p^\infty)}(\varepsilon)_{\mathrm{anti}}$ is affinoid perfectoid by Theorem 4.4.1. Since inverse limits of affinoid perfectoid spaces are affinoid perfectoid, we obtain an affinoid perfectoid cover of the topological space

$$|\mathcal{X}_{\Gamma(p^\infty)}(\varepsilon)_{\mathrm{anti}}| := \varprojlim_m |\mathcal{X}_{\Gamma(p^m) \cap \Gamma_0(p^\infty)}(\varepsilon)_{\mathrm{anti}}|.$$

We thus deduce the following.

THEOREM 4.4.3. *There exists a unique perfectoid space $\mathcal{X}_{\Gamma(p^\infty)}(\varepsilon)_{\mathrm{anti}}$ over $\mathbb{Q}_p^{\mathrm{cycl}}$ such that*

$$\mathcal{X}_{\Gamma(p^\infty)}(\varepsilon)_{\mathrm{anti}} \sim \varprojlim_m \mathcal{X}_{\Gamma(p^m) \cap \Gamma_0(p^\infty)}(\varepsilon)_{\mathrm{anti}}.$$

REMARK 4.4.4. We briefly indicate how to extend the results of this section to the minimal compactification of the Siegel modular variety and thus how to construct a perfectoid space $\mathcal{X}^*_{\Gamma(p^\infty)}(\varepsilon)_{\mathrm{anti}}$. When $K_p = \mathrm{GSp}_{2n}(\mathbb{Z}_p)$, recall that $X^*_{K_p}$ is the minimal (Baily–Borel–Satake) compactification of X_{K_p} over $\mathbb{Z}_{(p)}$ as constructed by Faltings and Chai [**CF90**]. This is a projective, but not necessarily smooth, scheme over $\mathbb{Z}_{(p)}$ and the line bundle ω extends canonically to an ample line bundle on $X^*_{K_p}$.

Both $\mathrm{Ha} \in \omega^{\otimes(p-1)}/p$ and the ordinary locus (defined as the complement of the vanishing locus of Ha) can be extended to an open dense subscheme of the minimal compactification $\overline{X}^*_{K_p}$. Indeed, the codimension of the boundary $\overline{X}^*_{K_p} \setminus \overline{X}_{K_p}$ of the minimal compactification is n: the boundary can be described in terms of smaller Siegel modular varieties and the relative dimension of the Siegel modular variety for GSp_{2m} over $\mathbb{Z}_{(p)}$ is $\frac{m(m+1)}{2}$. For $n \geq 2$, Koecher's extension principle (see [**Lan16**] for the most definitive version) guarantees that Ha extends canonically to the whole $\overline{X}^*_{K_p}$. The case $n = 1$, i.e. the case of modular curves, can be done in an ad hoc manner, for example using q-expansions.

The proof that the space $\mathcal{X}_{\Gamma_0(p^\infty)}(\varepsilon)_{\mathrm{anti}}$ is perfectoid extends to the minimal compactification and, in fact, one can prove something even stronger, namely that $\mathcal{X}^*_{\Gamma_0(p^\infty)}(\varepsilon)_{\mathrm{anti}}$ is actually an affinoid perfectoid space. This follows from the fact that ω extends to an ample line bundle, so one can find a global characteristic 0 lift of any sufficiently large pth power of Ha.

Going from level $\Gamma_0(p^\infty)$ to level $\Gamma(p^\infty)$ is much more subtle for the minimal compactification than for the good reduction locus. We cannot simply argue using Theorem 4.4.1, because the maps

$$\mathcal{X}^*_{\Gamma(p^m)}(\varepsilon)_{\mathrm{anti}} \to \mathcal{X}^*_{\Gamma_0(p^m)}(\varepsilon)_{\mathrm{anti}}$$

are ramified along the boundary. Instead, for every $m \in \mathbb{Z}_{\geq 1}$, we consider the congruence subgroups

$$\Gamma_1(p^m) := \left\{ g \in \mathrm{GSp}_{2n}(\mathbb{Z}_p) \mid g \equiv \begin{pmatrix} \mathrm{Id}_n & * \\ 0 & \mathrm{Id}_n \end{pmatrix} \pmod{p^m} \right\}$$

One first shows that there exists a unique affinoid perfectoid space $\mathcal{X}^*_{\Gamma_1(p^\infty)}(\varepsilon)_{\mathrm{anti}}$ over $\mathbb{Q}_p^{\mathrm{cycl}}$ such that

$$\mathcal{X}^*_{\Gamma_1(p^\infty)}(\varepsilon)_{\mathrm{anti}} \sim \varprojlim_m \mathcal{X}^*_{\Gamma_1(p^m)}(\varepsilon)_{\mathrm{anti}}.$$

This is the most subtle part of the argument and it uses *Tate's normalized traces* to extend the construction of the anticanonical tower via Theorem 4.4.1 over the boundary of the minimal compactification at level $\Gamma_1(p^m) \cap \Gamma_0(p^\infty)$. See [**Sch15**, §3.2] for more details.

Finally, the maps

$$\mathcal{X}^*_{\Gamma(p^m)}(\varepsilon)_{\mathrm{anti}} \to \mathcal{X}^*_{\Gamma_1(p^m)}(\varepsilon)_{\mathrm{anti}}$$

are finite étale for every $m \in \mathbb{Z}_{\geq 1}$ even over the boundary. This means that we can go from level $\Gamma_1(p^\infty)$ to level $\Gamma(p^\infty)$ using Theorem 4.4.1 as described above.

One concludes the following strenghtening of Theorem 4.4.3.

THEOREM 4.4.5. *There exists a unique affinoid perfectoid space $\mathcal{X}^*_{\Gamma(p^\infty)}(\varepsilon)_{\mathrm{anti}}$ over $\mathbb{Q}_p^{\mathrm{cycl}}$ such that*

$$\mathcal{X}^*_{\Gamma(p^\infty)}(\varepsilon)_{\mathrm{anti}} \sim \varprojlim_m \mathcal{X}^*_{\Gamma(p^m)}(\varepsilon)_{\mathrm{anti}}.$$

5. Perfectoid Shimura varieties and the Hodge–Tate period morphism

In this section, we construct the Hodge–Tate period morphism and use it to show that Siegel modular varieties (and other Shimura varieties) with infinite level at p are perfectoid. In § 5.1, we explain how to use Theorem 4.4.5 to show that there exists a perfectoid space $\mathcal{X}^*_{\Gamma(p^\infty)}$ over $\mathbb{Q}_p^{\mathrm{cycl}}$ such that

(5.0.1) $$\mathcal{X}^*_{\Gamma(p^\infty)} \sim \varprojlim_m \mathcal{X}^*_{\Gamma(p^m)}.$$

In § 5.2, we discuss the geometry of the Hodge–Tate period morphism in the Siegel case. Finally, in § 5.3, we discuss the Hodge–Tate period morphism for Shimura varieties of Hodge type.

5.1. Siegel modular varieties with infinite level at p are perfectoid.
Consider the inverse limit of topological spaces

$$|\mathcal{X}^*_{\Gamma(p^\infty)}| = \varprojlim_m |\mathcal{X}^*_{\Gamma(p^m)}|.$$

This must be the underlying topological space of a perfectoid space $\mathcal{X}^*_{\Gamma(p^\infty)}$ satisfying (5.0.1). The topological spaces at finite level $|\mathcal{X}^*_{\Gamma(p^m)}|$ are all *spectral spaces*, as they are the underlying topological spaces of quasi-compact and quasi-separated adic spaces, and the transition maps are *spectral maps*, since they underlie maps of adic spaces.[25] This implies that $|\mathcal{X}^*_{\Gamma(p^\infty)}|$ is itself a spectral topological space. The hard part is endowing this topological space with a perfectoid structure.

The perfectoid space $\mathcal{X}^*_{\Gamma(p^\infty)}(\varepsilon)_{\mathrm{anti}}$ covers a part of the topological space

$$|\mathcal{X}^*_{\Gamma(p^\infty)}| := \varprojlim_m |\mathcal{X}^*_{\Gamma(p^m)}|.$$

We will show that the entire topological space above underlies a perfectoid space, using the continuous $\mathrm{GSp}_{2n}(\mathbb{Q}_p)$ action on $|\mathcal{X}^*_{\Gamma(p^\infty)}|$[26] and the fact that the translates of $|\mathcal{X}^*_{\Gamma(p^m)}(\varepsilon)_{\mathrm{anti}}|$ under this action cover the entire space $|\mathcal{X}^*_{\Gamma(p^\infty)}|$. The rigorous way of proving this is via *the Hodge–Tate period morphism*, which has as target a flag variety $\mathscr{F}\ell$, which also has an action of $\mathrm{GSp}_{2n}(\mathbb{Q}_p)$. One of the most important properties of the Hodge–Tate period morphism is that it is equivariant for the action of $\mathrm{GSp}_{2n}(\mathbb{Q}_p)$ on both the Shimura variety at infinite level (or, for now, on the corresponding topological space) and on the flag variety $\mathscr{F}\ell$.

Recall that (V, ψ) denotes the split symplectic space of dimension $2n$ over $\mathbb{Q}$. Let Fl/$\mathbb{Q}$ be the flag variety parametrizing subspaces $W \subset V$ of dimension n which are totally isotropic under ψ. We consider the corresponding a dic space $\mathscr{F}\ell$ over $\mathbb{Q}_p$. The Hodge–Tate period morphism is first defined at the level of topological spaces:

$$|\pi_{\mathrm{HT}}| : |\mathcal{X}^*_{\Gamma(p^\infty)}| \to |\mathscr{F}\ell|$$

For simplicity, in these notes we will only describe the map on the good reduction locus $|\mathcal{X}_{\Gamma(p^\infty)}|$. For each pair (L, L^+), with $L/\mathbb{Q}_p^{\mathrm{cycl}}$ a complete non-archimedean field and $L^+ \subset L$ an open and bounded valuation subring, define

$$\mathcal{X}_{\Gamma(p^\infty)}(L, L^+) := \varprojlim_m \mathcal{X}_{\Gamma(p^m)}(L, L^+).$$

From this definition, one can check that $\mathcal{X}_{\Gamma(p^\infty)}(L, L^+)$ has a moduli interpretation in terms of abelian varieties A/L, equipped with a principal polarization, with a

---

[25] We can define a spectral topological space as any topological space that is homeomorphic to the underlying topological space of an affine scheme. For more on spectral spaces and spectral maps, in the context of adic spaces, see, for example, [**Wed**].

[26] This action can only be seen at level $\Gamma(p^\infty)$; at finite level one only has an action of $\mathrm{GSp}_{2n}(\mathbb{Z}_p)$. To see the action of $\mathrm{GSp}_{2n}(\mathbb{Q}_p)$ on $|\mathcal{X}_{\Gamma(p^\infty)}|$, it is easiest to first redefine the moduli problem in terms of abelian varieties up to isogeny, as in Example 2.3.6; then it is easy to see the group action on the p-part of the level structure.

K^p-level structure $\bar\eta^p$, and with a symplectic isomorphism $\eta_p : \mathbb{Z}_p^{2n} \xrightarrow{\sim} T_p A$. We have
$$|\mathcal{X}_{\Gamma(p^\infty)}| = \varinjlim_{(L, L^+)} \mathcal{X}_{\Gamma(p^\infty)}(L, L^+),$$
where the limit on the right hand side is not filtered, but each point comes from a unique minimal pair (L, L^+). The following is a reformulation of Lemma 3.3.4 of [**Sch15**], restricted to the good reduction locus.

LEMMA 5.1.1. *There exists a* $\mathrm{GSp}_{2n}(\mathbb{Q}_p)$-*equivariant, continuous map of topological spaces*
$$|\pi_{\mathrm{HT}}| : |\mathcal{X}_{\Gamma(p^\infty)}| \to |\mathscr{F}\ell|,$$
which is defined at the level of points by sending an abelian variety A/L together with a symplectic isomorphism
$$\eta_p : \mathbb{Z}_p^{2n} \xrightarrow{\sim} T_p A$$
to the (first piece of the) Hodge–Tate filtration $\mathrm{Lie}\, A \subset T_p A \otimes_{\mathbb{Z}_p} L \xrightarrow{\sim} L^{2n}$.

PROOF. First, define the map $|\pi_{\mathrm{HT}}|$ on points by the recipe in the statement of the lemma. Since $\mathrm{GSp}_{2n}(\mathbb{Q}_p)$ acts on the level structure η_p, the map $|\pi_{\mathrm{HT}}|$ is $\mathrm{GSp}_{2n}(\mathbb{Q}_p)$-equivariant by definition.

To show that there exists a map of topological spaces which agrees with $|\pi_{\mathrm{HT}}|$ on points, it is enough to work locally on $|\mathcal{X}_{\Gamma(p^\infty)}|$. It will therefore suffice to construct a cover of $|\mathcal{X}_{\Gamma(p^\infty)}|$ which is pulled back from a cover of $|\mathcal{X}_{K_p}|$. We work in the setting of Example 3.2.6, i.e. by considering the proper smooth morphism $\pi : \mathcal{A} \to \mathcal{X}_{K_p}$ of smooth adic spaces over $\mathrm{Spa}(\mathbb{Q}_p, \mathbb{Z}_p)$. The relative Hodge–Tate filtration of the universal abelian variety is encoded by the natural injection
$$R^1\pi_* \mathcal{O}_\mathcal{A} \otimes_{\mathcal{O}_{\mathcal{X}_{K_p}}} \widehat{\mathcal{O}}_{\mathcal{X}_{K_p}} \hookrightarrow R^1\pi_* \widehat{\mathcal{O}}_\mathcal{A} \simeq R^1\pi_* \widehat{\mathbb{Z}}_p \otimes_{\widehat{\mathbb{Z}}_p} \widehat{\mathcal{O}}$$
of sheaves on the flattened pro-étale site $(\mathcal{X}_{K_p})_{\mathrm{proét}}$.

Locally on $\mathcal{X}_{K_p}$, one can find a pro-finite étale cover $\widetilde{U} \to \mathcal{X}_{K_p}$ such that $\widetilde{U}$ is affinoid perfectoid. We show that it is possible to pull back $\widetilde{U}$ to an affinoid perfectoid space $\widetilde{U}_\infty$ such that $|\widetilde{U}_\infty|$ covers $|\mathcal{X}_{\Gamma(p^\infty)}|$. For each $m \geq 0$, the map $\mathcal{X}_{\Gamma(p^m)} \to \mathcal{X}_{K_p}$ is finite étale. Thus, we can form the pullback $\widetilde{U}_m := \widetilde{U} \times_{\mathcal{X}_{K_p}} \mathcal{X}_{\Gamma(p^m)}$ and, by Theorem 4.4.1, this is an affinoid perfectoid cover of $\mathcal{X}_{\Gamma(p^\infty)}$. We then take the inverse limit of the $\widetilde{U}_m$ as $m \to \infty$, which we can do for affinoid perfectoid spaces, and we obtain an affinoid perfectoid space $\widetilde{U}_\infty$. This is still an element of the flattened pro-étale site of $\mathcal{X}_{K_p}$.

We now evaluate the injection
$$R^1\pi_* \mathcal{O}_\mathcal{A} \otimes_{\mathcal{O}_{\mathcal{X}_{K_p}}} \widehat{\mathcal{O}}_{\mathcal{X}_{K_p}} \hookrightarrow R^1\pi_* \widehat{\mathcal{O}}_\mathcal{A} \simeq R^1\pi_* \widehat{\mathbb{Z}}_p \otimes_{\widehat{\mathbb{Z}}_p} \widehat{\mathcal{O}}$$
on $\widetilde{U}_\infty$. Since $R^1\pi_* \mathcal{O}_\mathcal{A}$ can be identified with $\mathrm{Lie}\,\mathcal{A}$, we get a totally isotropic submodule $(\mathrm{Lie}\,\mathcal{A}) \otimes_{\mathcal{O}_{c\mathcal{X}_{K_p}}} \mathcal{O}_{\widetilde{U}_\infty} \subset \mathcal{O}_{\widetilde{U}_\infty}^{2n}$, which defines a map of adic spaces
$$\widetilde{U}_\infty \to \mathscr{F}\ell.$$
The induced map on topological spaces is automatically continuous. By checking on points, one sees that this map factors through the restriction of $|\pi_{\mathrm{HT}}|$ to $|\widetilde{U}| \times_{|\mathcal{X}_{K_p}|} |\mathcal{X}_{\Gamma(p^\infty)}|$. Moreover, the map
$$|\widetilde{U}_\infty| \to |\widetilde{U}| \times_{|\mathcal{X}_{K_p}|} |\mathcal{X}_{\Gamma(p^\infty)}|$$

is both surjective and open, as it is a pro-finite étale cover and pro-finite étale maps are open. Thus, $|\pi_{\mathrm{HT}}|$ is continous. $\square$

REMARK 5.1.2. In fact, if we let $\mathcal{Z}_{\Gamma(p^m)}$ be the boundary of $\mathcal{X}^*_{\Gamma(p^m)}$, we can define the spectral topological space
$$|\mathcal{Z}_{\Gamma(p^\infty)}| := \varprojlim_m |\mathcal{Z}_{\Gamma(p^m)}|$$
and the construction of the map $|\pi_{\mathrm{HT}}|$ in Lemma 5.1.1 extends to the open Shimura variety $|\mathcal{X}^*_{\Gamma(p^\infty)}| \setminus |\mathcal{Z}_{\Gamma(p^\infty)}|$ with the same proof, thus we have a continuous, $\mathrm{GSp}_{2n}(\mathbb{Q}_p)$-equivariant map
$$|\pi_{\mathrm{HT}}| : |\mathcal{X}^*_{\Gamma(p^\infty)}| \setminus |\mathcal{Z}_{\Gamma(p^\infty)}| \to |\mathscr{F}\ell|.$$

Let $0 \leq \varepsilon < \frac{1}{2}$. Recall that $\mathcal{X}^*_{K_p}(\varepsilon) \subset \mathcal{X}^*_{K_p}$ is the locus where $|\mathrm{Ha}| \geq p^\varepsilon$. Let $|\mathcal{X}^*_{\Gamma(p^\infty)}(\varepsilon)| \subset |\mathcal{X}^*_{\Gamma(p^\infty)}|$ be the preimage of $|\mathcal{X}^*_{K_p}(\varepsilon)|$. We have
$$|\mathcal{X}^*_{\Gamma(p^\infty)}(\varepsilon)| = \mathrm{GSp}_{2n}(\mathbb{Z}_p)|\mathcal{X}^*_{\Gamma(p^\infty)}(\varepsilon)_{\mathrm{anti}}|.$$
This can be checked at finite level - for example at level $\Gamma_0(p)$, where the ε-neighborhood $\mathcal{X}^*_{\Gamma_0(p)}(\varepsilon)_{\mathrm{anti}} \subset \mathcal{X}^*_{\Gamma_0(p)}$ of the anticanonical locus is defined (recall that everything else is just pulled back from this level). In fact, by doing this, we see that we can replace $\mathrm{GSp}_{2n}(\mathbb{Z}_p)$ by finitely many translates of $|\mathcal{X}^*_{\Gamma(p^\infty)}(\varepsilon)_{\mathrm{anti}}|$ by elements of $\mathrm{GSp}_{2n}(\mathbb{Z}_p)$; thus, $|\mathcal{X}^*_{\Gamma(p^\infty)}(\varepsilon)|$ is quasi-compact. The key result is now the following (Lemma 3.3.10 of [**Sch15**]).

PROPOSITION 5.1.3. *There exist finitely many elements* $\gamma_1, \ldots, \gamma_k \in \mathrm{GSp}_{2n}(\mathbb{Q}_p)$ *such that*
$$|\mathcal{X}^*_{\Gamma(p^\infty)}| = \bigcup_{i=1}^k \gamma_i \cdot |\mathcal{X}^*_{\Gamma(p^\infty)}(\varepsilon)|.$$

PROOF. We sketch the main steps in the proof.
(1) First, one shows that if $|\pi_{\mathrm{HT}}|$ is the map in Remark 5.1.2, and $\mathscr{F}\ell(\mathbb{Q}_p)$ denotes the $\mathbb{Q}_p$-points of the adic space $\mathscr{F}\ell$, then
$$|\pi_{\mathrm{HT}}|^{-1}(\mathscr{F}\ell(\mathbb{Q}_p)) = \text{ closure of } |\mathcal{X}^*_{\Gamma(p^\infty)}(0)| \setminus |\mathcal{Z}_{\Gamma(p^\infty)}(0)|.$$
This is Lemma 3.3.6 of [**Sch15**]. The idea is that for an ordinary abelian variety, the Hodge–Tate filtration is $\mathbb{Q}_p$-rational and measures the relative position of the canonical subgroup.
(2) For $0 < \varepsilon < \frac{1}{2}$, one shows that there exists an open subset $U \subset \mathscr{F}\ell$ containing $\mathscr{F}\ell(\mathbb{Q}_p)$ and such that
$$|\pi_{\mathrm{HT}}|^{-1}(U) \subset |\mathcal{X}^*_{\Gamma(p^\infty)}(\varepsilon)| \setminus |\mathcal{Z}_{\Gamma(p^\infty)}(\varepsilon)|.$$
This is Lemma 3.3.7 of [**Sch15**]. Using induction, one reduces to the locus of good reduction. The proof then relies on Step 1 and on a compactness argument using the constructible topology on spectral spaces. For the compactness argument, one uses the continuity of the map
$$|\pi_{\mathrm{HT}}| : |\mathcal{X}_{\Gamma(p^\infty)}| \to |\mathscr{F}\ell|$$
and the fact that the space $|\mathcal{X}_{\Gamma(p^\infty)}|$ is spectral with quasi-compact open subset $|\mathcal{X}_{\Gamma(p^\infty)}(\epsilon)|$.

(3) One shows that there exist finitely many elements $\gamma_1, \ldots, \gamma_k \in \mathrm{GSp}_{2n}(\mathbb{Q}_p)$ such that
$$|\mathcal{X}^*_{\Gamma(p^\infty)}| \setminus |\mathcal{Z}_{\Gamma(p^\infty)}| = \bigcup_{i=1}^k \gamma_i \cdot (|\mathcal{X}^*_{\Gamma(p^\infty)}|(\varepsilon) \setminus |\mathcal{Z}_{\Gamma(p^\infty)}|(\varepsilon)).$$
This is Lemma III.3.9 of [**Sch15**]. This uses an open subset U as in Step 2; the quasi-compactness of $\mathscr{F}\ell$ implies that finitely many $\mathrm{GSp}_{2n}(\mathbb{Q}_p)$-translates of U cover $\mathscr{F}\ell$. The fact that $|\pi_{\mathrm{HT}}|$ is $\mathrm{GSp}_{2n}(\mathbb{Q}_p)$-equivariant allows one to conclude by taking preimages of everything.

(4) Finally, one shows that with the same $\gamma_1, \ldots, \gamma_k$ as above one has the desired equality
$$|\mathcal{X}^*_{\Gamma(p^\infty)}| = \bigcup_{i=1}^k \gamma_i \cdot |\mathcal{X}^*_{\Gamma(p^\infty)}|(\varepsilon).$$
This again relies on a compactness argument as in Step 2 above. The idea is that the right hand side is a quasi-compact open subset of $|\mathcal{X}^*_{\Gamma(p^\infty)}|$ which contains $|\mathcal{X}^*_{\Gamma(p^\infty)}| \setminus |\mathcal{Z}_{\Gamma(p^\infty)}|$ by Step 3 above. Any such subset must be the whole space. One concludes this by reducing to finite level, considering classical points, and again using a compactness argument for the constructible topology on a spectral space. $\square$

As a result, we see that $|\mathcal{X}^*_{\Gamma(p^\infty)}|$ is covered by finitely many translates of $|\mathcal{X}^*_{\Gamma(p^\infty)}(\varepsilon)_{\mathrm{anti}}|$, which is the underlying topological space of an affinoid perfectoid space. This proves the existence of the perfectoid space $\mathcal{X}^*_{\Gamma(p^\infty)}$. With a bit more work, one can also show that there exists a map of adic spaces
$$\pi_{\mathrm{HT}} : \mathcal{X}^*_{\Gamma(p^\infty)} \to \mathscr{F}\ell.$$
which agrees with the previously defined map $|\pi_{\mathrm{HT}}|$ on the underlying topological spaces.

REMARK 5.1.4. The closed subset $|\mathcal{Z}_{\Gamma(p^\infty)}| \subset |\mathcal{X}^*_{\Gamma(p^\infty)}|$ has an induced structure of a perfectoid space. If $\mathcal{Z}_{\Gamma(p^\infty)}$ denotes the boundary with the induced perfectoid structure, then the existence of the map of adic spaces
$$\pi_{\mathrm{HT}} : \mathcal{X}^*_{\Gamma(p^\infty)} \setminus \mathcal{Z}_{\Gamma(p^\infty)} \to \mathscr{F}\ell$$
follows by the same argument as in the proof of Lemma 5.1.1, using instead of $\widetilde{U}_\infty$ the affinoid perfectoid cover given by the disjoint union of finitely many copies of $\mathcal{X}^*_{\Gamma(p^\infty)}(\varepsilon)_{\mathrm{anti}} \setminus \mathcal{Z}_{\Gamma(p^\infty)}(\varepsilon)_{\mathrm{anti}}$. The tricky part is to show that the Hodge–Tate period morphism extends to the boundary. For this, one uses a version of Riemann's Hebbarkeitssatz for perfectoid spaces, concerning the extension of bounded functions from complements of Zariski closed subsets. See [**Sch15**, §2] for more details.

5.2. The Hodge–Tate period morphism in the Siegel case.
We summarize here some facts about the geometry of $\mathscr{F}\ell$ in the Siegel case. The flag variety admits the Plücker embedding
$$\mathscr{F}\ell \hookrightarrow \mathbb{P}^{\binom{2n}{n}-1}, W \mapsto \wedge^n W.$$

Any subset $J \subset \{1, 2, \ldots, 2n\}$ of cardinality n determines a homogeneous coordinate s_J on $\mathbb{P}^{\binom{2n}{n}-1}$. One can cover $\mathscr{F}\ell$ by open affinoid subsets $\mathscr{F}\ell_J$, which are defined by the conditions $|s_{J'}| \leq |s_J|$ for all $J' \subset \{1, 2, \ldots, 2n\}$ of cardinality n. These affinoid subsets are permuted transitively by the action of $\mathrm{GSp}_{2n}(\mathbb{Z}_p)$. For example, $\mathscr{F}\ell_{\{n+1,\ldots,2n\}}(\mathbb{Q}_p)$ parametrizes those totally isotropic direct summands $M \subset \mathbb{Z}_p^{2n}$ such that $M \oplus (\mathbb{Z}_p^n \oplus 0^n) \simeq \mathbb{Z}_p^{2n}$.

EXERCISE 5.2.1. *Show that the preimage of $\mathscr{F}\ell_{\{n+1,\ldots,2n\}}(\mathbb{Q}_p)$ under π_{HT} is given by the closure of $\mathcal{X}^*_{\Gamma(p^\infty)}(0)_{\mathrm{anti}}$.*

Since $\mathcal{X}^*_{\Gamma(p^\infty)}(0)_{\mathrm{anti}}$ is affinoid perfectoid, thus of the form $\mathrm{Spa}(R, R^+)$, and since taking the closure only adds higher rank points, which amounts to only changing the integral structure, i.e R^+, we see that the preimage of $\mathscr{F}\ell_{\{n+1,\ldots,2n\}}(\mathbb{Q}_p)$ under π_{HT} is affinoid perfectoid.

We claim that something stronger holds, namely the preimage of the whole of $\mathscr{F}\ell_{\{n+1,\ldots,2n\}}$ is affinoid perfectoid. To see this, note that the action of the diagonal element γ^{-1}, where $\gamma = (p, \ldots, p, 1, \ldots, 1) \in (\mathbb{Q}_p^\times)^n \times (\mathbb{Q}_p^\times)^n \subset \mathrm{GSp}_{2n}(\mathbb{Q}_p)$, contracts $\mathscr{F}\ell_{\{n+1,\ldots,2n\}}$ towards the point of $\mathscr{F}\ell_{\{n+1,\ldots,2n\}}(\mathbb{Q}_p)$ corresponding to $0^n \oplus \mathbb{Z}_p^n \subset \mathbb{Z}_p^{2n}$. In particular, the action of γ^{-1} contracts $\mathscr{F}\ell_{\{n+1,\ldots,2n\}}$ towards the image of the anticanonical locus $\mathcal{X}^*_{\Gamma(p^\infty)}(0)_{\mathrm{anti}}$ under π_{HT}. To make this precise, for any $0 < \varepsilon \leq \frac{1}{2}$, one can find some large integer N such that

$$\pi_{\mathrm{HT}}^{-1}\left(\gamma^{-N} \cdot \mathscr{F}\ell_{\{n+1,\ldots,2n\}}\right) \subset \mathcal{X}^*_{\Gamma(p^n)}(\varepsilon)_{\mathrm{anti}}$$

is a rational subset. This shows that $\gamma^{-N} \cdot \mathscr{F}\ell_{\{n+1,\ldots,2n\}}$ is affinoid perfectoid and thus that $\mathscr{F}\ell_{\{n+1,\ldots,2n\}}$ is itself affinoid perfectoid. Since the action of $\mathrm{GSp}_{2n}(\mathbb{Z}_p)$ permutes the cardinality n subsets J, we also see that the preimage of any $\mathscr{F}\ell_J$ under π_{HT} is affinoid perfectoid.

REMARK 5.2.2. The idea of using an element of $\mathrm{GSp}_{2n}(\mathbb{Q}_p)$ to contract a subset of $\mathcal{X}^*_{\Gamma(p^\infty)}$ towards the anticanonical locus seems quite fruitful. For example, this idea is used in [**Lud17**] to construct a perfectoid version of the Lubin-Tate tower at level $\Gamma_0(p^\infty)$.

EXAMPLE 5.2.3. For $n = 1$, the flag variety $\mathscr{F}\ell$ can be identified with the one-dimensional adic projective space $\mathbb{P}^1$. The Plücker embedding is the identity map. If (x_1, x_2) are the usual coordinates on $\mathbb{P}^1$, we see that $\mathscr{F}\ell = \mathbb{P}^1$ has a cover by two affinoid subsets $\mathscr{F}\ell_{\{2\}}$ and $\mathscr{F}\ell_{\{1\}}$, defined by the conditions $|x_1| \leq |x_2|$, respectively $|x_2| \leq |x_1|$. The image of the anticanonical locus under π_{HT} is given by $\{(\frac{x_1}{x_2}, 1) \in \mathbb{P}^1(\mathbb{Q}_p) \mid \frac{x_1}{x_2} \in \mathbb{Z}_p\}$ and the image of the canonical locus is the point $(1, 0) \in \mathbb{P}^1(\mathbb{Q}_p)$. The action of $\begin{pmatrix} p & 0 \\ 0 & 1 \end{pmatrix} \in \mathrm{GL}_2(\mathbb{Q}_p)$ expands the anticanonical locus towards the complement of the canonical point in $\mathbb{P}^1$.

To summarize the discussion in this section, we have the following result.

THEOREM 5.2.4.

(1) *For any sufficiently small tame level $K^p \subset \mathrm{GSp}_{2n}(\mathbb{A}_f^p)$, there exists a perfectoid space $\mathcal{X}^*_{\Gamma(p^\infty), K^p}$ over $\mathbb{Q}_p^{\mathrm{cycl}}$ such that*

$$\mathcal{X}^*_{\Gamma(p^\infty), K^p} \sim \varprojlim_m \mathcal{X}^*_{\Gamma(p^m), K^p}.$$

(2) *There exists a* $\mathrm{GSp}_{2n}(\mathbb{Q}_p)$-*equivariant map of adic spaces*
$$\pi_{\mathrm{HT}} : \mathcal{X}^*_{\Gamma(p^\infty),K^p} \to \mathscr{F}\ell$$
which agrees with the map defined explicitly on points in Lemma 5.1.1.

(3) *Let S be a finite set of bad primes for the tame level K^p. The map π_{HT} is equivariant with respect to the natural Hecke action of the abstract spherical Hecke algebra $\mathbb{T}^S$ on $\mathcal{X}^*_{\Gamma(p^\infty),K^p}$ and the trivial action of $\mathbb{T}^S$ on $\mathscr{F}\ell$.*

(4) *The map π_{HT} is "affinoid", in the following sense: for any subset $J \subset \{1,\ldots,2n\}$ of cardinality n, the preimage of $\mathscr{F}\ell_J$ under π_{HT} is affinoid perfectoid.*[27]

(5) *Let $\omega_{\mathscr{F}\ell} := (\wedge^n W_{\mathscr{F}\ell})^\vee$ be the natural ample line bundle on $\mathscr{F}\ell$. Recall that one also has the natural line bundle ω_{K^p} on $\mathcal{X}^*_{\Gamma(p^\infty),K^p}$, obtained by pullback from any finite level. There is a natural, $\mathrm{GSp}_{2n}(\mathbb{Q}_p)$-equivariant isomorphism*
$$\omega_{K^p} \simeq \pi^*_{\mathrm{HT}} \omega_{\mathscr{F}\ell}.$$
This isomorphism is also $\mathbb{T}^S$-equivariant.

5.3. Shimura varieties of Hodge type.

Let (G, X) be a Shimura datum of Hodge type. Let $K^p \subset G(\mathbb{A}_f)$ be a sufficiently small compact open subgroup. For any choice of compact open subgroup $K_p \subset G(\mathbb{Q}_p)$, we let $X_{K^p K_p}$ be the Shimura variety for G, at level $K^p K_p$, and defined over the reflex field E. Let C be a complete, algebraically closed extension of $\overline{\mathbb{Q}}_p$. We consider the adic space
$$\mathcal{X}_{K^p K_p} := \left(X_{K^p K_p} \times_{\mathrm{Spec}\, E} \mathrm{Spec}\, C\right)^{\mathrm{ad}}.$$

THEOREM 5.3.1 (Thm 4.1.1 of [**Sch15**]). *For any sufficiently small tame level K^p, there exists a perfectoid space $\mathcal{X}_{K^p}$ over $\mathrm{Spa}(C, \mathcal{O}_C)$ such that*
$$\mathcal{X}_{K^p} \sim \varprojlim_{K_p} \mathcal{X}_{K^p K_p}.$$

The proof goes by embedding the Shimura variety of Hodge type into a Siegel modular variety and using the fact that Siegel modular varieties with infinite level at p are perfectoid spaces, as explained in Section 5.1.

REMARK 5.3.2. There is also a version of this result for minimal compactifications. There is one subtlety, having to do with the fact that one does not necessarily have closed embeddings on the level of minimal compactifications. Because of this, one must consider a slightly modified space $X^*_{K^p K_p}$ at finite level, obtained by taking the scheme-theoretic image of $X_{K^p K_p}$ into the corresponding compactification of the Siegel modular variety. However, the map $X^*_{K^p K_p} \to X^{*}_{K^p K_p}$ is a universal homeomorphism, hence induces an isomorphism of diamonds by [**SW17**, Prop. 10.2.1] and the remarks following it. In particular, the spaces have the same étale cohomology. Because of this, we write $\mathcal{X}^*_{K^p}$ for the minimal compactification of the perfectoid Shimura variety $\mathcal{X}_{K^p}$. On the level of diamonds, it is the inverse limit of the diamonds corresponding to $\mathcal{X}^*_{K^p K_p}$.

One can define the Hodge–Tate period morphism more generally, for Shimura varieties of Hodge type, as done in Section 2 of [**CS17**] or even of abelian type [**She17**]. We contend ourselves here to discussing the Hodge–Tate period morphism

---

[27]This implies the following, apparently stronger, statement: for any $J \subset \{1,\ldots,2n\}$ the preimage of any rational open $U \subseteq \mathscr{F}\ell_J$ under π_{HT} is an affinoid perfectoid space.

for Shimura varieties of Hodge type, in order to give a sense of the role that the Shimura datum plays in the definition of a functorial p-adic period morphism and to illustrate the analogy with the complex picture described in Section 2.3. We will use Section 2 of [**CS17**] as a reference.

Let (G, X) be a Shimura datum of Hodge type, with Hodge cocharacter μ. Recall the parabolic subgroup $P_\mu \subset G \times_{\mathbb{Q}} E$ defined in § 2.3. This parabolic subgroup can be thought of as "the stabilizer of the Hodge–Tate filtration". We have a Hodge–Tate period morphism π_{HT}, which should be thought of as a p-adic analogue of the morphism π_{dR} from § 2.3. The following is part of Theorem 2.1.3 of [**CS17**].

THEOREM 5.3.3.

(1) *For any choice of tame level $K^p \subset G(\mathbb{A}_f)$, there is a morphism of adic spaces*

$$\pi_{\mathrm{HT}} : \mathcal{X}_{K^p} \to \mathscr{F}\ell_{G,\mu}.$$

This is functorial in the Shimura datum and agrees with the morphism constructed in Theorem 5.2.4 for Siegel modular varieties.

(2) *The map π_{HT} is equivariant with respect to the Hecke action of $G(\mathbb{Q}_p)$ on $\mathcal{X}_{K^p}$ and the natural action of $G(\mathbb{Q}_p)$ on $\mathscr{F}\ell_{G,\mu}$.*

(3) *The map π_{HT} is equivariant with respect to the action of Hecke operators away from p on $\mathcal{X}_{K^p}$ and the trivial action of these Hecke operators on $\mathscr{F}\ell_{G,\mu}$.*

PROOF. We say a few words about the proof. The main idea is to choose a symplectic embedding $(G, X) \hookrightarrow (\widetilde{G}, \widetilde{X})$, and keep track of Hodge tensors, the finite collection of elements $s_\alpha \in V^\otimes$ which are stabilized by $G \subset \widetilde{G}$. The relative p-adic étale cohomology

$$\mathcal{V}_p := R^1 \pi_{*,\text{ét}} \mathbb{Q}_p$$

of the abelian variety $\pi : \mathcal{A} \to \mathcal{X}_{K^p K_p}$ (restricted from the Siegel modular variety) is trivialized over $\mathcal{X}_{K^p}$. Moreover, under the trivialization, the p-adic realizations of Hodge tensors $s_{\alpha,p} \in \mathcal{V}_p^\otimes$ are identified with the $s_\alpha \in V^\otimes$. This can be rephrased as saying that the G-torsor of trivializations of $(\mathcal{V}_p, s_{\alpha,p})$ has a section over $\mathcal{X}_{K^p}$, which can be thought of as an object in the flattened pro-étale site of $\mathcal{X}_{K^p K_p}$.

The relative Hodge–Tate filtration gives rise to the Hodge–Tate period morphism; in order to show that this morphism factors through the appropriate flag variety $\mathscr{F}\ell_{G,\mu}$, it is enough to show that the G-torsor described above has a P_μ-structure. This amounts to showing that the p-adic realizations of Hodge tensors respect the Hodge–Tate filtration. The same argument automatically proves that the resulting morphism is independent of the choice of embedding $(G, X) \hookrightarrow (\widetilde{G}, \widetilde{X})$.

The latter statement can be seen as a consequence of the fact that the de Rham realizations of Hodge tensors respect the Hodge de Rham filtration, of the relationship between the Hodge–de Rham and Hodge–Tate filtrations described in Section 3, and of the fact that the de Rham and p-adic realizations of Hodge tensors are matched by the p-adic-de Rham comparison isomorphism. The latter result is known for abelian varieties defined over number fields and is due to Blasius [**Bla94**]. □

REMARK 5.3.4. For Shimura varieties of PEL type, the construction of the map π_{HT} in Theorem 5.3.3 is simpler, as one can keep track of the extra endomorphisms in the moduli problem and cut down to the desired flag variety directly.

EXAMPLE 5.3.5. Let $F/\mathbb{Q}$ be an imaginary quadratic field. We set $n = 2$ and we consider the corresponding quasi-split unitary similitude group $G/\mathbb{Q}$ as in Example 2.3.6. This has signature $(2,2)$ at infinity. The corresponding Shimura variety is of PEL type. Assume that $p = \mathfrak{p}\bar{\mathfrak{p}}$ splits in F. Let K be a complete nonarchimedean field which is an extension of $\mathbb{Q}_p^{\text{cycl}}$ and $K^+ \subset K$ an open and bounded valuation subring. For any abelian variety A/K parametrized by the Shimura variety for G, we can write its p-divisible group as a direct product

$$A[p^\infty] = A[\mathfrak{p}^\infty] \times A[\bar{\mathfrak{p}}^\infty].$$

The compatibility between the action of F on A by quasi-isogenies and the polarization λ means that conjugation in F is induced by the Rosati involution corresponding to λ. Therefore, $A[\bar{\mathfrak{p}}^\infty]$ is determined by $A[\mathfrak{p}^\infty]$. We understand the latter via the Hodge–Tate period morphism. The target $\mathscr{F}\ell_{G,\mu}$ of this morphism can be identified with the Grassmannian of 2-dimensional subspaces of a 4-dimensional vector space. This space can be described via the Plücker embedding into $\mathbb{P}^5$.

6. The cohomology of locally symmetric spaces: conjectures and results

In this section, we discuss some recent conjectures and results about the cohomology of locally symmetric spaces, including the case of Shimura varieties. This is an active area of research and there are many perspectives on it (coming from number theory, harmonic analysis, algebraic topology, representation theory). We will restrict ourselves to discussing the following two topics: the construction of Galois representations attached to systems of Hecke eigenvalues in the cohomology of locally symmetric spaces for GL_n/F, where F is a CM field, and vanishing conjectures and theorems for the completed cohomology of locally symmetric spaces.

In § 6.1 we state Scholze's main theorem on the existence of Galois representations and we give a brief sketch of the proof, which uses perfectoid Shimura varieties and the Hodge–Tate period morphism. In § 6.2, we define completed cohomology, state a conjecture of Calegari–Emerton, and discuss some recent results towards this in the case of Shimura varieties.

6.1. The construction of Galois representations. Let F be a CM field. For simplicity, assume that F is imaginary CM. We consider the symmetric space for GL_n/F (in other words, the symmetric space for the group $G = \text{Res}_{F/\mathbb{Q}}\text{GL}_n$) and we let $K \subset \text{GL}_n(\mathbb{A}_{F,f})$ be a neat compact open subgroup. The corresponding locally symmetric space X_K is a smooth orientable Riemannian manifold which does not admit the structure of an algebraic variety.

Let S' be the finite set of primes of F consisting of those primes above any ramified prime of $\mathbb{Q}$ and of those primes v where $K_v \subset \text{GL}_n(F_v)$ is not hyperspecial. [28] Choose a prime p for the coefficients that we will use throughout. Let

---

[28] Recall that a group scheme is reductive if it is smooth and affine, with connected reductive geometric fibers. If G/F_v is a reductive group, a hyperspecial subgroup of $G(F_v)$ is a subgroup that can be identified with the $\mathcal{O}_{F,v}$-points of some reductive model $\mathcal{G}$ of G over $\text{Spec } \mathcal{O}_{F,v}$. Such subgroups of $G(F_v)$ are maximal as compact open subgroups of $G(F_v)$.

$S = S' \cup \{v \text{ prime of } F, v \mid p\}$. If $v \notin S$, let

$$\mathbb{T}_v := \mathbb{Z}_p[\operatorname{GL}_n(\mathcal{O}_{F,v}) \backslash \operatorname{GL}_n(F_v) / \operatorname{GL}_n(\mathcal{O}_{F,v})]$$

be the Hecke algebra of bi-$\operatorname{GL}_n(\mathcal{O}_{F,v})$-invariant, compactly supported, $\mathbb{Z}_p$-valued functions on $\operatorname{GL}_n(\mathcal{O}_{F,v})$. (Recall that this is an algebra under the convolution of functions and that it is commutative.) For $v \notin S$ a prime of F and $1 \le i \le n$, we consider the double coset operator

$$T_{v,i} := [\operatorname{GL}_n(\mathcal{O}_{F,v}) \operatorname{diag}(\varpi_v, \ldots, \varpi_v, 1, \ldots, 1) \operatorname{GL}_n(\mathcal{O}_{F,v})]$$

with i occurrences of the uniformizer ϖ_v on the diagonal. Let $\mathbb{T}^S$ be the abstract Hecke algebra over $\mathbb{Z}_p$

$$\mathbb{T}^S := \otimes'_{v \notin S} \mathbb{T}_v,$$

which acts by correspondences on X_K and therefore also on $H^i_{(c)}(X_K, \mathbb{Z}/p^N\mathbb{Z})$.

Systems of Hecke eigenvalues occurring in $H^i_{\text{Betti}}(X_K, \mathbb{Z}/p^N\mathbb{Z})$ for some $N \in \mathbb{Z}_{\ge 1}$ will be interesting. Let

$$\mathbb{T}(K, i, N) := \operatorname{Im}\left(\mathbb{T}^S \to \operatorname{End}(H^i_{\text{Betti}}(X_K, \mathbb{Z}/p^N\mathbb{Z}))\right).$$

The goal will be to construct a Galois representation valued in $\mathbb{T}(K, i, n)$; we will not quite do this, but something that is close enough (at least for applications to modularity). The following is Corollary 5.4.3 of [**Sch15**].

THEOREM 6.1.1. *Let $\mathfrak{m} \subset \mathbb{T}(K, i, N)$ be a maximal ideal. Then there exists a unique continuous semisimple Galois representation*

$$\bar{\rho}_{\mathfrak{m}} : \operatorname{Gal}(\overline{F}/F) \to \operatorname{GL}_n(\overline{\mathbb{F}}_p)$$

such that for all $v \notin S$ $\bar{\rho}_{\mathfrak{m}}|_{\operatorname{Gal}(\overline{F}_v/F_v)}$ is unramified and the characteristic polynomial of $\bar{\rho}_{\mathfrak{m}}(\operatorname{Frob}_v)$ is equal to the image of

$$X^n - T_{v,1}X^{n-1} + \cdots + (-1)^i q_v^{i(i-1)/2} T_{v,i} X^{n-i} + \cdots + (-1)^n q_v^{n(n-1)/2} T_{v,n}.$$

in $\mathbb{T}(K, i, N)/\mathfrak{m}$.

One can show that the Galois representation $\bar{\rho}_{\mathfrak{m}}$ is actually valued in $\operatorname{GL}_n(\mathbb{F}_q)$, where $\mathbb{F}_q$ is the finite residue field of $\mathfrak{m}$. We say that $\mathfrak{m}$ is *non-Eisenstein* if $\bar{\rho}_{\mathfrak{m}}$ is (absolutely) irreducible. The following is Corollary 5.4.4 of [**Sch15**].

THEOREM 6.1.2. *Assume that $\mathfrak{m}$ is non-Eisenstein.*

Then there exists a nilpotent ideal $I \subset \mathbb{T}(K, i, N)$, of nilpotence degree bounded only in terms of $[F : \mathbb{Q}]$ and n, and a Galois representation

$$\rho_{\mathfrak{m}} : \operatorname{Gal}(\overline{F}/F) \to \operatorname{GL}_n(\mathbb{T}(K, i, N)_{\mathfrak{m}}/I)$$

such that for all $v \notin S$ a prime of F $\rho_{\mathfrak{m}}|_{\operatorname{Gal}(\overline{F}_v/F_v)}$ is unramified and the characteristic polynomial of $\rho_{\mathfrak{m}}(\operatorname{Frob}_v)$ is equal to the image of

$$X^n - T_{v,1}X^{n-1} + \cdots + (-1)^i q_v^{i(i-1)/2} T_{v,i} X^{n-i} + \cdots + (-1)^n q_v^{n(n-1)/2} T_{v,n}.$$

in $\mathbb{T}(K, I, N)_{\mathfrak{m}}/I$.

REMARK 6.1.3.

(1) The Galois representations $\bar{\rho}_{\mathfrak{m}}$ and $\rho_{\mathfrak{m}}$ are constructed by *p-adic interpolation* (in other words by keeping track of congruences modulo p^N for $N \in \mathbb{Z}_{\geq 1}$) from the Galois representations associated to (conjugate) self-dual, regular L-algebraic automorphic representations of GL_m/F for some $\mathfrak{m} \in \mathbb{Z}_{\geq 1}$.

These Galois representations were constructed in several steps by many people: Kottwitz, Clozel, Harris-Taylor, Shin, Chenevier-Harris [**Clo91, Kot92a, HT01, Shi11, CH09**], building on fundamental contributions by many others. In almost all cases, one uses a similar method to the one outlined in the introduction in the case of weight 2 modular forms, i.e. one uses the étale cohomology of certain *Shimura varieties*, which are higher-dimensional analogues of modular curves. However, we emphasize that the Galois representations in Theorem 6.1.2 are not "cut out" directly from the étale cohomology of some Shimura variety.

(2) Scholze's result recovers a theorem of Harris–Lan–Taylor–Thorne in characteristic 0, namely the existence of Galois representations for regular L-algebraic automorphic representations of GL_n/F, cf. [**HLTT16**]. It is possible to give a different proof of Theorem 6.1.2 (even in the torsion setting) as a result of Boxer's thesis [**Box15**], which uses integral models rather than perfectoid Shimura varieties and understands torsion in the *coherent* cohomology of Shimura varieties.

(3) Newton and Thorne [**NT16**] improve the bound on the nilpotence degree of I to $I^4 = 0$. In certain instances, it is possible to eliminate the nilpotent ideal completely.

(4) We have stated the main result for the trivial local system on X_K for simplicity. The analogous result also holds with coefficients in a local system on X_K corresponding to some irreducible algebraic representation of $\mathrm{Res}_{F/\mathbb{Q}} \mathrm{GL}_n$.

Let $\widetilde{G}$ be the unitary group defined in Example 2.3.11.

EXERCISE 6.1.4. *Show that $\mathrm{Res}_{F/\mathbb{Q}} \mathrm{GL}_n$ is the Levi subgroup of a maximal parabolic subgroup of $\widetilde{G}$. (This is the so-called* Siegel parabolic *of $\widetilde{G}$ and, up to conjugacy over $\mathbb{Q}$, coincides with the parabolic subgroup P_μ defined above.)*

Let $\widetilde{K} \subset \widetilde{G}(\mathbb{A}_f)$. This determines a Shimura variety of Hodge type $\widetilde{X}_{\widetilde{K}}$. There are two key steps for the proof of Theorem 6.1.2.

The first step is to show that any mod p^N system of Hecke eigenvalues in the étale cohomology of $\widetilde{X}_{\widetilde{K}}$ lifts to characteristic 0, to a system of Hecke eigenvalues associated to a cusp form on $\widetilde{G}$. This is proved in [**Sch15**, Thm. 4.3.1] for any Shimura variety of Hodge type. This can be seen by studying the geometry of the Hodge–Tate period morphism

$$\pi_{\mathrm{HT}} : \widetilde{\mathcal{X}}^*_{\widetilde{K}^p} \to \mathscr{F}\ell_{\widetilde{G},\mu}{}^{[29]}.$$

---

[29] In fact, it is enough to consider a bigger period domain, corresponding to a Siegel embedding $\widetilde{G} \hookrightarrow \mathrm{GSp}_{2n[F^+:\mathbb{Q}]}$.

We briefly mention the key ideas involved in the first step, i.e. in the proof of [**Sch15**, Thm. 4.3.1]. See *loc. cit.* for more details, as well as the survey [**Mor16**].

(1) First, Scholze applies a version of the primitive comparison isomorphism for rigid analytic varieties proved in [**Sch13**]. This reduces [**Sch15**, Thm. 4.3.1] to constructing Hecke congruences between $H^i_{\text{ét}}(\widetilde{\mathcal{X}}^*_{\widetilde{K}^p}, \mathcal{I}^+/p^N)$ and cusp forms for $\widetilde{G}$, where $\mathcal{I}^+ \subset \mathcal{O}^+_{\widetilde{\mathcal{X}}^*_{\widetilde{K}^p}}$ is the subsheaf of sections which vanish along the boundary.

(2) An important property of the morphism π_{HT} is that it is "affinoid" in the following sense: there exists a cover by affinoid open subsets
$$\mathcal{F}\ell_{\widetilde{G},\mu} = \bigcup_{i=1}^{M} U_i$$
(defined via an appropriate Plücker embedding) such that each preimage $V_i := \pi_{\text{HT}}^{-1}(U_i)$ is affinoid perfectoid. Moreover, since π_{HT} is Hecke equivariant, the open subsets $V_i \subset \widetilde{\mathcal{X}}^*_{\widetilde{K}^p}$ are stable under the action of Hecke operators away form p. For affinoid perfectoid spaces, the étale cohomology $H^i_{\text{ét}}(V_i, \mathcal{O}^+_{V_i})$ is almost zero for $i > 0$, cf. [**Sch13**, Lemma 4.10]; a version of this result also holds for sections that vanish along the boundary of $V_i \cap \widetilde{\mathcal{X}}^*_{\widetilde{K}^p}$. Using the open cover $(V_i)_{i=1}^M$ and the above observation, Scholze reduces [**Sch15**, Thm. 4.3.1] to constructing mod p^N Hecke congruences between the Cech cohomology of the cover $(V_i)_{i=1}^M$ and cusp forms for $\widetilde{G}$.

(3) The key ingredients for constructing these congruences are certain "fake Hasse invariants" pulled back from sections defined over the open cover of $\mathcal{F}\ell_{\widetilde{G},\mu}$. This is inspired by the way the classical Hasse invariant is used to construct Hecke congruences in the setting of modular forms; this method is used in [**DS74**] to construct Galois representations attached to cusp forms of weight 1 by constructing congruences to cusp forms of higher weight.

The second step is to relate the cohomology of X_K with $\mathbb{Z}/p^N\mathbb{Z}$-coefficients to the cohomology of $\widetilde{X}_{\widetilde{K}}$ with $\mathbb{Z}/p^N\mathbb{Z}$-coefficients, for some neat compact open subgroup $\widetilde{K} \subset \widetilde{G}(\mathbb{A}_f)$. This is done by studying the geometry of the Borel–Serre compactification of $\widetilde{X}_{\widetilde{K}}$, which we now consider simply as a locally symmetric space for $\widetilde{G}$. The rough idea is that the strata in the boundary of the Borel–Serre compactification are generalizations of locally symmetric spaces, more precisely locally symmetric spaces attached to the proper parabolic subgroups of $\widetilde{G}$. In the case of the Siegel parabolic P_μ, the corresponding locally symmetric spaces are torus bundles over the locally symmetric spaces for $\text{Res}_{F/\mathbb{Q}} \text{GL}_n$. Since this part of the argument does not involve p-adic geometry, we do not say more about it here; for more details about this step, see [**Sch15**, §5] and [**NT16**].

6.2. Vanishing of cohomology at infinite level. Completed cohomology, as introduced by Emerton in [**Eme06**], gives a way of defining p-adic automorphic forms for general reductive groups. Let $G/\mathbb{Q}$ be a connected reductive group with the corresponding tower of locally symmetric spaces $(X_K)_K$. Fix a *tame level*, i.e. a sufficiently small compact open subgroup $K^p \subset G(\mathbb{A}_f^p)$. The *completed cohomology* groups are defined as

$$\widetilde{H}^i(K^p) := \varprojlim_N \left(\varinjlim_{K_p} \left(H^i(X_{K^p K_p}, \mathbb{Z}/p^N \mathbb{Z}) \right) \right)$$

where K_p runs over all compact open subgroups of $G(\mathbb{Q}_p)$. For $N \in \mathbb{Z}_{\geq 1}$ we also define

$$\widetilde{H}^i(K^p, \mathbb{Z}/p^N \mathbb{Z}) := \varinjlim_{K_p} \left(H^i(X_{K^p K_p}, \mathbb{Z}/p^N \mathbb{Z}) \right).$$

The group $\widetilde{H}^i(K^p)$ is a p-adically complete $\mathbb{Z}_p$-module. If S' is the finite set of bad primes determined by the tame level K^p and $S = S' \cup \{p\}$, then $\widetilde{H}^i(K^p)$ has an action of the abstract Hecke algebra $\mathbb{T}^S$. Moreover, $\widetilde{H}^i(K^p)$ also has an action of the full group $G(\mathbb{Q}_p)$. This is induced from the action of c_g^* for $g \in G(\mathbb{Q}_p)$ on the directed system $\left(H^i((X_{K^p K_p}, \mathbb{Z}/p^N \mathbb{Z}) \right)_{K_p \subset G(\mathbb{Q}_p)}$, sending a class at level K_p to a class at level $K_p \cap g K_p g^{-1}$. As a representation of $G(\mathbb{Q}_p)$, one can prove that $\widetilde{H}^i(K^p)$ is *p-adically admissible*, which means that

(1) it is p-adically complete and separated, and the $\mathbb{Z}_p$-torsion subspace $\widetilde{H}^i(K^p)[p^\infty]$ is of bounded exponent, which means that $\widetilde{H}^i(K^p)[p^\infty] = \widetilde{H}^i(K^p)[p^M]$ for all sufficiently large integers M.
(2) each $\widetilde{H}^i(K^p)/p^N$, which is a smooth representation of $G(\mathbb{Q}_p)$, is also admissible as a representation of $G(\mathbb{Q}_p)$ (in the usual sense).

Recall that a *smooth* representation of $G(\mathbb{Q}_p)$ is one in which every vector has an open stabilizer in $G(\mathbb{Q}_p)$. It is not hard to show that $\widetilde{H}^i(K^p, \mathbb{Z}/p^N \mathbb{Z})$ are smooth representations of $G(\mathbb{Q}_p)$ for every $N \geq 1$. However, completed cohomology with $\mathbb{Z}_p$-coefficients is *not* a smooth representation of $G(\mathbb{Q}_p)$ - the smooth vectors in completed cohomology correspond to certain classical automorphic forms, which form a much smaller space than the space of all p-adic automorphic forms.

EXERCISE 6.2.1. *Using the fact that $\widetilde{H}^i(K^p)$ is p-adically admissible, show that the inverse system $\left(\widetilde{H}^i(K^p, \mathbb{Z}/p^N \mathbb{Z}) \right)_N$ satisfies Mittag-Leffler.*

REMARK 6.2.2.
(1) One can also make the definition for compactly-supported cohomology as well as for homology and Borel-Moore homology. See [**CE12**, **Eme14**] for more details on these and the relationships between them. See [**Eme14**] also for an overview of the role that completed cohomology plays in the p-adic Langlands program, in terms of both local and global aspects.
(2) We can identify the completed cohomology of tame level K^p with the cohomology of the perfectoid Shimura variety of tame level K^p. More precisely, there exists a natural, Hecke-equivariant comparison isomorphism

$$\widetilde{H}^i_{(c)}(K^p, \mathbb{Z}/p^N \mathbb{Z}) \xrightarrow{\sim} H^i_{\text{ét},(c)}(\mathcal{X}_{K^p}, \mathbb{Z}/p^N \mathbb{Z}).$$

(3) One can also make the following definition:

$$\widehat{H}^i(K^p) := \varprojlim_N \left(\varinjlim_{K_p} \left(H^i(X_{K^p K_p}, \mathbb{Z}_p) \right) / p^N \right),$$

which also has an action of the abstract Hecke algebra $\mathbb{T}^S$. Intuitively, the systems of Hecke eigenvalues (i.e. the maximal ideals of $\mathbb{T}^S$) in the support of $\widehat{H}^i(K^p)$ are those which can be p-adically interpolated from systems of Hecke eigenvalues in the support of $H^i(X_{K^pK_p}, \mathbb{Z}_p)$ for some finite level K_p, i.e. systems of Hecke eigenvalues corresponding to classical automorphic forms. The difference between $\widetilde{H}^i(K^p)$ and $\widehat{H}^i(K^p)$ can be expressed as a limit over *torsion classes* occurring in the cohomology of locally symmetric spaces at finite level, as seen in Exercise 6.2.3 below.

EXERCISE 6.2.3. *Consider the short exact sequence of sheaves*

$$0 \to \mathbb{Z}_p \xrightarrow{\cdot p^N} \mathbb{Z}_p \to \mathbb{Z}/p^N\mathbb{Z} \to 0$$

on $X_{K^pK_p}$ for every K_p. By analyzing the cohomology long exact sequence, prove that we have an injection

$$\widehat{H}^i(K^p) \hookrightarrow \widetilde{H}^i(K^p)$$

and describe its cokernel in terms of torsion classes, i.e. in terms of the groups $H^i(X_{K^pK_p}, \mathbb{Z}_p)[p^N]$.

REMARK 6.2.4. In particular, if the groups $H^i(X_{K^pK_p}, \mathbb{Z}_p)[p^N]$ are zero for all $N \in \mathbb{Z}_{\geq 1}$ and all compact-open K_p, then we have an isomorphism $\widehat{H}^i(K^p) \xrightarrow{\sim} \widetilde{H}^i(K^p)$. This happens, for example, if G is a definite unitary group, so that the locally symmetric spaces are finite sets of points. This also happens in the case of modular curves. However, we will be primarily concerned with a general $G = \operatorname{Res}_{F/\mathbb{Q}} \operatorname{GL}_n$, in which case the groups $H^i(X_{K^pK_p}, \mathbb{Z}_p)$ are known to contain torsion. For example, Bergeron–Venkatesh conjecture in [**BV13**] that the size of the torsion subgroup in Betti cohomology grows exponentially with the index of K when $l_0 = 1$.

Here are some further important properties of completed cohomology:

(1) The Hochschild–Serre spectral sequence can be used to recover cohomology at finite level from completed cohomology. More precisely, if $K_p \subset G(\mathbb{Q}_p)$ is a compact-open subgroup, then we have spectral sequences

$$E_2^{i,j} = H^i(K_p, \widetilde{H}^j(K^p, \mathbb{Z}/p^N\mathbb{Z})) \Rightarrow H^{i+j}(X_{K^pK_p}, \mathbb{Z}/p^N\mathbb{Z})$$

and

$$E_2^{i,j} = H^i(K_p, \widetilde{H}^j(K^p)) \Rightarrow H^{i+j}(X_{K^pK_p}, \mathbb{Z}_p),$$

where $H^i(K_p, \ )$ denotes the continuous group cohomology of K_p. See [**Eme06**, Prop. 2.1.11] for a proof, which is slightly subtle with $\mathbb{Z}_p$-coefficients.[30]

(2) One can work with cohomology at finite level with coefficients in a local system $\mathcal{V}_\xi$ corresponding to some algebraic representation ξ of G and the completed cohomology groups one obtains match up. More precisely, assume that ξ is an algebraic representation of G defined over $\mathbb{Q}_p$ (for simplicity, otherwise we would introduce a field of coefficients E which is

---

[30]Since K_p is a compact locally $\mathbb{Q}_p$-analytic group, the category of p-adically admissible representations of K_p over $\mathbb{Z}_p$ has enough injectives. Therefore, the continuous cohomology groups $H^i(K_p, \ )$ can be identified with the derived functors of the functor "taking K_p-invariants" on the category of p-adically admissible K_p-representations.

a finite extension of $\mathbb{Q}_p$). Let $V_\xi^\circ \subset V_\xi$ be a $\mathbb{Z}_p$-lattice stable under the action of $\mathcal{G}(\mathbb{Z}_p)$. The local system $\mathcal{V}_\xi^\circ$ is defined as follows:

$$\mathcal{V}_\xi^\circ := G(\mathbb{Q}) \setminus \left(X \times G(\mathbb{A}_f)/K \times V_\xi^\circ \right).$$

The completed cohomology groups corresponding to the local system $\mathcal{V}_\xi^\circ$ are defined as

$$\widetilde{H}^i(K^p, \mathcal{V}_\xi^\circ) := \varprojlim_N \left(\varinjlim_{K_p} \left(H^i(X_{K^p K_p}, \mathcal{V}_\xi^\circ/p^N) \right) \right),$$

where K_p runs over compact-open subgroups of $\mathcal{G}(\mathbb{Z}_p)$. Then we have a natural, Hecke-equivariant isomorphism of p-adically admissible representations of $\mathcal{G}(\mathbb{Z}_p)$

$$\widetilde{H}^i(K^p, \mathcal{V}_\xi^\circ) \xrightarrow{\sim} V_\xi^\circ \otimes_{\mathbb{Z}_p} \widetilde{H}^i(K^p).$$

(3) Let $(V_\xi^\circ)^\vee$ denote the $\mathbb{Z}_p$-dual of V_ξ°, endowed with the contragredient action of $\mathcal{G}(\mathbb{Z}_p)$. Let $K_p \subset \mathcal{G}(\mathbb{Z}_p)$ be a compact-open subgroup. By combining the first two items, one obtains a *control theorem* for completed cohomology in the form of a spectral sequence

$$E_2^{i,j} = \operatorname{Ext}^i_{\mathbb{Z}_p[\![K_p]\!]}\left((V_\xi^\circ)^\vee, \widetilde{H}^j(K^p) \right) \Longrightarrow H^{i+j}(X_{K^p K_p}, \mathcal{V}_\xi^\circ).$$

Motivated by heuristics coming from the Langlands program, Calegari and Emerton made several conjectures about the usual and compactly-supported completed cohomology of general locally symmetric spaces: see [**CE12**, Conj. 1.5] (which is, however, stated in terms of homology and Borel–Moore homology). In the case of tori, their conjectures are equivalent to Leopoldt's conjecture, cf. [**Hil10**].

A consequence of the Calegari–Emerton conjectures is the following. Let l_0 be the invariant defined in § 2.1 and set $q_0 := \frac{1}{2}(\dim_\mathbb{R} X - l_0)$. Then for every sufficiently small $K^p \subset G(\mathbb{A}_f^p)$ we expect that $\widetilde{H}^i_{(c)}(K^p) = 0$ for $i > q_0 + l_0$. Even this weaker conjecture is wide open for general locally symmetric spaces. Some small-dimensional examples can be studied by hand, such as arithmetic hyperbolic 3-manifolds. In the case of Shimura varieties, where $l_0 = 0$ and q_0 is equal to the dimension of the Shimura variety, Scholze recently proved the following result, cf. [**Sch15**, Thm. 4.2.2].

THEOREM 6.2.5. *Let (G, X) be a Shimura datum of Hodge type. Let $K^p \subset G(\mathbb{A}_f^p)$ be a sufficiently small compact open subgroup and let $N \in \mathbb{Z}_{\geq 1}$. We have*

$$\widetilde{H}^i_c(K^p, \mathbb{Z}/p^N\mathbb{Z}) = 0$$

for all $i > d = \dim_\mathbb{C} X$.

SKETCH OF PROOF OF THEOREM 6.2.5. The primitive comparison theorem of [**Sch13**] (applied at finite level, then taking a direct limit) gives an almost isomorphism between $\widetilde{H}^i(K^p, \mathbb{F}_p) \otimes_{\mathbb{F}_p} \mathcal{O}_C/p$ and $H^i_{\text{ét}}(\mathcal{X}^*_{K^p}, \mathcal{I}^+/p)$, where $\mathcal{I}^+ \subseteq \mathcal{O}^+_{\mathcal{X}^*_{K^p}}$ is the subsheaf sections which vanish along the boundary. For any affinoid perfectoid cover of $\mathcal{X}^*_{K^p}$, the étale cohomology of the restriction of the sheaf $\mathcal{I}^+$ is almost equal to 0 in degree $i > 0$. This allows us to replace $H^i_{\text{ét}}(\mathcal{X}^*_{K^p}, \mathcal{I}^+/p)$ with $H^i_{\text{an}}(\mathcal{X}^*_{K^p}, \mathcal{I}^+/p)$. The vanishing theorem now follows from a theorem of Scheiderer [**Sch92**], who shows that the cohomological dimension of a spectral space of Krull dimension d is at most d. □

REMARK 6.2.6. The same idea can be used to prove the analogous vanishing theorem for the cohomology $\varinjlim_r H^i(X_r, \mathbb{F}_p)$, where $X \subset \mathbb{P}^N$ is a projective variety, and X_r is the pullback of X under the morphism

$$\mathbb{P}^N \to \mathbb{P}^N, (x_0 : x_1 : \ldots x_N) \mapsto (x_0^{p^r} : x_1^{p^r} : \ldots x_N^{p^r}).$$

A direct geometric proof of this vanishing theorem, which does not use p-adic geometry, was recently given by Esnault in [**Esn18**] and her result also holds for X of characteristic $\ell \neq p$. It would be interesting to see whether Theorem 6.2.5 or even Theorem 6.2.7 below admit more direct proofs.

We now explain a strenghtening of Theorem 6.2.5 in the particular case of Siegel modular varieties. Let (G, X) denote the Siegel datum in Example 2.3.4. Let $K^p \subset G(\mathbb{A}_f^p)$ be sufficiently small and for every $m \in \mathbb{Z}_{\geq 1}$ recall the compact open subgroups

$$\Gamma_1(p^m) := \left\{ g \in \mathrm{GSp}_{2n}(\mathbb{Z}_p) \mid g \equiv \begin{pmatrix} \mathrm{Id}_n & * \\ 0 & \mathrm{Id}_n \end{pmatrix} \pmod{p^m} \right\}.$$

Define $\widetilde{H}^i_{(c)}(K^p, \mathbb{Z}/p^N\mathbb{Z})_{\Gamma_1} := \varinjlim_m H^i_{(c)}(X_{K^p\Gamma_1(p^m)}, \mathbb{Z}/p^N\mathbb{Z})$.

THEOREM 6.2.7. *We have $\widetilde{H}^i_c(K^p, \mathbb{Z}/p^N\mathbb{Z})_{\Gamma_1} = 0$ for all $i > d = \dim_{\mathbb{C}} X$.*

This is a part of [**CGH+18**, Thm. 1.1.2]; the result in *loc. cit.* is slightly more general, as it also considers certain Shimura varieties attached to quasi-split unitary groups as in Examples 2.3.6 and 2.3.11. In fact, we expect an even more general statement to be true, but we leave the question of extending Theorem 6.2.7 to other Shimura varieties to be explored in future work.

SKETCH OF PROOF OF THEOREM 6.2.7. Assume for simplicity that $n = 1$, in which case $G = \mathrm{GL}_2$ and we are working with modular curves. Set $N_0 := \begin{pmatrix} 1 & \mathbb{Z}_p \\ 0 & 1 \end{pmatrix}$; this is the group of $\mathbb{Z}_p$-points of the unipotent radical N of the standard (upper-triangular) Borel subgroup B of G. As in the proof of Theorem 6.2.5, we start by applying the primitive comparison theorem of [**Sch13**]. We then analyze the Leray spectral sequence for the morphism of sites

$$|\pi_{\mathrm{HT}/N_0}| : \left(\mathcal{X}^*_{\Gamma_1(p^\infty)}\right)_{\text{ét}} \to |\mathbb{P}^{1,\mathrm{ad}}|/N_0,$$

using it to compute the cohomology $H^i_{\text{ét}}(\mathcal{X}^*_{\Gamma_1(p^\infty)}, \mathcal{I}^+/p)$.[31] We consider the stratification of $\mathbb{P}^{1,\mathrm{ad}}$ into B-orbits: there is an open cell $\mathbb{A}^{1,\mathrm{ad}}$, containing $\pi_{\mathrm{HT}}(\mathcal{X}^*_{\Gamma(p^\infty)}(\varepsilon)_{\text{anti}})$ as an open subset, and a closed cell, which consists of one point ∞, the image under π_{HT} of the canonical tower over the ordinary locus. Since $N_0 \subset B(\mathbb{Q}_p)$, this stratification descends to the quotient $|\mathbb{P}^{1,\mathrm{ad}}|/N_0$. We see the following two phenomena:

(1) The fibers of π_{HT/N_0} over rank 1 points of $|\mathbb{A}^{1,\mathrm{ad}}|/N_0$ are affinoid perfectoid spaces. Indeed, the action of $\begin{pmatrix} p & 0 \\ 0 & 1 \end{pmatrix} \in \mathrm{GL}_2(\mathbb{Q}_p)$, expands the anticanonical locus in $\mathbb{P}^{1,\mathrm{ad}}$ to cover all of $\mathbb{A}^{1,\mathrm{ad}}$ and the anticanonical locus is affinoid perfectoid already at level $\Gamma_0(p^\infty)$; see also [**Lud17**] for a version of this argument.

(2) The fiber of π_{HT/N_0} over ∞/N_0 is not affinoid perfectoid, but it has a "$\mathbb{Z}_p$-cover"[32] which is affinoid perfectoid. Indeed, at full infinite level,

---

[31] In order to make this rigorous, we need to consider $\mathcal{X}^*_{\Gamma_1(p^\infty)}$ as a diamond and use the étale cohomology of diamonds as developed in [**Sch17**].

[32] For the precise meaning of "$\mathbb{Z}_p$-cover", see [**CGH+18**, Thm. 2.2.7].

the Weyl group element $\begin{pmatrix} 0 & 1 \\ 1 & 0 \end{pmatrix} \in \mathrm{GL}_2(\mathbb{Q}_p)$ takes the anticanonical locus in $\mathbb{P}^{1,\mathrm{ad}}$ to an open neighborhood of ∞. The cohomological dimension of this fiber is then bounded by the cohomological dimension of $\mathbb{Z}_p$ for continuous group cohomology.

While $R^0|\pi_{\mathrm{HT}/N_0}|_*(\mathcal{I}^+/p)$ is supported on the whole of $|\mathbb{P}^{1,\mathrm{ad}}|/N_0$, we deduce that $R^1|\pi_{\mathrm{HT}/N_0}|_*(\mathcal{I}^+/p)$ is almost zero outside a closed spectral subspace of $|\mathbb{P}^{1,\mathrm{ad}}|/N_0$ of dimension 0 and $R^2|\pi_{\mathrm{HT}/N_0}|_*(\mathcal{I}^+/p)$ is almost zero everywhere. We conclude by appealing to [**Sch92**] to compute the terms in the Leray spectral sequence.

The case of GSp_{2n} for arbitrary n is similar, using the generalized Bruhat stratification of the algebraic flag variety $\mathrm{Fl}_{G,\mu}$ into P_μ-orbits. Setting $N_0 := \cap_m \Gamma_1(p^m)$, we obtain a stratification

$$|\mathscr{F}\ell_{G,\mu}|/N_0 = \bigsqcup |\mathscr{F}\ell_{G,\mu}^w|/N_0$$

into locally closed strata, indexed by certain Weyl group elements w. Moreover, for every point $x \in |\mathscr{F}\ell_{G,\mu}^w|/N_0$, we show that

$$R^i \pi_{\mathrm{HT}/N_0,*}(\mathcal{I}^+/p)_x^a = 0 \text{ for } i > d - \dim \mathscr{F}\ell_{G,\mu}^w.$$

This, together with [**Sch92**], is enough to prove the desired vanishing for Siegel modular varieties. □

References

[BCDT01] Christophe Breuil, Brian Conrad, Fred Diamond, and Richard Taylor, *On the modularity of elliptic curves over* **Q**: *wild 3-adic exercises*, J. Amer. Math. Soc. **14** (2001), no. 4, 843–939, DOI 10.1090/S0894-0347-01-00370-8. MR1839918

[BG14] Kevin Buzzard and Toby Gee, *The conjectural connections between automorphic representations and Galois representations*, Automorphic forms and Galois representations. Vol. 1, London Math. Soc. Lecture Note Ser., vol. 414, Cambridge Univ. Press, Cambridge, 2014, pp. 135–187, DOI 10.1017/CBO9781107446335.006. MR3444225

[Bla94] Don Blasius, *A p-adic property of Hodge classes on abelian varieties*, Motives (Seattle, WA, 1991), Proc. Sympos. Pure Math., vol. 55, Amer. Math. Soc., Providence, RI, 1994, pp. 293–308. MR1265557

[Box15] George Andrew Boxer, *Torsion in the Coherent Cohomology of Shimura Varieties and Galois Representations*, ProQuest LLC, Ann Arbor, MI, 2015. Thesis (Ph.D.)–Harvard University. MR3450451

[BS73] A. Borel and J.-P. Serre, *Corners and arithmetic groups*, Comment. Math. Helv. **48** (1973), 436–491, DOI 10.1007/BF02566134. Avec un appendice: Arrondissement des variétés à coins, par A. Douady et L. Hérault. MR0387495

[BV13] Nicolas Bergeron and Akshay Venkatesh, *The asymptotic growth of torsion homology for arithmetic groups*, J. Inst. Math. Jussieu **12** (2013), no. 2, 391–447, DOI 10.1017/S1474748012000667. MR3028790

[BW00] A. Borel and N. Wallach, *Continuous cohomology, discrete subgroups, and representations of reductive groups*, 2nd ed., Mathematical Surveys and Monographs, vol. 67, American Mathematical Society, Providence, RI, 2000. MR1721403

[CE12] Frank Calegari and Matthew Emerton, *Completed cohomology—a survey*, Non-abelian fundamental groups and Iwasawa theory, London Math. Soc. Lecture Note Ser., vol. 393, Cambridge Univ. Press, Cambridge, 2012, pp. 239–257. MR2905536

[CF90] Gerd Faltings and Ching-Li Chai, *Degeneration of abelian varieties*, Ergebnisse der Mathematik und ihrer Grenzgebiete (3) [Results in Mathematics and Related Areas (3)], vol. 22, Springer-Verlag, Berlin, 1990. With an appendix by David Mumford. MR1083353

[CG18] Frank Calegari and David Geraghty, *Modularity lifting beyond the Taylor-Wiles method*, Invent. Math. **211** (2018), no. 1, 297–433, DOI 10.1007/s00222-017-0749-x. MR3742760

[CGH+18] A. Caraiani, D. R. Gulotta, C.-Y. Hsu, C. Johansson, L. Mocz, E. Reinecke, and S.-C. Shih, *Shimura varieties at level $\Gamma_1(p^\infty)$ and Galois representations*, ArXiv e-prints (2018).
[CH09] Gaëtan Chenevier and Michael Harris, *Construction of automorphic Galois representations, II*, Camb. J. Math. **1** (2013), no. 1, 53–73, DOI 10.4310/CJM.2013.v1.n1.a2. MR3272052
[Clo91] Laurent Clozel, *Représentations galoisiennes associées aux représentations automorphes autoduales de GL(n)* (French), Inst. Hautes Études Sci. Publ. Math. **73** (1991), 97–145. MR1114211
[CS17] Ana Caraiani and Peter Scholze, *On the generic part of the cohomology of compact unitary Shimura varieties*, Ann. of Math. (2) **186** (2017), no. 3, 649–766, DOI 10.4007/annals.2017.186.3.1. MR3702677
[Del71] Pierre Deligne, *Travaux de Shimura* (French), Séminaire Bourbaki, 23ème année (1970/71), Exp. No. 389, Springer, Berlin, 1971, pp. 123–165. Lecture Notes in Math., Vol. 244. MR0498581
[Del79] Pierre Deligne, *Variétés de Shimura: interprétation modulaire, et techniques de construction de modèles canoniques* (French), Automorphic forms, representations and L-functions (Proc. Sympos. Pure Math., Oregon State Univ., Corvallis, Ore., 1977), Proc. Sympos. Pure Math., XXXIII, Amer. Math. Soc., Providence, R.I., 1979, pp. 247–289. MR546620
[Del82] Pierre Deligne, James S. Milne, Arthur Ogus, and Kuang-yen Shih, *Hodge cycles, motives, and Shimura varieties*, Lecture Notes in Mathematics, vol. 900, Springer-Verlag, Berlin-New York, 1982. MR654325
[DI87] Pierre Deligne and Luc Illusie, *Relèvements modulo p^2 et décomposition du complexe de de Rham* (French), Invent. Math. **89** (1987), no. 2, 247–270, DOI 10.1007/BF01389078. MR894379
[DS74] Pierre Deligne and Jean-Pierre Serre, *Formes modulaires de poids 1* (French), Ann. Sci. École Norm. Sup. (4) **7** (1974), 507–530 (1975). MR0379379
[Eme06] Matthew Emerton, *On the interpolation of systems of eigenvalues attached to automorphic Hecke eigenforms*, Invent. Math. **164** (2006), no. 1, 1–84, DOI 10.1007/s00222-005-0448-x. MR2207783
[Eme14] Matthew Emerton, *Completed cohomology and the p-adic Langlands program*, Proceedings of the International Congress of Mathematicians—Seoul 2014. Vol. II, Kyung Moon Sa, Seoul, 2014, pp. 319–342. MR3728617
[Esn18] H. Esnault, *Cohomological dimension in pro-p-towers*, ArXiv e-prints (2018).
[Fal02] Gerd Faltings, *Almost étale extensions*, Astérisque **279** (2002), 185–270. Cohomologies p-adiques et applications arithmétiques, II. MR1922831
[FC90] Gerd Faltings and Ching-Li Chai, *Degeneration of abelian varieties*, Ergebnisse der Mathematik und ihrer Grenzgebiete (3) [Results in Mathematics and Related Areas (3)], vol. 22, Springer-Verlag, Berlin, 1990. With an appendix by David Mumford. MR1083353
[Fra98] Jens Franke, *Harmonic analysis in weighted L_2-spaces* (English, with English and French summaries), Ann. Sci. École Norm. Sup. (4) **31** (1998), no. 2, 181–279, DOI 10.1016/S0012-9593(98)80015-3. MR1603257
[Hil10] Richard Hill, *On Emerton's p-adic Banach spaces*, Int. Math. Res. Not. IMRN **18** (2010), 3588–3632, DOI 10.1093/imrn/rnq042. MR2725506
[HLTT16] Michael Harris, Kai-Wen Lan, Richard Taylor, and Jack Thorne, *On the rigid cohomology of certain Shimura varieties*, Res. Math. Sci. **3** (2016), Paper No. 37, 308, DOI 10.1186/s40687-016-0078-5. MR3565594
[HT01] Michael Harris and Richard Taylor, *The geometry and cohomology of some simple Shimura varieties*, Annals of Mathematics Studies, vol. 151, Princeton University Press, Princeton, NJ, 2001. With an appendix by Vladimir G. Berkovich. MR1876802
[Kot92a] Robert E. Kottwitz, *On the λ-adic representations associated to some simple Shimura varieties*, Invent. Math. **108** (1992), no. 3, 653–665, DOI 10.1007/BF02100620. MR1163241
[Kot92b] Robert E. Kottwitz, *Points on some Shimura varieties over finite fields*, J. Amer. Math. Soc. **5** (1992), no. 2, 373–444, DOI 10.2307/2152772. MR1124982

[Lan13a] Kai-Wen Lan, *Arithmetic compactifications of PEL-type Shimura varieties*, London Mathematical Society Monographs Series, vol. 36, Princeton University Press, Princeton, NJ, 2013. MR3186092

[Lan13b] Kai-Wen Lan, *Arithmetic compactifications of PEL-type Shimura varieties*, London Mathematical Society Monographs Series, vol. 36, Princeton University Press, Princeton, NJ, 2013. MR3186092

[Lan16] Kai-Wen Lan, *Higher Koecher's principle*, Math. Res. Lett. **23** (2016), no. 1, 163–199, DOI 10.4310/MRL.2016.v23.n1.a9. MR3512882

[Lud17] Judith Ludwig, *A quotient of the Lubin-Tate tower*, Forum Math. Sigma **5** (2017), e17, 41, DOI 10.1017/fms.2017.15. MR3680340

[Mil90] J. S. Milne, *Canonical models of (mixed) Shimura varieties and automorphic vector bundles*, Automorphic forms, Shimura varieties, and L-functions, Vol. I (Ann Arbor, MI, 1988), Perspect. Math., vol. 10, Academic Press, Boston, MA, 1990, pp. 283–414. MR1044823

[Mil04] J.S. Milne, *Introduction to shimura varieties*, Available online at http://www.jmilne.org/math/xnotes/svi.pdf (2004).

[Mor16] Sophie Morel, *Construction de représentations galoisiennes de torsion [d'après Peter Scholze]* (French), Astérisque **380, Séminaire Bourbaki. Vol. 2014/2015** (2016), Exp. No. 1102, 449–473. MR3522182

[NT16] James Newton and Jack A. Thorne, *Torsion Galois representations over CM fields and Hecke algebras in the derived category*, Forum Math. Sigma **4** (2016), e21, 88, DOI 10.1017/fms.2016.16. MR3528275

[PR94] Vladimir Platonov and Andrei Rapinchuk, *Algebraic groups and number theory*, Pure and Applied Mathematics, vol. 139, Academic Press, Inc., Boston, MA, 1994. Translated from the 1991 Russian original by Rachel Rowen. MR1278263

[Sch92] Claus Scheiderer, *Quasi-augmented simplicial spaces, with an application to cohomological dimension*, J. Pure Appl. Algebra **81** (1992), no. 3, 293–311, DOI 10.1016/0022-4049(92)90062-K. MR1179103

[Sch12a] Peter Scholze, *Perfectoid spaces*, Publ. Math. Inst. Hautes Études Sci. **116** (2012), 245–313, DOI 10.1007/s10240-012-0042-x. MR3090258

[Sch12b] Peter Scholze, *Perfectoid spaces: a survey*, Current developments in mathematics 2012, Int. Press, Somerville, MA, 2013, pp. 193–227. MR3204346

[Sch13] Peter Scholze, *p-adic Hodge theory for rigid-analytic varieties*, Forum Math. Pi **1** (2013), e1, 77, DOI 10.1017/fmp.2013.1. MR3090230

[Sch15] Peter Scholze, *On torsion in the cohomology of locally symmetric varieties*, Ann. of Math. (2) **182** (2015), no. 3, 945–1066, DOI 10.4007/annals.2015.182.3.3. MR3418533

[Sch17] P. Scholze, *Etale cohomology of diamonds*, ArXiv e-prints (2017).

[She17] Xu Shen, *Perfectoid Shimura varieties of abelian type*, Int. Math. Res. Not. IMRN **21** (2017), 6599–6653, DOI 10.1093/imrn/rnw202. MR3719474

[Shi11] Sug Woo Shin, *Galois representations arising from some compact Shimura varieties*, Ann. of Math. (2) **173** (2011), no. 3, 1645–1741, DOI 10.4007/annals.2011.173.3.9. MR2800722

[SW17] Peter Scholze and Jared Weinstein, *Berkeley lectures on p-adic geometry*, http://www.math.uni-bonn.de/people/scholze/Berkeley.pdf.

[TW95] Richard Taylor and Andrew Wiles, *Ring-theoretic properties of certain Hecke algebras*, Ann. of Math. (2) **141** (1995), no. 3, 553–572, DOI 10.2307/2118560. MR1333036

[Wed] T. Wedhorn, *Adic spaces*, Preliminary version, available online at https://www2.math.uni-paderborn.de/fileadmin/Mathematik/People/wedhorn/Lehre/AdicSpaces.pdf.

[Wed99] Torsten Wedhorn, *Ordinariness in good reductions of Shimura varieties of PEL-type* (English, with English and French summaries), Ann. Sci. École Norm. Sup. (4) **32** (1999), no. 5, 575–618, DOI 10.1016/S0012-9593(01)80001-X. MR1710754

[Wil95] Andrew Wiles, *Modular elliptic curves and Fermat's last theorem*, Ann. of Math. (2) **141** (1995), no. 3, 443–551, DOI 10.2307/2118559. MR1333035

Email address: a.caraiani@imperial.ac.uk

IMPERIAL COLLEGE LONDON, 180 QUEEN'S GATE, KENSINGTON, LONDON SW7 2AZ

Selected Published Titles in This Series

242 Bhargav Bhatt, Ana Caraiani, Kiran S. Kedlaya, Peter Scholze, and Jared Weinstein, Perfectoid Spaces, 2019

237 Dusa McDuff, Mohammad Tehrani, Kenji Fukaya, and Dominic Joyce, Virtual Fundamental Cycles in Symplectic Topology, 2019

236 Bernard Host and Bryna Kra, Nilpotent Structures in Ergodic Theory, 2018

235 Habib Ammari, Brian Fitzpatrick, Hyeonbae Kang, Matias Ruiz, Sanghyeon Yu, and Hai Zhang, Mathematical and Computational Methods in Photonics and Phononics, 2018

234 Vladimir I. Bogachev, Weak Convergence of Measures, 2018

233 N. V. Krylov, Sobolev and Viscosity Solutions for Fully Nonlinear Elliptic and Parabolic Equations, 2018

232 Dmitry Khavinson and Erik Lundberg, Linear Holomorphic Partial Differential Equations and Classical Potential Theory, 2018

231 Eberhard Kaniuth and Anthony To-Ming Lau, Fourier and Fourier-Stieltjes Algebras on Locally Compact Groups, 2018

230 Stephen D. Smith, Applying the Classification of Finite Simple Groups, 2018

229 Alexander Molev, Sugawara Operators for Classical Lie Algebras, 2018

228 Zhenbo Qin, Hilbert Schemes of Points and Infinite Dimensional Lie Algebras, 2018

227 Roberto Frigerio, Bounded Cohomology of Discrete Groups, 2017

226 Marcelo Aguiar and Swapneel Mahajan, Topics in Hyperplane Arrangements, 2017

225 Mario Bonk and Daniel Meyer, Expanding Thurston Maps, 2017

224 Ruy Exel, Partial Dynamical Systems, Fell Bundles and Applications, 2017

223 Guillaume Aubrun and Stanisław J. Szarek, Alice and Bob Meet Banach, 2017

222 Alexandru Buium, Foundations of Arithmetic Differential Geometry, 2017

221 Dennis Gaitsgory and Nick Rozenblyum, A Study in Derived Algebraic Geometry, 2017

220 A. Shen, V. A. Uspensky, and N. Vereshchagin, Kolmogorov Complexity and Algorithmic Randomness, 2017

219 Richard Evan Schwartz, The Projective Heat Map, 2017

218 Tushar Das, David Simmons, and Mariusz Urbański, Geometry and Dynamics in Gromov Hyperbolic Metric Spaces, 2017

217 Benoit Fresse, Homotopy of Operads and Grothendieck–Teichmüller Groups, 2017

216 Frederick W. Gehring, Gaven J. Martin, and Bruce P. Palka, An Introduction to the Theory of Higher-Dimensional Quasiconformal Mappings, 2017

215 Robert Bieri and Ralph Strebel, On Groups of PL-homeomorphisms of the Real Line, 2016

214 Jared Speck, Shock Formation in Small-Data Solutions to 3D Quasilinear Wave Equations, 2016

213 Harold G. Diamond and Wen-Bin Zhang (Cheung Man Ping), Beurling Generalized Numbers, 2016

212 Pandelis Dodos and Vassilis Kanellopoulos, Ramsey Theory for Product Spaces, 2016

211 Charlotte Hardouin, Jacques Sauloy, and Michael F. Singer, Galois Theories of Linear Difference Equations: An Introduction, 2016

210 Jason P. Bell, Dragos Ghioca, and Thomas J. Tucker, The Dynamical Mordell–Lang Conjecture, 2016

209 Steve Y. Oudot, Persistence Theory: From Quiver Representations to Data Analysis, 2015

For a complete list of titles in this series, visit the
AMS Bookstore at www.ams.org/bookstore/survseries/.